D1121723
UNBC
LIBRARY

# Chemical Oceanography

# Chemical Oceanography

Authors

**Frank J. Millero, Ph.D.**
Rosenstiel School of Marine
and Atmospheric Science
University of Miami
Miami, Florida 33149

**Mary L. Sohn, Ph.D.**
Department of Chemistry
Florida Institute of Technology
Melbourne, Florida 32901

**CRC Press**
**Boca Raton Ann Arbor London**

**Library of Congress Cataloging-in-Publication Data**

Millero, Frank J.
Chemical oceanography/Frank J. Millero and Mary L. Sohn
p. cm.
Includes bibliographical references and index.
ISBN 0-8493-8840-6
1. Chemical oceanography. I. Sohn, Mary L., 1952–
II. Title.
GC111.2.M55 1991 91-15798
551.46′01—dc20 CIP

Developed by Telford Press

This book represents information obtained from authentic and highly regarded sources. Reprinted material is quoted with permission, and sources are indicated. A wide variety of references are listed. Every reasonable effort has been made to give reliable data and information, but the author and the publisher cannot assume responsibility for the validity of all materials or for the consequences of their use.



Direct all inquiries to CRC Press, Inc., 2000 Corporate Blvd., N.W., Boca Raton, Florida 33431.



International Standard Book Number 0-8493-8840-6

Library of Congress Card Number 91-15798

Printed in the United States of America 2 3 4 5 6 7 8 9 0

*Frank J. Millero wishes to dedicate this book to his wife, Judy, who has accepted his dedication to science.*

# Preface

This book is the result of a course that I have taught over the last 20 years at the University of Miami. Shortly after I arrived in Miami in 1966 from a short stint in industry, I was asked to teach a graduate course in chemical oceanography. At the time I knew little about the oceans and found it necessary to do a lot of reading. Luckily J. P. Riley and G. Skirrow (1965) had just edited a two-volume book on chemical oceanography. The book contained chapters written by a number of authors that described the basic areas of the field. I used these volumes as a source for my first course and the present book is patterned after these volumes. My early reading included the classical text *The Oceans* by Sverdrup, Johnson, and Fleming (1942); the book by Harvey (1961) *The Chemistry and Fertility of Seawater,* and the book *Oceanography* edited by M. Sears (1961). I used the text *Marine Chemistry* by Horne (1969) and the text *Introduction to Marine Chemistry* by Riley and Chester (1971) for some time, but both became outdated. The volumes of *The Sea: Ideas and Observations* edited by Hill and others (1963, 1974, 1977) were used as a source of supplemental material. The second edition by Riley and Skirrow (1975) and later volumes by Riley and Chester (1976, 1978, 1983) and by Riley (1989) are still used to keep the course up to date.

In 1980 I started to teach an undergraduate course in chemical oceanography and found the need for a textbook. The text *Chemical Oceanography* by Broecker (1971) was used for a time, but it did not cover the classical areas that were more familiar to me. The more recent book *Tracers in the Sea* by Broecker and Peng (1982) was better, but the undergraduate students found it difficult to follow. Although graduate students can probe the volumes of *Chemical Oceanography* and current literature references, undergraduate students need a more fundamental approach to the field. The present text is my attempt to put together my views of the field. It is largely descriptive and is biased by my strong mechanistic feelings as a physical chemist. Since it is meant to be an undergraduate text, I have not given all the original references. I have mentioned by name many of those colleagues of mine that have contributed to the field — I apologize for any omissions. Since my background and interests in organic chemistry are rather meager, I asked Mary Sohn to join me as co-author. She has been very helpful in correcting some of my mistakes and wrote the chapter on organic chemistry.

As with most endeavors, I owe thanks to a number of people: my many students, both graduate and undergraduate who have struggled through my courses; Mrs. Rita Marvez who has typed and re-typed the many drafts of this book; Mrs. Sara Sotolongo who has drawn many of the figures; and Kara Kern, who carefully proofed the final copy. The support of the National Science Foundation and the Office of Naval Research of my research work should also be acknowledged. These agencies provided me with research funds to probe the thermodynamics and kinetics of processes occurring in the oceans.

I am sure the book has some flaws, but I hope it will prove useful to those wishing to get an introduction to the field of chemical oceanography. Some of the books that have shaped my views of the field are given below.

*The Oceans,* H. U. Sverdrup, M. W. Johnson, and R. H. Fleming, Prentice-Hall, Englewood Cliffs, New Jersey (1942).

*The Chemistry and Fertility of Seawater,* H. W. Harvey, Cambridge University Press, London (1955)

*Oceanography,* M. Sears, Ed., Pub. 67, A.A.A.S., Washington, D.C. (1961)

*The Sea: Ideas and Observations,* M. N. Hill, Ed., Vol. 2, John Wiley & Sons, New York (1963).

*The Sea: Ideas and Observations,* E. D. Goldberg, Ed., Vol. 5, John Wiley & Sons, New York (1974).

*The Sea: Ideas and Observations,* E. D. Goldberg, I. N. McCave, J. J. O'Brien, and J. H. Steel, Eds., Vol. 6, John Wiley & Sons, New York (1977).

*Marine Chemistry,* R. A. Horne, Wiley-Interscience, New York (1960).

*Introduction to Marine Chemistry,* J. P. Riley and R. Chester, Academic Press, New York (1971).

*Chemical Oceanography,* W. S. Broecker, Harcourt, Brace and Jovanovich Inc., New York (1971).

*Chemical Oceanography,* J. P. Riley and G. Skirrow, Vols. 1 and 2, 1st ed., Academic Press, New York (1965).

*Chemical Oceanography,* J. P. Riley and G. Skirrow, Vols. 1 to 4, 2nd ed., Academic Press, New York (1975).

*Chemical Oceanography,* J. P. Riley and R. Chester, Vols. 5 and 6, 2nd ed., Academic Press, New York (1976).

*Chemical Oceanography,* J. P. Riley and R. Chester, Vol. 7, 2nd ed., Academic Press, New York (1978).

*Chemical Oceanography,* J. P. Riley and R. Chester, Vol. 8, 2nd ed., Academic Press, New York (1983).

*Chemical Oceanography,* J. P. Riley, Vols. 9 and 10, 2nd ed., Academic Press, New York (1989).

*Tracers in the Sea,* W. S. Broecker and T. H. Peng, Eldigio Press, Palisades, N.Y. (1982).

# Table of Contents


1. **Descriptive Oceanography** .......... 1

2. **The Composition of the Major Components of Seawater** .......... 59

3. **Minor Elements of Seawater** .......... 115

4. **Ionic Interactions** .......... 157

5. **Dissolved Gases Other than $CO_2$** .......... 225

6. **The Carbonate System** .......... 267

7. **Micronutrients in the Oceans** .......... 321

8. **Organic Compounds** .......... 353

9. **Processes in the Oceans** .......... 415

**Index** .......... 511

# 1 Table of Contents

**Descriptive Oceanography** .......... 3
*1. Physical Characteristics of the Oceans* .......... 5
*2. Distribution of Temperature and Salinity for Ocean Waters* .......... 11
*3. Circulation and Water Masses of the Oceans* .......... 25
*3.1. Atlantic Ocean Waters* .......... 27
*3.2. Southern Ocean Waters* .......... 33
*3.3. Pacific Ocean Waters* .......... 39
*3.4. Indian Ocean Waters* .......... 43
*3.5. Arctic and Adjacent Seas* .......... 44
*3.6. The Age of Water Masses* .......... 49
*3.7. Closed Basins* .......... 50
*3.7.1. Mediterranean Sea* .......... 50
*3.7.2. Red Sea* .......... 51
*3.7.3. Estuaries* .......... 52
*Further Reading* .......... 55

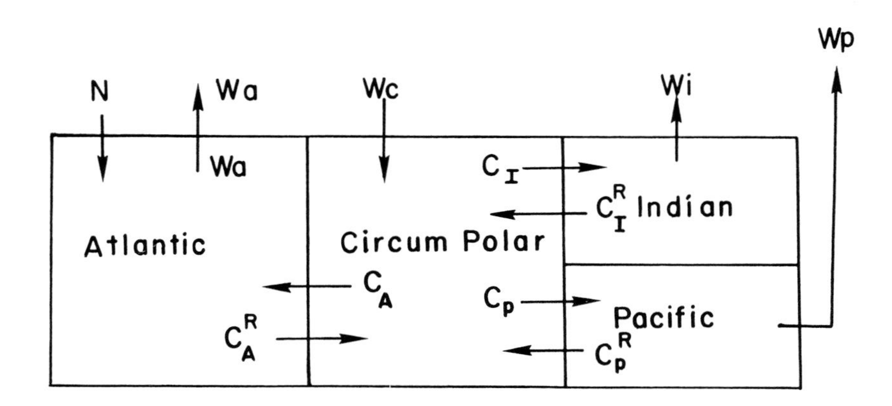


| | | |
|---|---|---|
| N | North Atlantic Water Formation | (14 Sv) |
| Wa | Atlantic Upwelling Rates | (10 Sv) |
| $C_A$ | Transport rate from Circumpolar to Atlantic | ( 7 Sv) |
| $C_A^R$ | Transport Rate from Atlantic to Circumpolar | (11 Sv) |
| $C_I - C_I^R$ | Net Transport from Circumpolar to Indian | (20 Sv) |
| $C_p - C_p^R$ | Net Transport from Circumpolar to Pacific | (25 Sv) |
| Wc | Bottom Water Formed in Circumpolar | (41 Sv) |
| Wi | Indian Ocean Upwelling | (20 Sv) |
| Wp | Pacific Ocean Upwelling | (25 Sv) |

Box model for the deep water circulation of the world ocean waters.

# 1 Descriptive Oceanography

Before we discuss the chemistry of the oceans, it is useful to review some descriptive features of the oceans. Oceanography is the scientific study of the oceans and can be divided into four major areas:

1. Physical oceanography — The study of the physics of the oceans and their interaction with the atmosphere.
2. Biological oceanography — The study of the biology of the oceans.
3. Geological oceanography — The study of the geology and geophysics of the oceans.
4. Chemical oceanography — The study of the chemistry of the oceans.

The basic goal of oceanography is to obtain a clear and systematic description of the oceans. The reasons for developing our knowledge of the oceans are to obtain sources of food, chemicals, and energy; at the same time, the oceans control the climate of the earth and are also a sink for industrial and radioactive wastes, both processes we would like to better understand.

The goal of the physical oceanographer is to obtain a *systematic and quantitative* description of ocean waters and their movements. The ocean currents circulate continuously with small scale variations due to tides and waves generated by winds and earthquakes. This physical study of the oceans has been approached by:

1. The descriptive approach — Observations are made of specific features and reduced to a single characterization of the general features.
2. The dynamic approach — The known laws of physics are applied to the oceans. An attempt is made to solve the mathematical equations for the motion of a body acted on by forces.

In this brief review we will be concerned with the descriptive physical oceanography of the oceans. We feel it is important to understand these descriptive features since the movement of the waters of the oceans affects the biogeochemical processes. Much of the material covered in this chapter is taken from the books by Pickard and Emery (1982), Dietrich and Kalle (1963), and Tchernia (1980). These texts should be referred to for further details.

The field of physical oceanography in its modern context was initiated with the collections of physical data on surface currents and winds which were made by Matthew Fontaine in 1855. The H.M.S. Challenger expedition (1872 to 1876) was the first to gather oceanographic data from around the world. The Meteor expedition (1925 to 1927) yielded the first details of the physics of the Atlantic ocean waters. In recent years we have had

1. The Geophysical Year of the Ocean (GYO) from 1957 to 1958.
2. The International Indian Ocean Expedition (INDEX) from 1962 to 1965.
3. The Global Atmospheric Research Program (GARP).
4. The GARP Atlantic Tropical Experiment (GATE) in 1974.
5. The Middle Ocean Dynamics of Eddies (MODE) in 1973.
6. The Geochemistry Sections Study (GEOSECS) from 1970 to 1980.

The earlier theoretical studies of surface tides were made by Newton and Laplace and those of waves were made by Gerstner and Stokes. The Scandinavian meteorologists Bjerknes, Ekman, and Helland-Hansen developed dynamic oceanography. Recent studies have examined coastal processes, western boundary currents like the Gulf Stream, small scale fluctuations, eddies or rings, bottom water movement, and the use of tracers to study large scale mixing processes. The next ten years will be concerned with the World Ocean Climate Experiment (WOCE). This study will be concerned with a global study of ocean currents and water masses and their interactions with the atmosphere. Future oceanography will have a strong component dealing with satellites and remote sensing techniques.

## 1. PHYSICAL CHARACTERISTICS OF THE OCEANS

The physical characteristics of the oceans and adjacent seas are given in Table 1.1. Some facts to keep in mind when examining this table are that 71% of the surface of the earth is oceans ($361 \times 10^6$ $km^3$); the highest mountain on land is 8,848 m (Everest) while the deepest ocean trench is 11,022 m (Mariana); the Southern Hemisphere has the largest percentage of water (80.9%); the various depth zones of the oceans are given (Table 1.2) in percentages. The world oceans (including adjacent seas) are made up of 50% by the Pacific, 29% by the Atlantic, and 21% by the Indian. The major depth zone of the ocean (74%) lies between 3 to 6 km.

As will be discussed later, 50% of ocean waters have a temperature range between 1.3 and 3.8°C and a salinity (grams of salt in 1 kg of seawater) between 34.6 and 34.8. The mean depth of the oceans is 3.7 km, the mean temperature is 3.5°C, and the mean salinity is 34.7.

The classification of the large scale topographic features of the world oceans is shown in Figure 1.1.* The principal features of the ocean floor are shown in Figure 1.2. This figure is an oversimplification of the actual bottom, but gives the major features. Continental margins show a wide range of unique features and a number of studies have been made on seamounts that occur in deep sea basins. In recent years the detailed structure has been examined for the mid-ocean ridge systems (Figure 1.3).

The *shore* is the part of the land mass close to the sea that has been modified by the sea. The *continental shelf* is seaward from the shore and has a gradient or slope of 1 to 500 and an average width of 65 km. The *break in slope* is the outer limit of the shelf and has a slope of about 1 to 20 and an average depth of 130 m. The *continental slope* is about 400 m from the shelf to the deep sea bottom. In some places it extends as much as 9000 m vertically over a relatively short horizontal distance (e.g., off the west coast

* Figures 1.1 and 1.17 from Dietrich et al., 1980; Figures 1.2, 1.5 to 1.12, 1.14 to 1.16, 1.23 to 1.27, 1.31, 1.33, 1.36, and 1.38 from Pickard and Emery, 1982.

**TABLE 1.1**
**Area, Volume, Mean, and Maximum Depths of the Oceans and Their Adjacent Seas**

| Sea | Area[a] (106 km$^2$) | Volume[a] (106 km$^3$) | Depth | |
|---|---|---|---|---|
| | | | Mean[a] (m) | Maximum[b] (m) |
| Oceans without adjacent seas | | | | |
| Pacific Ocean | 166.24 | 696.19 | 4,188 | 11,022[a] |
| Atlantic Ocean | 84.11 | 322.98 | 3,844 | 9,219[b] |
| Indian Ocean | 73.43 | 284.34 | 3,872 | 7,455[c] |
| Total | 323.78 | 1,303.51 | 4,026 | — |
| Mediterranean seas | | | | |
| Arctic[d] | 12.26 | 13.70 | 1,117 | 5,449 |
| Austral-Asiatic[e] | 9.08 | 11.37 | 1,252 | 7,440 |
| American | 4.36 | 9.43 | 2,164 | 7,680 |
| European[f] | 3.02 | 4.38 | 1,450 | 5,092 |
| Total | 28.72 | 38.88 | 1,354 | — |
| Intracontinental Mediterranean seas | | | | |
| Hudson Bay | 1.23 | 0.16 | 128 | 218 |
| Red Sea | 0.45 | 0.24 | 538 | 2,604 |
| Baltic Sea | 0.39 | 0.02 | 55 | 459 |
| Persian Gulf | 0.24 | 0.01 | 25 | 170 |
| Total | 2.31 | 0.43 | 184 | — |
| Marginal seas | | | | |
| Bering Sea | 2.26 | 3.37 | 1,491 | 4,096 |
| Sea of Okhotsk | 1.39 | 1.35 | 971 | 3,372 |
| East China Sea | 1.20 | 0.33 | 275 | 2,719 |
| Sea of Japan | 1.01 | 1.69 | 1,673 | 4,225 |
| Gulf of California | 0.15 | 0.11 | 733 | 3,127 |
| North Sea | 0.58 | 0.05 | 93 | 725[g] |
| Gulf of St. Lawrence | 0.24 | 0.03 | 125 | 549 |
| Irish Sea | 0.10 | 0.01 | 60 | 272 |
| Remaining seas | 0.30 | 0.15 | 470 | — |
| Total | 7.23 | 7.09 | 979 | — |
| Oceans, including adjacent seas | | | | |
| Pacifc Ocean | 181.34 | 714.41 | 3,940 | 11,022[a] |
| Atlantic Ocean | 106.57 | 350.91 | 3,293 | 9,219[b] |
| Indian Ocean | 74.12 | 284.61 | 3,840 | 7,455[c] |
| World ocean | 362.03 | 1,349.93 | 3,729 | 11,022[a] |

[a] Vitiaz Depth in the Mariana Trench.
[b] Milwaukee Depth in the Puerto Rico Trench.
[c] Planet Depth in the Sunda Trench.
[d] Consisting of Artic Ocean, Barents Sea, Canadian Archipelago, Baffin Bay, and Hudson Bay.
[e] Including Andaman Sea.
[f] Including Black Sea.
[g] In the Skagerrak area.

Data from Dietrich et al. (1980).

**TABLE 1.2**
**Depth Zones in the Oceans**

| Ocean area[a] | Depth zone (km) 0—0.2 | 0.2—1 | 1—2 | 2—3 | 3—4 | 4—5 | 5—6 | 6—7 | 7—8 | 8—9 | 9—10 | 10—11 | % of world ocean |
|---|---|---|---|---|---|---|---|---|---|---|---|---|---|
| Pacific Ocean[b] | 1.631 | 2.583 | 3.250 | 6.856 | 21.796 | 34.987 | 26.884 | 1.742 | 0.188 | 0.063 | 0.019 | 0.001 | 45.919 |
| Austral Asiatic Mediterranean[c] | 51.913 | 9.255 | 10.433 | 12.151 | 6.698 | 7.780 | 1.636 | 0.076 | 0.058 | 0 | 0 | 0 | 2.509 |
| Bering Sea | 46.443 | 5.975 | 7.623 | 10.330 | 29.629 | 0 | 0 | 0 | 0 | 0 | 0 | 0 | 0.625 |
| Sea of Okhotsk | 26.475 | 39.479 | 22.383 | 3.403 | 8.260 | 0 | 0 | 0 | 0 | 0 | 0 | 0 | 0.384 |
| East China Sea[d] | 81.305 | 11.427 | 5.974 | 1.239 | 0.055 | 0 | 0 | 0 | 0 | 0 | 0 | 0 | 0.332 |
| Sea of Japan | 23.498 | 15.176 | 19.646 | 20.096 | 21.551 | 0.033 | 0 | 0 | 0 | 0 | 0 | 0 | 0.280 |
| Gulf of California | 46.705 | 20.848 | 25.891 | 6.556 | 0 | 0 | 0 | 0 | 0 | 0 | 0 | 0 | 0.042 |
| Atlantic Ocean[b] | 7.025 | 5.169 | 4.295 | 8.590 | 19.327 | 32.452 | 22.326 | 0.738 | 0.067 | 0.012 | 0 | 0 | 23.909 |
| Arctic Mediterranean[e] | 47.083 | 17.427 | 9.317 | 11.153 | 12.834 | 2.195 | 0 | 0 | 0 | 0 | 0 | 0 | 3.386 |
| American Mediterranean | 23.443 | 10.674 | 13.518 | 15.313 | 20.796 | 13.440 | 2.572 | 0.193 | 0.051 | 0 | 0 | 0 | 1.203 |
| European Mediterranean[f] | 22.868 | 20.814 | 18.362 | 30.326 | 7.426 | 20.204 | 0 | 0 | 0 | 0 | 0 | 0 | 0.834 |
| Baltic Sea | 99.832 | 0.168 | 0 | 0 | 0 | 0 | 0 | 0 | 0 | 0 | 0 | 0 | 0.105 |
| Indian Ocean[b] | 3.570 | 2.685 | 3.580 | 10.029 | 25.259 | 36.643 | 16.991 | 1.241 | 0.001 | 0 | 0 | 0 | 20.282 |
| Red Sea | 41.454 | 43.058 | 14.920 | 0.568 | 0 | 0 | 0 | 0 | 0 | 0 | 0 | 0 | 0.125 |
| Persian Gulf | 100.000 | 0 | 0 | 0 | 0 | 0 | 0 | 0 | 0 | 0 | 0 | 0 | 0.066 |
| World ocean | 7.492 | 4.423 | 5.376 | 8.497 | 20.944 | 31.689 | 21.201 | 1.232 | 0.105 | 0.032 | 0.009 | 0.001 | 100.000 |

[a] As a percentage of the surface of each ocean.
[b] Without adjacent seas.
[c] Including Andaman Sea.
[d] Including Yellow Sea.
[e] Consisting of Arctic Ocean, Barents Sea, Canadian Archipelago, Baffin Bay, and Hudson Bay.
[f] Including Black Sea.

From Dietrich et al. (1980).

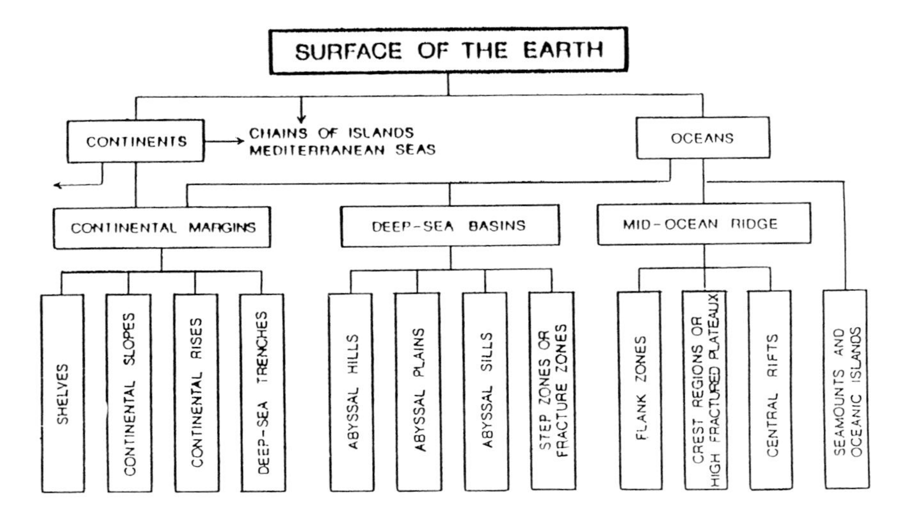


FIGURE 1.1. Divisions of the surface of the earth.

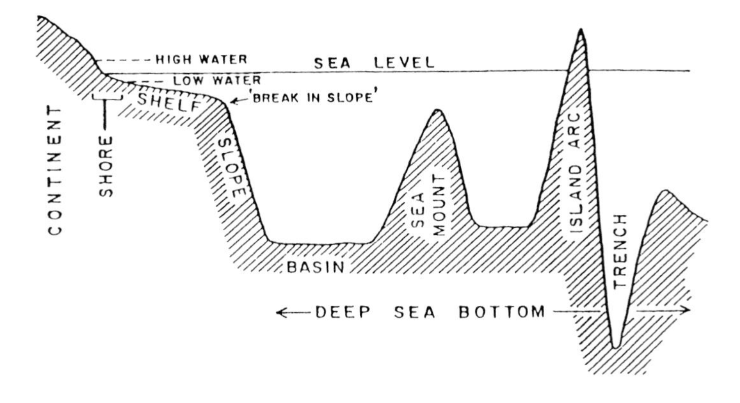


FIGURE 1.2. Structure of the ocean bottom.

of America). The *deep sea bottom* represents the most extensive area (76% of ocean basins) and is from 3 to 6 km in depth. This area is not completely flat but contains welts, furrows, swells, and basins.

The bottom sediments contain material formed in the sea (pelagic) and brought from the land by rivers and the atmosphere (nonpelagic). The components of sediments can be divided into a detrital fraction transported as a solid and a non-detrital or authigenic fraction transported as dissolved matter. The sediment can be divided into four types (Figure 1.4):

1. Hydrogenous — Formed by reactions (precipitation and adsorption) in the water.

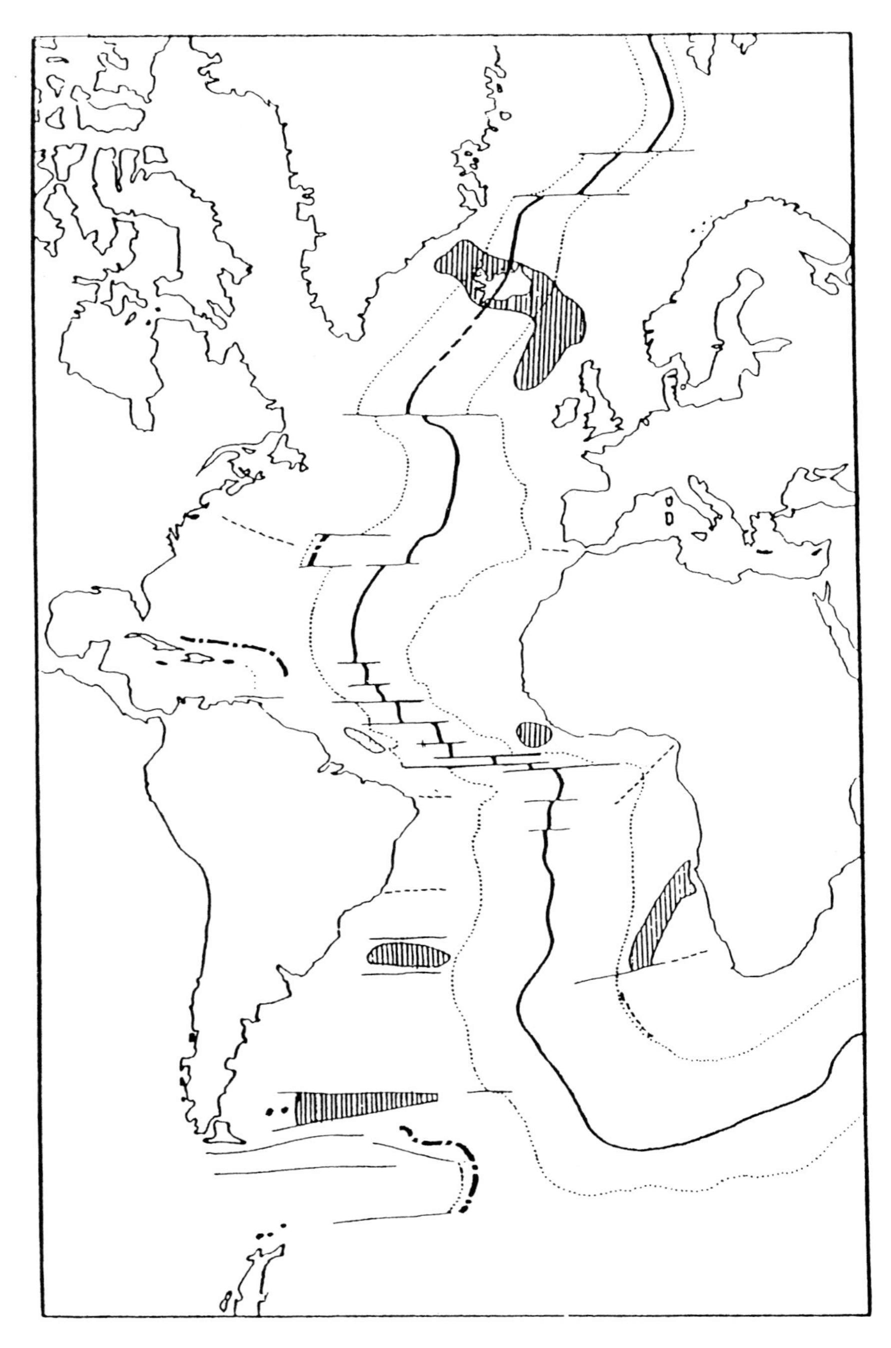

FIGURE 1.3. Active ridge systems in the oceans.

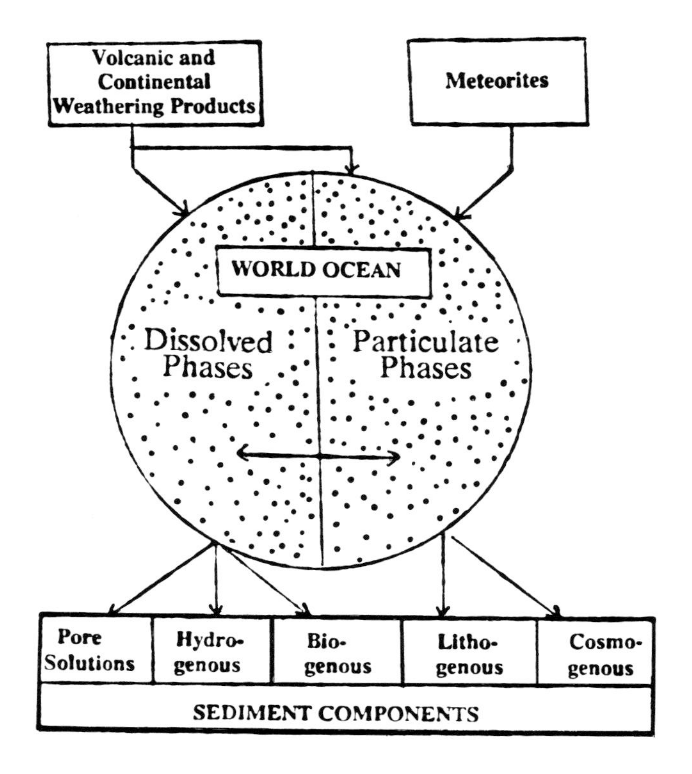


FIGURE 1.4. Types of ocean sediments.

2. Biogenous — Produced by living organisms from the parts of shells and skeletons.
3. Lithogenous — From the weathering of the earth surface and transported by rivers and winds to the oceans.
4. Cosmogenous — From extraterrestrial sources.

Examples of the various types include:

1. Hydrogenous — Aragonite as "whitings" in the Bahama Banks; manganese nodules in the deep Pacific; sulfates and phosphates.
2. Biogenous — Calcite from foraminifera and cocoliths; aragonite from pteropods; silica from radiolarium and diatoms.
3. Lithogenous — Clay minerals and quartz as rock fragments transported by wind, rivers, glacial waters, and volcanic sources.
4. Cosmogenous — Ferric meteorites from outer space.

## 2. DISTRIBUTION OF TEMPERATURE AND SALINITY FOR OCEAN WATERS

Much of our descriptive knowledge of the physics of the oceans comes from an examination of the properties of seawater from place to place. Our knowledge of the distribution of various properties comes from collections of measurements made at an oceanographic or hydrographic station. Properties such as temperature, salinity (the approximate grams of sea salt in 1 kg of seawater), oxygen, nutrients, etc. are measured as a function of depth. These measurements at a given station are plotted as a function of the depth (which can be related to the pressure, 10 m = 1 bar) and are called a vertical profile. Profiles from a string of stations can be combined to form a vertical section. The common values of a given property are connected to give a contour of the property of interest as a function of depth and distance. Similar contours of common properties can be made of surface waters and waters at a fixed depth (e.g., 4000 m). The contoured lines can be used to elucidate the horizontal and vertical flow of waters due to sinking, upwelling, mixing, precipitation, freezing, etc. The basic assumption made is that the ocean waters have similar average properties over a period of time. The variation with time (temporal) can be observed from anchor stations, fixed buoys, floating instrument packages and, for surface waters, from satellites.

The vertical distribution of density controls vertical mixing in the oceans. Since the density of seawater is directly related to the temperature and salinity of ocean waters, it is useful to examine the typical distributions of these properties. The surface distribution of ocean waters is zonal with the lines of constant temperature (isotherms) running parallel with latitude (Figures 1.5 and 1.6). Near the west coast of South America and Africa, lower temperature surface waters exist due to upwelling. Open ocean waters vary from −2°C (close to the freezing point) in polar regions to 28°C near the equator, though the temperature in confined areas can be as high as 40°C. The annual variations of temperature are 2°C at the equator and poles. The variations are 8°C at 40N or S and as high as 15°C in coastal surface waters. The diurnal variations are 0.3°C in the open oceans and 2 to 3°C in shallow waters.

Typical temperature profiles are shown in Figure 1.7. Below the surface the water can be divided into three zones. The upper zone from 5 to 200 m has similar temperatures to the surface waters. Below this mixed layer, from 200 to 1000 m, the temperature decreases rapidly with depth (the zone with

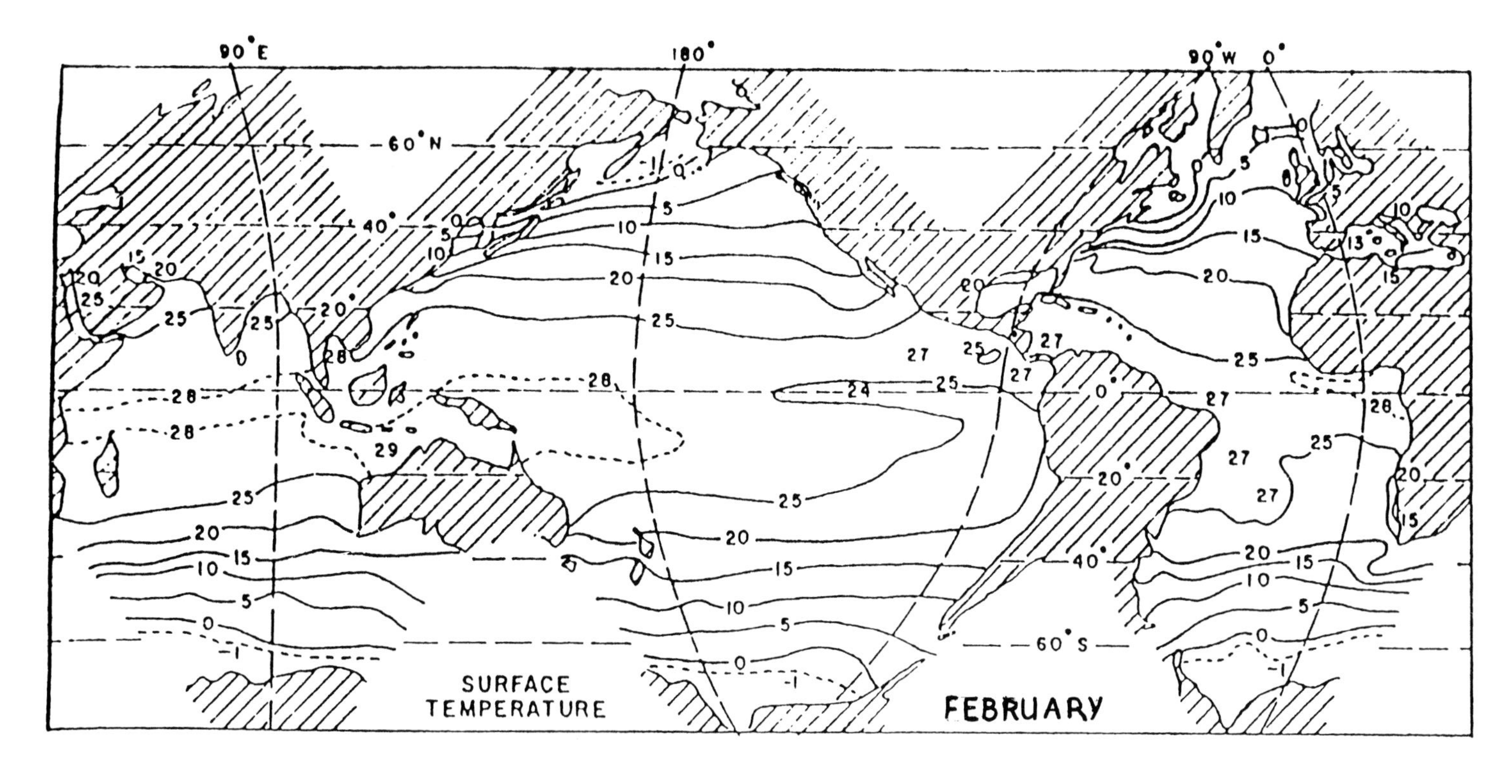


FIGURE 1.5. Temperature of ocean surface waters (February).

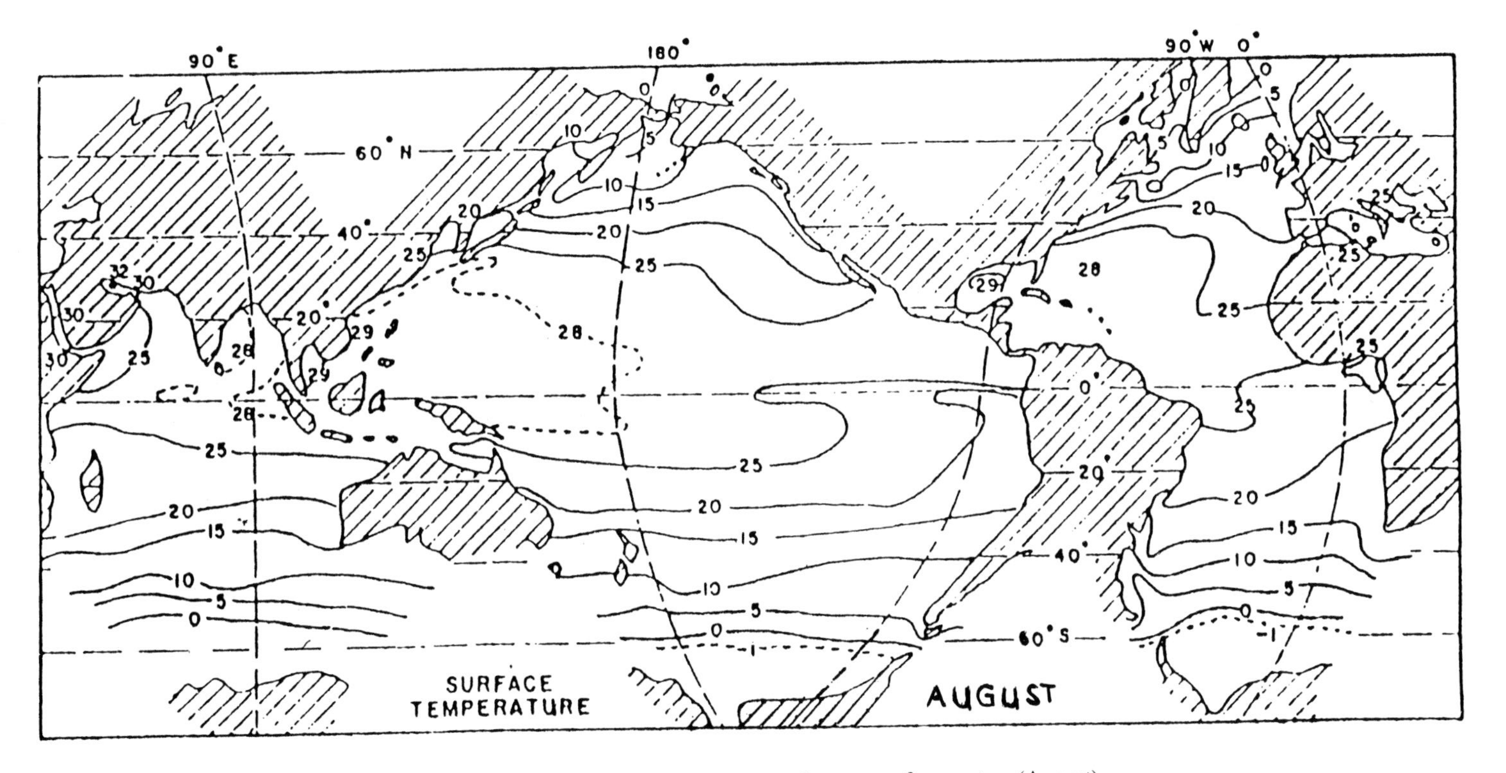


FIGURE 1.6. Temperature of ocean surface waters (August).

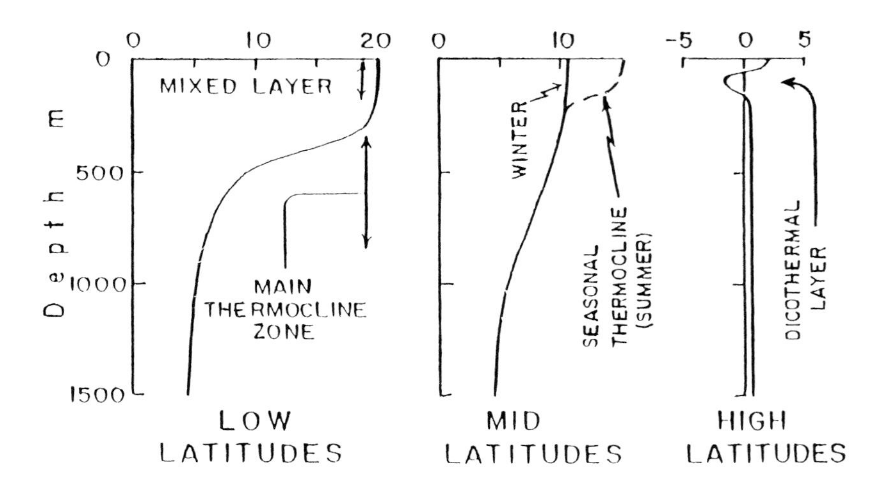


FIGURE 1.7. Typical temperature profiles in the ocean; temperature in °C.

the maximum decrease is called the thermocline). In the deep zone the temperature changes slowly with depth. In low latitude waters the surface mixed layer temperatures will be about 20°C. The deep waters in this region will have temperatures between 5 and 2°C and be separated from the mixed layer by a permanent thermocline. In mid latitudes, the surface mixed layer will be about 15°C in the summer and will decrease to about 5 to 10°C in the winter. This change in temperature results in a seasonal thermocline in the summer months which disappears in the winter. This growth and decay of the thermocline is shown in Figure 1.8. The near vertical distribution of temperature in the winter results in the mixing of the waters to a great depth and the replenishment of nutrients to the surface waters.

In high latitudes the surface temperatures are much lower. The main thermocline may not be present and only a seasonal thermocline may occur. Between 50 to 100 m a dicothermal layer can develop. This layer of cold water at $-1.6$°C is sandwiched between the warmer surface and deep waters. The stability is maintained by an increase of salinity with depth through the layer.

The temperature of deep waters increases to a depth of about 3000 m; however, in deep trenches, the *in situ* temperature increases slowly with depth due to the effect of an increase in pressure. If seawater of $S = 35$ and $t = 5$°C were lowered to 4000 m adiabatically (not allowing heat to exchange with the surrounding), the temperature would increase to 5.45°C due to compression. An appropriate decrease in temperature of 0.45°C would occur if the waters were brought from 4000 m to the surface due to expansion. The

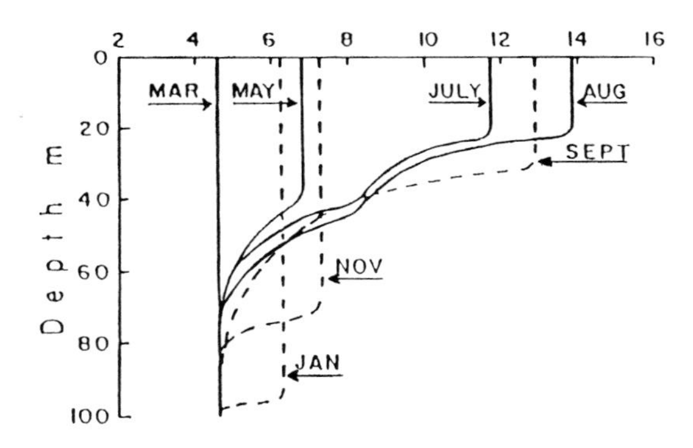


FIGURE 1.8. Growth and decay of the thermocline; temperature in °C.

*in situ* temperature at great depths is higher than the potential temperature ($\theta$) which has been corrected for the effect of pressure on the volume of seawater. An example of the importance of potential temperature is shown for the temperature data in the Mindanao trench (Figure 1.9). The *in situ* temperature of the deep waters is 2°C, which is higher than the waters above the trench. This leads to the calculated deep densities being higher than the waters above the trench. This is an unstable density structure (Table 1.3) and would result in the uplifting of these waters. The potential temperatures, however, show a smooth decrease with depth and the densities increase slowly with depth, as expected for a stable water column.

By connecting the constant temperature of various profiles (isotherms) it is possible to produce sections of temperatures for the major ocean basins. The resulting sections for the major oceans are shown in Plates 1, 2, and 3.* The deep waters have similar temperatures (2°C) while the surface waters show a great deal of structure. The convergence zones are clearly demonstrated in the South Atlantic, Pacific, and Indian oceans. The upwelling in the equatorial regions is also clearly shown.

The typical surface salinities of the oceans are shown in Figure 1.10. These surface values are affected by physical processes occurring in the waters. The salinity will increase due to evaporation and freezing and the salinity will decrease due to precipitation of rain, river runoff, and the melting of ice. The difference between the evaporation and precipitation occurring at different latitudes controls the surface salinities. This is demonstrated in Figure 1.11. The decrease in the surface salinities near the equator are caused by

* Plates 1 to 3 follow page 48.

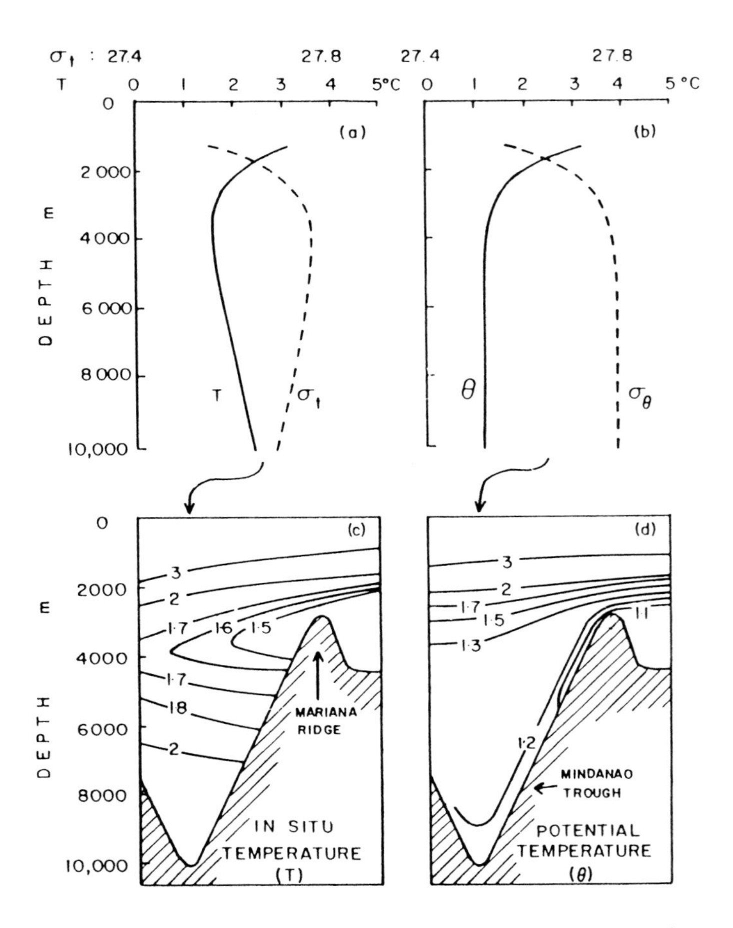


FIGURE 1.9. *In situ* and potential temperature in deep sea trench.

**TABLE 1.3**
***In Situ* and Potential Temperature in the Mindanao Trench**

| Depth | Salinity | Temperature (°C) *In situ* | Temperature (°C) Potential | $\varphi_T$ | Potential |
|---|---|---|---|---|---|
| 1455 | 34.58 | 3.20 | 3.09 | 27.55 | 27.56 |
| 2470 | 34.64 | 1.82 | 1.65 | 27.72 | 27.73 |
| 3470 | 34.67 | 1.52 | 1.31 | 27.76 | 27.78 |
| 4450 | 34.67 | 1.65 | 1.25 | 27.76 | 27.78 |
| 6450 | 34.67 | 1.93 | 1.25 | 27.74 | 27.79 |
| 8450 | 34.69 | 2.23 | 1.22 | 27.72 | 27.79 |
| 10035 | 34.67 | 2.48 | 1.16 | 27.69 | 27.79 |

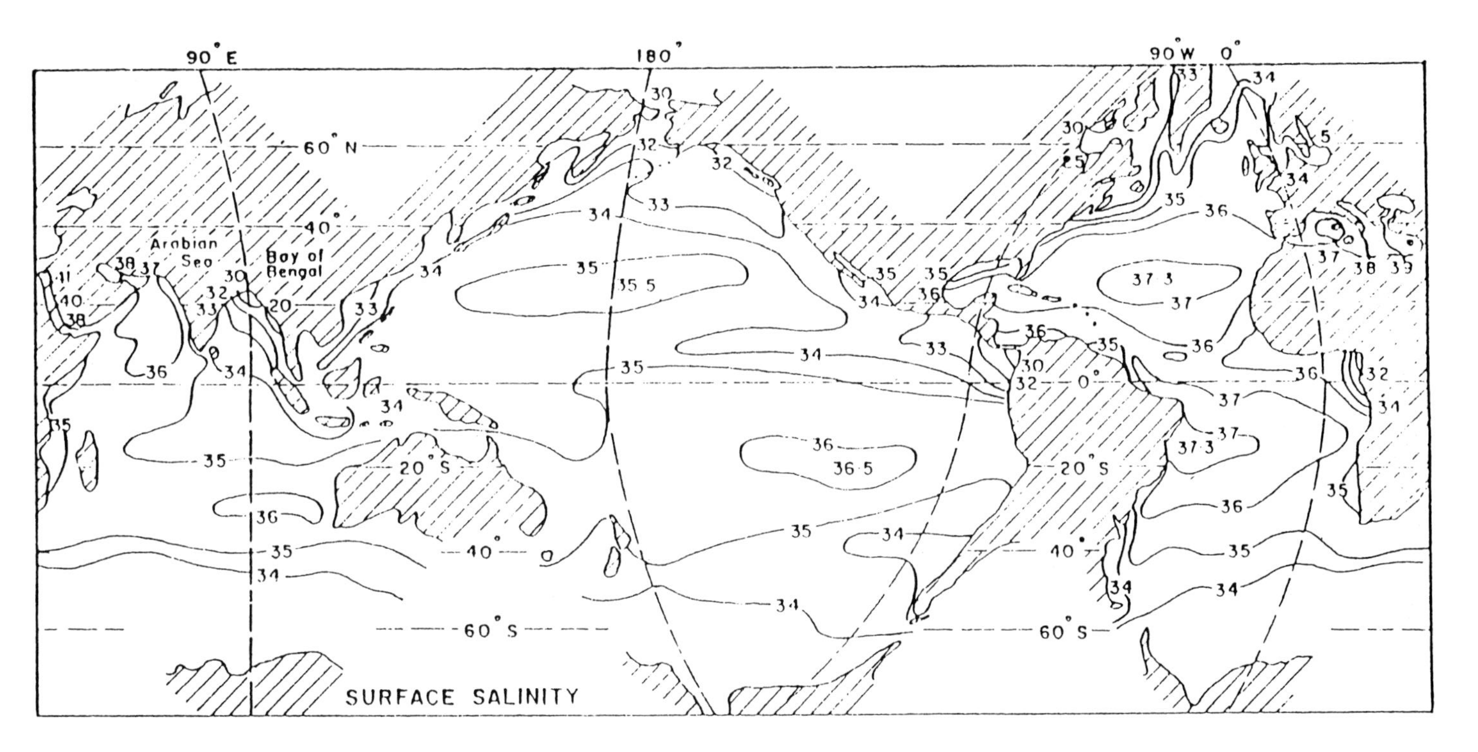


FIGURE 1.10. Salinity of ocean surface waters.

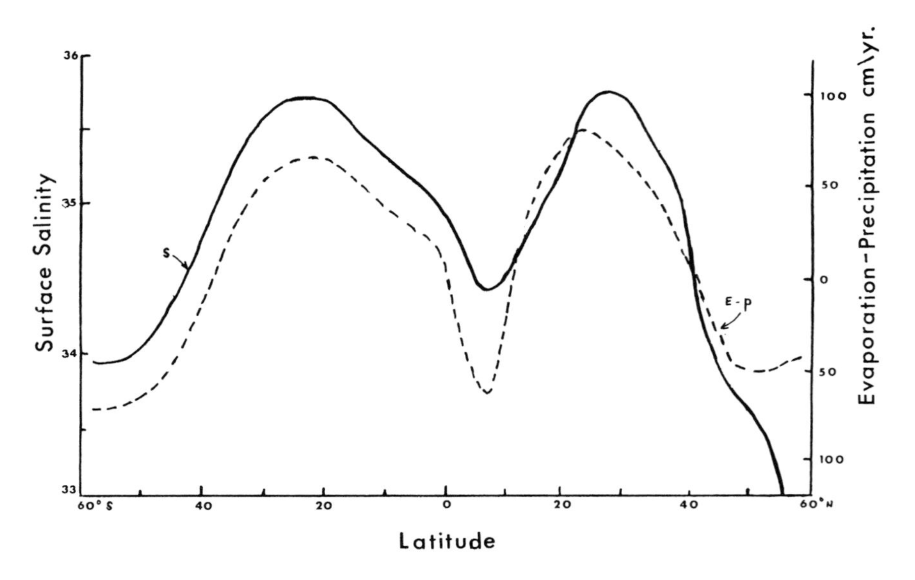


FIGURE 1.11. Surface salinity compared to evaporation minus precipitation (cm/year).

greater precipitation; the increase in the mid latitudes (30N and S) is caused by higher rates of evaporation. The annual variation of salinity in the open oceans is 0.5 while the values range from S = 33 to 37. Higher values occur in regions of high evaporation such as the Mediterranean (S = 39) and Red Seas (S = 41). Values as high as S = 300 can be attained in closed lagoons during the summer in arid areas. The salinity of the North Atlantic waters is higher (S = 37.3) than North Pacific waters (S = 35.5).

The higher salinity of the North Atlantic waters is important because it leads to a higher density when the waters are colder. This in turn leads to the formation of deep waters in the North Atlantic but not in the North Pacific. Recently, Bruce Warren (Woods Hole) has given a clever answer to why the North Atlantic waters have a higher salinity than the North Pacific. The higher salinity is due to the fact that evaporation rates are about twice as high in the North Atlantic. This overcomes the higher river input into the Atlantic. The precipitation in both oceans is about the same. The lower evaporation rates in the North Pacific are due to the lower surface temperature. Cold waters have a lower specific humidity which reduces the evaporation. This thermal effect is due to differences in the winds in the two oceans. In the Pacific the maximum westerlies of the wind field are further south. This limits the ef-

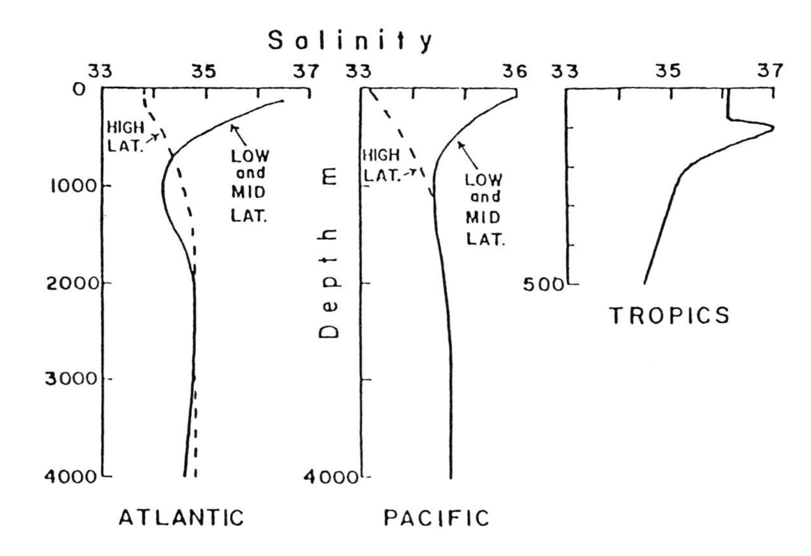


FIGURE 1.12. Typical salinity profiles in the ocean.

fectiveness of the winds to transfer warm subtropical water to the North. This may be a response to large scale factors. The density of the North Pacific is thus reduced by decreasing the temperature. The inflow of water into the North Pacific matches the Ekman-induced upwelling in the subpolar gyre. The surface waters of the Pacific have a lower density due to the decrease in salinity caused by lower evaporation. In the North Atlantic the densities are higher due to the decrease in temperature. This allows for deep water formation.

The typical vertical distributions of salinity in the Atlantic, Pacific, and tropics are shown in Figure 1.12. There is a marked salinity minimum at 600 to 1000 m in the Atlantic. This is due to the movement of sub-Antarctic intermediate water to the north. The maximum near 100 to 200 m in the tropics is close to the top of the thermocline. It results from water sinking at the tropical salinity maximum and flowing toward the equator. In high latitudes the surface salinity values are low due to the melting of ice. The salinity generally increases with depth to 2000 m with no subsurface maximum. In low and mid latitudes the surface salinities are higher due to evaporation being greater than rainfall. In coastal waters with river runoff there is a strong rapid increase in salinity (halocline) between the fresh river water and deep saline water. The salinity of deep water below 4000 m is quite uniform (34.6 to 34.9) while the temperature ranges from $-1$ to 2°C.

Sections of salinity for the Atlantic, Pacific, and Indian oceans are shown in Plates 4, 5, and 6.* The section in the intermediate waters (400 to 1500 m) clearly shows the northern movement of the sub-Antarctic intermediate water. The Atlantic waters show the southern movement of North Atlantic deep water and the remaining core of high salt water between 20 to 30 N which may be due to Mediterranean waters. The North Pacific waters show the southern movement of sub-Arctic intermediate water but no formation of deep water. In the Indian Ocean no cold, deep waters are formed in the north due to the Asian land mass. A salty intermediate water, however, is formed from the waters coming from the Red Sea.

A summary of the temperature and salinity of the world ocean waters is shown in Figure 1.13. The stippled area represents the ranges for 99% of the waters (t = −2 to 32°C and S = 33 to 37); the solid area represents 75% of the waters (t = 0 to 6°C and S = 34 to 35). The weighted mean values are t = 3.5°C and S = 34.7 with a mean depth of 3.73 km (373 bar pressure).

The temperature and salinity of ocean waters control their density. The change in the density with depth is important because it determines the water static stability. When the stability is high, vertical mixing is minimized. The density of ocean waters is normally characterized by the symbol $\sigma_T$ which is called Sigma-T. This quantity is related to the relative density (d) of seawater by

$$\sigma_T = (d - 1)\ 10^3 \tag{1}$$

where $d = \rho/\rho^0(\mathrm{Max})$, $\rho$ is the density of seawater (kg $m^{-3}$), and $\rho^0(\mathrm{Max})$ = 999.974 kg $m^{-3}$ is the maximum density of water at 3.98°C. The density of seawater as a function of salinity (S), temperature (t°C), and pressure (P, bar) can be calculated from the International Equation of State given below in Table 1.4. A typical density profile is shown in Figure 1.14. The density normally increases with increasing depth. Minimum energy is obtained when the less dense water is at the surface. The density of surface waters shown in Figure 1.15 is strongly affected by temperature except in polar regions where salinity becomes important. In coastal waters, fjords, and estuarines, salinity often controls the density at all depths. In equatorial and tropical regions there is a shallow upper layer of nearly uniform density, then a layer where the density increases rapidly. This depth region is called the pycnocline. Below this region the density increases slowly with depth. Most deep waters

* Plates 4 to 6 follow page 48.

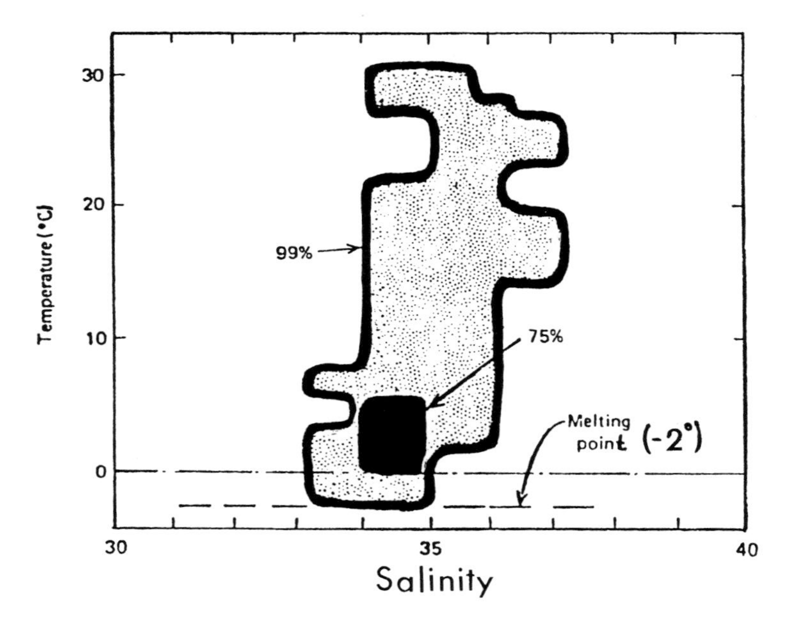


FIGURE 1.13. Temperature versus salinity for ocean waters.

have a $\sigma_T$ near 27.9. At higher latitudes the surface values of $\sigma_T = 27$ so that the increase of density with increasing depth is much smaller. As shown earlier in Table 1.3, for deep ocean waters the values of $\sigma_T$ at the adiabatic temperature should be used to examine the stability of deep waters (i.e., $\sigma_\Theta$).

The geographic distribution of density for the Atlantic in a north-south section is shown in Figure 1.16. The upper layers tend to be concave upward, showing an increase from the equator to the pole. Below 2000 m the values of $\sigma_\Theta$ range from 27.6 to 27.9 for most deep waters. It should be pointed out that the flow of ocean waters tends to follow constant density surfaces (isopycnals). One can frequently trace the origin of deep waters back to the region of formation at the surface.

Although the density of a given water can be arrived at by a number of possible combinations of S and t, one finds that only a limited number of combinations occur and these occur in different regions. Thus, it is possible to recognize a water mass by its characteristic combination of water properties. By plotting temperature against salinity (T-S plots) for individual oceanographic stations it is possible to develop a characteristic diagram. The T-S relation for the major ocean waters is shown in Figure 1.17. It is not necessary to plot density on this diagram since it depends upon T and S. Each point on this diagram corresponds to a particular combination of T and S and a given

## TABLE 1.4
## The International Equation of State for Seawater ($m^3\ kg^{-1}$)

$$v^P = v^0(1 - P/K)$$
$$\rho = 1/\rho^0(1 - P/K)$$

where

$$\rho^0 = 999.842594 + 6.793952 \times 10^{-2}t - 9.095290 \times 10^{-3}t^2 + 1.001685 \times 10^{-4}t^3 - 1.120083 \times 10^{-6}t^4 + 6.536336 \times 10^{-9}t^5 + (8.24493 \times 10^{-1} - 4.0899 \times 10^{-3}t + 7.6438 \times 10^{-5}t^2 - 8.2467 \times 10^{-7}t^3 + 5.3875 \times 10^{-9}t^4)\ S + (-5.72466 \times 10^{-3} + 1.0227 \times 10^{-4}t - 1.6546 \times 10^{-6}t^2)S^{3/2} + 4.8314 \times 10^{-4}\ S^2$$

$$K = 19652.21 + 148.4206t - 2.327105t^2 + 1.360477 \times 10^{-2}t^3 - 5.155288 \times 10^{-5}t^4 + S(54.6746 - 0.603459t + 1.09987 \times 10^{-2}t^2 - 6.1670 \times 10^{-5}t^3) - S^{3/2}(7.944 \times 10^{-2} + 1.6483 \times 10^{-2}t - 5.3009 \times 10^{-4}t^2) + P[3.239908 + 1.43713 \times 10^{-3}t + 1.16092 \times 10^{-4}t^2 - 5.77905 \times 10^{-7}t^3 + S(2.2838 \times 10^{-3} - 1.0981 \times 10^{-5}t - 1.6078 \times 10^{-6}t^2) + S^{3/2}(1.91075 \times 10^{-4})] + P^2[8.50935 \times 10^{-5} - 6.12293 \times 10^{-6}t + 5.2787 \times 10^{-8}t^2 + S(-9.9348 \times 10^{-7} + 2.0816 \times 10^{-8}t + 9.1697 \times 10^{-10}t^2)]$$

| Check values: | S | t | P | $\nu$ | K |
|---|---|---|---|---|---|
| | 35 | 5°C | 0b | 1027.67547 | 22185.93358 |
| | | | 1000 | 1069.48914 | 25577.49819 |

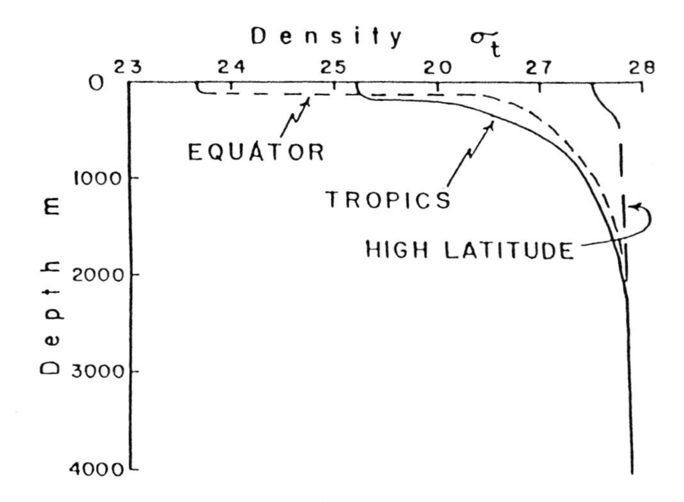


FIGURE 1.14. Typical Sigma-t profiles in the ocean.

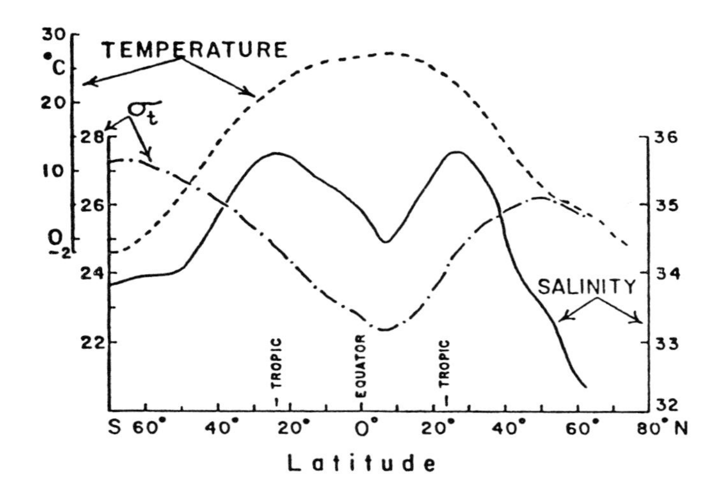


FIGURE 1.15. Comparison of Sigma-t to temperature and salinity for surface waters.

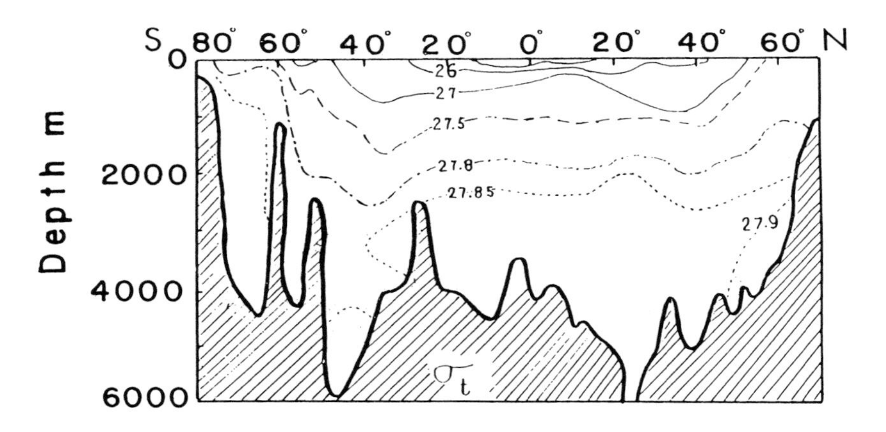


FIGURE 1.16. North-south section of Sigma-t in the Atlantic Ocean.

density. The same density, however, can be attained by different combinations of T and S. These combinations lie on a smooth curve on the T-S diagram and can be shown as the dotted lines shown in Figure 1.17. When discussing a T-S diagram a water type is represented by a point and a water mass by a line. These are ideal definitions and the actual data shows some scatter. Climatic processes at the surface form water types and a water mass results from the mixing of two or more water types. Since surface waters are not

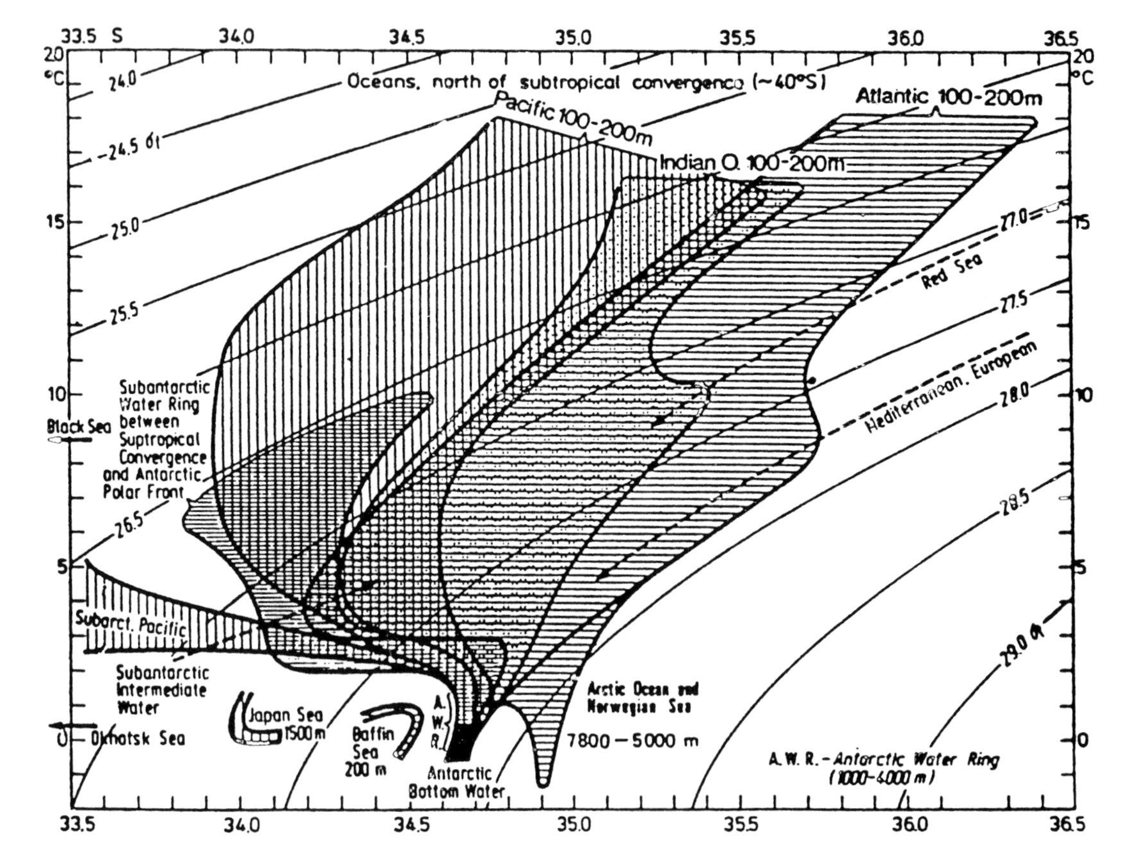


FIGURE 1.17. Temperature salinity (T-S) diagram for waters of the ocean.

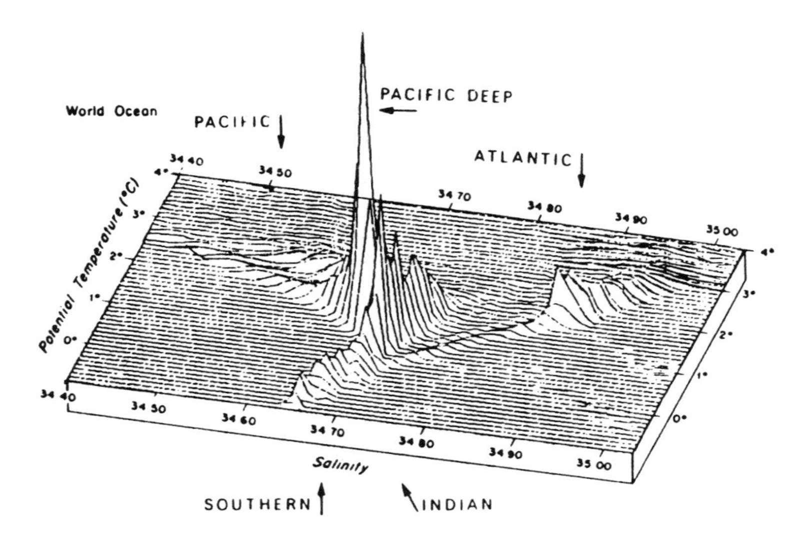


FIGURE 1.18. Three-dimensional distribution of ocean waters as function of potential temperature, salinity, and specific volume.

normally conservative, they are omitted from T-S diagrams. The T-S diagrams have two disadvantages: (1) they give a poor indication of the distribution with depth of the water property because depth along a T-S diagram is not linear; and (2) they also give no indication of the spatial distribution of water properties. This is more clearly demonstrated using vertical or horizontal sections.

It is also possible to consider the volume of ocean waters with given properties of T and S. Montgomery was the first to suggest that T-S-V diagrams could be useful. A three dimensional T-S-V diagram for waters of the ocean is shown in Figure 1.18.

## 3. CIRCULATION AND WATER MASSES OF THE OCEANS

The energy from the sun is responsible for the circulation of ocean waters. This circulation can be divided into two types:

1. Wind driven circulation — Due to the wind blowing over the surface waters. This affects the upper few hundred meters of the oceans yielding primarily horizontal circulation.

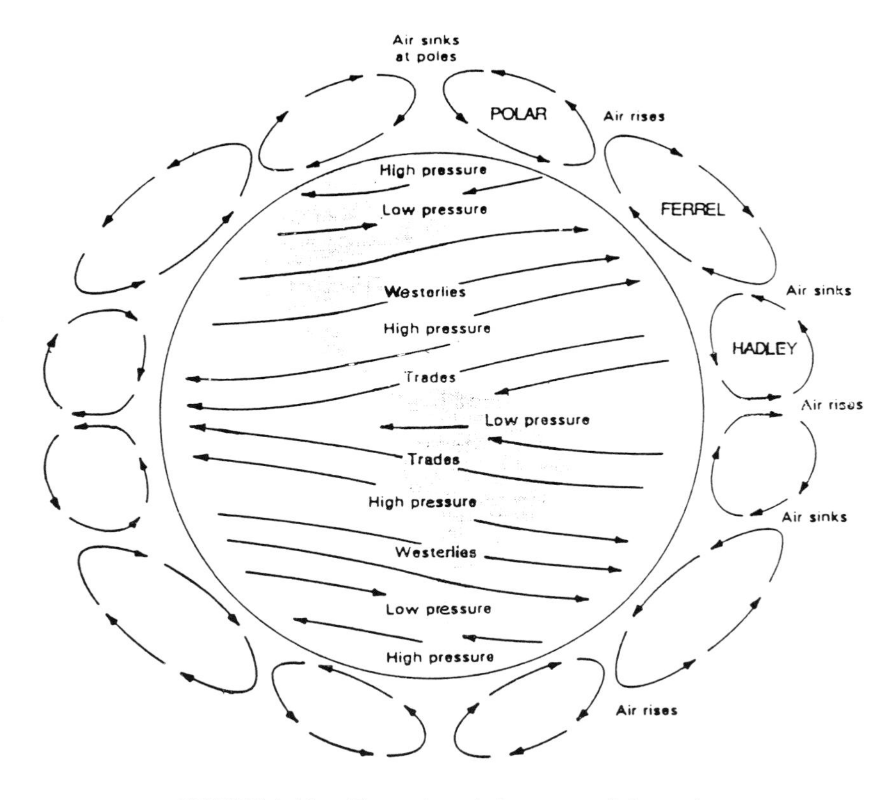


FIGURE 1.19. The major wind systems of the earth.

2. Thermohaline circulation — Movement of water due to density differences caused by changes in T and S. This leads to deep sea and vertical circulation. The processes that result in mixing are eddy diffusion caused by concentration gradients and advection caused by currents. The pycnocline acts as a barrier to thermohaline mixing in the oceans.

The surface circulation of waters in the various oceans is similar and due to the distribution of the major winds shown in Figure 1.19. The trade winds and the westerlies combine to lead to a clockwise gyre in the northern hemisphere and a counterclockwise gyre in the southern hemisphere. In the northern hemisphere the waters have a right hand component due to the spinning of the earth. This so-called "coriolis effect" has a left hand component in the southern hemisphere. In the North Atlantic and the Pacific the currents are narrower and faster on the west side of the ocean. This westward intensification results in the Gulf Stream in the North Atlantic and the Kuroshio current in the North Pacific. When the trade winds from the north or south cross the equator, the coriolis effect causes a countercurrent in a direction opposite to that of the major gyres.

The surface ocean currents of the oceans are divided into a number of separate currents which are shown in Plate 7.*

## 3.1. Atlantic Ocean

### Waters

As shown in Figure 1.20 the upper water circulation of the Atlantic ocean consists of two gyres: (1) a clockwise one in the North, and (2) a counter-clockwise one in the South. In the North the gyre starts with the North Equatorial Current that is driven by the North East trade winds. It is joined by the South Equatorial Current and becomes the Antilles Current off the West Indies. The Florida Current comes from waters that go through the Yucatan and Gulf of Mexico (called the Gulf of Mexico Loop Current) and escapes between Florida and Cuba into the Atlantic. The Gulf Stream is the northern extension of the Florida Current which joins with the Antilles Current. The flow of the Florida Current waters is about $26 \times 10^6$ $cm^3$ $sec^{-1}$ (26 Sv, Sverdrups). The Antilles Current has a flow rate of about 12 Sv and the flow of the Gulf Stream off the Chesapeake Bay is about 80 Sv. The North Atlantic Current is a continuation of the Gulf Stream. The Labrador Current waters flow south out of the Labrador Sea. This cold, low salinity water mixes with the warm Gulf Stream water causing a complicated mixing regime. Part of the North Atlantic Current continues toward the Arctic and becomes the Norwegian Current. The remainder goes south to complete the gyre. The waters flowing off the coast of Africa are called the Canary Current.

In the South Atlantic the counterclockwise gyre starts near the equator due to the stress of the South East trade winds causing the South Equatorial Current to move toward Central America. Part goes north and part goes south. The south component is called the Brazil Current. This current turns east at the subtropical convergence and becomes part of the Circumpolar Current. It turns north off the coast of Africa and becomes the Benguela Current. The Brazil Current is warm and salty water, while the Benguela Current is cold and less saline. The Falkland Current flows north from the Drake passage up the coast of South America. It separates the Brazil Current from the coast to about 25S.

One of the most studied currents in the Atlantic is the Gulf Stream. Recent

* Plate 7 follows page 48.

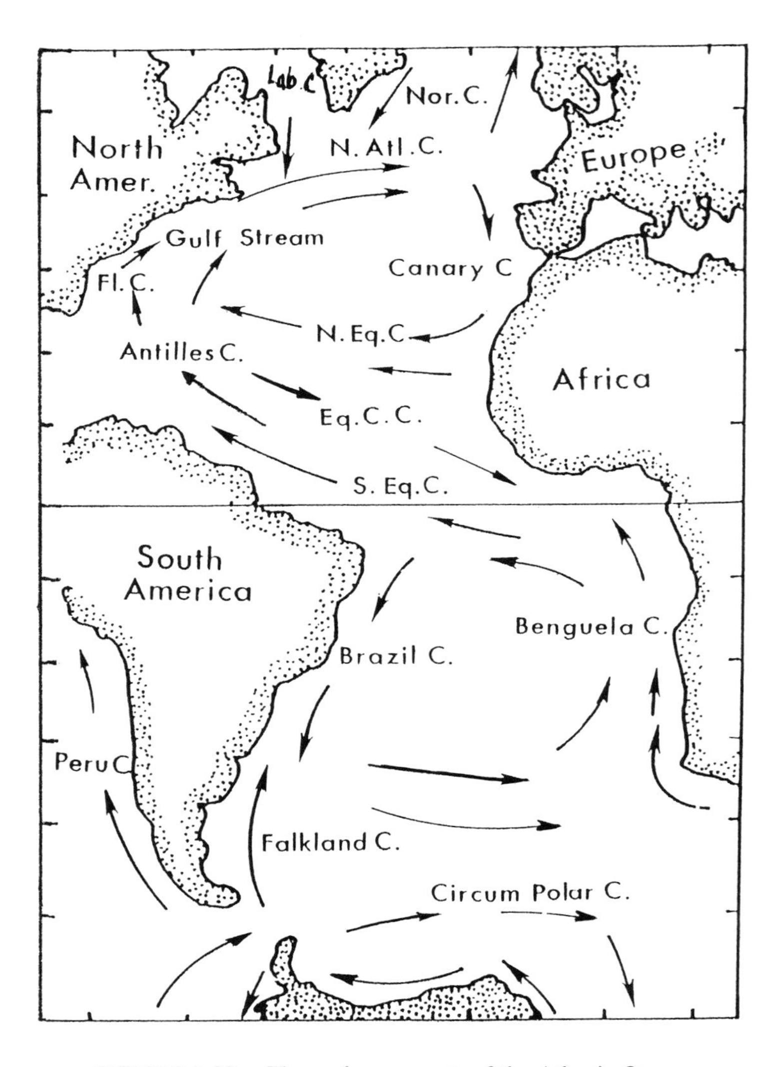


FIGURE 1.20. The surface currents of the Atlantic Ocean.

studies have been made on the formation of Gulf Stream rings. These rings are formed from the meandering of the Gulf Stream (Figure 1.21). Two types of rings are formed; cold core rings and warm core rings. The cold core rings are formed from the waters north of the Gulf Stream. This colder slope water forms a counterclockwise gyre of water that moves to the south. These rings (*Science,* 212, 1091, 1981) can exist for 2 to 4 years, but usually rejoin the Gulf Stream after 6 to 12 months. The warm core rings are formed from the closure of a ring of warm Sargasso Sea water. This clockwise gyre of warm water moves toward the north.

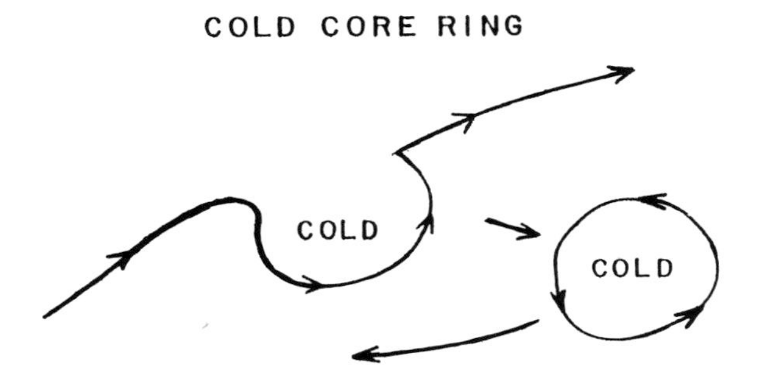


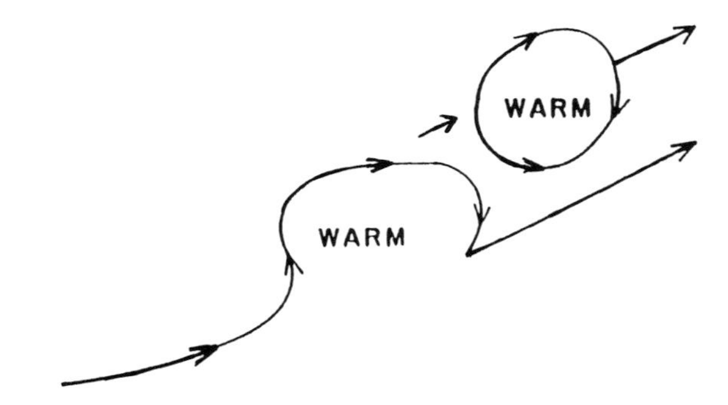


FIGURE 1.21. The formation of Gulf Stream rings.

The cold core rings are rich in nutrients and support high productivity with an enrichment of animal life. The structure of a typical ring is shown in Figure 1.22. The cold, rich nutrient waters are brought toward the surface due to the circulation. This increases the productivity of the rings. By using satellites it is possible to study the formation and destruction of rings by remote sensing techniques. This satellite information (Plate 8*) is also very useful in positioning ships in a ring and tracking a ring as a function of time.

The water masses in the Atlantic have been studied for some time. Two expeditions, the Meteor in 1925 to 27 and the International Geophysical Year in 1957 to 58, supplied the earlier data while, more recently, the GEOSECS and TTO (Transient Tracers in the Ocean) programs have provided physical and chemical data on the Atlantic. The T-S diagram for water masses in the Atlantic are given in Figure 1.23. The upper waters below the surface are

* Plate 8 follows page 48.

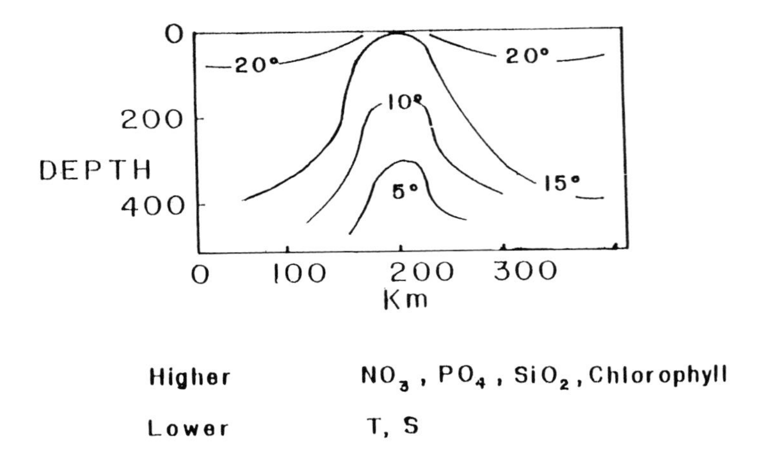


FIGURE 1.22. The temperature structure of a Gulf Stream ring.

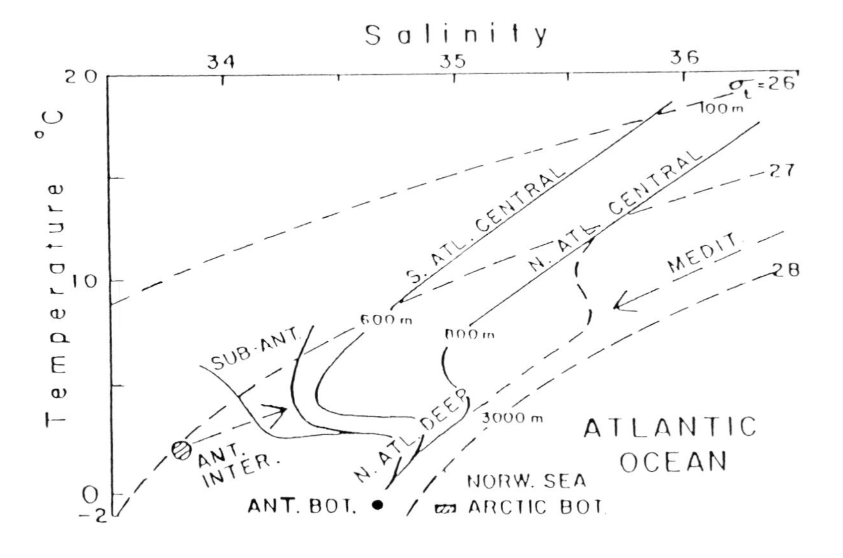


FIGURE 1.23. The T-S diagram for the Atlantic Ocean.

called Atlantic Central Waters. They extend to depths of 300 m on either side of the equator and deepen to 600 to 900 m at mid latitudes. These waters are thought to be formed by the sinking of waters on the equatorward side of the subtropical convergences. The cooler, less saline water sinks at higher latitudes. The more saline warmer water sinks at lower latitudes and flows toward the equator. In the South Atlantic the Central Water terminates at a depth where it merges with the Antarctic Intermediate water. In the North

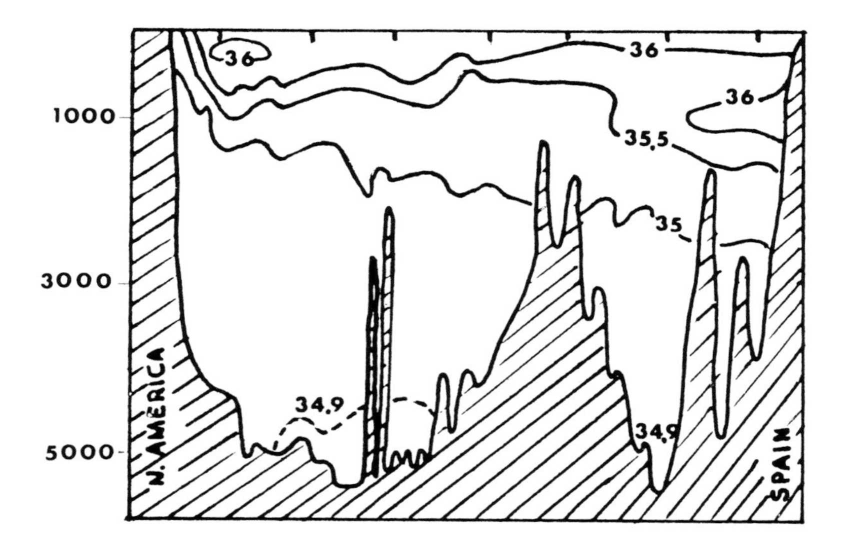


FIGURE 1.24. East-west section of salinity in Atlantic Ocean.

Atlantic, the Arctic Intermediate water is not as dominant. The Central water merges into high latitudes in the North Atlantic and into Mediterranean water at low latitudes. The effect of the Mediterranean water can be demonstrated by examining the horizontal section of salinity at a depth of 1000 m (Figure 1.24).

The temperature and salinity section for the Atlantic shown earlier (as well as the $O_2$ section to be shown later) demonstrated that the great bulk of Atlantic water is deep water. The major source of this North Atlantic deep water is the Norwegian Sea. This water flows over sills between Scotland, Iceland, and Greenland and sinks into the deep Atlantic. This flow is not continuous, but occurs in pulses. The salinity is 34.9 and the temperature is between 2 to 3°C for most of the North Atlantic deep waters.

The bottom waters in the western and eastern basins of the South Atlantic are different because of the separation due to the Mid-Atlantic ridge (Figure 1.25). On the west side the bottom water has temperatures down to 0.4°C, while on the east side the minimum is 2.4°C. The bottom waters are prevented from flowing into the eastern basin by the Walfish Ridge which has a sill depth of 3500 m. The salinity difference is not dramatic (34.7 on the west and 34.9 on the east).

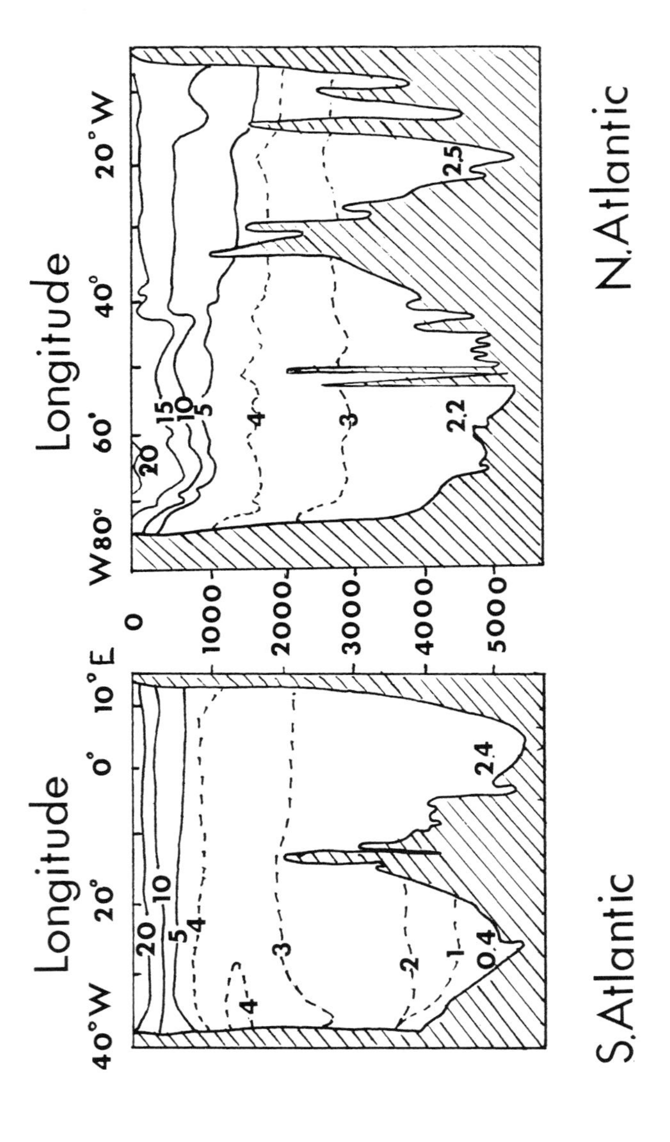


FIGURE 1.25. East-west section of temperature in Atlantic Ocean.

The recent tritium measurements by Ostlund and co-workers in Miami have clearly demonstrated the speed of formation of North Atlantic deep waters. Their results for tritium in the North Atlantic obtained from the earlier GEOSECS and more recent TTO program are shown in Figure 1.26. The measurements show the deep southward penetration of tritium, which has a half-life $t_{1/2}$ = 12 years and entered North Atlantic waters after the 1960 atomic bomb tests. The TTO measurement made about ten years after the GEOSECS work show the further movement of North Atlantic deep waters.

A summary of the balance of water masses for the South Atlantic is shown in Figure 1.27. The southward flow consists of 17 Sv of surface waters and 18 Sv of deep waters. The northward flow consists of 23 Sv of upper gyre waters, 9 Sv of Antarctic intermediate waters, and 3 Sv of Antarctic bottom waters.

Another feature of the circulation of the Atlantic waters that should be mentioned is the vertical circulation in equatorial regions. This circulation is shown in Figure 1.28. There is a general upwelling that is revealed in the profiles of salinity and temperature described earlier. These features bring cold, rich nutrient waters near to the surface and support primary productivity near the equator. Recent studies have also examined the countercurrents near the equatorial regions. More will be said about this when discussing the Pacific Ocean waters.

### 3.2. Southern Ocean Waters

Surface water circulation in the southern oceans is controlled by the circumpolar and polar gyres that are driven, respectively, by the west wind drift (clockwise) and east wind drift (counterclockwise) (Figure 1.29). The surface circulation is strongly affected by a number of convergence zones shown in Figure 1.30. Going north from Antarctica the surface sea temperature increases slowly until a region is reached where it rapidly increases by 2 to 3°C. This convergence zone is called the Antarctic convergence. This zone occurs at about 50S in the Atlantic and Indian oceans and 60S in the Pacific ocean. The region from the continent to the Antarctic convergence is called the Arctic zone. The surface temperatures of waters in this zone are −2 to 1°C in the winter and −1 to 4°C in the summer. Salinities are less than 34.5 due to the melting of ice and the colder surface waters moving north sink in this region and continue to flow north. This water mass is called Antarctic intermediate water and this mixes with the North Atlantic deep water as it

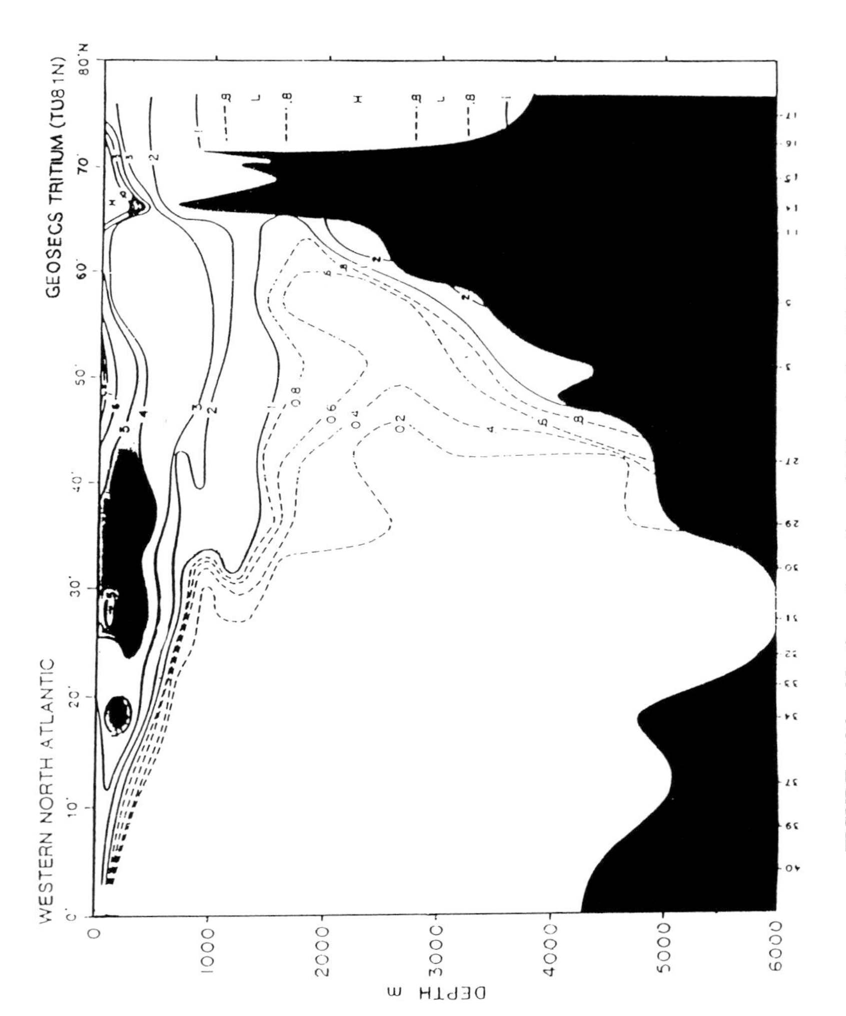


FIGURE 1.26. North-south section of tritium in the Atlantic Ocean.

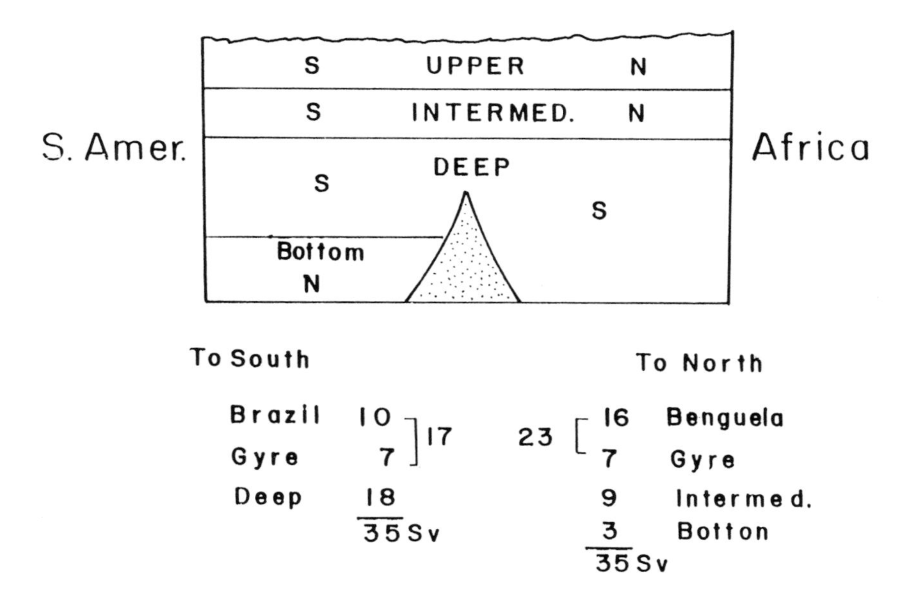


FIGURE 1.27. The balance of water masses in the South Atlantic.

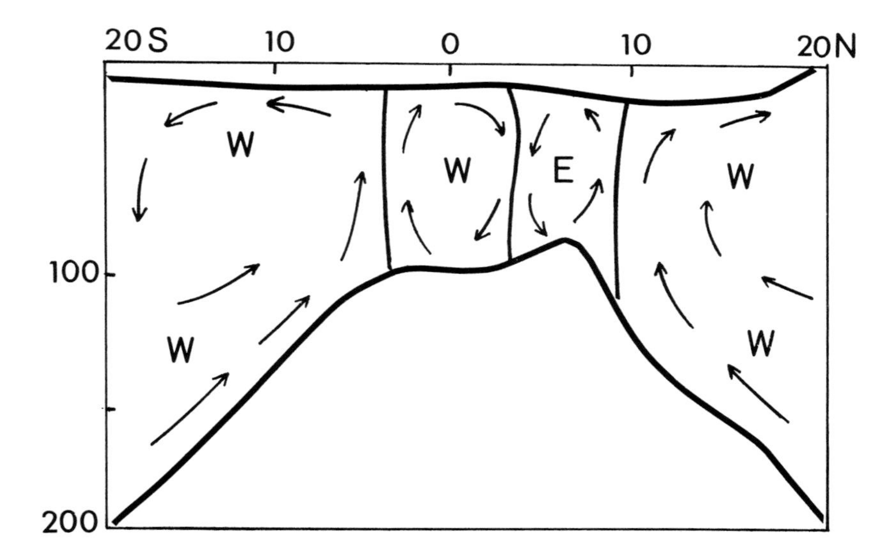


FIGURE 1.28. Vertical circulation in the equatorial Atlantic.

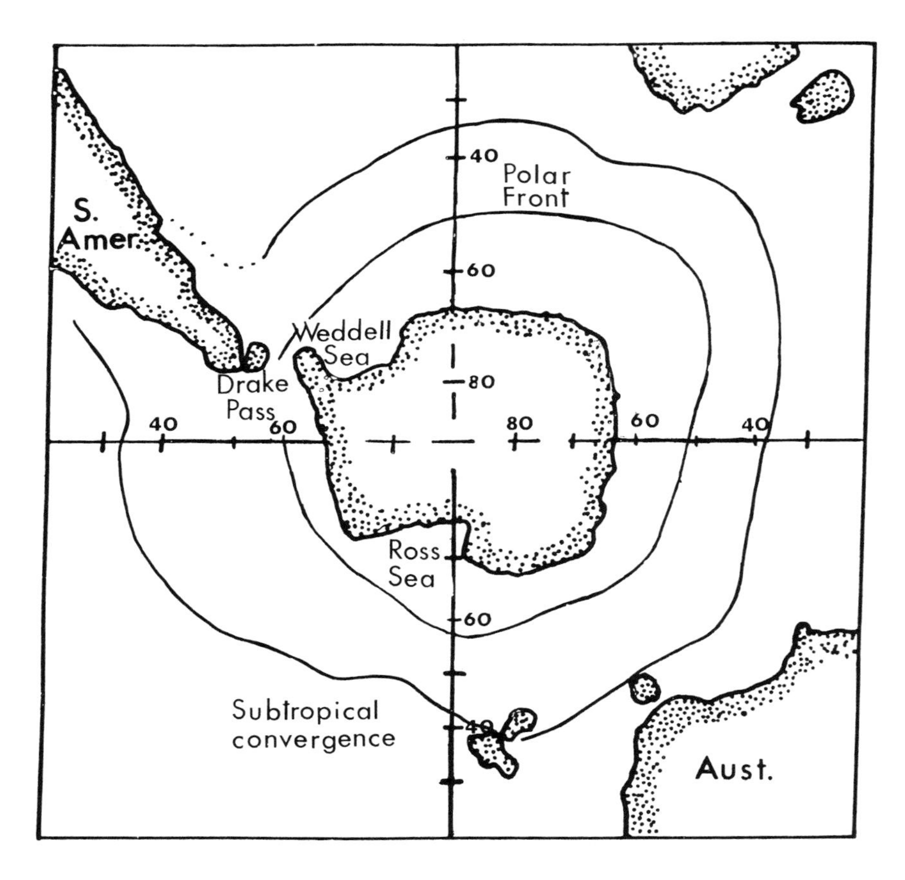


FIGURE 1.29. The surface currents of the southern oceans.

moves north. It is about 500 m thick and has t = 2°C and S = 33.8. It has a $\sigma_T$ = 27.0 and forms a low salinity tongue with its center at 800 to 1000 m at 40°C. It also has a high $O_2$ concentration (220 to 300 $\mu M$) when it leaves the surface.

If one continues north from the Antarctic convergence, the temperature rises slowly to a region where it rises rapidly by about 4°C with a salinity increase of 0.5. This region is called the subtropical convergence. It occurs at about 40N around most of the Antarctic, but it is poorly defined in the eastern South Pacific. The area between the Antarctic and subtropical convergence is called the subantarctic zone. It has surface temperatures of 4 to 10°C in the winter and up to 14°C in the summer. The salinity is 33.9 to 34.9 in the winter and as low as 33 in the summer as ice melts.

The strong circumpolar current has a northern component due to the

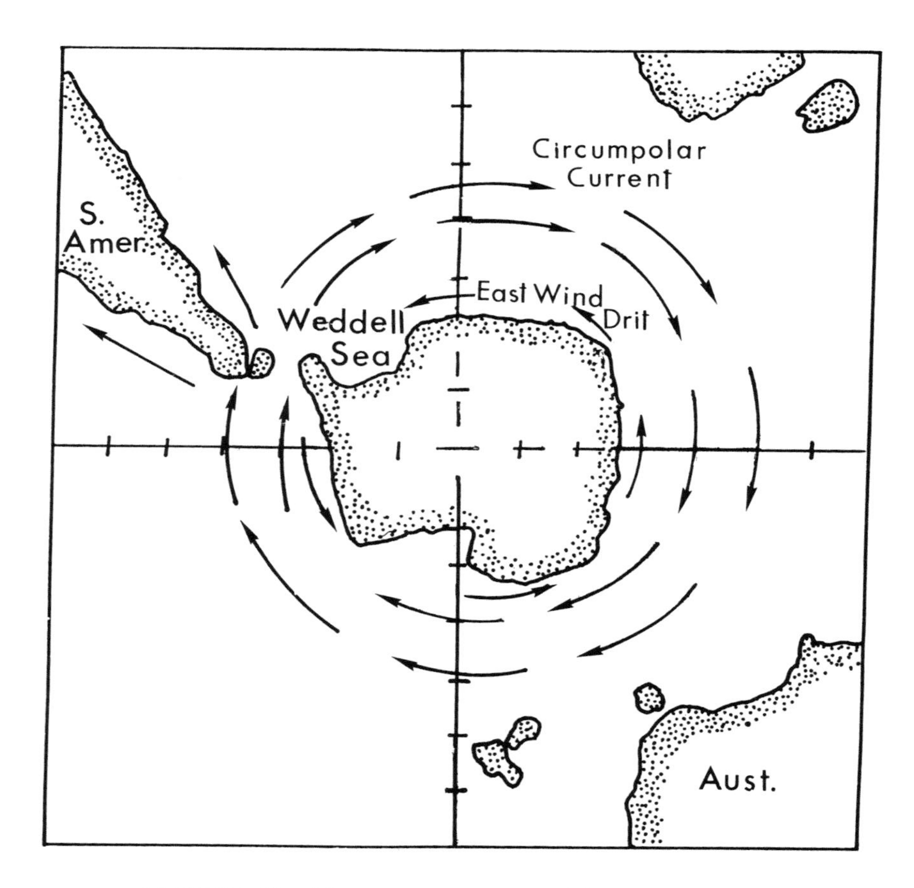


FIGURE 1.30. The convergence zones of the southern oceans.

coriolis force (from the rotation of the earth). The speed of this current varies from 4 cm $s^{-1}$ in the Arctic zone to 15 cm $s^{-1}$ north of the Antarctic convergence. It has the largest volume flow of all the water masses (150 to 190 Sv).

A sketch of the water masses in the southern oceans is shown in Figure 1.31. The circumpolar water extends from the surface to 400 m. At 400 m it has a t = 2.5°C and a t = 0°C at the bottom. The salinity is 34.7 for the circumpolar water. It is made up of North Atlantic deep water with some mixing with surface and intermediate waters.

Antarctic bottom water is formed in the Weddell and Ross Seas. It is a mixture of circumpolar water and shelf water (t = −1.9°C, the freezing point of seawater; S = 34.6; and $\sigma_T$ = 27.9, the highest in the southern ocean). The bottom water is formed due to its high density ($\sigma_T$ = 27.9) and

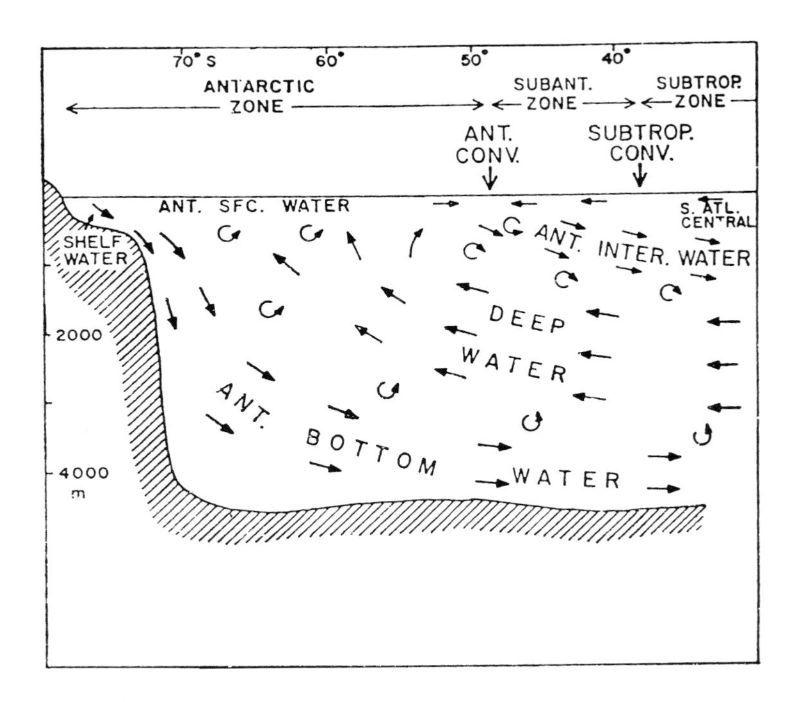


FIGURE 1.31. South-north section of water masses in the southern oceans.

flows down the continental slope northward in the Atlantic, Indian, and Pacific oceans. As it moves north it mixes with deep water and the temperature rises to 1.5°C and the salinity to 34.7 to 34.8.

As mentioned earlier, Antarctic intermediate water forms at the Antarctic convergence. It includes Antarctic surface water and circumpolar water. It moves north in the Atlantic, Indian, and Pacific oceans mixing with deep waters and surface waters.

The deep water in the southern oceans comes from the North Atlantic deep water which is circulated as part of the circumpolar current. It moves toward the north in the Indian and Pacific oceans and lies between 1500 and 3000 m in the Subantarctic region (t = 2 to 3°C and S = 34.6 to 34.8). This "warm" deep water is sandwiched between the colder intermediate and bottom waters. The North Atlantic origin of deep waters was suggested by earlier oceanographers due to the similarities of the temperature and salinity (t = 2°C and S = 34.9). These waters lose $O_2$ and gain nutrients as they make their tour from the North Atlantic to the North Pacific.

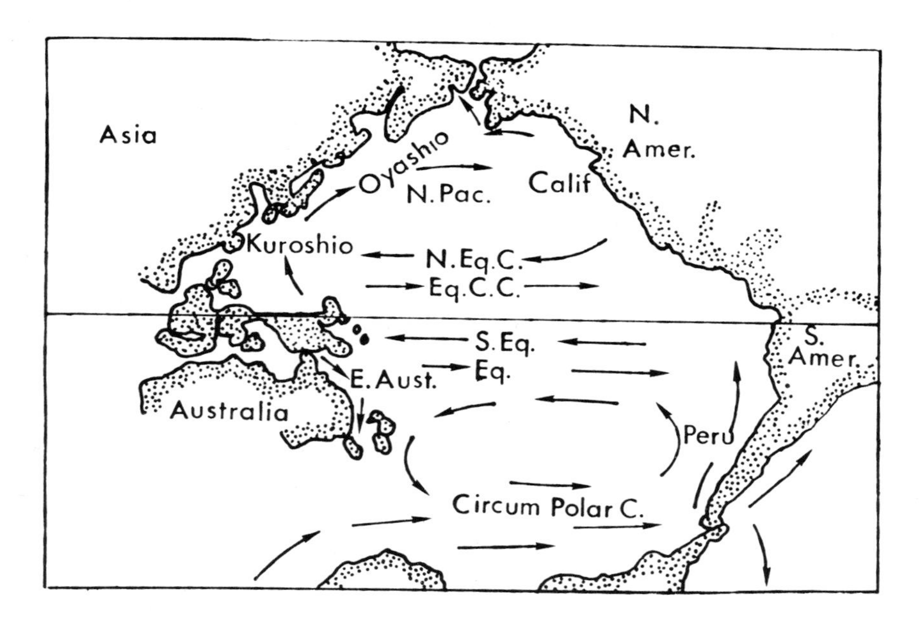


FIGURE 1.32. The surface currents of the Pacific Ocean.

## 3.3. Pacific Ocean Waters

Surface water circulation in the Pacific is similar to the Atlantic (Figure 1.32). In the North Pacific there is a clockwise gyre and in the South Pacific there is a counterclockwise gyre. The equatorial currents are more distinguishable in the Pacific due to its larger size. The North and South equatorial currents flow toward the west and represent part of the two major gyres. Both are driven by the trade winds. The North and South equatorial countercurrents flow toward the east. The North Equatorial Current has surface speeds of 25 to 30 cm $s^{-1}$ and volume flows of 45 Sv. The equatorial countercurrent has speeds of 35 to 60 cm $s^{-1}$ except in the Spring (March to April) when it decreases to 20 cm $s^{-1}$. The countercurrents extend to depths of 1500 m and have volume transports as high as 60 Sv. At the equator there is an undercurrent called the Cromwell Current which is below the sea surface (200 m in the west to 40 m in the east). It is only 0.2 km thick but 300 km wide and has a length of 14,000 km. The speeds are up to 170 cm $s^{-1}$ and the maximum volume transports are 70 Sv; the average is 40 Sv. This current was discovered by Buchanan in 1886 during the Challenger expedition, but was forgotten and rediscovered by Cromwell, Montgomery, and Stroup in 1952.

No major transport of water from the South to North Pacific Ocean occurs near the equator in the Pacific. The North Pacific does have a strong western boundary current like the Atlantic. It is called the Kuroshio Current and Extension. Its volume flow is about 65 Sv. It also meanders like the Gulf Stream and eddies or rings are formed. The California Current in the North and Peru or Humboldt Current in the South Pacific result in upwelling along the coasts. Both currents bring cold water toward equatorial regions and influence the climate in these areas. The Peru Current turns west between January and March at a few degrees south of the equator. The north equatorial countercurrent follows it south and brings warm water near the coast. This phenomenon is called ''El Niño'' because it occurs near Christmas. This increase in temperature kills fish and increases evaporation from the ocean and precipitation on the land, resulting in floods. In recent years a number of studies have been made on the ''El Niño'' in attempts to predict its behavior due to processes occurring in the Eastern Pacific.

The water masses in the Pacific are more complicated than the Atlantic due to its size. The salinity of surface waters shows a maximum in the tropics like the Atlantic, but the values in the North Pacific are lower than in the North Atlantic. The salinities in the South Pacific are greater than in the North Pacific, but less than the South Atlantic. As discussed earlier, this is related to the high evaporation rates in the Atlantic. The surface temperatures of the Pacific waters are highest near the equator and in the Western basin due to the circulation from the North and South equatorial currents. The lower temperatures off the coast of Peru are due to the upwelling. The winds off the coast move the surface water off the shelf and they are replaced with cold, nutrient rich intermediate waters (500 m). The waters between 100 to 800 m in the Pacific are called central waters.

The waters between 20N to 10N are the most saline waters in the Pacific. The surface waters are separated from the Pacific equatorial waters by a strong thermocline. Vertical transport is prohibited, the discontinuation layer being 150 to 200 m in the west and 50 m or less in the east. It can reach the surface off the west coast of Central America (called the Costa Rican thermal dome). A salinity maximum occurs in the equatorial waters as with the Atlantic. At about 800 m the salinity goes through a minimum due to the flow, South, of Arctic intermediate waters and the flow, North, of Antarctic intermediate waters. The Pacific subarctic waters ($S = 33.5$ to $34.5$ and $t = 2$ to $4°C$) are more extensive than the Atlantic subarctic waters. It is formed at the subarctic convergence (western Pacific) between the more saline Kuroshio extension and the cold Oyashio waters.

The Antarctic intermediate waters in the Pacific (t = 2.2°C and S = 33.8) form at the Antarctic convergence. It is limited in its northern extension by the Pacific equatorial waters. In the Atlantic, the equatorial waters are not as clearly defined and the intermediate waters go across the equator.

The North Pacific intermediate waters are delineated by a salinity minimum at 800 m in the west and 300 m in the east. These waters appear to circulate in a clockwise gyre similar to surface waters. The high $O_2$ values indicate that the waters are replenished with surface waters. Reid (SCRIPPS) has pointed out, however, that the density of these waters differs from those at the surface. He thus concludes that these intermediate waters attain their properties below the surface.

The deep Pacific waters (2000 m to the bottom) have t = 1.1 to 2.2°C and S = 34.65 to 34.75. The salinity increases or remains constant with depth unlike the Atlantic where a salinity maximum is observed. The reason for the uniformity is that no deep water is formed in the Pacific; it is a sink for North Atlantic (and other) deep waters.

In the South Pacific, bottom waters are supplied from the Antarctic circumpolar waters. The temperature of these bottom waters increases from 0.9° in the Antarctic to 1.5°C in the North Pacific. This is thought to be due to heat flow from the interior of the earth. The deep and bottom water movements in the Pacific are very slow and only a little exchange occurs between the South and North Pacific. The similarities in the properties (t = 1.5°C and S = 34.7) in the deep waters of the Pacific and Indian oceans form the largest water mass of the world oceans and are called oceanic or common water. It is largely a mixture formed from North Atlantic deep water and Antarctic bottom waters. As will be discussed later, the deep Pacific waters have much lower $O_2$ and higher nutrients than deep Atlantic waters. This is due to the waters being older and due to the high productivity of some of the surface waters. The Pacific has no source of high salinity water like the Mediterranean Sea in the Atlantic and the Red Sea in the Indian oceans.

The T-S diagrams for the major water masses in the Pacific are shown in Figure 1.33. The water masses are more complicated than in the Atlantic because of its size. The upper water masses below the surface have a variety of properties. Between 100 to 800 m the North and South Pacific waters and the Pacific Equatorial water masses dominate. The equatorial water extends over the entire ocean and has a very uniform T-S relationship. The surface waters are separated from the equatorial waters by a strong thermocline which inhibits vertical transfer. The depth of the discontinuity layer decreases from 150 to 200 m in the west to 50 m or less in the east. The layer sometimes

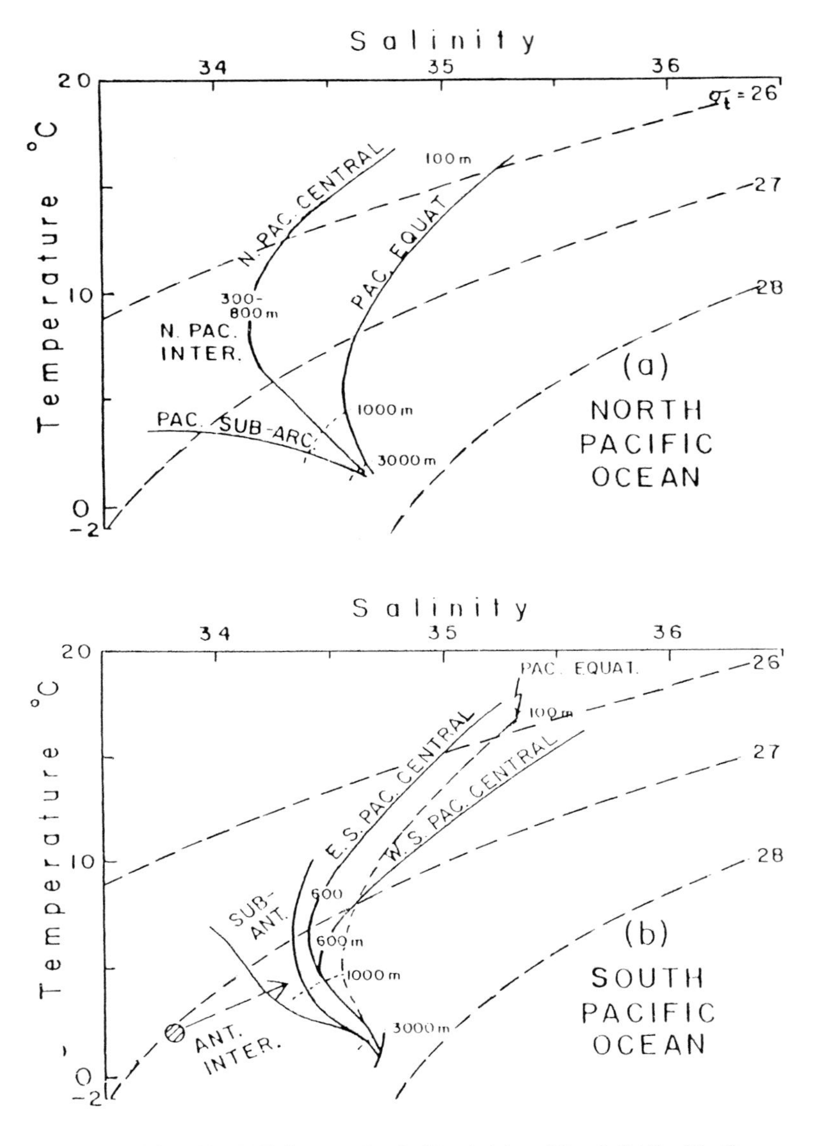


FIGURE 1.33. The T-S diagram for the North (a) and South (b) Pacific Ocean.

reaches the surface near the American coast and is called the Costa Rica thermal dome.

The equatorial undercurrent in the Pacific, called the Cromwell Current, has been studied for some time. It was discovered by Cromwell, Montgomery, and Stroup in 1954. It caused a great deal of excitement because it had not been predicted by theory. The current was discovered by tracking drogues

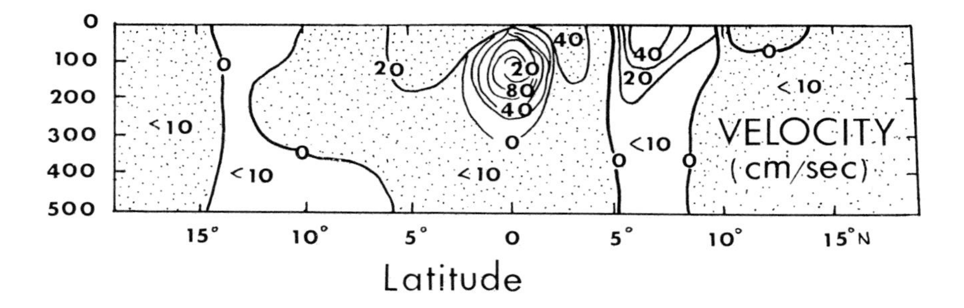


FIGURE 1.34. North-south velocity section in the Pacific Ocean.

that were released at different depths. The submerged current was observed as a narrow undercurrent above the thermocline. The technique was similar to the methods used by Buchanan in the 19th century to study the Atlantic undercurrent. The general features of the Cromwell Current are shown in Figure 1.34. The shaded area in the velocity section indicates eastward flow. The speeds are as high as 120 cm $sec^{-1}$ and estimated transport is about 40 Sv. The south equatorial current flows above the undercurrent and separates it from the surface countercurrents to the north and south. It should be pointed out that the vertical circulation in the equatorial Pacific is similar to the Atlantic (Figure 1.28).

### 3.4. Indian Ocean Waters

The surface currents in the Indian Ocean are shown in Figure 1.35. The Indian differs from the Atlantic and Pacific in its limited northern extent due to the Asian land mass. The subtropical convergence at 40S is considered to be the southern boundary of the Indian Ocean. In the South Indian Ocean the counterclockwise gyre is similar to the South Atlantic and Pacific oceans. It is bound on the south by the Polar Current. Part of this current flows north along the west coast of Australia and is augmented by a coastal current that flows toward the west along the south of Australia. Part of this current flows west and becomes the South Equatorial current. The southward current along the south coast of Africa is called the Agulhas Current, completing the gyre. Part of this current flows west into the Atlantic to form the Benguela Current.

The equatorial current systems in the Indian Ocean differ from the Atlantic and Pacific due to seasonal changes in the winds. From November to March the winds and currents are similar to the other oceans except they are further south. The South Equatorial Current (20 to 8S) flows west throughout the

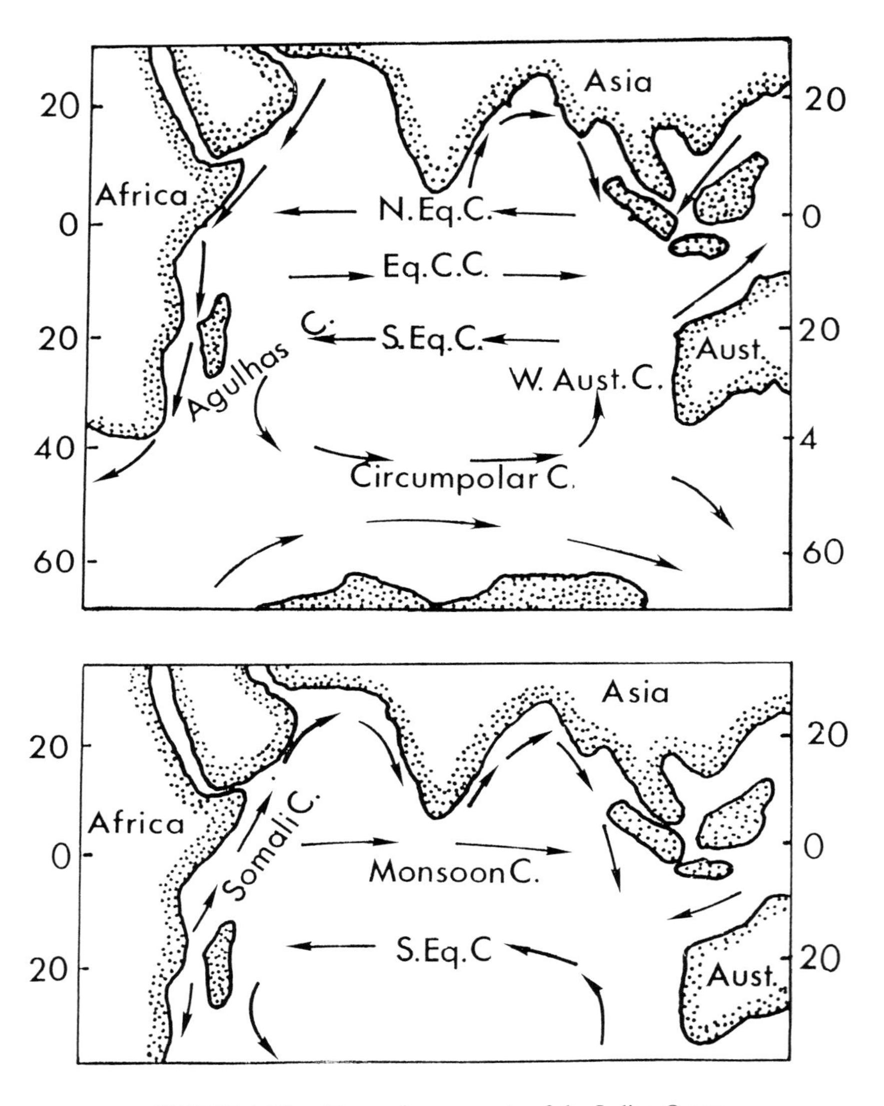


FIGURE 1.35. The surface currents of the Indian Ocean.

year. The equatorial countercurrent (8S to 2S) flows toward the east. The North Equatorial Current (2S to 10N) flows to the east and is maintained by the North East tradewinds.

In April the winds change and from May to September the North East tradewinds are replaced by winds from the southwest called the South West monsoon winds. These are really a continuation across the equator of the South East tradewinds. The North Equatorial Current is replaced by a south-

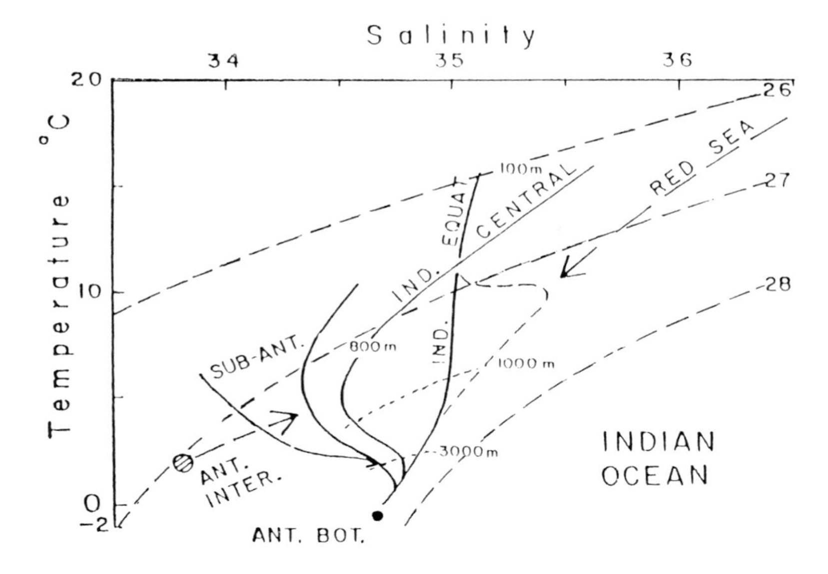


FIGURE 1.36. The T-S diagram for the Indian Ocean.

west Monsoon Current. The equatorial countercurrent also disappears or becomes undistinguishable from the Monsoon Current. During the monsoon period the South Equatorial Current turns north supplying the Somali Current which flows north along the east coast of Africa.

The temperature and salinity of waters near the equator in the Indian Ocean are similar to those in the Atlantic and Pacific. Waters in the Bay of Bengal have lower salinities (S = 31 to 34) due to river runoff, particularly during the monsoon season. Waters in the Arabian Sea have a high salinity (S = 36.5) due to evaporation.

The T-S diagrams of the major water masses in the Indian Ocean are shown in Figure 1.36. The equatorial waters have a S = 34.9 to 35.2 and are located 10N at a depth of 100 to 2000 m. The central waters are similar to South Atlantic central water (800 to 1000 m). The Antarctic intermediate waters are characterized by the usual salinity minimum. It is formed at the Antarctic convergence as in the Atlantic and Pacific. The deep waters in the Indian Ocean have t = 1 to 3°C and S = 34.6 to 34.8. This deep water has similar properties to Pacific deep waters and is the remnant of North Atlantic deep water. The Antarctic bottom water in the south Indian Ocean has similar properties to bottom waters formed in the Atlantic and Pacific.

In the northern and western Indian Ocean a warm saline water mass flows out of the Red Sea to a depth of 1000 to 1500 m. The sill depth of the Red

Sea is about 125 m. Since the evaporation exceeds precipitation in the Red Sea and there is little river runoff, the salinity gets very high (42.5) as does the temperature. This salty water sinks and flows out of the Red Sea across the sill. The water that leaves the Red Sea has different properties than deep Red Sea waters. Inside the sill the t = 24°C and S = 39.8; outside the sill the t = 15°C and S = 36. Surface inflow occurs similar to the Mediterranean Sea.

### 3.5. Arctic and Adjacent Seas

The surface circulation in the Arctic and adjacent seas is shown in Figure 1.37. The Norwegian Current is a continuation of the North Atlantic current that turns north and flows into the Norwegian Sea. The East Greenland Current flows southwestward along the Greenland coast. It is composed of waters from the Arctic Sea, and some of these waters mix with Norwegian Current waters. The speed of the current is about 30 cm $s^{-1}$. The ridge from Greenland to Scotland is 1000 m and prevents deeper Atlantic water from entering the Norwegian and Arctic Seas. The East Greenland Current carries ice from the Arctic and into the North Atlantic.

In the Norwegian and Greenland Seas, the surface waters consist of the Greenland gyre (above 1500 m, t = $-1.1$ to $-1.7$°C and S = 34.86 to 34.90 and below this depth S = 34.92 and t $< -1.1$°C). The deep waters of the Norwegian gyre have the same salinity, but the temperatures are greater than $-0.95$°C (similar to the bottom water of the Arctic basin). The Norwegian Sea thus forms a barrier to the passage of the colder Greenland gyre water. The deep waters of both seas have high $O_2$ concentrations indicating that its formation takes place by the cooling and sinking of winter waters.

The West Greenland Current is the result of East Greenland Current water that flows around the tip of Greenland into the Labrador Sea. The Baffin Bay Current flows southward and becomes the Labrador Current, which flows into the Atlantic. It has a low salinity (S = 30 to 34) and a low temperature (less than 0°C). The volume flow of the Labrador Current has been estimated to be 5.6 Sv. This can be compared to the East Greenland Current with a flow of 7.5 Sv. The difference (1.9 Sv) is the estimate of outflow as deep water into the Atlantic.

The Arctic Sea is divided into two basins separated by the Lomonosov ridge between Greenland and Siberia. The Canadian basin has a maximum depth of 3800 m while the Eurasian basin has a maximum depth of 4200 m. The sill depth separating the basins is 1200 to 1400 m and the main connection

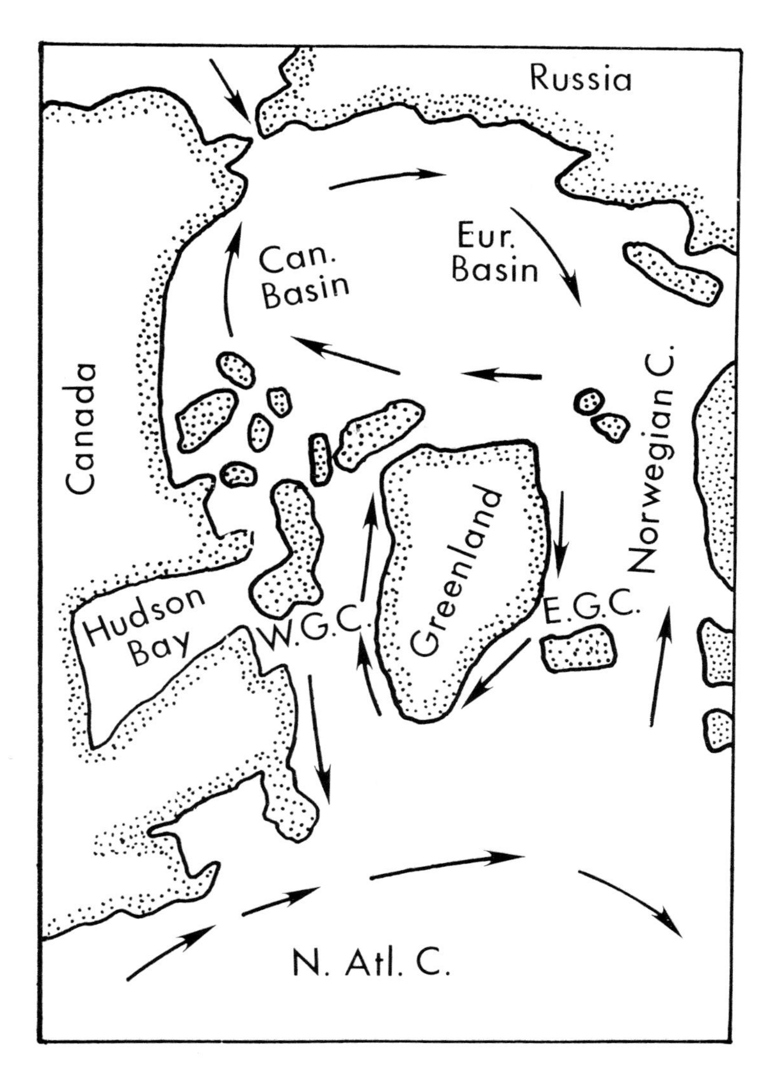


FIGURE 1.37. The surface currents of the Arctic Sea.

of the Eurasian basin with the Atlantic is through Greenland and Spitzbergen which has a sill depth of 1500 m. The Bering Strait has a sill depth of 50 m; thus, the Canadian basin has little exchange with the North Pacific.

The upper circulation of waters in the Arctic was obtained from the records of ships held in ice. The clockwise circulation in the Canadian basin leads out to the East Greenland Current. This is augmented by the counterclockwise circulation in the Eurasian basin.

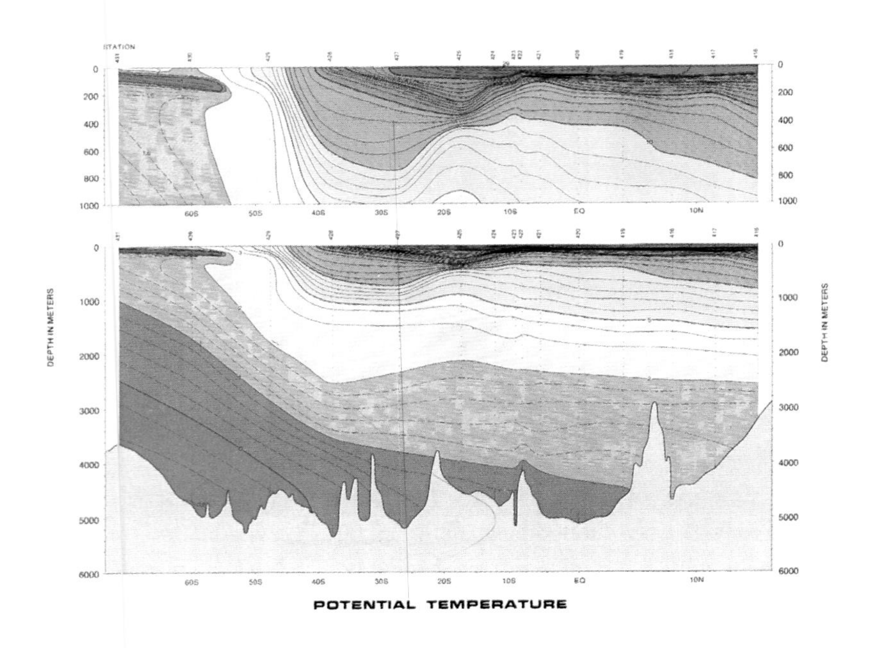


PLATE 3. North-south section of potential temperature in Indian Ocean.

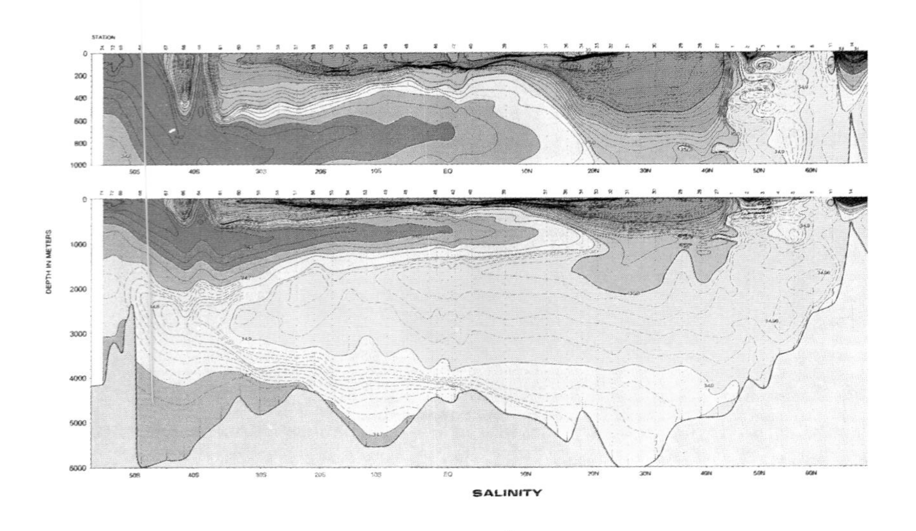


PLATE 4. North-south section of salinity in Atlantic Ocean.

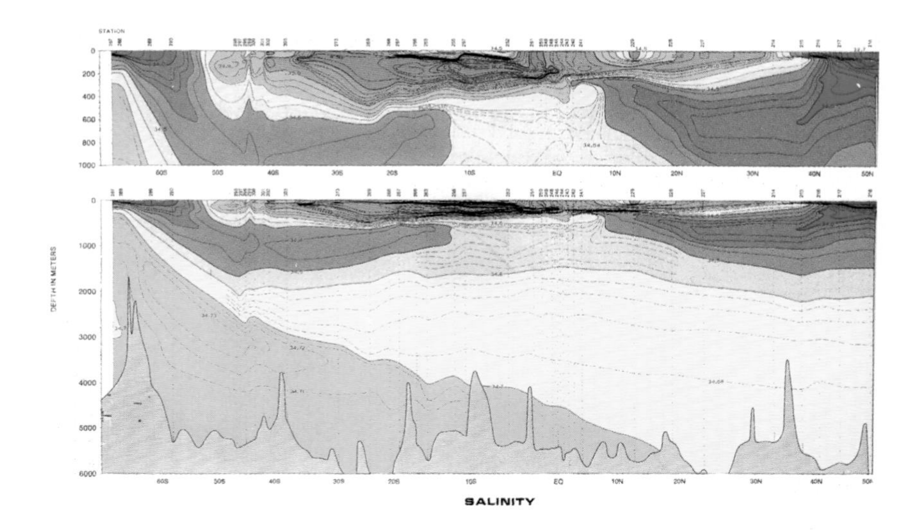


PLATE 5. North-south section of salinity in Pacific Ocean.

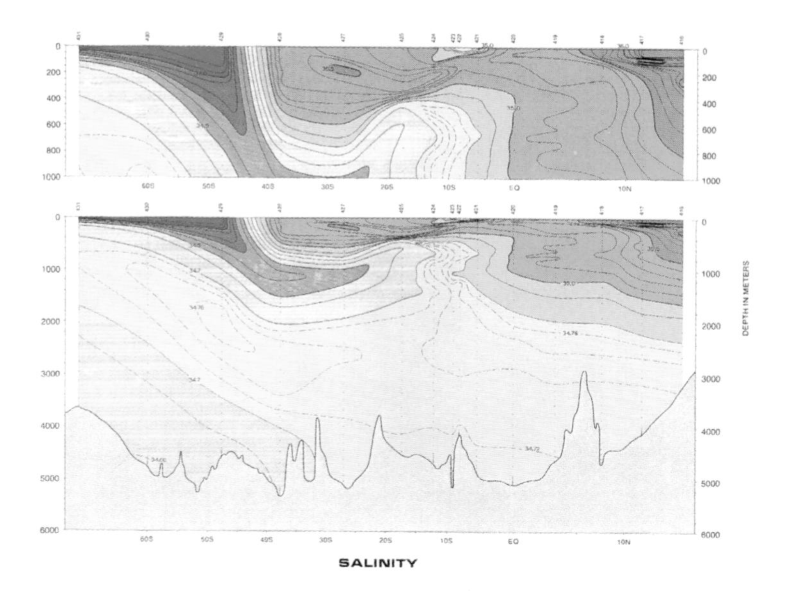


PLATE 6. North-south section of salinity in Indian Ocean.

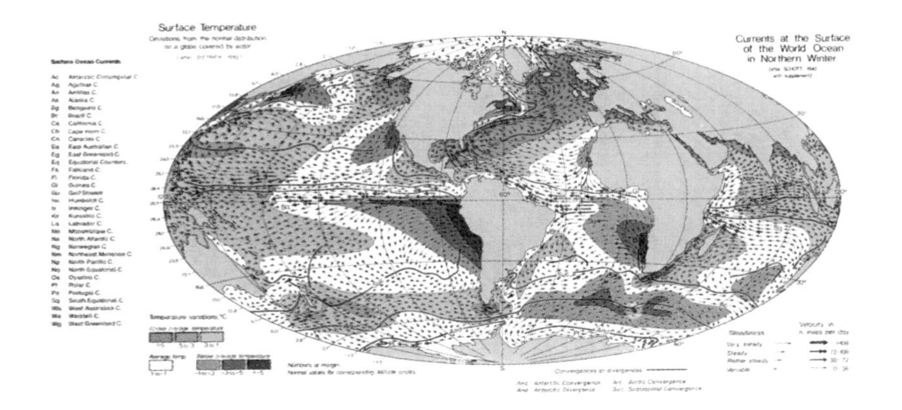


PLATE 7. The surface currents of the oceans.

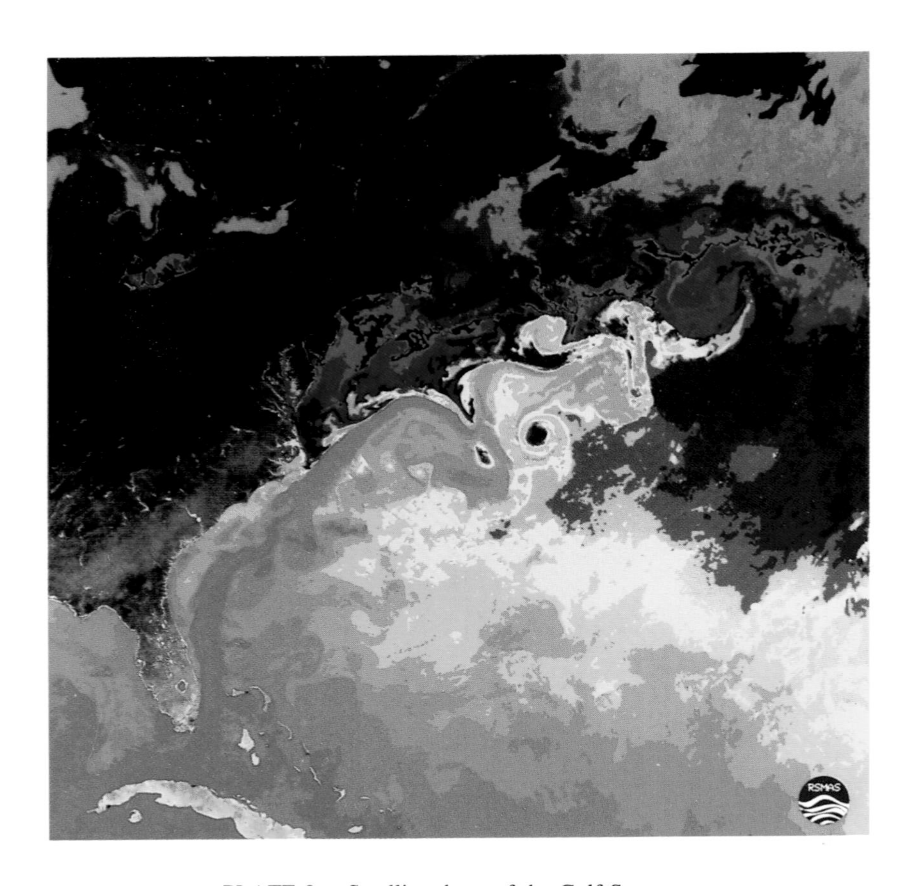


PLATE 8. Satellite photo of the Gulf Stream.

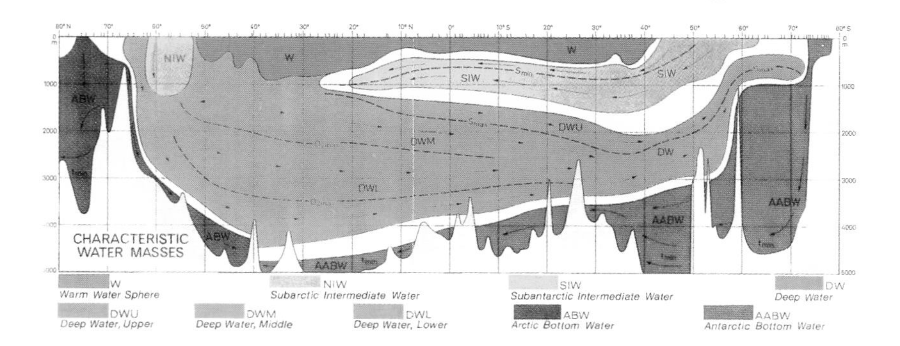


PLATE 9. Characteristic water masses in the Alantic Ocean.

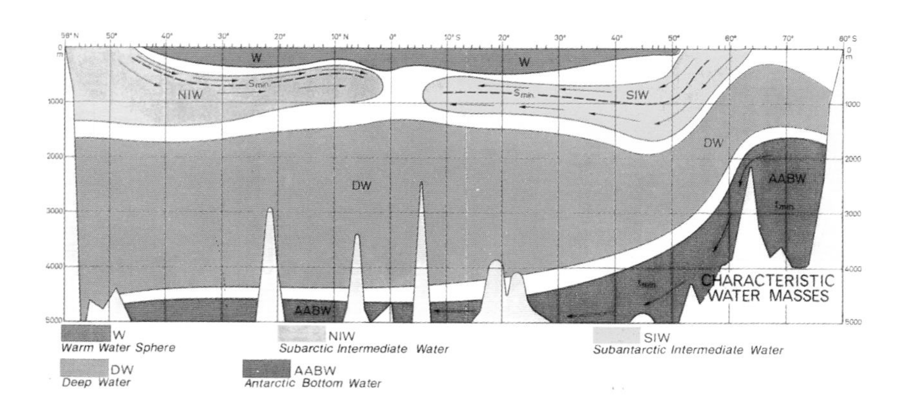


PLATE 10. Characteristic water masses in the Pacific Ocean.

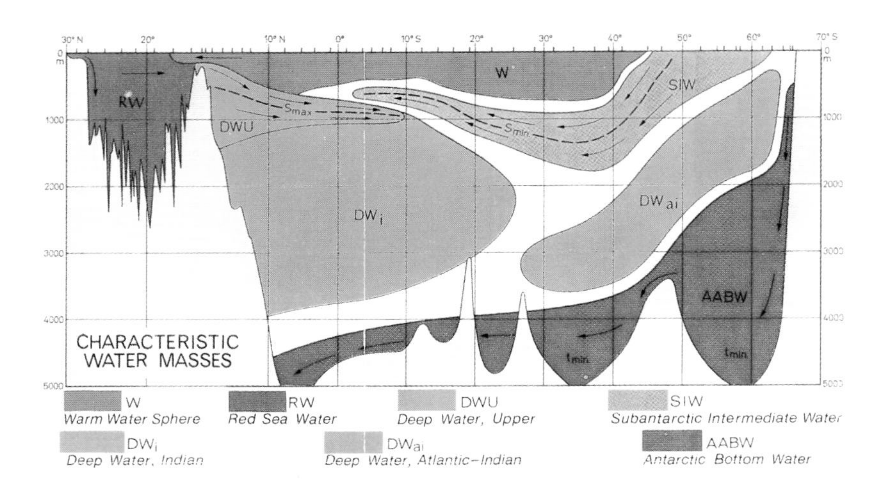


PLATE 11. Characteristic water masses in the Indian Ocean.

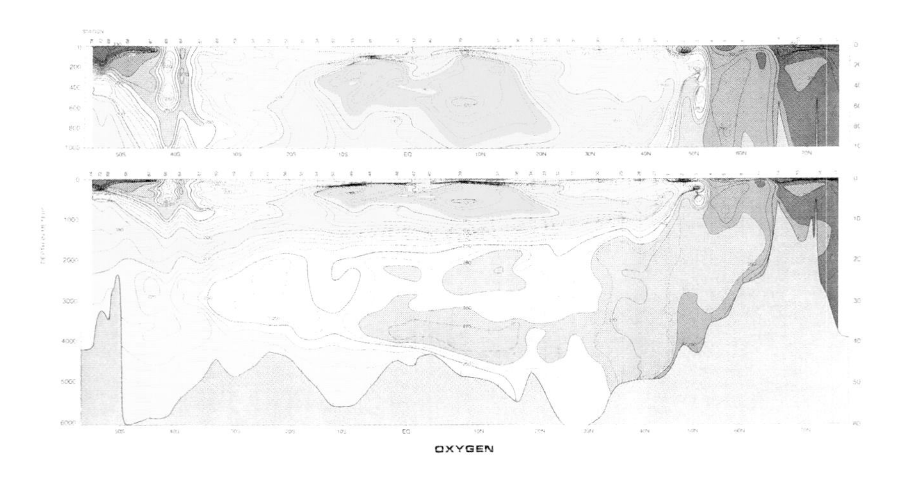


PLATE 12. A section of oxygen in the Atlantic Ocean.

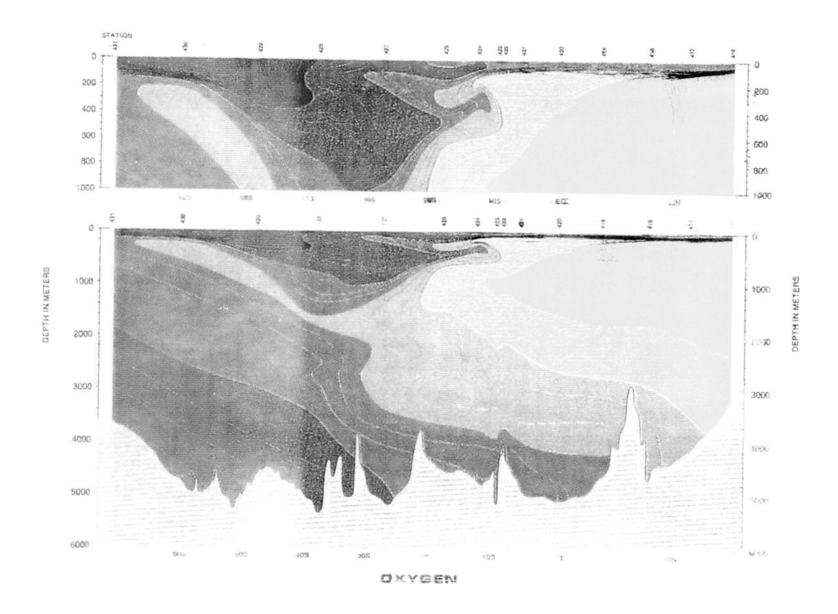


PLATE 13. A section of oxygen in the Indian Ocean.

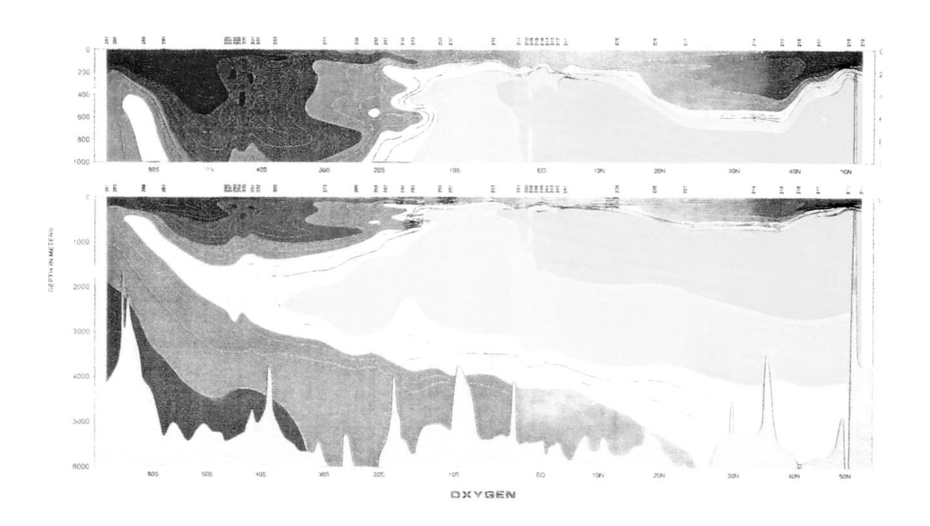


PLATE 14. A section of oxygen in the Pacific Ocean.

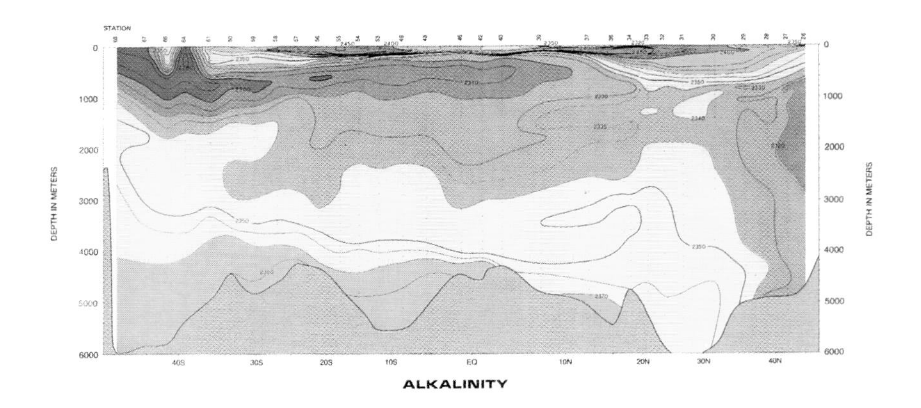


PLATE 15. A section of total alkalinity in the Atlantic Ocean.

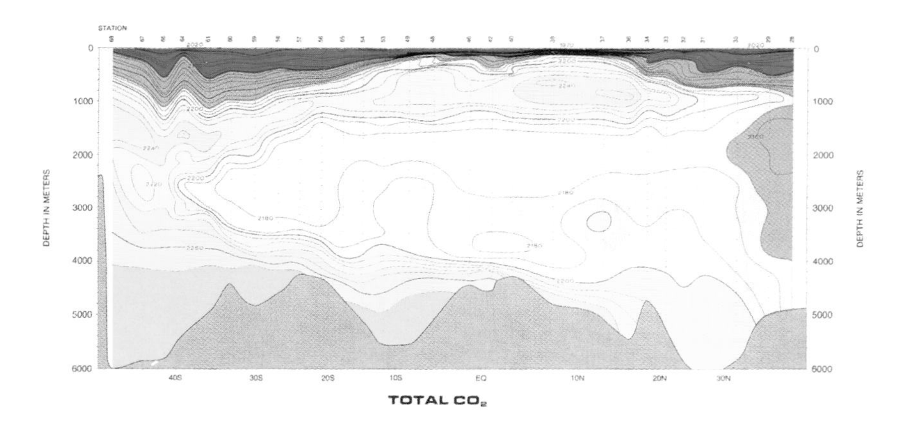


PLATE 16. A section of total carbon dioxide in the Atlantic Ocean.

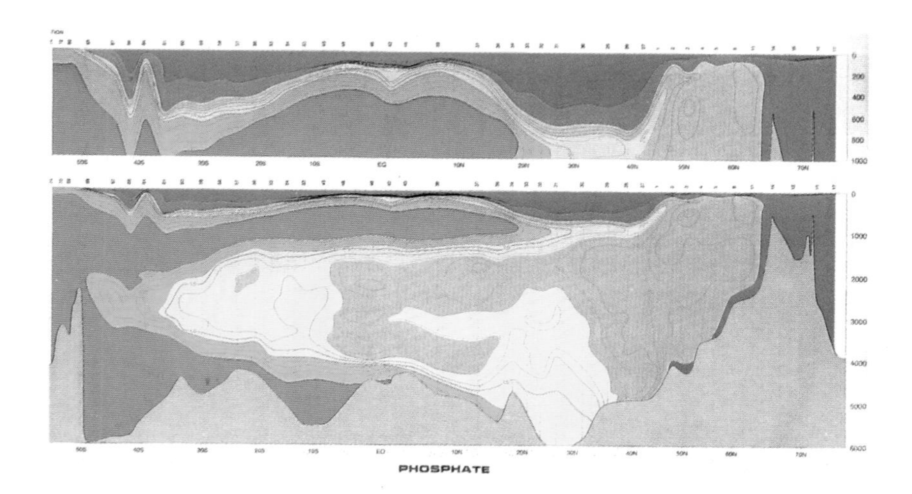


PLATE 17. Section of phosphate in the Atlantic Ocean.

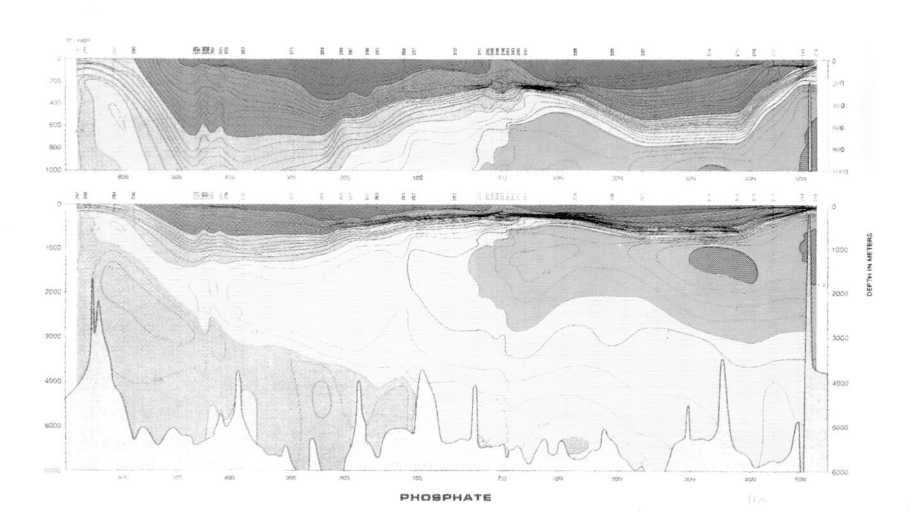


PLATE 18. Section of phosphate in the Pacific Ocean.

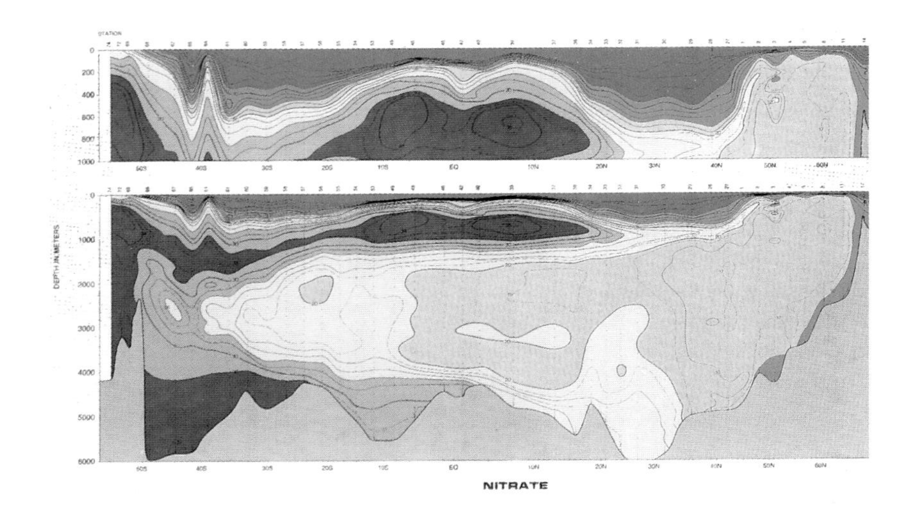


PLATE 19. Section of nitrate in the Atlantic Ocean.

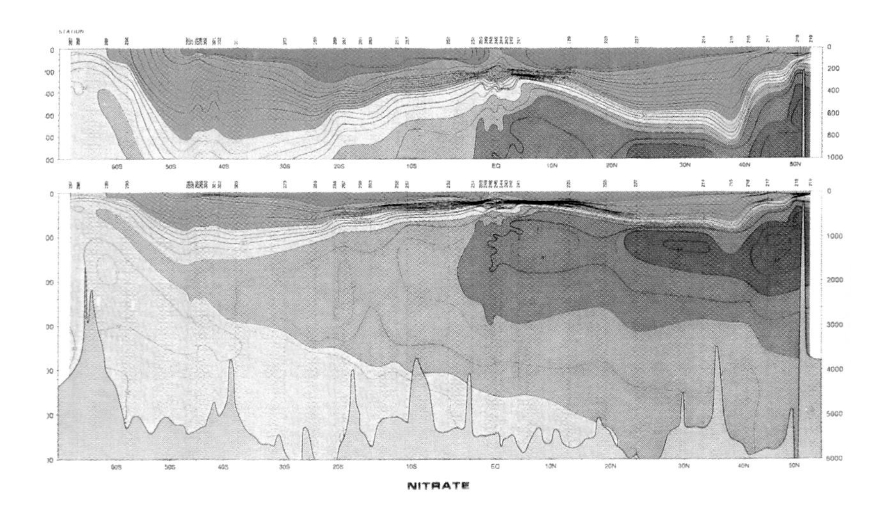


PLATE 20. Section of nitrate in the Pacific Ocean.

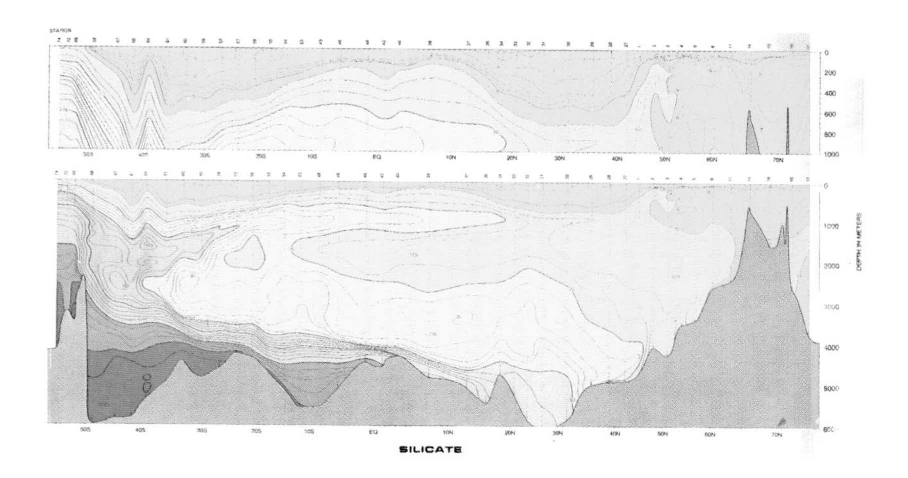


PLATE 21. Section of silicate in the Atlantic Ocean.

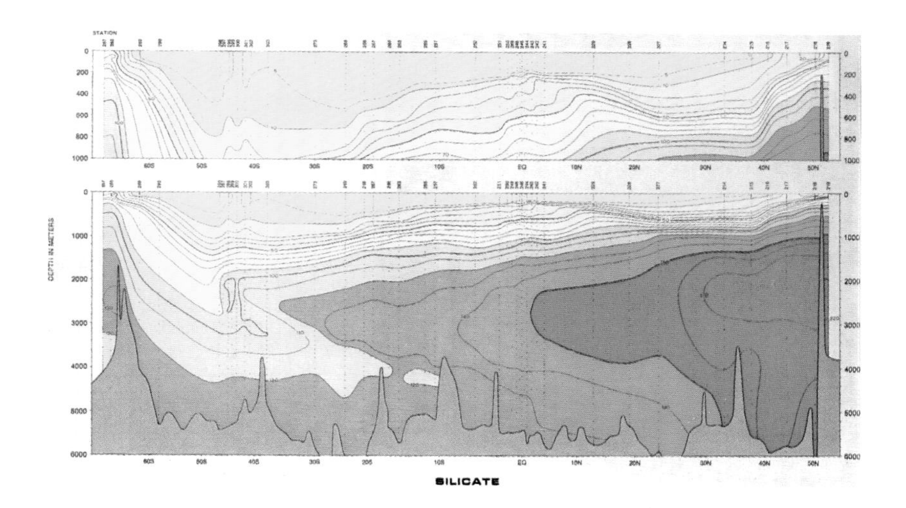


PLATE 22. Section of silicate in the Pacific Ocean.

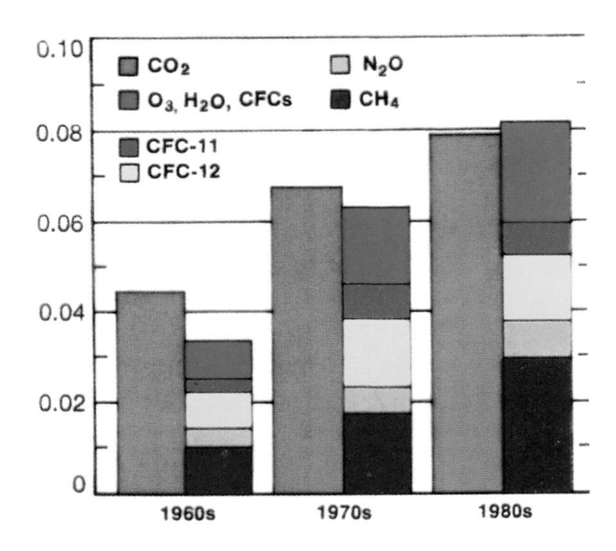


PLATE 23. Temperature change due to $CO_2$ and various greenhouse gases.

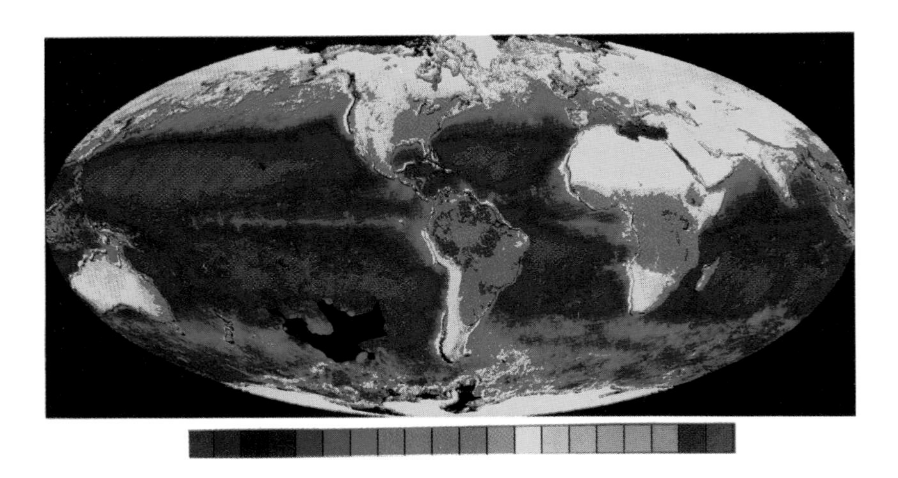

PLATE 24. World pigment map produced by NASA Goddard Space Flight Centre and the University of Miami from CZCS satellite data. This composite image of the ocean chlorophyll concentration was produced from 31,352 4-km resolution CZCS scenes from November 1978 to June 1981. The scale indicates increasing pigment concentrations ranging from $<.05$ mg/m$^3$ (violet at left) to 30 mg/m$^3$ (red-brown at right).

Recently, Ostlund and co-workers (Miami) have determined the deep water exchange in the Arctic Sea. They found the deep water exchange in the Eurasian basin is between 10 to 100 years, while it is about 700 years in the Canadian basin. Less than 10 to 15% of the deep waters originate from the shelves.

The polar ice cap covers 70% of the Arctic Sea. It is always present and extends from the Pole to about the 1000 m isobath. Some of the ice melts in the summer and the thickness decreases to a few meters. Open spaces called polynyas may form and ridges (rafting) occur in the fall. In the winter the thickness is about 3 m with hummocks that increase to a height of 10 m above sea level and up to 40 m meters below sea level. The polar cap ice is not always the same; up to one third is carried away by the East Greenland Current. Pack ice covers about 25% of the Arctic area outside the polar cap. It is lighter than cap ice and is carried further south than polar cap ice. This ice floe causes problems for navigation. Ice that forms from the land out from shore is called fast ice. It normally breaks up and melts in the summer. Icebergs are pieces of glaciers that come from the west coast of Greenland. It is different from pack ice and can extend to 70 m above sea level (as much as 500 m long). The volume below sea level varies from 1:1 to 7:1 depending upon its shape. The drift of icebergs is influenced by water currents while pack ice is influenced by winds. The pack ice moves to the right in the Northern Hemisphere due to the coriolis force. This was known by sailors long before it was noted by Nansen or suggested by Ekman.

## 3.6. The Age of Water Masses

A summary of the major water masses in the world oceans is given in Plates 9, 10, and 11.* These figures are based upon the water properties of various ocean waters. For a number of years many workers have attempted to estimate the age of the deep water masses in the world oceans. The recent $^{14}C$ work from the GEOSECS program has given us a new look at the deep water circulation in the world oceans (Ostlund et al.). Most of the deep water formation takes place in the North Atlantic. The average residence time for water in the North Atlantic is about 246 years which is the youngest of the oceans. The major source of this North Atlantic deep water is from the Norwegian Sea. This deep water flows over the sills between Greenland and Iceland. The temperature and salinity of this water is, respectively, 2°C and

* Plates 9 to 11 follow page 48.

**TABLE 1.5**[a]

| Caption | Various Waters | Volume |
|---|---|---|
| N | North Atlantic water formation | 14 Sv |
| Wa | Atlantic upwelling rates | 10 Sv |
| $C_A$ | Transport rate from Circumpolar to Atlantic | 7 Sv |
| $O_A^R$ | Transport rate from Atlantic to Circumpolar | 11 Sv |
| $C_I - C_I^R$ | Net transport from Circumpolar to Indian | 20 Sv |
| $C_P - C_P^R$ | Net transport from Circumpolar to Pacific | 29 Sv |
| Wc | Bottom water formed in Circumpolar | 41 Sv |
| Wi | Indian Ocean Upwelling | 20 Sv |
| Wp | Pacific Ocean upwelling | 25 Sv |

[a] See Figure 1.39.

35.0. This water flows to the south and, after mixing with bottom waters, into the South Pacific and South Indian oceans. It is, thus, the major water mass of deep ocean basins. From $^{14}C$ data, new estimates have been made of the time it takes for these waters to make a complete cycle. These estimates are given in Table 1.5. The estimates come from using a simple box model to account for the $^{14}C$ measurements (half-life: $t_{1/2}$ = 5800 years). A sketch of the box model with the values for the estimated rates of exchange are shown in Figure 1.39. The net flow of water from the Atlantic to the Circumpolar is 4 Sv, while the net flow from the Circumpolar to the Indian is 20 Sv and to the Pacific 25 Sv. The upwellings in the Atlantic (10 Sv), Indian (20 Sv), and Pacific (25 Sv) are balanced by the downwelling of North Atlantic waters (14 Sv) and Antarctic bottom water formed in the Circumpolar region (41 Sv). The 500 years estimate for the replacement time for deep Pacific waters is much shorter than earlier estimates (1000 to 1500 years).

## 3.7. Closed Basins

### 3.7.1. Mediterranean Sea

The Mediterranean Sea is divided into a western and eastern basin separated by a sill of 400 m which extends from Sicily to North Africa. The maximum depths are 3400 m in the western basin and 4200 m in the eastern basin. The sill depth between the Mediterranean and the Atlantic is 330 m.

The Mediterranean Sea is an example of a closed basin that has a higher rate of evaporation than the rate of precipitation and river input. This results

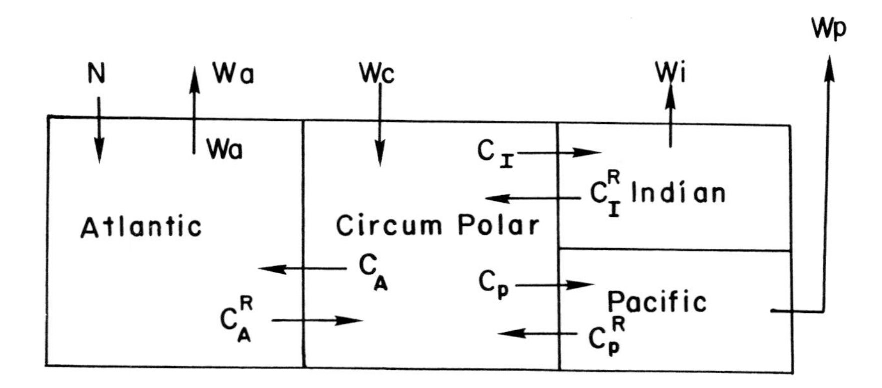


FIGURE 1.39. Box model for the deep water circulation of the world ocean waters.

in high salinities (S = 39) being formed in the surface layers. These waters sink in the winter and form saline deep waters. Off the coast of Turkey, Lavantine intermediate waters are formed with t = 15°C and S = 39.1. These waters flow toward the west at 200 to 600 m and out through the Straits of Gibraltar into the Atlantic. The Mediterranean water that flows into the Atlantic has a t = 13°C and S = 37.3 due to mixing with the incoming Atlantic waters.

The deep and bottom waters are formed during the winter off the Riviera in the western basin (t = 12.6°C and S = 38.4) and in the South Adriatic Sea in the eastern basin (t = 13.3°C and S = 38.65). Since the deep waters are replenished, they have high concentrations of oxygen.

### 3.7.2. Red Sea

The average depth of the Red Sea is 1000 m, with a maximum depth of 2200 m. It is separated from the Indian Ocean by a sill of 100 m depth. Because of the very dry climate, evaporation exceeds precipitation by about 200 cm/year. There is no river runoff into the Red Sea. This leads to salinities as high as 42.5, combined with temperatures of 30°C in the summer and 18°C in the winter. The deep water below the sill depth has a uniform t = 21.7°C and S = 40.6. This water is formed in the winter in the north. The intermediate water that flows out over the sill starts with t = 25°C and S = 39.8 and becomes t = 15°C and S = 36 outside the sill. This subsurface flow is compensated by a surface inflow of Indian Ocean water. This is similar to the Mediterranean Sea. The Red Sea water that flows over the sill is an identifiable feature in the Northern Indian Ocean.

In the northern Red Sea hot brine pools were discovered in some deep trenches. Temperatures of 58°C and salinities of 320 have been found. The compositions of the solutions indicate that the brines were formed by hot seawaters dissolving some evaporite deposits.

### 3.7.3. Estuaries

Conditions in coastal waters of the ocean are quite different in many ways from those in the open oceans. Variations with location and time are generally larger. These differences are caused by river runoff, tidal currents, and the effects of coastal boundaries on circulation. Tidal currents twice a day change the volume of water in a harbor or bay and promote vertical mixing. This breaks down any stratification in the water and can cause a resuspension of river sediments. River runoff reduces the salinity of the surface waters. The runoff normally has seasonal fluctuations that frequently follow the local precipitation or the summer melt of snow fields and glaciers. Most of the extreme conditions in coastal waters occur in or near river mouths or in estuaries.

An estuary can have a wide meaning in oceanography. The simple definition is the tidal region at the mouth of a large river. Pritchard's definition is "a semi-enclosed body of coastal water having a free connection with the sea" and containing a measurable amount of sea salt. One can classify an estuary into two major types:

1. A positive estuary where the salinity is less than the ocean due to precipitation and river runoff being greater than evaporation (e.g., the Baltic Sea).
2. A negative estuary where the salinity is greater than the ocean due to evaporation being greater than precipitation and river runoff (e.g., the Red and Mediterranean Seas).

It is also possible to define an estuary in terms of its shape. Three such divisions are:

1. A coastal plain estuary that results from land subsidence or a rise of sea level flooding a river valley (e.g., the St. Lawrence and Chesapeake Bays).
2. A deep basin estuary that has a sill toward the sea which is shallower than the basin and the outside sea (e.g., fjords of Norway and Canada).
3. A bar built estuary that has a narrow channel between the shore and a bar built up nearby by waves and sedimentation (e.g., Mexican lagoons).

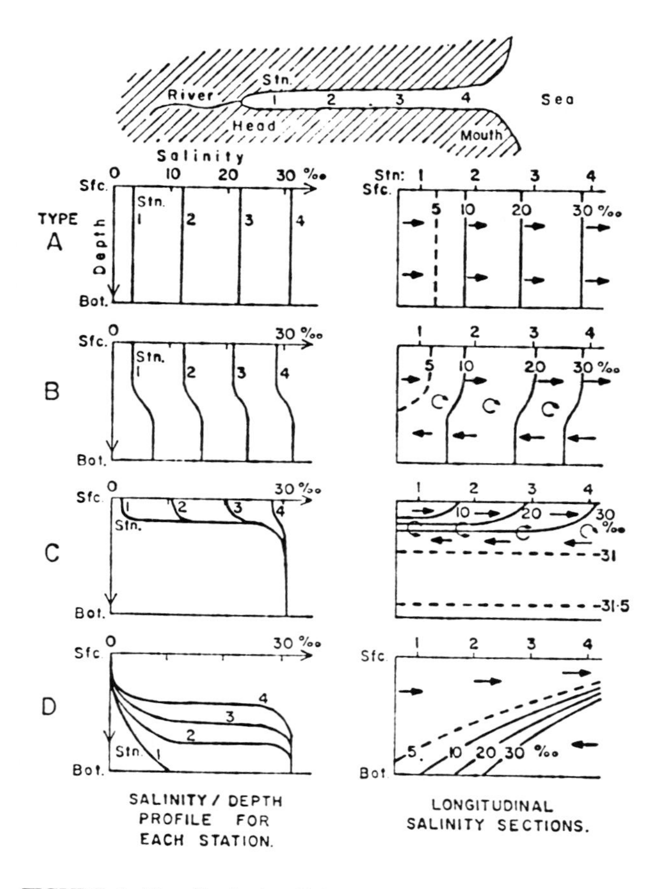


FIGURE 1.40. Typical salinity depth profiles and sections for estuaries.

Finally, it is possible to classify an estuary based upon the distribution of the water properties. Four types that are shown in Figure 1.40 are:

A. A vertically mixed estuary is a shallow basin where the water is well mixed. This leads to a similar salinity from the surface to the bottom at any particular place along the estuary. The salinity increases as the estuary flows from the head to the mouth. The Servern River in England is an example of this type of estuary.

B. A slightly stratified estuary is also shallow with the salinity increasing from the head to the mouth. The water, however, is in two layers. The upper layer is a little less saline than the deep layer. There is a net seaward flow in the upper layer and an inward flow in the deep layer.

Vertical mixing of the two layers results in the increase of salinity from the head to mouth in both layers. The James River in the Chesapeake Bay is an example of this type of estuary.

C. The highly stratified estuary is typical for fjords. The upper layer increases in salinity from zero at the head to the value close to seawater at the mouth. The deep water has an almost uniform salinity. There is a net outflow in the surface layer and inflow in the deeper water. A very strong halocline exists between the two layers. The depth of the interface is substantially the same throughout the estuary. The circulation in this type of estuary is dependent upon the sill depth, the river runoff, and the outside water density distribution. If the sill depth is too shallow, the inflow of deep saline water will not occur. This leads to the bottom waters becoming stagnant and perhaps devoid of oxygen. This situation occurs in Norwegian fjords but not all deep basin estuaries. The rate of renewal is affected by the river runoff and outside density structure (both of which can be seasonal).

D. The salt wedge estuary is named after its salinity structure. The saline water intrudes from the sea as a wedge below the river water. This situation occurs for rivers with a large volume transport such as the Mississippi or Fraser Rivers. The wedge is really very thin so that the isohalines are almost horizontal. There is a strong horizontal gradient of salinity at the bottom as in the slightly stratified estuary and a strong vertical gradient. The surface waters are largely fresh due to the strong river outflow.

This rather brief survey provides a discussion of descriptive aspects of estuarine systems. More will be said about the chemistry of estuaries in the next chapter.

## FURTHER READING

Bowden, K. F., Oceanic and Estuarine Mixing Processes, chap. 1, *Chemical Oceanography,* Vol. 1, 2nd ed., J. P. Riley and G. Skirrow, Eds., Academic Press, New York, 1-41 (1975).

Campbell, J. A., The Geochemical Ocean Sections Study-GEOSECS, chap. 44, *Chemical Oceanography,* Vol. 8, 2nd ed., J. P. Riley and R. Chester, Eds., Academic Press, New York, 89-155 (1983).

Deacon, D., *Scientist and the Sea 1650-1900,* Academic Press, New York (1971).

Dietrich, G., K. Kalle, W. Krauss and G. Siedler, *General Oceanography,* 2nd ed., Wiley and Sons, New York (1980).

Neumann, G., *Ocean Currents,* Elsevier, New York (1968).

Pickard, G. L. and W. J. Emery, *Descriptive Physical Oceanography,* 4th ed. Pergamon Press, Oxford (1982).

Tchernia, P., *Descriptive Regional Oceanography,* Pergamon Press, New York (1980).

# 2 Table of Contents

**The Composition of the Major Components of Seawater** .......... 59

1. *Introduction* .......... 59
2. *Method of Determination* .......... 62
3. *Composition and Stoichiometry of Average Seawater* .......... 65
4. *The Concept of Salinity* .......... 70
5. *Methods of Determining Salinity* .......... 78
6. *Causes of the Major Components Not Being Conservative* .......... 83
   - 6.1. *Estuaries* .......... 83
   - 6.2. *Evaporation in Isoalted Basins* .......... 101
   - 6.3. *Admixture with Brines* .......... 101
   - 6.4. *Precipitation and Dissolution* .......... 102
   - 6.5. *Submarine Volcanism* .......... 103
   - 6.6. *Exchange Between Atmosphere and Sea* .......... 104
   - 6.7. *Anoxic Basins* .......... 108
   - 6.8. *Freezing* .......... 108
   - 6.9. *Interstitial Waters* .......... 109
7. *Isotopic Variations* .......... 109
   - 7.1. *Deuterium* .......... 111
   - 7.2. *$^{18}O$ Results* .......... 111
   - 7.3. *Isotopes of Sulfur* .......... 112

*Further Reading* .......... 113

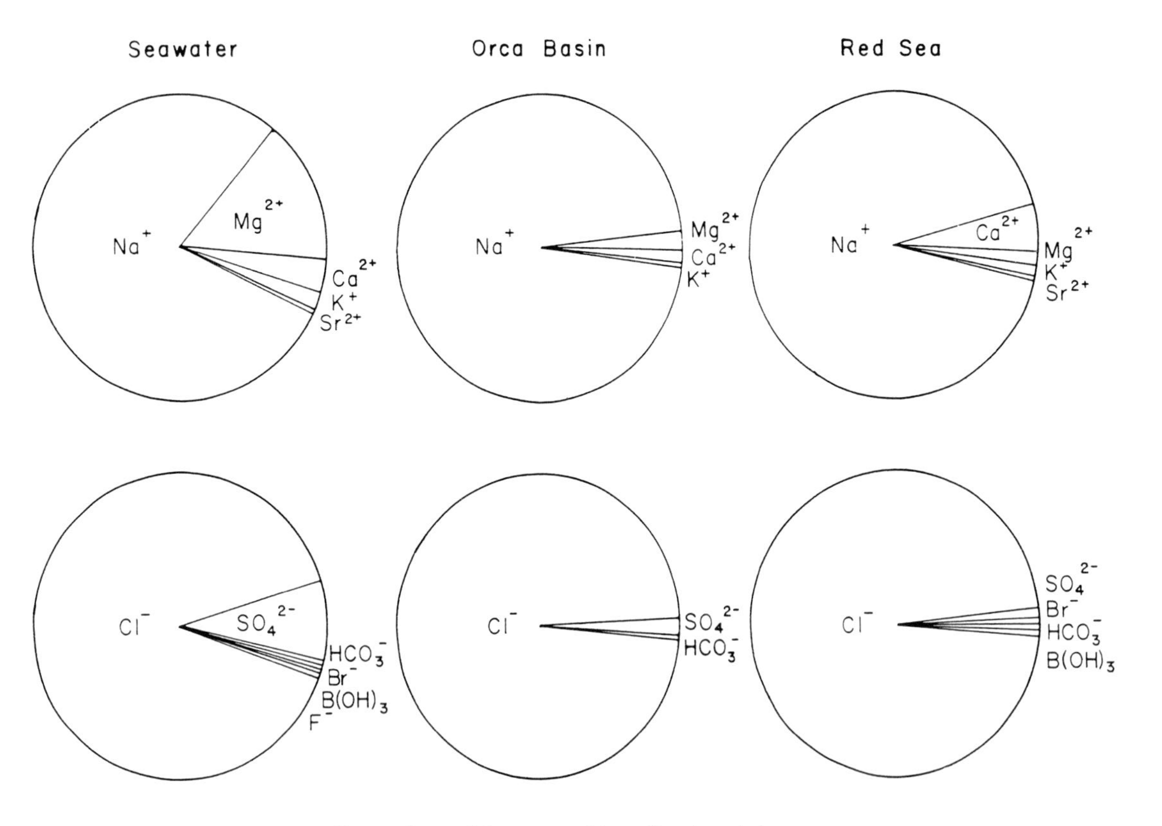


Comparison of the composition of various brines.

# 2 The Composition of the Major Components of Seawater

## 1. INTRODUCTION

Seawater is composed of a number of different components that can be divided into the following phases:

1. Solids (material that does not pass through a 0.45 μm filter)
   a. Particulate organic material (plant detritus)
   b. Particulate inorganic material (minerals)
2. Gases
   a. Conservative ($N_2$, Ar, Xe)
   b. Nonconservative ($O_2$ and $CO_2$)
3. Colloids (passes through 0.45 μm filter, but is not dissolved)
   a. Organic
   b. Inorganic
4. Dissolved Solutes
   a. Inorganic solutes
      1. Major ( $> 1$ ppm )
      2. Minor ( $< 1$ ppm )
   b. Organic solutes

In this chapter we will discuss the various dissolved major components of seawater. We will attempt to answer questions such as:

1. What are the major dissolved components of seawater?
2. How are they measured?
3. How are they distributed in the oceans?
4. What processes caused them to arrive at the present composition and how are they regulated?

These are just a few of the questions one might hope to answer. As we progress through the topic, we will develop new questions that require further study.

The earliest chemical analysis of seawater was made by Bergman in 1779. Since that time many workers have determined one or more elements in various ocean waters. Marcet in 1819 was the first to suggest that the relative composition of sea salt is nearly constant. Marcet analyzed water samples from the Arctic, Atlantic, Mediterranean, Black, Baltic, China, and White seas. Although his techniques were rather crude and the atomic weights were only approximately known, his conclusions are correct; ''all the species of seawater contain the same ingredients all over the world, these bearing very nearly the same proportion to each other, so that they differ only as to the total amount of their saline contents''. This simple suggestion has been called by Dana Kester (U.R.I.) the first law of chemical oceanography. Other earlier workers made analyses of seawaters; however, the first extensive investigations of the major inorganic components of seawater were made by Forchhammer in 1865. He determined $Cl^-$, $SO_4^{2-}$, $Mg^{2+}$, $Ca^{2+}$, and $K^+$ directly and $Na^+$ by difference. He made these measurements on several hundred surface waters from all parts of the world. He noted that the ratios of the major constituents (greater than 1 ppm by weight) were nearly constant with only very slight variations. These major dissolved components make up 99.9% of the soluble ionic species of seawater. His results have been criticized on the grounds that he used only surface samples and that his analytical methods were inaccurate. In 1884 Dittmar analyzed 70 seawater samples collected at various depths for the major oceans during the cruise of the H.M.S. Challenger (1873 to 1876). Although his results are not as extensive as one might wish and the samples were stored for as long as two years, he did adopt analytical techniques that are still used today. He also checked his experimental techniques on synthetic samples to examine the reliability of his measurements. His results agreed fairly well with the work of Forchhammer. He did find small variations in

**TABLE 2.1**
**Relative Composition of Major Components of Seawater**

| Solute | $g_i$/Cl(‰) A | B | C | D |
|---|---|---|---|---|
| $Na^+$ | 0.5556 | 0.5555 | 0.5567 | 0.55653 |
| $Mg^{2+}$ | 0.06695 | 0.06692 | 0.06667 | 0.06626 |
| $Ca^{2+}$ | 0.02106 | 0.02126 | 0.02128 | 0.02127 |
| $K^+$ | 0.0200 | 0.0206 | 0.0206 | 0.02060 |
| $Sr^{2+}$ | 0.00070 | 0.00040 | 0.00042 | 0.00041 |
| $Cl^-$ | 0.99894 | — | — | 0.99891 |
| $SO_4^{2-}$ | 0.1394 | — | 0.1400 | 0.14000 |
| $HCO_3^-$ | 0.00735 | — | — | 0.00586 |
| $Br^-$ | 0.00340 | — | 0.003473 | 0.00347 |
| $CO_3^{2-}$ | — | — | — | 0.00060 |
| $B(OH)_4^-$ | — | — | — | 0.00034 |
| $F^-$ | $7 \times 10^{-5}$ | — | — | 0.000067 |
| $B(OH)_3$ | 0.00137 | | | 0.00105 |
| | 1.81484 | | | 1.81537 |

A: Dittmar as recalculated by Lyman and Fleming (*J. Mar. Res.*, 3, 134, 1940).
B: Cox and Culkin (*Deep Sea Res.*, 13, 789, 1966.)
C: Riley and Tongadai (*Chem. Geol.*, 2, 263, 1967); Morris and Riley (*Deep Sea Res.*, 13, 699, 1966).
D: Millero (*Ocean Sci. Eng.*, 7, 403, 1982).

$Mg^{2+}$, $K^+$, $SO_4^{2-}$, $Ca^{2+}$, and $Na^+$. The values of $Ca^{2+}$ for deep waters were found to be 0.3% higher than surface waters which was five times higher than the analytical error. The results of Dittmar were recalculated in 1940 by Lyman and Fleming using modern atomic weights (Table 2.1).

Since the work of Dittmar, many workers have studied and still are studying the composition of natural waters. In 1965 Culkin made a thorough review of all the measurements made on the major components of seawater since the work of Dittmar. About the same time as this review appeared an extensive study of the major components was being carried out by Cox, Culkin, and Riley as part of an international study of the salinity of seawater. Since the relative composition of ocean waters is nearly constant, it is possible to characterize the composition by measuring only one component that is easy to measure and is conservative in its behavior. A conservative component is one which is unreactive and changes from place to place due to the addition or subtraction of water. The constituent originally selected to characterize a given sample of seawater or other natural water is chlorinity, Cl (‰).

The chlorinity was originally defined as the chlorine equivalent to the total halide concentration in parts per thousand by weight (g Cl/kg seawater) measured by titration with $AgNO_3$. Since changes occurred in the atomic weights of Ag and Cl, the chlorinity was redefined in 1937 by Jacobsen and Knudsen as "the chlorinity is the mass in grams of pure silver necessary to precipitate the halogens in 328.5233 grams of seawater." This definition gives the value of Cl (‰) = 0.3285233 Ag (‰), where Ag (‰) is the grams of silver per kilogram of seawater. The 1969 atomic weights yield ATW(Cl)/ATW(Ag) = 35.453/107.868 = 0.328670. Thus, the true chlorinity or chlorinity equivalent is equal to 0.328670/0.3284233 = 1.00045 times the Cl (‰).

Before we discuss the more recent measurements of the major components of seawater, it is appropriate to briefly discuss the methods used in their determination.

## 2. METHODS OF DETERMINATION

The methods used to determine the major components of natural waters are given in detail elsewhere (Culkin, 1965). In this section we will briefly discuss some of the popular methods that are presently being used.

**Chlorinity** — As discussed earlier the chlorinity of seawater is determined by adding $AgNO_3$ to precipitate AgCl.

$$Ag^+ + \text{Seawater} \rightarrow AgCl(s) + AgBr(s) \quad (1)$$

Potassium chromate is added as an indicator. After the halides are removed, silver chromate is precipitated

$$2Ag^+ + CrO_4^{2-} \rightarrow Ag_2CrO_4 \text{ (red solid)} \quad (2)$$

A number of workers have used other indicators and modern workers use an automated potentiometric end point (using a Ag,AgCl electrode). Since the heat of precipitation of AgCl is quite large and exothermic (heat is given off), we have used a titration calorimeter to determine the end point. The fact that the heat of solution is endothermic (heat taken up) yields a very sharp endpoint. The $AgNO_3$ is added by using a constant flow burette and the time for the endpoint to occur is recorded.

No matter what method is used to determine the endpoint, standard seawater of known chlorinity is used to standardize the $AgNO_3$ solution. Since chlorinity is a conservative component of seawater, it is important to measure or determine its value when studying the other major components of natural waters.

Approximate values of Cl (‰) can be determined from conductivity and density measurements. The measurements of Cox, Culkin, and Riley in 1967 yielded the equation

$$Cl(‰) = -0.050 + 15.66367R_{15} + 7.08943R_{15}^2 - 5.91110R_{15}^3 + 3.31363R_{15}^4 - 0.73240R_{15}^5 \quad (3)$$

for various ocean waters where $R_{15}$ is the conductivity ratio at 15°C of the sample relative to standard seawater. More reliable values for the chlorinity of ocean waters can be obtained from the density or the conductivity ratio of seawater as a function of the salinity determined using the new practical salinity scale and the new International Equation of State for Seawater.

**Sulfate** — The most widely used method to determine $SO_4^{2-}$ in seawater is the addition of $BaCl_2$ and subsequent precipitation of $BaSO_4$, which is weighed. Since other salts coprecipitate (e.g., $Ca^{2+}$), errors can result from using this technique. Measurements relative to standard seawater of known concentration can yield very precise measurements. The endpoint of the titration of seawater with $BaCl_2$ can be determined using calorimetry, potentiometry, or conductivity.

**Bromine** — Bromine is normally determined after being coprecipitated with $Cl^-$ by determination of weight loss after $Br_2$ is liberated. This requires large samples since the concentration of $Br^-$ is quite low. The $Br_2$ is liberated by adding chromic acid or potassium permanganate. The liberated $Br_2$ can be determined colorimetrically or by titration.

**Fluorine** — Fluorine is analyzed in natural waters by using colorimetric methods or using a specific ion electrode. The details are given elsewhere (Kremling, 1976).

**Bicarbonate and carbonate** — The concentration of $HCO_3^-$ and $CO_3^{2-}$ are determined by measuring at least two parameters of the carbonate system (pH; the total alkalinity, $A_T$; the total carbon dioxide, $CO_2$; or the partial pressure of $CO_2$ [$PCO_2$]. In most routine work, pH and $A_T$ are used to characterize the carbonate system. More details of these measurements will be discussed later in Chapter 6.

**Boric acid and borate** — Boron exists mainly as boric acid in seawater. At a pH of 8, borate ion is 25% of total boron. Total boron is determined as boric acid by forming a complex with mannitol and glycol. It is then converted to a strong acid and titrated with a base. It is first necessary to destroy the organic boron compounds by oxidation with permanganate. Colorimetric techniques using a colored indicator such as boric acid-curcumin complex are much quicker.

**Magnesium** — The classical method for determining $Mg^{2+}$ is a gravimetric determination of the precipitate formed following the addition of ammonium phosphate (after $Ca^{2+}$ is removed). The magnesium ammonium phosphate is converted to magnesium pyrophosphate and weighed. In recent years, volumetric methods have been used. EDTA (ethylenediammine N,N,N′,N′ tetraacetic acid) has been used to titrate $Mg^{2+}$ after $Ca^{2+}$ has been removed. The total equivalents of divalent ions can be determined by using an ion exchange resin or an EDTA titration. $Mg^{2+}$ is determined by difference. Problems in the determination of the endpoint were discussed by Carpenter and Manella in 1973.

**Calcium** — The classical gravimetric method used to determine $Ca^{2+}$ is precipitation as the oxalate. This can be used as a method of separating $Ca^{2+}$ and $Mg^{2+}$. Calcium oxalate is weighed as is or ignited to $CaCO_3$ or CaO. As with all precipitation techniques, coprecipitation problems are encountered and various steps have to be taken to obtain a true precipitate. Most modern measurements of $Ca^{2+}$ have been made by using volumetric methods. EGTA — [ethylene glycol-bis (2-aminoethyl)- N,N,N′,N′ tetraacetic acid] — is used. The endpoint is detected potentiometrically or colorimetrically using metallochromic indicators.

**Potassium** — Early measurements of potassium were made after removal of divalent ions. The total cations can be determined by ion exchange methods (titrating the released protons with base). The divalent ions are removed and determined by EDTA titration. Potassium can be determined by precipitation with chloroplatinate, $K_2PtCl_6$, which is sparingly soluble in 80% ethanol (the Na salt is very soluble). Since potassium tetraphenylboron is not very soluble, it can be weighed after precipitation by adding sodium tetraphenyl-boron.

**Sodium** — Since the concentration of sodium is quite high, it cannot be accurately determined by direct methods. The classical direct determination is made by precipitation with zinc uranyl acetate which is sparingly soluble.

The most reliable method to determine $Na^+$ is by difference. This can be done on individual samples by determining the total cations (equal to the total equivalent anions) and subtracting the equivalent concentrations of $K^+$, $Mg^{2+}$, $Ca^{2+}$, and $Sr^{2+}$. Since seawater should have an equivalent concentration of cations and anions, the preferred method is to determine the $Na^+$ after the other major components are determined (i.e., by difference).

## 3. COMPOSITION AND STOICHIOMETRY OF AVERAGE SEAWATER

To examine the thermodynamic properties of seawater, we (Millero, 1982) found it necessary to select a reasonably reliable composition for average seawater ( $S = 35$, $pH = 8.1$, and $t = 25°C$ ). Before we discuss these results, we will briefly examine the measurements made by Cox, Culkin, and Riley and co-workers on the major components of seawater. The studies made by these workers were for the major cations $Na^+$, $K^+$, $Mg^{2+}$, $Ca^{2+}$, and $Sr^{2+}$. These results are given in Table 2.1 along with the earlier results of Dittmar as recalculated by Lyman and Fleming. With the exception of $Ca^{2+}$, both studies demonstrated little or no depth dependence for the major cations (Table 2.2). The value of g Ca/Cl(‰) increases by 0.3 to 0.5% in agreement with the earlier findings of Dittmar. More will be said about this later. In the major oceans the values $g_i$/Cl(‰) are reasonably constant (Table 2.3). Small variations do occur in some areas, e.g., Ca/Cl (‰) in the Red Sea.

The values of $g_i$/Cl(‰) that we feel are representative of average seawater are given in Table 2.1 and Table 2.4. The values for $Ca^{2+}$, $K^+$, and $Sr^{2+}$ are taken from the work of Culkin and Cox (1966) and Riley and Tongadai (1967); the values for $SO_4^{2-}$ and $Br^-$ were taken from the work of Morris and Riley (1966); the value for $F^-$ was taken from Warner (N.R.L.); the value of $Cl^-$ was determined from the chloride equivalent by subtracting the equivalent $Br^-$; and the value of $Na^+$ was determined by difference (i.e., by assuming the cation and anion equivalents were equal, Table 2.5). The results for $HCO_3^-$, $CO_3^{2-}$, $B(OH)_3$, and $B(OH)_4^-$ were determined for seawater of $pH = 8.1$ using B/Cl $= 0.000230$ (Culkin, 1965) and $A_T$/Cl $= 0.00735$ (Lyman and Fleming, 1940). The value for $Mg^{2+}$ was taken from Carpenter and Manella (1973). By appropriate manipulation of the values of $g_i$/Cl(‰), it is possible to determine the stoichiometry of the various components of seawater as well as average seawater having a Cl(‰) $= 19.374$. Table 2.4

**TABLE 2.2**
**The Variation of $g_i$/Cl(‰) with Depth**

| Depth | Na A | Na B | K A | K B | Mg A | Mg B | Ca A | Ca B | $10^6$ Sr A | $10^6$ Sr B |
|---|---|---|---|---|---|---|---|---|---|---|
| 0—100 m | .5554 | .5566 | .0206 | .0206 | .06691 | .06666 | .02124 | .02123 | 42 | 39 |
| 700—1500 m | .5557 | .5567 | .0206 | .0206 | .06691 | .06668 | .02128 | .02131 | 43 | 40 |
| > 1500 m | .5555 | .5568 | .0206 | .0206 | .06695 | .06667 | .02130 | .02133 | 43 | 40 |
| MEAN | .5555 | .5567 | .0206 | .0206 | .06692 | .06667 | .02126 | .02128 | 42 | 40 |
| Std. error | .0007 | .0007 | .0002 | .0001 | .00004 | .00007 | .00004 | .00006 | 2 | 2 |
| Number | 49 | 93 | 54 | 84 | 66 | 91 | 66 | 81 | 38 | 58 |

A: Culkin and Cox (1966).
B: Riley and Tongudai (1967).

**TABLE 2.3**
**Values of $g_i$/Cl(‰)l for Various Oceans**

| Ocean | Na A | Na B | K A | K B | Mg A | Mg B | Ca A | Ca B | $10^5$ Sr A | $10^5$ Sr B |
|---|---|---|---|---|---|---|---|---|---|---|
| Atlantic | .5552 | .5567 | .0206 | .0206 | .06691 | .06671 | .02127 | .02123 | 39 | 42 |
| Pacific | .5555 | .5568 | .0206 | .0206 | .06681 | .06668 | .02128 | .02126 | 40 | 42 |
| N. Seas | .5553 | .5566 | .0205 | .0205 | .06690 | .06669 | .02121 | .02121 | 39 | 41 |
| S. Seas | .5567 | .5561 | .0206 | .0207 | .06691 | — | .02130 | .02126 | 40 | 42 |
| Indian | .5554 | .5569 | .0207 | .0206 | .06696 | .06661 | .02124 | .02122 | 40 | 43 |
| Mediterranean | .5557 | .5568 | .0206 | .0206 | .06685 | .06668 | .02131 | .02123 | 39 | 44 |
| Red | .5563 | — | .0206 | — | .06685 | — | .02115 | — | 38 | — |
| Persian Gulf | .5557 | — | .0208 | — | .06645 | — | .02123 | — | 38 | — |
| Baltic | .5554 | — | .0205 | — | .06694 | — | .02127 | — | 38 | — |
| North | .5541 | — | .0206 | — | .06703 | — | .02118 | — | 40 | — |
| Std. S.W. | .5562 | — | .0205 | — | .06690 | — | .02122 | — | 39 | — |
| MEAN | .5555 | .5567 | .0206 | .0206 | .06692 | .06667 | .02126 | .02128 | 40 | — |

A: Culkin and Cox (1966).
B: Riley and Tangudai (1967).

gives the values of $n_i$/Cl where $n_i$ is the number of moles, $e_i$/Cl where $e_i$ is equivalents and $I_i$/Cl where $I_i$ is the ionals of component i. By appropriate summation the total mass is given by

$$g_T = 1.81537 \, Cl(‰) \quad (4)$$

**TABLE 2.4**
**Composition of 1 kg of Natural Seawater as a Function of Chlorinity**

| Species | $g_i$/Cl | $n_i$/Cl | $e_i$/Cl | $n_iZ_i^2$/Cl |
|---|---|---|---|---|
| $Na^+$ | 0.55653 | 0.0242077 | 0.0242077 | 0.0242077 |
| $Mg^{2+}$ | 0.06626 | 0.0027262 | 0.0054524 | 0.0109048 |
| $Ca^{2+}$ | 0.02127 | 0.0005307 | 0.0010614 | 0.0021228 |
| $K^+$ | 0.02060 | 0.0005268 | 0.0005368 | 0.0005268 |
| $Sr^{2+}$ | 0.00041 | 0.0000047 | 0.0000094 | 0.0000187 |
| $Cl^-$ | 0.99891 | 0.0281756 | 0.0281756 | 0.0281756 |
| $SO_4^{2-}$ | 0.14000 | 0.0014575 | 0.0029149 | 0.0058298 |
| $HCO_3^-$ | 0.00586 | 0.0000960 | 0.0000960 | 0.0000960 |
| $Br^-$ | 0.00347 | 0.0000434 | 0.0000434 | 0.0000434 |
| $CO_3^{2-}$ | 0.00060 | 0.0000100 | 0.0000200 | 0.0000400 |
| $B(OH)_4^-$ | 0.00034 | 0.0000043 | 0.0000043 | 0.0000043 |
| $F^-$ | 0.00006 | 0.0000035 | 0.0000035 | 0.0000035 |
| 1/2 = | 0.0288932 | 0.0312577 | 0.0359867 | |
| $B(OH)_3$ | 0.00105 | 0.0000170 | 0.0000170 | |
| | 1.815367 | 0.0289102 | 0.0312577 | |

From Millero (1982), pH = 8.1 and t = 25°C.

Since the nonelectrolyte boric acid is part of the major components, the total moles and equivalents of ‘‘sea salt’’ are determined from

$$n_T = {}^1/_2 \Sigma n_i + n_B \tag{5}$$

$$e_T = {}^1/_2 \Sigma e_i + n_B \tag{6}$$

We are, thus, treating ‘‘sea salt’’ as a solution formed by mixing an electrolyte and a nonelectrolyte. This leads to

$$n_T = 0.0289102 \text{ Cl} \tag{7}$$

$$e_T = 0.0312747 \text{ Cl} \tag{8}$$

The total ionic strength ($I_T = 1/2 \Sigma n_iZ_i^2$) of sea salt per kg of solution is given by

$$I_T = 0.0359867 \text{ Cl} \tag{9}$$

**TABLE 2.5**
**$Na^+$ Value for Seawater (pH = 8.1)**

| | g/Cl | e/Cl | M′(eq. wt.) |
|---|---|---|---|
| Anions | | | |
| $Cl^-$ | 0.99891 | 0.0281756 | 35.4530 |
| $SO_4^{2-}$ | 0.14000 | 0.0029149 | 48.0288 |
| $HCO_3^-$ | 0.00586 | 0.0000960 | 61.0172 |
| $Br^-$ | 0.00347 | 0.0000434 | 79.9040 |
| $CO_3^{2-}$ | 0.00060 | 0.0000200 | 30.0046 |
| $B(OH)_2^-$ | 0.00034 | 0.0000043 | 78.8396 |
| $F^-$ | 0.000067 | 0.0000035 | 18.9984 |
| | $\Sigma$ = | 0.0312577 | |
| Cations | | | |
| $Mg^{2+}$ | 0.06626 | 0.0054524 | 21.1525 |
| $Ca^{2+}$ | 0.02127 | 0.0010614 | 20.0400 |
| $K^+$ | 0.02060 | 0.0005268 | 39.1020 |
| $Sr^{2+}$ | 0.00041 | 0.0000094 | 43.6200 |
| | $\Sigma$ = | 0.0070500 | |

e(Na)/Cl = 0.312577 − 0.0070500 = 0.0242077

g(Na)/Cl = e(Na)/Cl × FW = 0.0242077 × 22.9898 = 0.55653

The weight fraction ($X_i = g_i/g_T$), mole fraction ($N_i = n_i/n_T$), equivalent fraction ($E_i = e_i/e_T$), and the molecular weight ($M_i$) of the components of seawater are given in Table 2.6. The mean molecular ($M_T$) and equivalent weight ($M_T'$) of seawater are given by

$$M_T = N_i M_i = 62.793 \tag{10}$$

$$M_T' = E_i M_i' = 58.046 \tag{11}$$

Thus far, all of the concentrations have been given for one kg of solution. These values can be converted to molarity (moles/l or mol $dm^{-3}$) by multiplying by the density. The molality can be determined by dividing by the grams of water in 1 kg of seawater

$$g_{H_2O} = 1000 - 1.81537 \text{ Cl} \tag{12}$$

**TABLE 2.6**
**The Molecular Weight, Weight Fraction, Mole Fraction, Equivalent Fraction, and Ionic Strength Fraction of the Major Solutes in Water**

| Species | $M_i$ | $X_i$ | $N_i$ | $E_i$ | $I_i$ |
|---|---|---|---|---|---|
| $Na^+$ | 22.9898 | 0.306566 | 0.837341 | 0.774035 | 0.672685 |
| $Mg^{2+}$ | 24.3050 | 0.036500 | 0.094299 | 0.174339 | 0.303023 |
| $Ca^{2+}$ | 40.0800 | 0.011717 | 0.018357 | 0.033938 | 0.058988 |
| $K^+$ | 39.1020 | 0.011348 | 0.018222 | 0.016844 | 0.014639 |
| $Sr^{2+}$ | 87.6200 | 0.000226 | 0.000163 | 0.000301 | 0.000520 |
| $Cl^-$ | 35.4530 | 0.550252 | 0.974590 | 0.900907 | 0.782945 |
| $SO_4^{2-}$ | 96.0576 | 0.077119 | 0.050415 | 0.093203 | 0.161999 |
| $HCO_3^-$ | 61.0172 | 0.003228 | 0.003321 | 0.003070 | 0.002668 |
| $Br^-$ | 79.9040 | 0.001911 | 0.001501 | 0.001388 | 0.001206 |
| $B(OH)_3$ | 61.8322 | 0.000578 | 0.000588 | 0.000544 | — |
| $CO_3^{2-}$ | 60.0092 | 0.000331 | 0.000346 | 0.000639 | 0.001112 |
| $B(OH)_4^-$ | 78.8396 | 0.000187 | 0.000149 | 0.000137 | 0.000119 |
| $F^-$ | 18.9984 | 0.000037 | 0.000121 | 0.000112 | 0.000097 |

The total molarity ($c_T$), molality ($m_T$), equivalent molarity or normality ($N_T$), volume ionic strength ($I_V$), and weight ionic strength ($I_m$) are given by

$$c_T = 0.289102 \text{ Cl} \times \rho \tag{13}$$

$$m_T = 28.9102 \text{ Cl}/[1000 - 1.81537 \text{ Cl}] \tag{14}$$

$$N_T = 0.031275 \text{ Cl} \times \rho \tag{15}$$

$$e_T = 0.031275 \text{ Cl}/[1000 - 1.81537 \text{ Cl}] \tag{16}$$

$$I_V = 0.035987 \text{ Cl} \times \rho \tag{17}$$

$$I_m = 0.035987 \text{ Cl}/[1000 - 1.81537 \text{ Cl}] \tag{18}$$

where $\rho$ is the density in g $cm^{-3}$. These equations can be converted into functions of the salinity (S) using the approximate relationship

$$S = 1.80655 \text{ Cl} \tag{19}$$

**TABLE 2.7**
**Composition of 1 kg of Seawater with Cl = 19.374‰**

| Species | $g_i$ | $n_i$[a] | $e_i$[a] | $I_i$ |
|---|---|---|---|---|
| $Na^+$ | 10.7822 | 0.46900 | 0.46900 | 0.46900 |
| $Mg^{2+}$ | 1.2837 | 0.05282 | 0.10563 | 0.21126 |
| $Ca^{2+}$ | 0.4121 | 0.01028 | 0.02056 | 0.04112 |
| $K^+$ | 0.3991 | 0.01021 | 0.01021 | 0.01021 |
| $Sr^{2+}$ | 0.0079 | 0.00009 | 0.00018 | 0.00036 |
| $Cl^-$ | 19.3529 | 0.54587 | 0.54587 | 0.54587 |
| $SO_4^{2-}$ | 2.7124 | 0.02824 | 0.05648 | 0.11296 |
| $HCO_3^-$ | 0.1135 | 0.00186 | 0.00186 | 0.00186 |
| $Br^-$ | 0.0672 | 0.00084 | 0.00084 | 0.00084 |
| $CO_3^-$ | 0.0116 | 0.00019 | 0.00039 | 0.00078 |
| $B(OH)_4^-$ | 0.0013 | 0.00007 | 0.00007 | 0.00007 |
| | $^1/_2\ \Sigma$ = 0.55978 | $^1/_2\ \Sigma$ = 0.60559 | $^1/_2\ \Sigma$ = 0.69717 | |
| $B(OH)_3$ | 0.0203 | 0.00033 | 0.00033 | |
| | $g_T$ = 35.1709 | $n_T$ = 0.56011 | $m_T$ = 0.60592 | $I_T$ = 0.69717 |

[a] To convert to molar units multiply by the density. To convert to molal units multiply by $X_{H_2O}$ = 0.96483.

The composition of average seawater of Cl = 19.374 is given in Table 2.7. These results can be used to aid in the preparation of 1 kg of artificial seawater. A recipe that can be used to make up artificial seawater is given in Table 2.8. It should be kept in mind that the trace metal impurities in most reagent grade salts are higher than in natural seawater. Thus, care must be taken when using artificial seawater solutions in biological studies.

## 4. THE CONCEPT OF SALINITY

Salinity was originally conceived as a measure of the mass of dissolved salts in a given mass of seawater. The experimental determination of the salt content of seawater by drying and weighing presents some difficulties. At the temperatures necessary to drive off the last traces of $H_2O$, the bicarbonates and carbonates are decomposed to oxides $[O_2^-]$.

$$2HCO_3^- \rightarrow [O_2^-] + H_2O + 2CO_2 \quad (20)$$

**TABLE 2.8**
**Preparation of 1 kg of Artificial Seawater of S = 35.00**

Gravimetric Salts

| Salt | g/kg | mol/kg | Mol wt |
|---|---|---|---|
| NaCl | 23.9849 | 0.41040 | 58.4428 |
| $Na_2SO_4$ | 4.0111 | 0.02824 | 142.0372 |
| KCl | 0.6986 | 0.00937 | 74.5550 |
| $NaHCO_3$ | 0.1722 | 0.00205 | 84.0070 |
| KBr | 0.1000 | 0.00084 | 119.0060 |
| $B(OH)_3$ | 0.0254 | 0.00041 | 61.8322 |
| NaF | 0.0029 | 0.00007 | 41.9882 |
| Total | 28.9951 | | |

Volumetric Salts

| | | | |
|---|---|---|---|
| $MgCl_2$ | 0.05282 | 5.0290 | 95.211 |
| $CaCl_2$ | 0.01028 | 1.1409 | 110.986 |
| $SrCl_2$ | 0.00009 | 0.0143 | 158.526 |

Use 1 *M* $MgCl_2$, $CaCl_2$, and $SrCl_2$ (standardize by $AgNO_3$ titration). 52.8 ml of 1 *M* $MgCl_2$, 10.3 ml of 1 *M* $CaCl_2$, and 0.1 ml of 1 *M* $SrCl_2$ are needed. The densities of these solutions are 1.017 g/ml and 1.013 g/ml, respectively, for $MgCl_2$, $CaCl_2$, and $SrCl_2$ solutions at 1 *M*. The grams of water of each solution are given by

$$H_2O = g_{SOLN} - g_{SALT} = ml \times density - g_{SALT}$$

Addition of Water

g $H_2O$ to add = 1000 − g $H_2O$ from $MgCl_2$, $CaCl_2$, and $SrCl_2$

$$CO_3^{2-} \rightarrow [O_2^-] + CO_2 \quad (21)$$

$Br_2$, some $Cl_2$ gas, and $B(OH)_3$ are also vaporized. For example, heating $MgCl_2$ solutions to dryness yields HCl gas. The $Cl_2$ and $Br_2$ lost can be determined by titrating with $AgNO_3$ before and after heating. Earlier workers

(Marcet, Forchhammer and Dittmar) found difficulty in determining the salinity by evaporation.

A complete chemical analysis of seawater is the only reliable way to determine the true salinity of seawater ($S_T$, in parts per thousand). This method, however, is too time consuming for routine studies. The early work of Forchhammer gave

$$S_T = 1.812 \text{ Cl } (‰) \tag{22}$$

while Dittmar found

$$S_T = 1.8056 \text{ Cl } (‰) \tag{23}$$

and Lyman and Fleming found

$$S_T = 1.8148 \text{ Cl } (‰) \tag{24}$$

Our analysis of the major components of seawater gives

$$S_T = 1.81537 \text{ Cl } (‰) \tag{25}$$

For average seawater of Cl (‰) = 19.374, we obtain $S_T$ = 35.171. This can be compared to a value of S = 35.00 obtained from the Knudsen equation (Equation 26). As shown in Table 2.9, the composition data given in Table 2.4 gives a calculated salinity of 35.001 if the Mg/Cl ratio of Carpenter and Manella is used.

In 1899 the International Council for the Exploration of the Sea (ICES) named Knudsen as chairman of a commission appointed to investigate and advise on the definition of the salinity and density of seawater. On the basis of an evaporation method, Forch, Knudsen, and Sorensen defined the salinity as, "the weight in grams of dissolved inorganic salts in one kilogram of seawater, when all bromides and iodides are replaced by an equivalent quantity of chlorides, and all the carbonates are replaced by an equivalent quantity of oxides". The sample is dried to constant weight at 480°C. The Cl and Br lost is allowed for by adding a weight of Cl equivalent to the loss during drying. Salinity is, thus, the weight of dissolved solids minus the weight loss of $HCO_3^-$ and $CO_3^{2-}$ and minus the difference between $Br_2$ and its equivalent $Cl_2$. The determination has rarely been used because the method is too difficult for routine work. One can prevent the loss of HCl by adding NaF before

**TABLE 2.9**
**Calculation of the Salinity for Average Seawater from Composition Data**

| Before Evaporation | | After Evaporation | |
|---|---|---|---|
| g $HCO_3^-$ | 0.1135 | g | 0.0149 |
| g $CO_3^{2-}$ | 0.0116 | g | 0.0031 |
| g $Br^-$ | 0.0672 | g Cl | 0.0298 |
| | 0.1923 | | 0.0478 |

| | | |
|---|---|---|
| g of salts loss from $HCO_3^-$, $CO_3^{2-}$, and $Br^-$: | | |
| 0.1923 − 0.0478 | = | 0.1445 |
| g of $B(OH)_3$ lost | = | 0.0255 |
| Total salts lost | = | 0.1700 |

| | A | B |
|---|---|---|
| $g_T$ | 35.1616 | 35.1709 |
| Loss of salts | 0.1700 | 0.1700 |
| | 34.9916 | 35.0009 |

A: Using average Mg/Cl(‰) ratio from Culkin and Cox (1966) and Riley and Tongudai (1967). B: Using Mg/Cl(‰) ratio from Carpenter and Manella (1972).

evaporation. In 1964 Morris and Riley made further improvements to the evaporation technique. Based on the assumption that the relative composition of seawater was constant, the commission defined chlorinity (given earlier) and suggested that it could be used as a measure of salinity. Measurements of the chlorinity and evaporation salinity were made on nine samples of seawater (two from the Baltic, two from the Atlantic, four from intermediate Baltic-North Sea waters, and a Red Sea water). The results were found to fit the equation

$$S\ (‰) = 1.805\ Cl\ (‰) + 0.030 \tag{26}$$

The standard deviation was 0.01‰ and the largest deviation was 0.022 ‰. This formula was used in oceanography for about 65 years. As discussed earlier, this definition is based on the 1902 atomic weights. The commission specified that titration results should be determined by using tables produced using Copenhagen "Normal" seawater as a standard. In order to free chlorinity from its dependence on "Normal" seawater stored at Copenhagen, the new definition of chlorinity was given in 1937. This definition is still used although standard seawater, once called "Copenhagen Seawater", of known salinity is now provided by the Institute of Ocean Sciences (IOC) at Wormley, England.

One of the problems with this salinity-chlorinity relationship is that when $Cl(‰) = 0$, the value of salinity is 0.03 ‰. Low salinity samples for example, from the Baltic have little chloride. As will be discussed later the form of the Knudsen equation is quite general for a given estuarine system formed by mixing river water with seawater. Both the intercept and slope vary from location to location and from time to time.

With the development of precise conductivity bridges in the 1950s, it became possible to determine conductivity salinities to ±0.003 ‰. In 1961 nonthermostated commercial conductivity bridges became available. All these bridges gave conductivity ratios between the sample and standard seawater ($R = C_{sample}/C_{std}$) and used standard seawater to calibrate the bridges. Although standard seawater was calibrated for chlorinity, it was not meant to be a standard for conductivity. At about the same time the old definition of salinity came under question because of the uncertain accuracy of gravimetric salinity definitions and because of the small number of samples used. A panel called the Joint Panel for Oceanographic Tables and Standards (JPOTS) sponsored by UNESCO, ICES, IAPSO, and SCOR was appointed to develop a conductivity standard for salinity.

A number of samples were collected from around the world. These samples were analyzed for chemical composition, chlorinity, and conductivity ratios. The JPOTS panel decided to revise the old relationship between salinity and chlorinity as follows

$$S\ (‰) = 1.80655\ Cl\ (‰) \qquad (27)$$

which is equivalent to the original definition at $S = 35$ or $Cl = 19.374$. Near $S = 35$ the two equations are identical, however, at $S = 32$ or 38 the difference is 0.003 ‰.

Cox, Culkin, and Riley developed a relationship between the conductivity ratio at 15°C ($R_{15} = C_{15}$ [sample]/$C_{15}$ [std seawater]) and the chlorinity of samples collected throughout the world. Samples deeper than 200 m were deleted (due to the $Ca^{2+}$ effect on $R_{15}$). Since most of the deep samples had

S = 34.8, a discontinuity occurred when the results were deleted. The polynomial was adjusted by adding 0.0018 to make S = 35 have R = 1.0. The Cl(‰) as a function of $R_{15}$ was converted to salinity using Cl = S/1.80655. The resultant equation was

$$S\ (‰) = -0.08996 + 28.2970\ R_{15} + 12.80832\ R_{15}^2 - 10.67869\ R_{15}^3 + 5.98624\ R_{15}^4 - 1.32311\ R_{15}^5 \quad (28)$$

Unfortunately this relationship was referred to as the new definition of salinity while in fact this polynomial merely expresses Cl in terms of $R_{15}$ and is strictly valid at S = 35 which is equivalent to the old definition.

About the same time (1969) as the various organizations recommended the acceptance of the so-called redefinition of salinity, *in situ* salinometers became commercially available. Since the new definition only went to 10°C, it became necessary to use either extrapolated values or the equations of Brown and Allentoft based on the dilution of seawater with pure water. Many workers attempted to use equations that joined these two studies.

A number of samples of standard seawater bottled from 1962 to 1975 were examined for conductivity, salinity, and density by our laboratory. All the conductance measurements were made relative to the same standard seawater ($P_{64}$–1973). The results were in good agreement (±0.0012 average deviation) with measurements made by Poisson. The differences in the conductivity salinities and the salinities determined from S = 1.80655 Cl are shown in the following Figure 2.1.

The differences range from −0.002 to 0.008. The maximum spread of 0.0098 in salinity represents the maximum error one would obtain in using the various standard seawater samples. No systematic correlations of the differences were found with the age of the standard. These results pointed out the need to characterize the conductivity of seawater relative to a KCl standard. In order to examine the causes of these differences the relative densities of the same samples were measured. These results were then compared to the calculated values obtained using conductivity and chlorinity derived salinities. The differences had a mean of $\pm 2 \times 10^{-6}$ g cm$^{-3}$ and showed no measurable differences. Thus, the myth that conductivity salinities are better than chlorinity-derived salinities to determine density does not hold true for seawater collected in the North Atlantic over a period of 38 years.

One of the samples ($P_{49}$) had densities 35 to 40 $\times 10^{-6}$ g cm$^{-3}$ higher than the other samples. This is equivalent to 0.06 in salinity. Since the

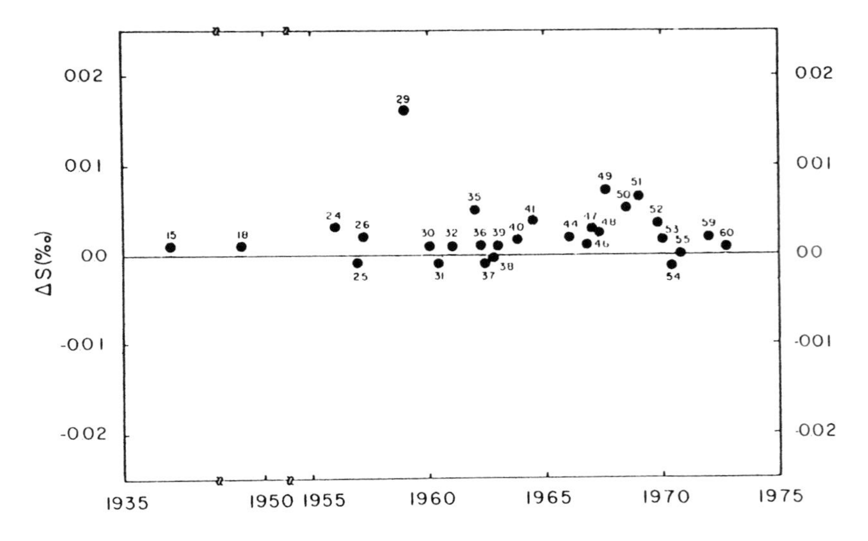


FIGURE 2.1. Difference in conductivity-salinity and chlorinity-salinity as a function of time.

conductive salinity is only 0.0073 higher than the $P_{64}$ sample, this increase in density was attributed to higher concentrations of the nonelectrolyte $SiO_2$ that occurred during closing of the sample or during storage. In general these results indicate that standard seawater can be used as a conductivity and density standard to $\pm 0.002$ in S and $\pm 2 \times 10^{-6}$ g cm$^{-3}$ in $\rho$. The deviations of 0.01 in conductivity salinity can be avoided by using seawater that has been calibrated relative to a KCl solution. The errors in calculated density due to changes in the composition of seawater that are due to real changes in composition cannot be avoided. More will be said about this later. All of the samples had a higher conductance salinity of $0.0021 \pm 0.0005$ than the $P_{64}$ sample used as a reference.

In 1975 the JPOTS committee that had been considering a new equation of state of seawater suggested that a background paper be prepared on this matter. It was concluded that a revision was needed in the definition of salinity and the committee recommended the Practical Salinity Scale of 1978. This new scale breaks the Cl-S relationship in favor of a salinity-conductivity ratio relationship. All waters of the same conductivity ratio have the same salinity (even though the composition may differ). Since salinity is normally used to determine a physical property like density, this was thought to be the best method for determining the effect of changes in ionic composition. This is not always the case since nonelectrolytes like $SiO_2$ are not detected by conductivity.

A standard seawater of salinity S = 35.000 (no units or ‰ are needed) has, by definition, a conductivity ratio of 1.0 at 15°C with a KCl solution containing a mass of 32.4356 g of KCl in a mass of 1 kg of solution. This value was determined as an average of three independent laboratory studies. The salinity dependence of the conductivity ratio was determined by measuring the conductivity (C) ratio at various temperatures of S = 35.000 seawater weight evaporated or diluted with water. The final equation is given by

$$S = a_0 + a_1R_t^{1/2} + a_2R_t + a_3R_t^{3/2} + a_4R_t^2 + a_5R_t^{5/2} + \Delta S \quad (29)$$

where

$$\Delta S = \frac{(t - 15)}{(1 + k(t - 15))} b_0 + b_1R_t^{1/2} + b_2R_t + b_3R_t^{3/2} + b_4R_t^2 + b_5R_t^{5/2} \quad (30)$$

$a_0 = 0.0080$ $b_0 = 0.0005$

$a_1 = -0.1692$ $b_1 = -0.0056$ $k = 0.0162$

$a_2 = 25.3851$ $b_2 = -0.0066$

$a_3 = 14.0941$ $b_3 = -0.0375$

$a_4 = -7.0261$ $b_4 = 0.0636$

$a_5 = 2.7081$ $b_5 = -0.0144$

$\Sigma a_i = 35.000$ $\Sigma b_i = 0.0000$

and $R_t = C(S,t,0)/C(35,t,0)$ at atmospheric pressure ($p = 0$).

As part of this new practical scale of salinity, equations were given to reduce *in situ* conductivity, temperature, and depth (pressure) data to salinity. These measurements give a conductivity ratio

$$R = C(S,t,P)/C(35,15,0) \quad (31)$$

$$R = \frac{C(S,t,P)}{C(S,t,0)} \times \frac{C(S,t,0)}{C(35,t,0)} \times \frac{C(35,t,0)}{C(35,15,0)} \quad (32)$$

$$R = R_P \cdot R_t \cdot r_t \tag{33}$$

The desired $R_t$ is determined from $R_t = R/r_t(1 + \alpha)$ where $R_p = (1 + \alpha)$ and

$$(1 + \alpha) = \frac{A_1P + A_2P^2 + A_3P^3}{(1 + B_1t + B_2t^2 + B_3R + B_4tR)} \tag{34}$$

$A_1 = 2.070 \times 10^{-5}$ $\quad B_1 = 3.426 \times 10^{-2}$

$A_2 = -6.370 \times 10^{-10}$ $\quad B_2 = 4.464 \times 10^{-4}$

$A_3 = 3.989 \times 10^{-15}$ $\quad B_3 = 4.215 \times 10^{-1}$

$B_4 = -3.107 \times 10^{-3}$

The value of $r_t$ is given by

$$r_t = c_0 + c_1t + c_2t^2 + c_3t^3 + c_4t^4 \tag{35}$$

where

$c_0 = 6.7660 \times 10^{-1}$

$c_1 = 2.00564 \times 10^{-2}$

$c_2 = 1.104259 \times 10^{-4}$

$c_3 = -6.9698 \times 10^{-7}$

$c_4 = 1.0031 \times 10^{-9}$

## 5. METHODS OF DETERMINING SALINITY

As discussed earlier, the salinity of seawater is normally determined by a conductance measurement using the practical salinity scale. Although this method yields very precise results, they are not always accurate since the conductance responds only to ionic components. In practice, any physical

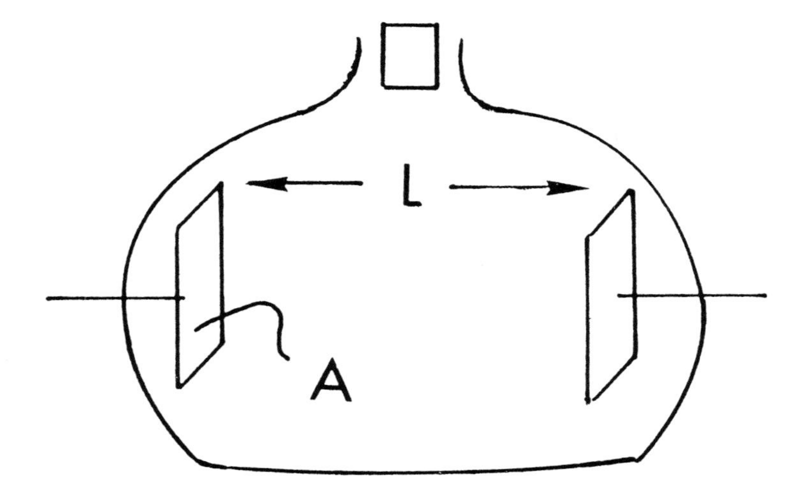


FIGURE 2.2. Typical conductance cell (A = area and L = length).

property (density, refractive index, sound speed, etc.) at a fixed temperature and pressure can be used to determine the salinity. Before we discuss these other methods we will examine the most popular method of measuring salinity by conductivity.

The conductivity is a measure of the mobility of ions to carry a charge through a solution. The specific conductivity is defined as

$$L_{sp} = k/R = \ell/AR \tag{36}$$

where R is the resistance and the cell constant k equals the distance ($\ell$) between the electrodes (usually platinum) divided by the cross-sectional area (A) of the electrodes. The equivalent conductance $\Lambda$ is defined as the conductance of 1 gram equivalent of the electrolyte

$$\Lambda = 1000\ L_{sp}/N \tag{37}$$

where N is the normality [eq/liter]. Since conductance is a function of temperature, it is necessary to make the measurements at a constant or known temperature for precisions of $\pm 0.001‰$ in salinity. Another problem is that DC current cannot be used due to polarization effects on the electrodes. Alternating current systems are normally used in conductivity measurements. The cell contains at least two platinum electrodes with a cross-sectional area A and a distance between the cells of $\ell$ (Figure 2.2).

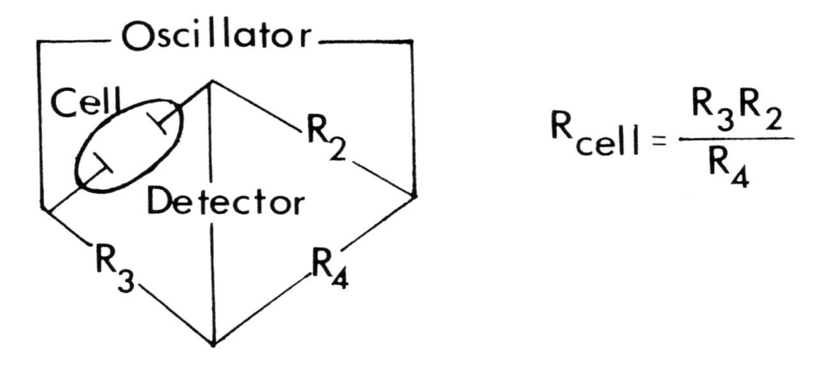


FIGURE 2.3. Wheatstone Bridge used to measure resistance of a solution in a conductance cell.

In the direct method, the cell is one arm of a Wheatstone bridge (Figure 2.3). The resistor $R_3$ is adjusted until the detection shows a zero reading. The resistance of the cell is given by

$$R_{cell} = R_3R_2/R_4 \tag{38}$$

$$L_{sp} = k/R_{cell} \tag{39}$$

The cell constant is determined using a standard seawater of known conductance relative to KCl. The conductance ratio of the sample of seawater is determined by

$$R_t = L_{sp}\ (\text{sample})/L_{sp}\ (\text{standard seawater}) \tag{40}$$

$$= R_{seawater}/R_{sample} \tag{41}$$

Modern salinometers like the Autosal system are nulled to a given conductivity ratio for the standard seawater and give a digital reading of $R_T$ at a given temperature. Laboratory measurements in physical chemistry are made with the following conditions:

1. Oscillation frequencies of 500 to 5000 cps.
2. Cell resistance of 1000 Ω.
3. Temperature bath of oil (to minimize capacity effects).
4. Measurements are made below room temperature (to avoid bubbles).
5. Short wires are used to minimize lead resistance.
6. A good detector is used to null the bridge and avoid capacitance effects.

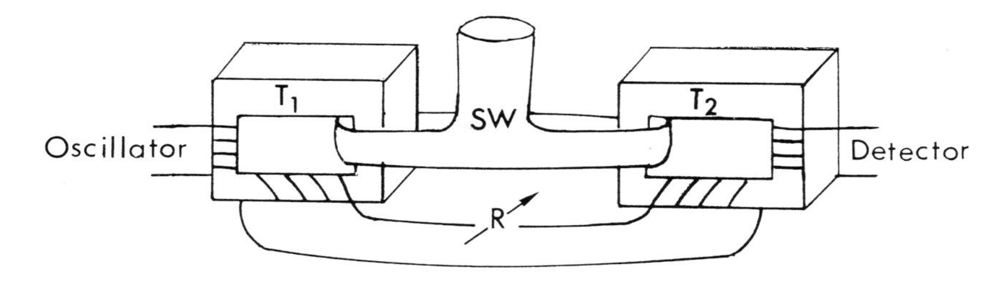


FIGURE 2.4. Inductance conductance cell.

In the inductive conductance bridges, a transformer bridge (Figure 2.4) is used. The variable resistor R is adjusted to give a zero reading using standard seawater on the signal sent from the oscillator using a detector. Although conductivity salinometers yield reliable values of R near a salinity of 35.0, it is good practice to calibrate the system using weight diluted seawater. For solutions of natural waters with low salinities or salinities above 42.0 it is convenient to use other physical measurements such as density, refractive index, or sound speeds. In our laboratory we use a vibrating flow densimeter to measure the density of natural waters relative to pure water. A sketch of the densimeter is shown in Figure 2.5. The measurements are normally made by passing water or sample through a vibrating tube encased in a constant temperature jacket ($\pm 0.01$°C). The density of the solution is proportional to the square of the period of the vibration

$$\rho = A + B\,\tau^2 \qquad (42)$$

The difference between the density of the solution ($\rho$) and pure water ($\rho_o$) is related by

$$\rho - \rho_o = B\,(\tau^2 - \tau_o^2) \qquad (43)$$

where $\tau$ and $\tau_o$ are, respectively, the periods of vibration for the solution and water. The system can be calibrated using $N_2$ and $H_2O$ or standard seawater and $H_2O$. The instrument constant (B) is determined from this calibration. At t = 25°C the values of $\rho - \rho_o$ are related to the practical salinity by

$$S = 1.3343\,(\rho - \rho_o)10^3 + 2.155306 \times 10^{-4} \times [(\rho - \rho_o)10^3]^2 - 1.171160 \times 10^{-5}\,[(\rho - \rho_o)10^3]^3 \quad (\sigma = 0.0012) \qquad (44)$$

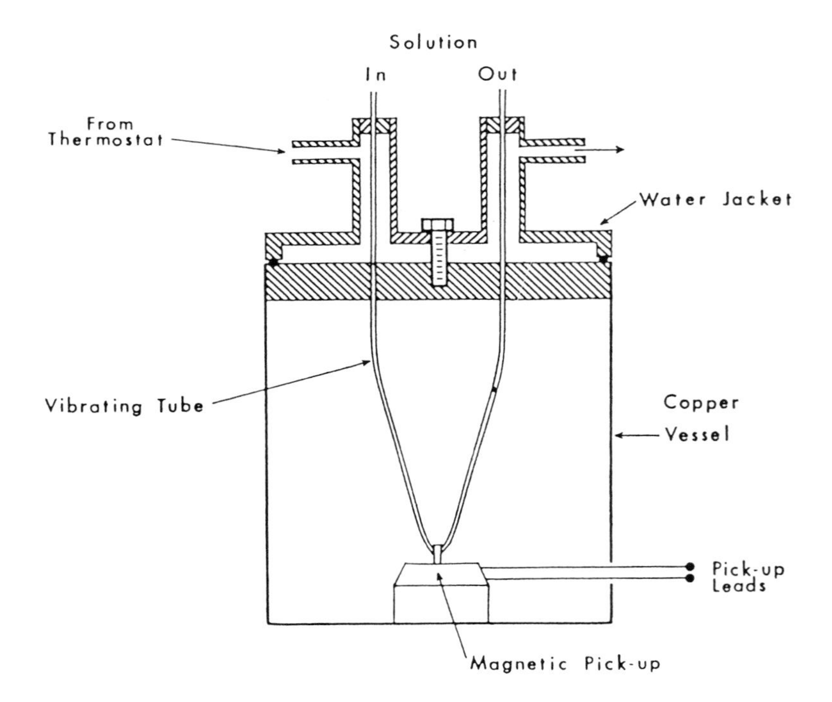


FIGURE 2.5. Sketch of vibrating densimeter.

From measurements of sound speeds (U) relative to pure water ($U_o$) the salinity can be determined from

$$S = 0.91712 (U - U_o) + 7.670097 \times 10^{-4} \times (U - U_o)^2 - 1.107557 \times 10^{-5} (U - U_o)^3$$

$$(\sigma = 0.0031) \qquad (45)$$

From measurements of the refractive index at 546.227 nm minus the value for standard seawater ($\Delta n$) the salinity can be determined from

$$S = 35.00 + 5.3302 \times 10^{-3} \Delta n + 2.274 \times 10^{-5} \Delta n^2 + 3.9 \times 10^6 \Delta n^3 + 10.59 \Delta n (t - 20) + 2.5 \times 10^2 \Delta n^2 (t - 20)^2$$

$$(\sigma = 0.0055) \qquad (46)$$

In routine measurements it is possible to determine relative densities to $\pm 3 \times 10^{-6}$ g cm$^{-3}$, sound speeds to $\pm 0.03$ m sec$^{-1}$ and refractive indices to

**TABLE 2.10**
**Precision in Salinity Determined by Using Various Methods**

1. Composition studies of major components ± 0.01
2. Evaporation to dryness ± 0.01
3. Chlorinity ± 0.002
4. Density ± 0.004
5. Conductivity ± 0.001
6. Sound speeds ± 0.03
7. Refractive index ± 0.05

±0.00001. These precisions are equivalent to errors in salinity of ±0.004, ±0.028, and ±0.053, respectively, for density, sound speeds, and refractive indices. In Table 2.10 we have summarized the precision of the salinities determined by using various methods.

## 6. CAUSES OF MAJOR COMPONENTS NOT BEING CONSERVATIVE

Although the major components of seawater are relatively constant, a number of factors can cause them to be nonconservative. In this section some of the areas will be examined (estuaries, anoxic basins and sediments, hydrothermal vents, evaporated basins) where processes (such as precipitation, dissolution, evaporation, freezing, and oxidation) can change the composition of some major components of seawaters.

### 6.1 Estuaries

The concentrations of salts in rivers are controlled by the nature of the rocks being weathered and soil types, yielding ground waters which differ in chemical composition. The total solids in most rivers are less than 200 mg $kg^{-1}$ or 0.2‰, however, the ratios of $SO_4^{2-}$, $HCO_3^-$, $K^+$, $Mg^{2+}$, and $Ca^{2+}$ to $Cl^-$ are usually greater than in seawater. The relative composition of rivers entering the oceans on various continents is shown in Figure 2.6. The composition of average world river water is compared with normal seawater in Figure 2.7. Although the total dissolved solids vary from 70 to 200 ppm, the equivalent fractions of the major components of most rivers are similar. The major cations are $Ca^{2+}$, $Mg^{2+}$, and $Na^+$, and the major anions are $HCO_3^-$,

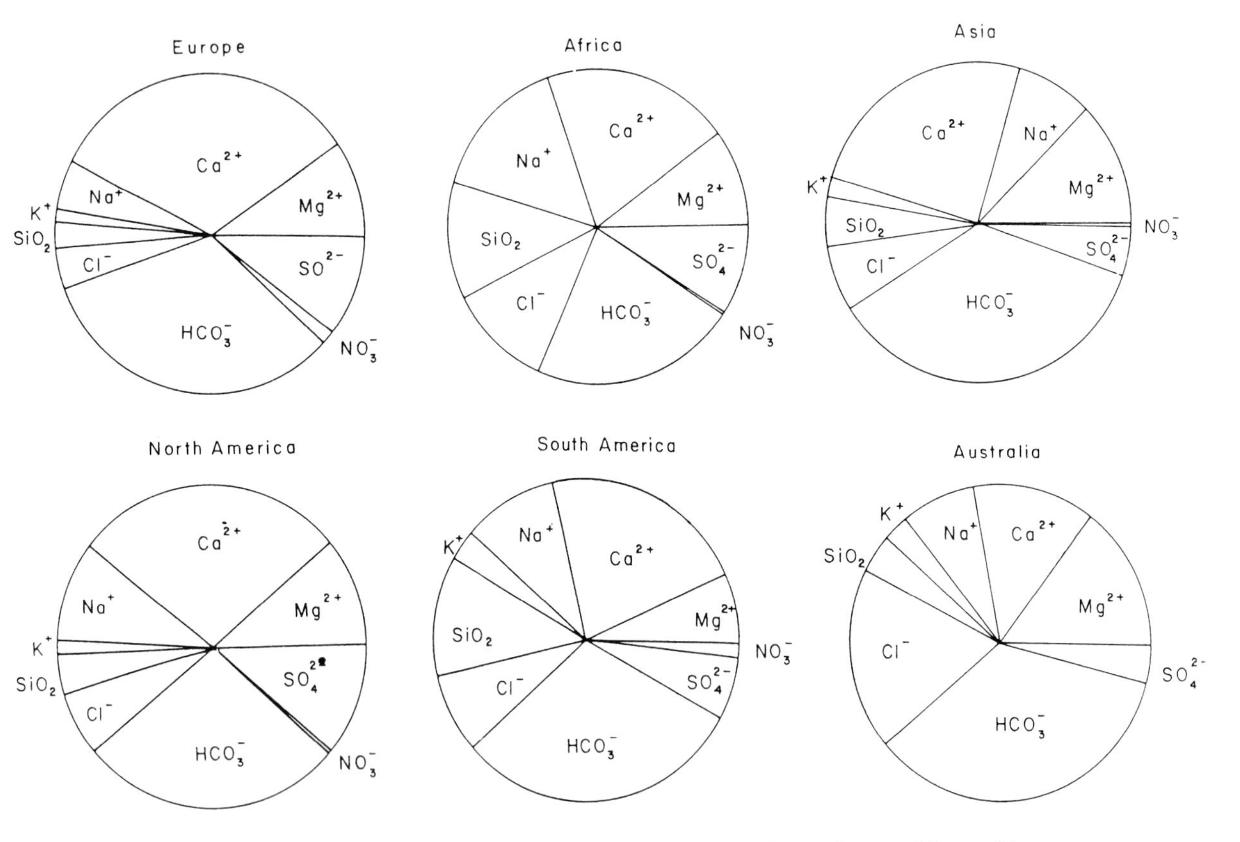


FIGURE 2.6. The major components of various rivers on the continents of the world.

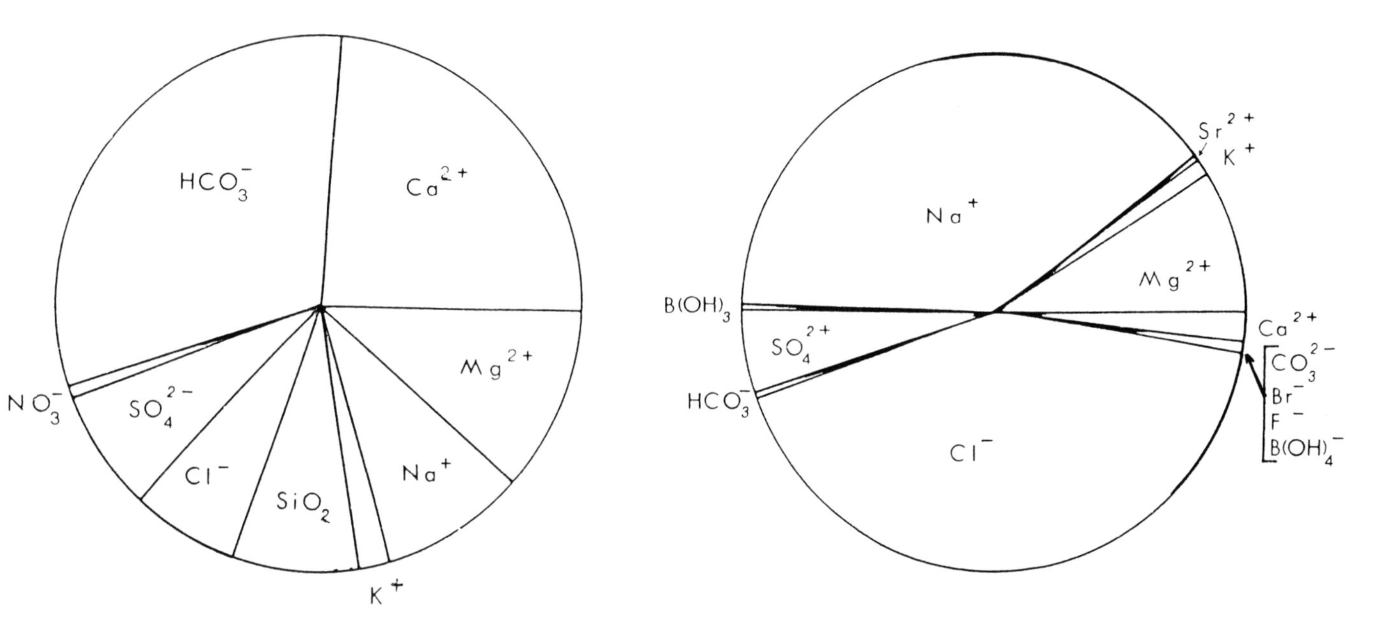


FIGURE 2.7. The equivalent fraction of the major constituents of the average world river and seawater.

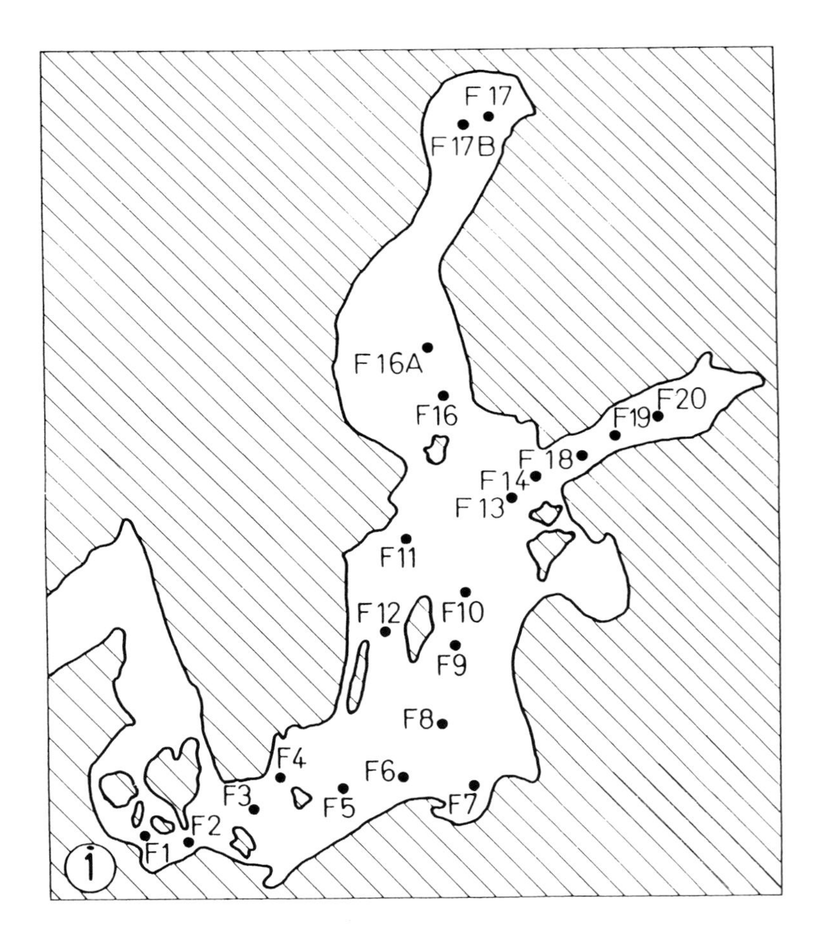


FIGURE 2.8. Stations in the Baltic Sea.

$SO_4^{2-}$, and $Cl^-$. Most of the NaCl is recycled from sea salt aerosols. The $SiO_2$ is predominantly in the unionized form $Si(OH)_4$ at the pH of most rivers (7.3 to 8.0). The major constituents of world river water are $Ca^{2+}$ and $HCO_3^-$ (from weathering of $CaCO_3$).

The formation of the Baltic sea estuary (Figure 2.8) has been analyzed by mixing average Baltic river water with seawater. The Baltic is a stratified positive estuary with a large salinity and temperature gradient (Figure 2.9). The nutrients $NO_3^-$ and $PO_4^{3-}$ are quite high in the deep water due to the oxidation of plant material (Figure 2.10). The $O_2$ levels are quite low in the deep water due to this oxidation (Figure 2.11). As discussed earlier, the dissolved constituents of an estuary are a linear function of $Cl^-$. The chemical data of Kremling (Kiel) for the major constituents are shown in the Figures

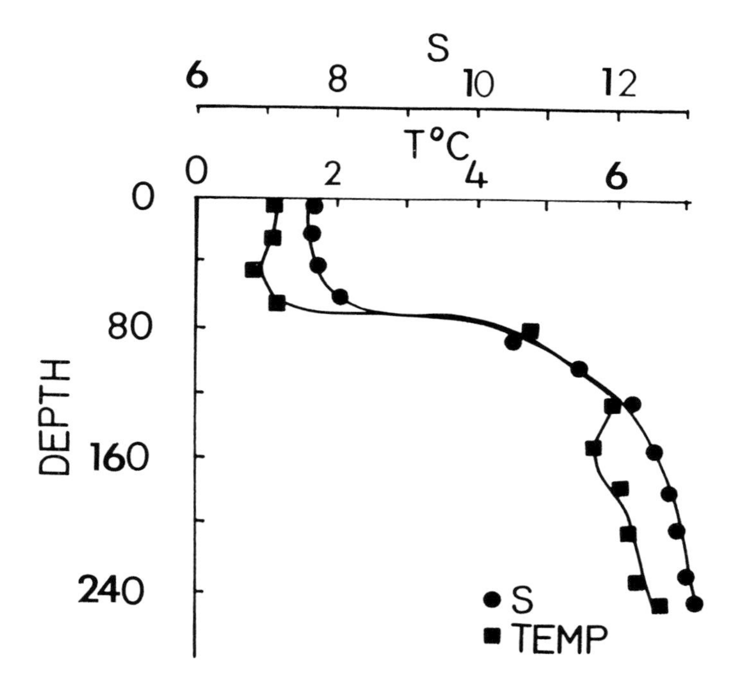


FIGURE 2.9. Salinity and temperature profiles for the Baltic Sea.

2.12 and 2.13 in which the grams of each component ($g_i$) per kg of seawater is plotted against the chlorinity. A summary of the intercepts (equal to $g_R$) are shown in Table 2.11. The total grams of salts (g $kg^{-1}$) is given by

$$g_T = g_R + [(35.171 - g_R)/19.374] \text{ Cl } (‰) \tag{47}$$

where $g_R$ is the grams of river salts. Since the slopes are related to the river concentration, more reliable values of $g_R$ can be obtained from

$$g_R = \frac{[g_E - g_{SW}]\ 19.374}{19.374 - \text{Cl } (‰)} \tag{48}$$

and

$$g_{SW} = k_i \times \text{Cl } (‰) \tag{49}$$

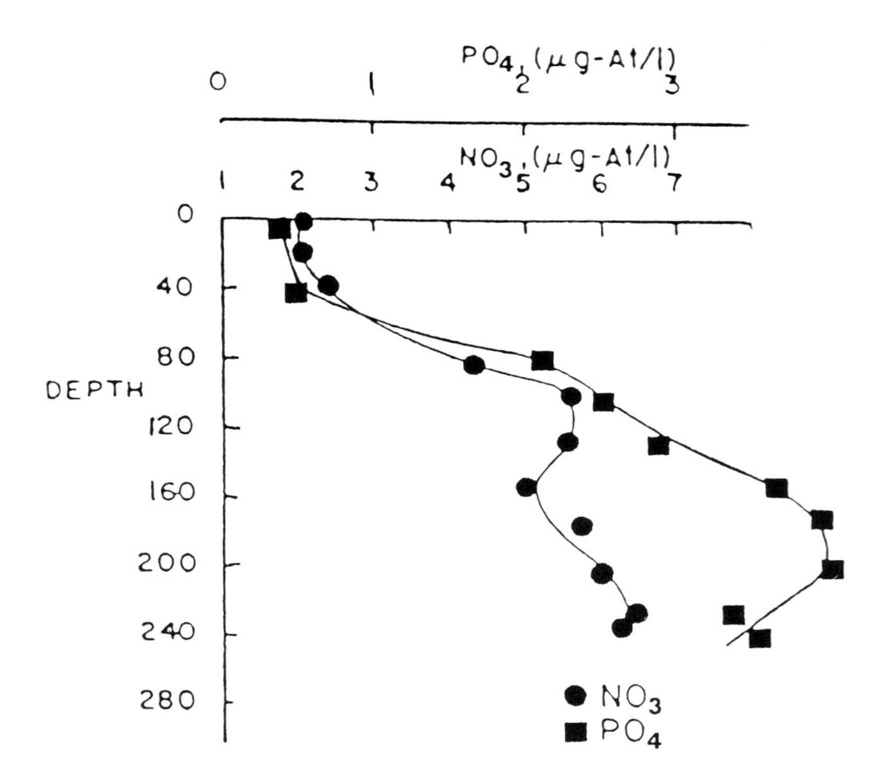


FIGURE 2.10. Phosphate and nitrate profiles for the Baltic Sea.

Where $k_i$ is the average ratio of $g_i$/Cl for seawater (Table 2.4). The values of $g_R$ determined from this equation have been determined (Table 2.12) from all the measurements of Kremling. The resulting values are compared in Table 2.11 to the earlier results of Lyman and Fleming. The $g_T$ has increased from 0.073 to 0.121 over the last 60 years. This increase is largely due to $Ca^{2+}$ and $HCO_3^-$. The $SO_4^{2-}$ has decreased apparently due to the waters becoming more anoxic (Figure 2.14). A comparison of Baltic river water and World river water is shown in Figure 2.15. Since the flow rate of the major rivers entering the Baltic has decreased (Figure 2.16), the increase in $g_T$ can be attributed to less rainfall diluting the ground water (which has a fixed concentration of $Ca^{2+}$ and $HCO_3^-$). It is interesting to note that the flow rate is cyclic (with a cycle nearing that of the sun spot cycle of 11 years). This would lead to cycling of the input of solids into the estuary and cause a pulsing rate for the renewal of the deep waters (from the North Sea). One could make the far-out postulation that the banding in the Baltic Sea sediments of FeS is

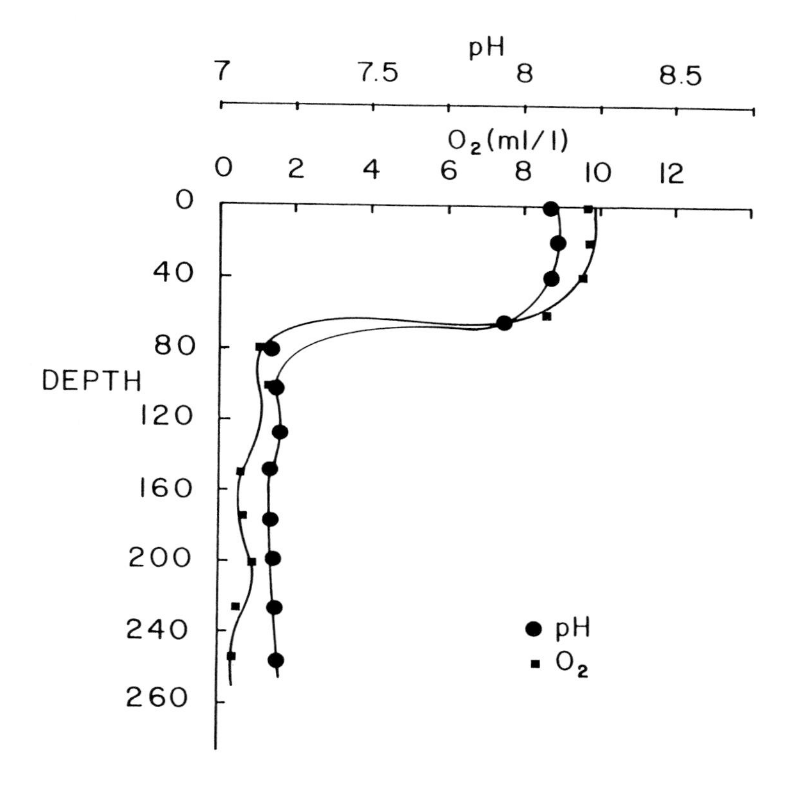


FIGURE 2.11. Oxygen and pH profiles for the Baltic Sea.

caused by the pulsing flow rate of the rivers which is caused, in turn, by sun spot activity. The examination of a deep basin in the Baltic (Figure 2.17) as a function of time shows a pulsing (oxic to anoxic); however, the periods are quite short.

The practical salinity scale is frequently used to characterize the composition of estuarine waters. Since the composition of these waters is different than those used in setting up the scale (i.e., seawater diluted with pure water), it is appropriate to discuss its limitations. A typical estuarine solution can be formed by mixing ''World'' river with average seawater (Table 2.13). If $SiO_2$ is left out of the river end member, the total grams of salts in 1 kg of solution is related to Cl by

$$g(E) = 0.092 + 1.80271\ Cl \tag{50}$$

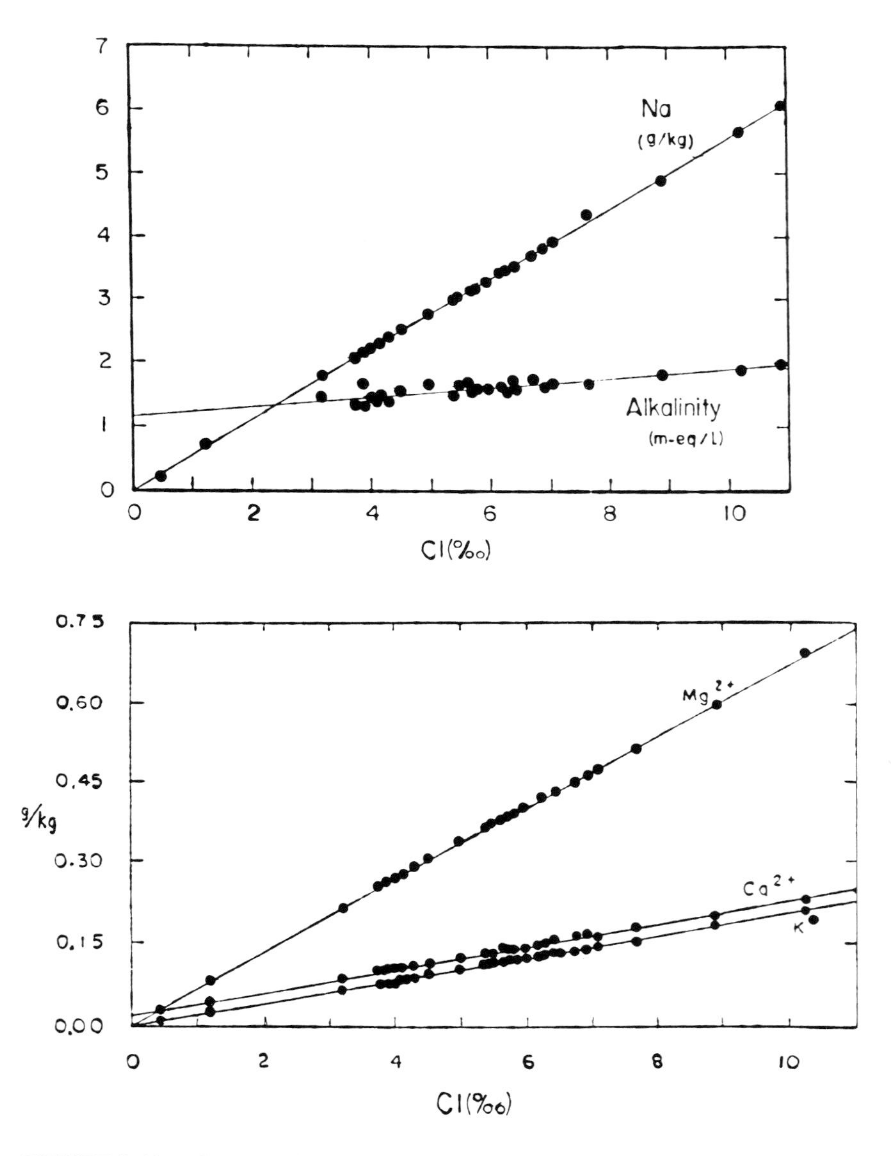


FIGURE 2.12. Concentrations of $Na^+$ and alkalinity ($HCO_3^-$) and of $Mg^{2+}$, $Ca^{2+}$, and $K^+$ as a function of the chlorinity in the Baltic Sea.

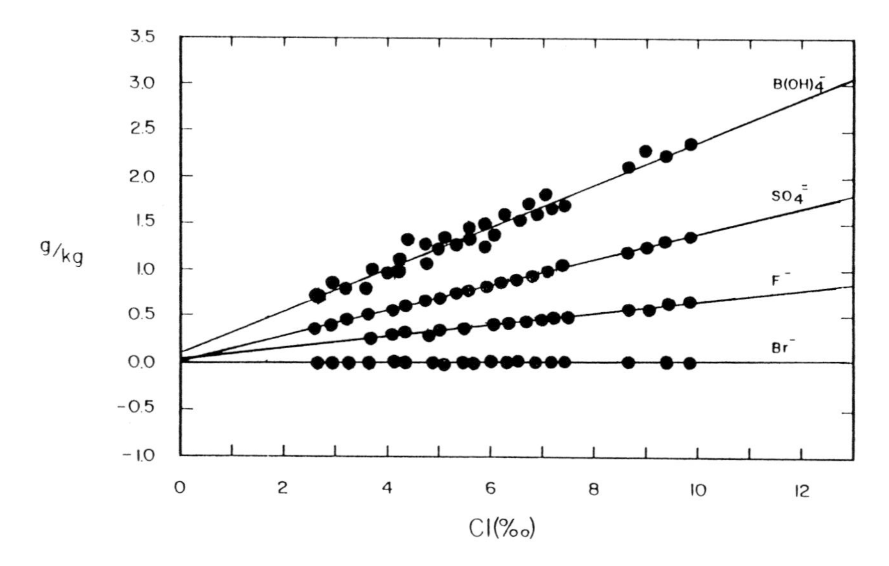


FIGURE 2.13. Concentrations of $B(OH)_4^-$, $SO_4^{2-}$, $F^-$ and $Br^-$ as a function of chlorinity in the Baltic Sea.

**TABLE 2.11**
**Composition of Baltic Waters[a]**

| Solute | L & F[b] | Linear | Individual | Best |
|---|---|---|---|---|
| $Na^+$ | — | 11.3 ± 23.4 | 0.1 ± 7.0 | 5.2 ± 3.8 |
| $Mg^{2+}$ | 2.0 | 3.8 ± 2.2 | 3.0 ± 0.5 | 3.0 ± 0.5 |
| $Ca^{2+}$ | 15.4 | 19.5 ± 2.3 | 20.8 ± 0.6 | 20.8 ± 0.6 |
| $K^+$ | — | 0.1 ± 1.4 | −0.5 ± 0.5 | 0 ± 0.5 |
| $Sr^{2+}$ | — | — | — | — |
| $Cl^-$ | — | — | — | — |
| $SO_4^{2-}$ | 6.1 | 6.8 ± 3.4 | 3.9 ± 1.0 | 3.9 ± 1.0 |
| $HCO_3^-$ | 49.3 | 80.2 ± 4.8 | 86.9 ± 4.2 | 86.9 ± 4.2 |
| $Br^-$ | — | −0.2 ± 0.1 | −0.8 ± 0.1 | 0.0 ± 0.8 |
| $B(OH)_4^-$ | — | — | — | — |
| $F^-$ | — | 0.07 ± 0.03 | 0.05 ± 0.02 | 0.8 ± 0.1 |
| $B(OH)_3$ | — | 0.7 ± 0.1 | 0.8 ± 0.1 | 0.1 ± 0.03 |
| | 72.8 | 122.3 ± 37.7 | 114.3 ± 14.0 | 120.7 ± 11.5 |

[a] $g_T = 0.120 + 1.8092$ Cl (‰), Millero (1967 data).
[b] $g_T = 0.073 + 1.8110$ Cl (‰), Lyman and Fleming (1900 data).

**TABLE 2.12**
**Composition of River Water Entering the Baltic**

| | $g_i(R)10^3$, g kg$^{-1}$ | | |
|---|---|---|---|
| **Ion** | **Surface waters** | **Deep waters** | **Weighted average** |
| $Na^+$ | 5.4 ± 8.3 | 5.3 ± 10.1 | 5.3 ± 9.3 |
| $K^+$ | −0.8 ± 1.5 | −0.9 ± 1.8 | −0.9 ± 1.7 |
| $Ca^{2+}$ | 21.1 ± 1.1 | 20.9 ± 1.3 | 21.0 ± 1.2 |
| $Mg^{2+}$ | 2.9 ± 1.3 | 3.0 ± 2.1 | 3.0 ± 1.8 |
| $HCO_3^-$ | 84.6 ± 4.7 | 88.1 ± 4.7 | 86.5 ± 4.7 |
| $SO_4^{2-}$ | 6.0 ± 3.8 | 3.3 ± 5.0 | 4.4 ± 4.5 |
| $Br^-$ | −0.7 ± 0.2 | −1.1 ± 0.6 | −0.9 ± 0.4 |
| $F^-$ | 0.06 ± 0.02 | 0.06 ± 0.03 | 0.06 ± 0.03 |
| $B(OH)_3$ | 0.8 ± 0.4 | 0.8 ± 0.4 | 0.8 ± 0.4 |
| T | 119.4 ± 21.3 | 118.5 ± 26.0 | 119.4 ± 24.0 |

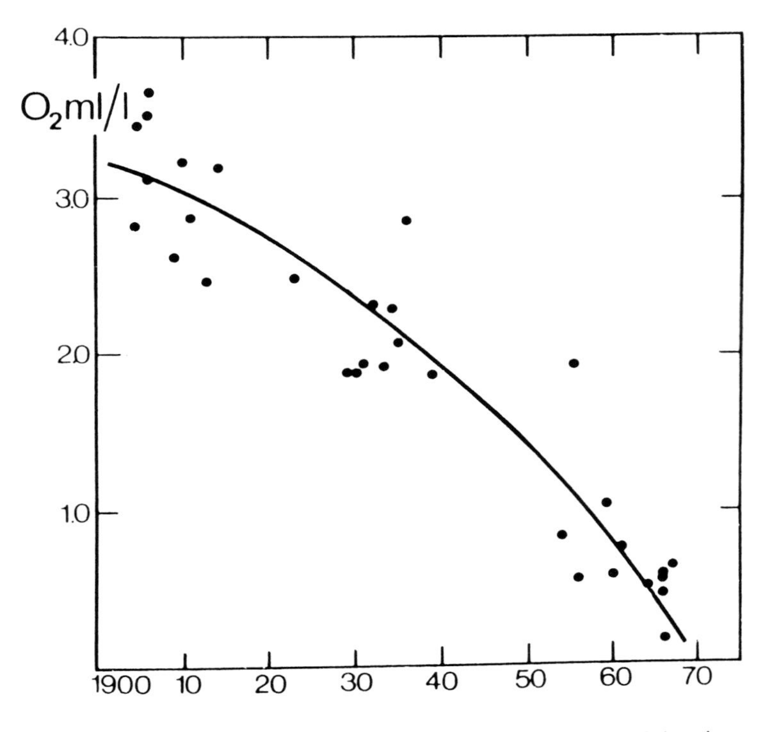


FIGURE 2.14. The concentration of oxygen as a function of time in a deep basin of the Baltic Sea.

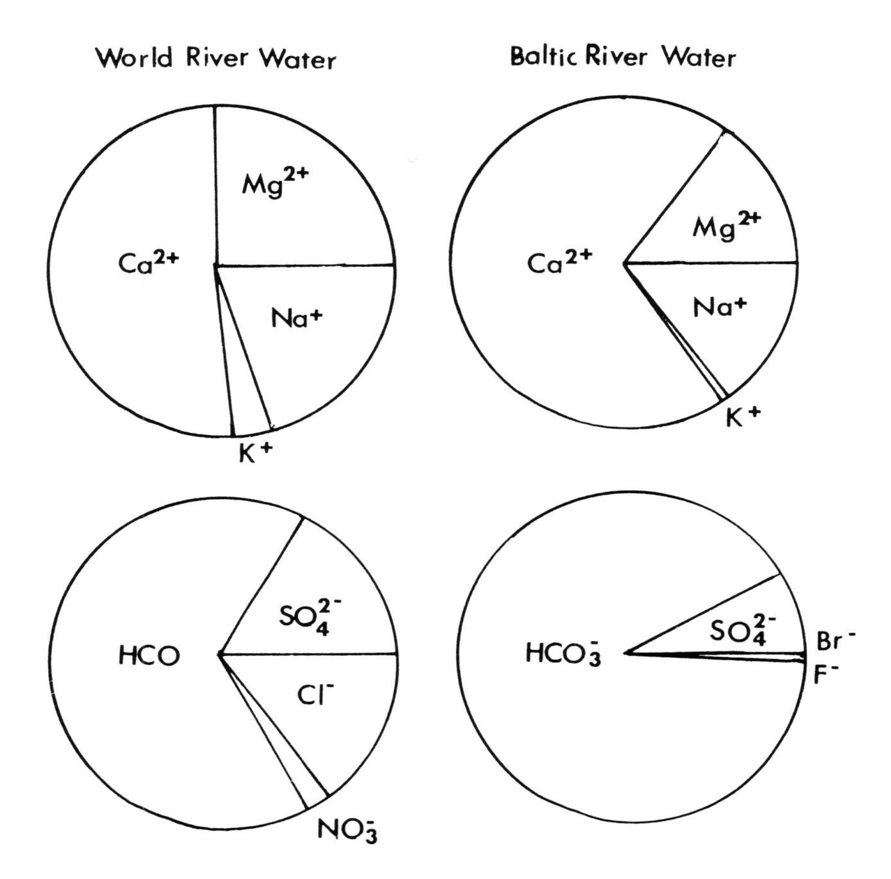


FIGURE 2.15. Comparison of the major sea salts in world and Baltic River waters.

Since salts are lost when seawater is evaporated, the true salinity ($S_T = g_T/1.00488$) is given by

$$S_T = 0.092 + 1.80183\ Cl \tag{51}$$

The values of the practical conductivity salinity of the estuarine mixtures were found to be related by

$$S_{COND} = 0.044 + 1.803898\ Cl \tag{52}$$

while the values of salinity determined by density were given by

$$S_{DENS} = 0.092 + 1.80186\ Cl \tag{53}$$

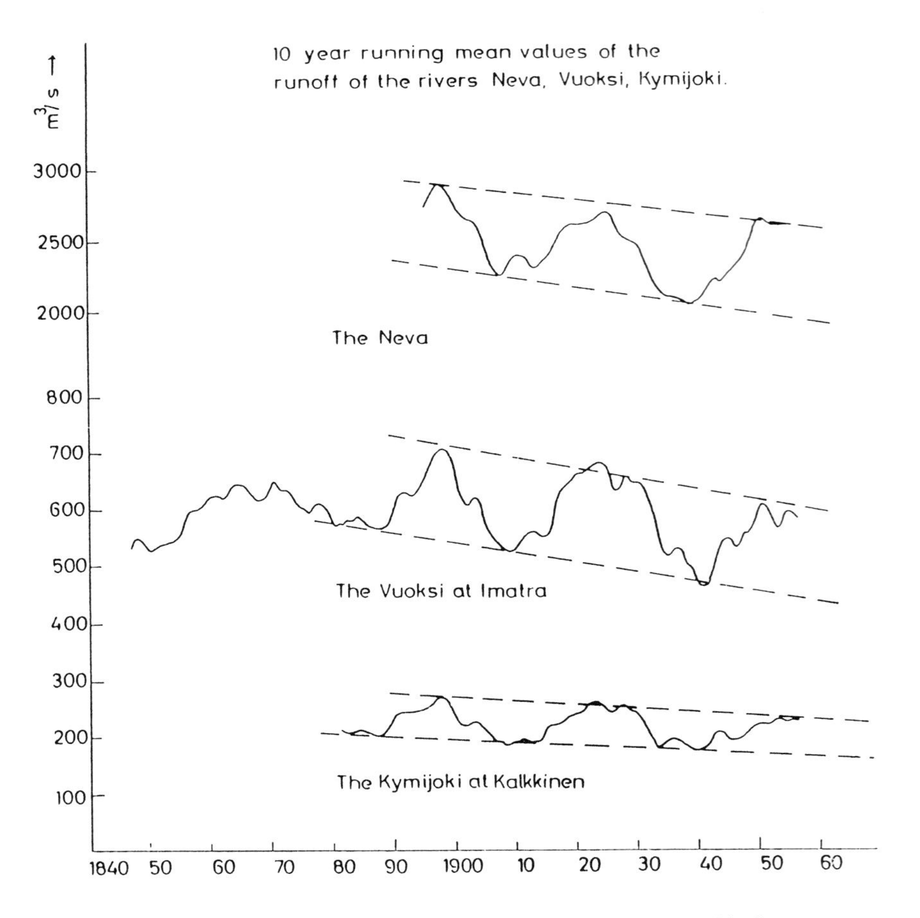


FIGURE 2.16. Mean values of runoff of rivers into the Baltic Sea.

The differences between $S_T$ and $S_{COND}$ or $S_{DENS}$ are shown below (Figure 2.18). The values of salinity determined by density are in excellent agreement with the true salinity. These results indicate that density derived salinities are more reliable for typical estuarine solutions than conductivity values. The lower values of $S_{COND}$ are related to the differences in the equivalent conductance of the main components of seawater and "World" river water (Table 2.14). The conductances of $Na^+$ and $Cl^-$ are larger than $Mg^{2+}$, $Ca^{2+}$, and $HCO_3^-$; thus, at a given Cl, the conductance of seawater diluted with pure water is greater than that of estuarine waters. Below Cl = 2.0, the conductivity of river water is 0.94 ± 0.02 lower than seawater diluted to the same Cl. This can be compared to $\Lambda°_{RW}/\Lambda°_{SW}$ = 0.88 calculated from infinite dilution conductivity data (Table 2.14). Part of this difference is related to the limi-

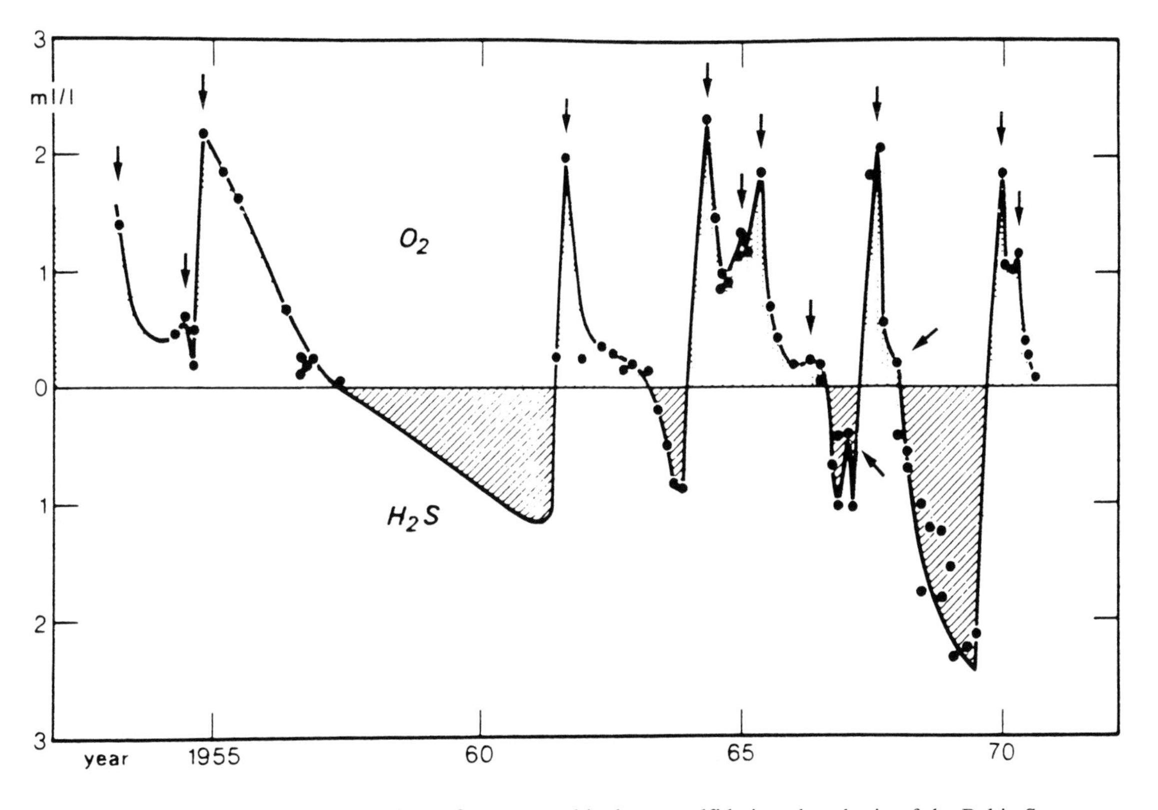


FIGURE 2.17. The concentrations of oxygen and hydrogen sulfide in a deep basin of the Baltic Sea.

**TABLE 2.13**
**The Composition of 1 l of World or Average River Water**

| Species | $g_i$ $10^6$ | $n_i$ $10^3$ | $e_i$ $10^3$ | $T_i$ $10^3$ |
|---|---|---|---|---|
| $Na^+$ | 6.5 | 0.283 | 0.283 | 0.283 |
| $Mg^{2+}$ | 4.1 | 0.169 | 0.337 | 0.674 |
| $Ca^{2+}$ | 15.0 | 0.374 | 0.749 | 1.496 |
| $K^+$ | 2.3 | 0.059 | 0.059 | 0.059 |
| $Cl^-$ | 7.8 | 0.220 | 0.220 | 0.220 |
| $SO_4^{2-}$ | 11.2 | 0.117 | 0.233 | 0.466 |
| $HCO_3^-$ | 58.4 | 0.950 | 0.950 | 0.950 |
| $CO_3^{2-}$ | — | 0.002 | 0.004 | 0.008 |
| $NO_3^-$ | 1.0 | 0.016 | 0.016 | 0.016 |
| $Si(OH)_3O^-$ | — | 0.005 | 0.005 | 0.005 |
| | | $1/2\Sigma = 1.086$ | $1/2\Sigma = 1.428$ | $1/2\Sigma = 2.089$ |
| $Si(OH)_4$ | 20.5 | 0.213 | 0.213 | — |
| | $g_T = 126.8$ | $n_T = 1.299$ | $e_T = 1.641$ | $I_T = 2.089$ |

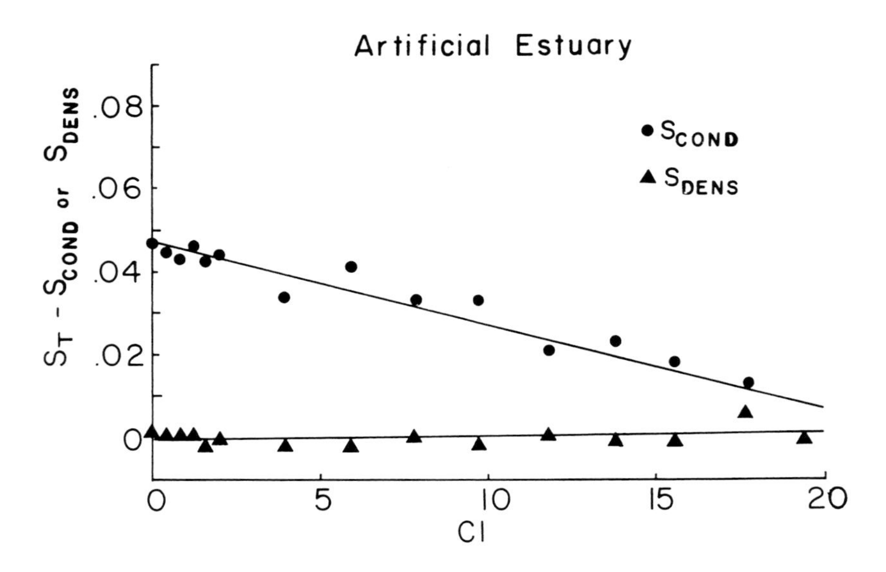


FIGURE 2.18. Comparison of the salinity determined from conductivity and density for estuarine waters.

**TABLE 2.14**
**Calculation of the Infinite Dilution Equivalent Conductance of Various Waters at 25°C**

| | | $E_i\ \Lambda^0_i$ [a] | | |
|---|---|---|---|---|
| **Ion** | $\Lambda^0_i$ | **Seawater** | **World river water** | **St. Lawrence River** |
| $Ca^{2+}$ | 59.51 | 46.97 | 31.41 | 35.40 |
| $Mg^{2+}$ | 53.50 | 9.32 | 12.71 | 13.15 |
| $Na^+$ | 50.10 | 38.71 | 9.67 | 7.24 |
| $K^+$ | 73.50 | 1.24 | 3.06 | 0.91 |
| $HCO_3^-$ | 44.50 | 0.14 | 29.79 | 25.19 |
| $SO_4^{2-}$ | 89.02 | 7.46 | 13.14 | 13.89 |
| $Cl^-$ | 76.35 | 68.77 | 11.83 | 18.61 |
| $NO_3^-$ | 71.46 | 0.20 | 0.81 | — |
| | $\Lambda^0$ = | 127.85[b] | 112.2 | 114.41 |

[a] $E_i$ is the equivalent function of species.
[b] $\Lambda^0 = \Sigma^{E}{}_i,\ \Lambda^0_i$.

tation of the practical salinity scale to S = 2.000. The scale can be extended to lower salinities by using the equation

$$S = S_{PSS} - \frac{a_0}{1 + 1.5X + X^2} - \frac{b_0\, f(t)}{1 + Y^{1/2} + Y + Y^{3/2}} \qquad (54)$$

where $S_{PSS}$ is the value determined from the Practical Salinity Scale given earlier and the other parameters are given by

$$a_0 = 0.0080$$

$$b_1 = 0.0005$$

$$X = 400\ R_t$$

$$Y = 100\ R_t$$

$$f(t) = (t - 15)/[1 - 0.0162\ (t - 25)]$$

where $r = R_{15}/0.0032$. The addition of this term allows one to calculate reliable values of $S_T$ for seawater diluted with pure water to dilute solutions. It also should prove useful in characterizing the salinity of rivers and dilute lakes.

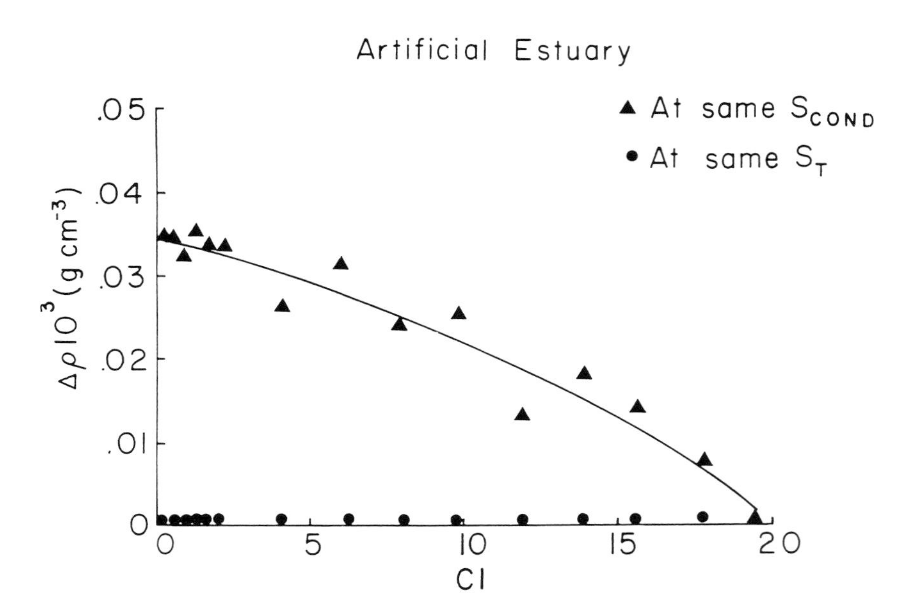


FIGURE 2.19. Comparison of the measured and calculated densities determined from conductivity-salinity and composition for estuarine waters.

These results indicate that it is possible to make reasonable estimates of the true salinity for an estuarine system if the composition of the river end member is known. For careful work, laboratory mixing experiments should be made, especially when the composition of the river end member is not known. As mentioned earlier one should calibrate the salinometer with seawater diluted with pure water to be sure it is working properly in dilute solutions. Since the reason for measuring the salinity is frequently to determine a physical property like density, it is interesting to examine the errors involved in using $S_{COND}$. A comparison is given below in Figure 2.19.

The maximum error is $35 \times 10^{-6}$ g $cm^{3-}$ in density. If the comparisons are made at the same $S_T$ the differences are within $\pm 3 \times 10^{-6}$ g $cm^{3-}$. It should be pointed out that salinities derived from density will be reliable only when the river salts have an effect on density similar to that of sea salts. This is true for average river waters, but not for every river. St. Lawrence River waters have a density $14 \times 10^{-6}$ g $cm^{3-}$ larger than seawater diluted to the same Cl.

These results can be compared to the measurements for the St. Lawrence system. The measurements of Poisson (University of Paris) are shown below in Figure 2.20. The differences between $S_T$ and $S_{COND}$ are linear and they approach 0.02 to 0.04 in dilute solutions similar to the model results. The

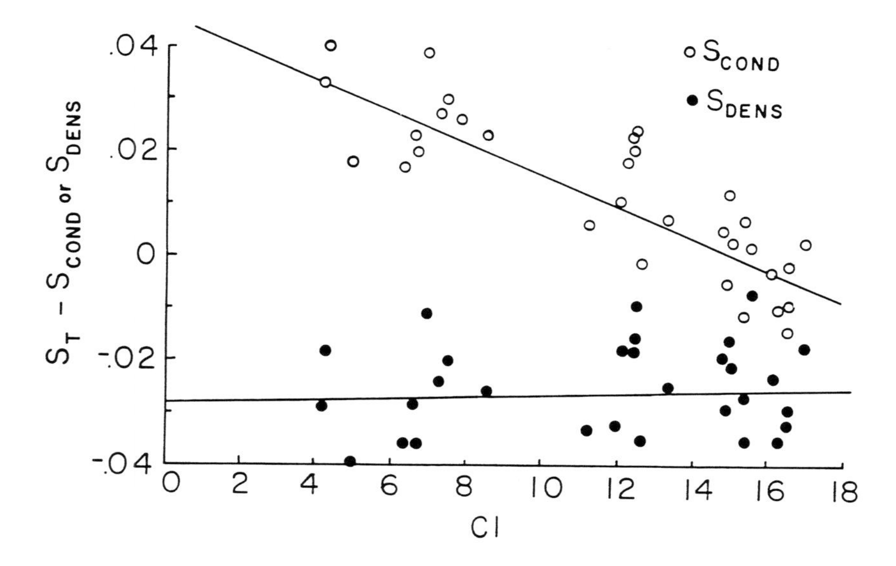


FIGURE 2.20. Total salinity minus salinity determined from density or conductivity for estuarine waters.

composition data gives $\Lambda°_{RW}/\Lambda°_{SW} = 0.89$ (Table 2.14) which gives $S_T - S_{COND} = 0.02$ in reasonable agreement with the extrapolated value.

The values of $S_T - S_{DENS}$ for the St. Lawrence are $-0.03 \pm 0.01$ over the entire chlorinity range. Partly, this is related to the density differences between the river and seawater diluted to the same Cl. The linear offset could also be caused by the presence of a dissolved nonelectrolyte of similar concentration over the entire salinity range. The negative values at high chlorinities are caused by differences in the seawater end members. This is more clearly demonstrated in the following Figure 2.21 of $S_{DENS}$ minus $S_{COND}$ vs. Cl. At high chlorinities the differences approach 0.02 instead of zero as found in the Baltic estuarine system.

The divergence is more clearly demonstrated when one compares the differences in the measured and calculated densities using $S_{COND}$ (Figure 2.22). For the Saguenay River and Gulf of St. Lawrence the measured densities agree to $\pm 10 \times 10^{-6}$ g $cm^{-3}$ with the calculated values. At high salinities the deviations for the St. Lawrence estuary are as high as $20 \times 10^{-6}$ g $cm^{-3}$. The extrapolated deviations for this estuary of $50 \times 10^{-6}$ g $cm^{-3}$ are higher than those found for our model results. The expected differences due to $S_T - S_{COND} = 0.04$ would be $35 \times 10^{-6}$ g $cm^{3-}$, if the river salts were the same. As stated earlier, this is not the case (the extra 15 ppm is due to these differences).

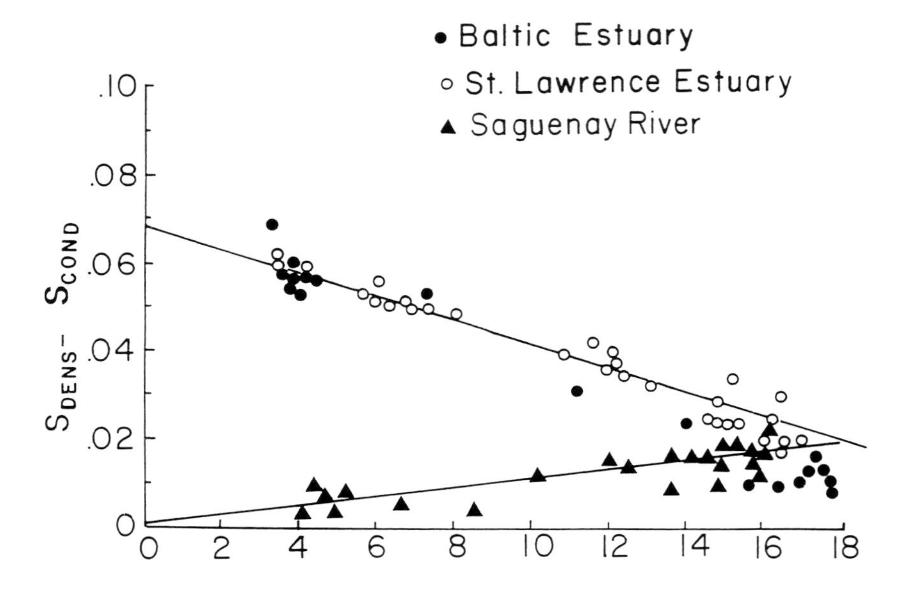


FIGURE 2.21. Density-salinity minus conductivity-salinity for various estuarine waters.

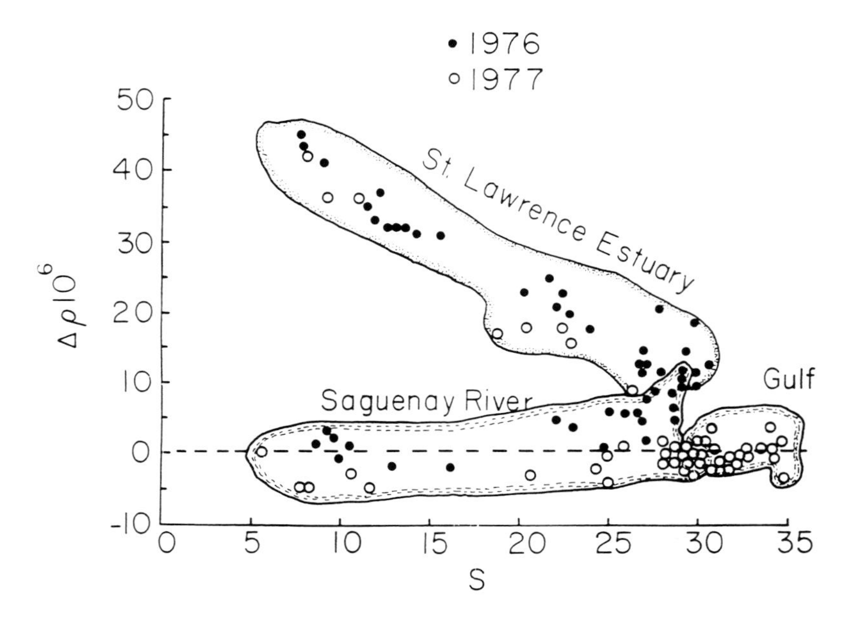


FIGURE 2.22. Comparison of the measured and calculated densities of estuarine waters.

These results indicate that, although dilute St. Lawrence River waters have a conductivity similar to world river waters, the densities are different. Thus, for careful estuarine work it is necessary to measure both density and conductivity as a function of chlorinity to fully characterize the system. If errors of $\pm 0.04$ in salinity and $\pm 50 \times 10^{-6}$ g cm$^{-3}$ in density can be tolerated, the Practical Salinity Scale and the International Equation of State of Seawater can be used for estuarine systems without a knowledge of their detailed ionic composition.

### 6.2. Evaporation in Isolated Basins

The formation of evaporites in hot arid areas (Stassfurt Deposits) were first studied by Van Hoff in 1912 (Borchert, 1965). During the slow evaporation of seawater, the precipitation of salts changes the composition of the solution. We have studied the changes in the composition of the major components of seawater in Mexican lagoons during evaporation to a salinity of close to 200 or Cl equal to 100. Our results are shown below in Figures 2.23 and 2.24. At a Cl = 40 (S = 72) there is a loss of $Ca^{2+}$ and $SO_4^{2-}$, probably as $CaSO_4 \cdot 2H_2O$ (gypsum). The loss of $HCO_3^-$ indicates that $CaCO_3$, as aragonite may also be precipitated. The small decreases in $Mg^{2+}$ and $K^+$ could be due to coprecipitation. Thermodynamic calculations indicate that the precipitation of $CaSO_4$ in the lagoon waters (Figure 2.25) occurs close to the predicted chlorinity. If seawater is evaporated further, other salts will precipitate (Table 2.15)

In laboratory experiments Na, K, and $MgSO_4$ are formed in stage IV rather than K-MgCl (the exact minerals formed depend on the temperature of evaporation). In nature $SO_4^{2-}$ is lost in brines giving $H_2S$, thus natural evaporate beds are almost devoid of sulfates. The most important solid phases in stage IV are KCl and $MgCl_2 \cdot 6H_2O$. Recently, Weare (University of California, San Diego) has used a thermodynamic model to examine the predicted phases precipitated during the evaporation of seawater. Their results are given in Table 2.16.

### 6.3. Admixture With Brines

Fissures in the ocean floor in the Red Sea yield hot, highly saline water (45 to 58°C, S = 225 to 326‰) at 2000 m. The composition of these waters are quite different from average seawater (Figure 2.26). Confined brines can also be formed by the dissolution of evaporites at low temperatures. The Orca

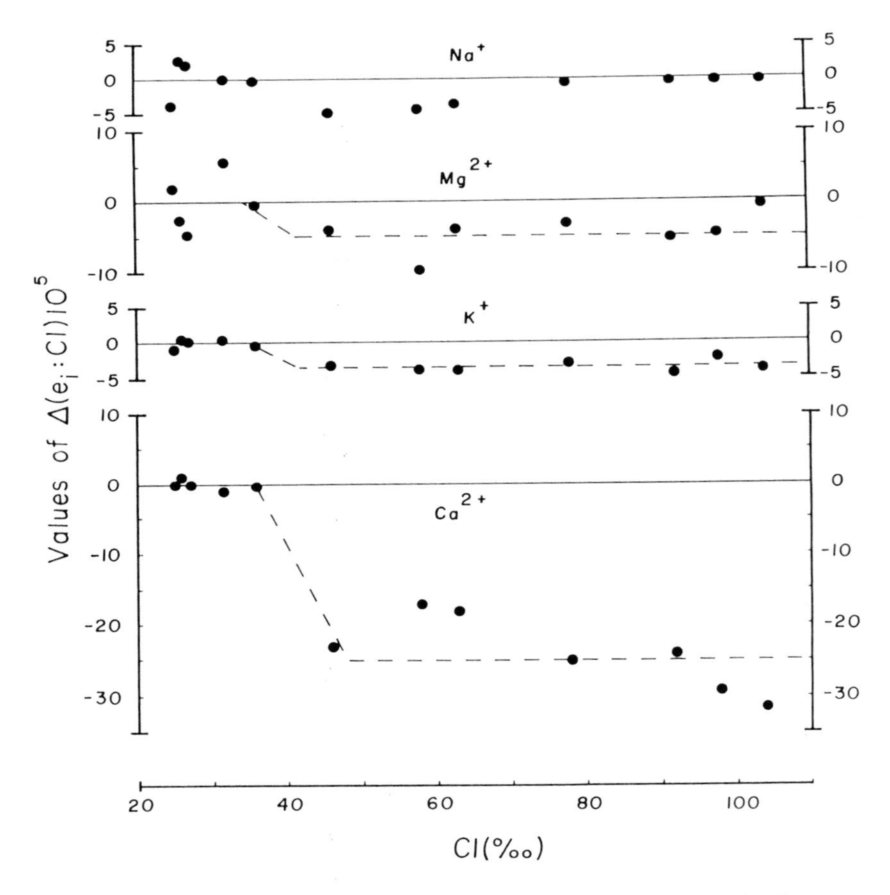


FIGURE 2.23. Values of the equivalents for cations divided by chlorinity for evaporating Mexican lagoon waters.

basin, for example, in the Gulf of Mexico is formed by this method (Figure 2.27). These waters also have a composition different from average seawater (Figure 2.28). Mixtures formed with these brines will have values of $g_i$/Cl governed by the two end numbers.

### 6.4. Precipitation and Dissolution

The dissolution of $CaCO_3$ in the form of aragonite and calcite in the deep oceans causes the $Ca^{2+}$ concentration to increase by about 1% in the deep Pacific. Recent measurements from the North Pacific are shown in Figure 2.26. The normalized values of $Ca^{2+}$ increase by as much as 1.3% in these waters. A deficiency of $Ca^{2+}$ occurs when aragonite is precipitated on the Bahama Banks and in the Red Sea (Wilson, 1970).

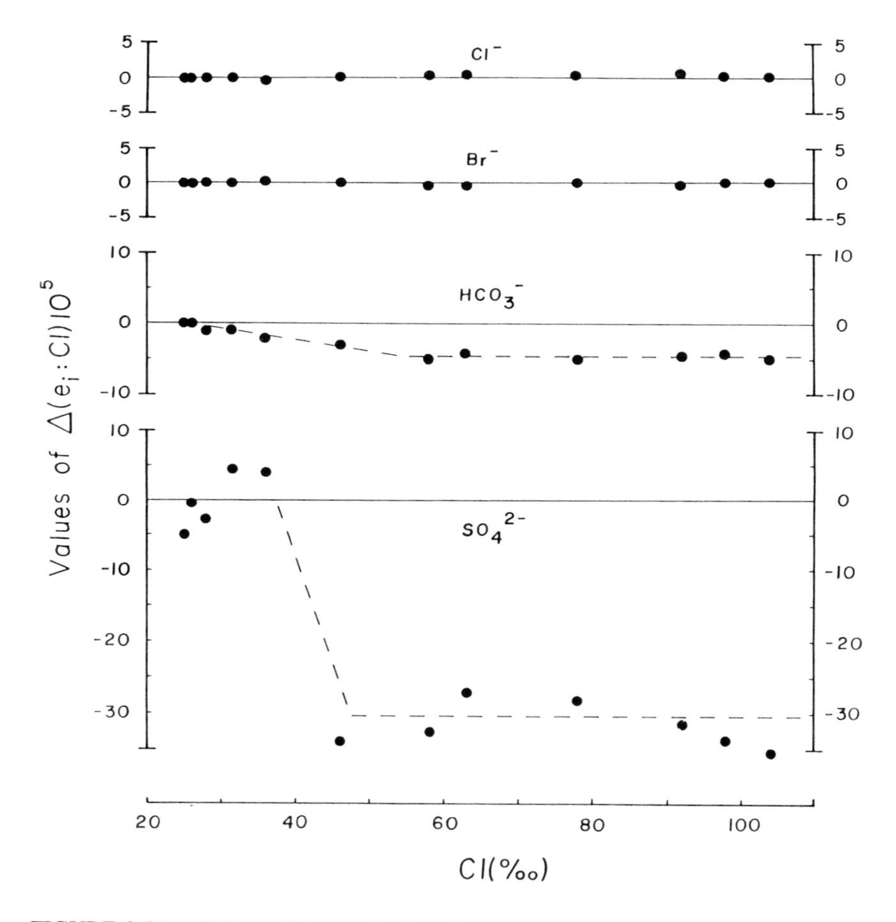


FIGURE 2.24. Values of the equivalents for anions divided by chlorinity for evaporating Mexican lagoon waters.

## 6.5. Submarine Volcanism

Molten magma has little effect on most of the major components of seawater. High F/Cl ratios, 8.0 to 9.0 $\times$ $10^{-5}$ (normal would be 6.7 $\times$ $10^{-5}$) were found near the Mid-Atlantic Ridge possibly due to the injection of volcanic gases. Brewer (MBARI) suggested that the excess F may be present in colloidal form, since the F ion electrode results have shown the same concentrations as surface samples (it could also be complexed with Ca or other trace metals). Recent studies in hydrothermal vents have shown changes in some of the major constituents (Si and Ca increase; Mg, K, B, and $SO_4^{2-}$ decrease). More will be said about these vent systems later.

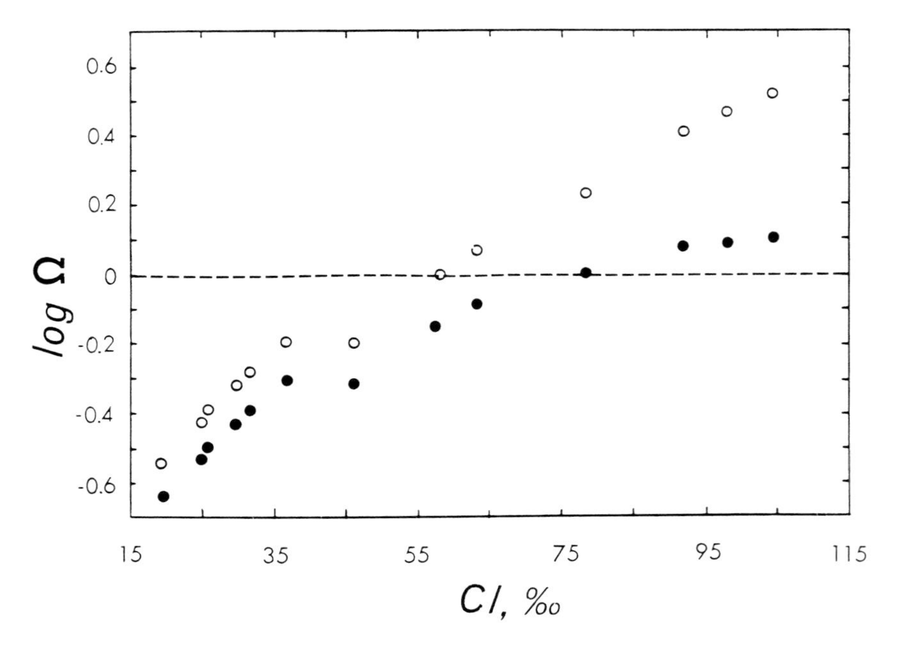


FIGURE 2.25. Values of the log of the saturation index (measured $CaSO_4$/calc.$CaSO_4$) for evaporating Mexican lagoon waters.

**TABLE 2.15**
**Sequence of Salts Formed from Evaporation of Seawater**

| Stage | Density | Wt % Liquid | Solid | % of total solid |
|---|---|---|---|---|
| | 1.026 | 100 | | |
| I | 1.140 | 50 | $CaCO_3$ + $MgCO_3$ | 1 |
| II | 1.214 | 10 | $CaSO_4$ (gypsum) | 3 |
| III | 1.236 | 3.9 | NaCl (halite) | 70 |
| IV | — | — | Na-Mg-K $SO_4$ and KCL, $MgCl_2$ | 26 |

## 6.6. Exchange Between Atmosphere and Sea

Sea and atmospheric bubbles give off 10 × $10^9$ tons of ions into the atmosphere each year. Most return directly or indirectly to the oceans. Considerable fractionation occurs. The bubbles gather dissolved species and particles to their surfaces which may take up certain ions selectively. White cap aerosols of ions and organics have a size of 0.1 to 20 μm and may cause

## TABLE 2.16
## Predicted Sequence of Salts Precipitated from Evaporation of Seawater

Equilibrium Evaporation Seawater

| Segment | First appearance of | C.F. | % $H_2O$ left | I | $^aH_2O$ | Facies |
|---|---|---|---|---|---|---|
| a | G + sol. | 3.62 | 27.63 | 2.6 | 0.929 | pene-saline |
| b | A + sol. | 9.82 | 10.18 | 6.6 | 0.772 | |
| c | A + H + sol. | 10.82 | 9.24 | 7.2 | 0.744 | saline |
| d | A + H + Gl + sol. | 13.15 | 7.60 | 7.5 | 0.738 | |
| e | | 29.17 | 3.43 | 9.1 | 0.714 | |
| f | A + H + Gl + Po + sol. | 38.50 | 2.60 | 10.1 | 0.697 | |
| g | A + H + Po + sol. | 44.76 | 2.23 | 10.7 | 0.685 | |
| h | A + H + Po + Ep + sol. | 73.56 | 1.36 | 13.0 | 0.590 | super-saline |
| | A + H + Po + Hx + sol. | 85.05 | 1.18 | 13.8 | 0.567 | |
| | A + H + Po + Ki + sol. | 102.40 | 0.98 | 14.9 | 0.498 | |
| i | A + H + Po + Ki + Car + sol. | 117.11 | 0.85 | 15.15 | 0.463 | |
| | A + H + Ki + Car + sol. | 159.74 | 0.63 | 15.33 | 0.457 | |
| | A + H + Ki + Car + Bi + sol. | 246 | 0.41 | 17.40 | 0.338 | |

C.F.: concentration factor. For seawater, C.F. = 1.
I: ionic strength.
$\alpha H_2O$: activity of $H_2O$.

**Mineral Abbreviations**: A, anhydrite, $CaSO_4$; Ap, aphthitalite, $NaK_3(SO_4)_2$; Ant, antarcticite, $CaCl_2{\cdot}6H_2O$; Bi, bischofite, $MgCl_2{\cdot}6H_2O$; Bl, bloedite, $Na_2Mg(SO_4)_2{\cdot}4H_2O$; Car, carnallite, $KMgCl_3{\cdot}6H_2O$; Ep, epsomite, $MgSO_4{\cdot}7H_2O$; G, gypsum, $CaSO_4{\cdot}2H_2O$; Gl, glauberite, $Na_2Ca(SO_4)_2$; H, halite, NaCl; Hx, hexahydrite, $MgSO_4{\cdot}6H_2O$; Ka, kainite, $KMgClSO_4{\cdot}3H_2O$; Ki, kieserite, $MgSO_4{\cdot}H_2O$; Le, leonite, $K_2Mg(SO_4)_2{\cdot}4H_2O$; Pic, picromerite, $K_2Mg(SO_4)_2{\cdot}6H_2O$; Po, polyhalite, $K_2MgCa_2(SO_4)_4{\cdot}2H_2O$; Syl, sylvite, KCl; Syn, syngenite, $K_2Ca(SO_4)_2{\cdot}6H_2O$; Tc, tachyhydrite, $Mg_2CaCl_2{\cdot}12H_2O$; Th, thenardite, $Na_6SO_4$.

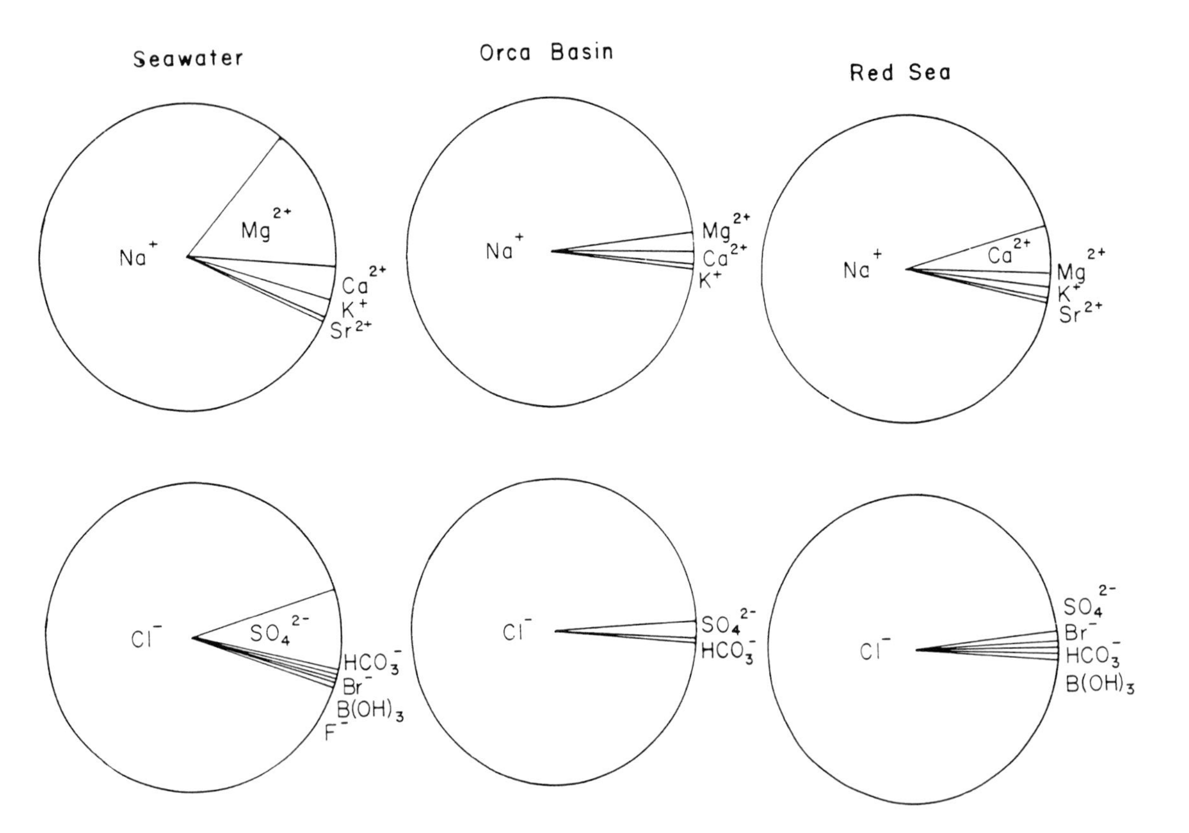


FIGURE 2.26. Comparison of the composition of various brines.

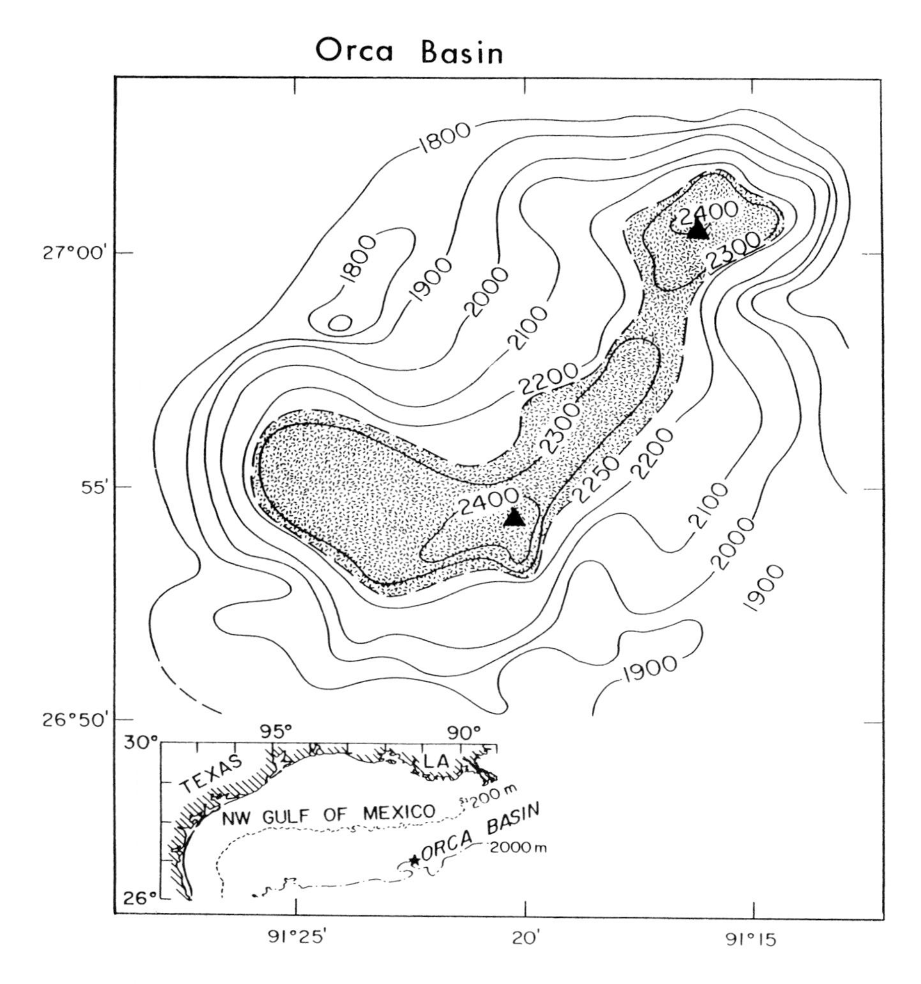


FIGURE 2.27. Location of the Orca Basin in the Gulf of Mexico.

nucleation of rain. Mainly $Na^+$, $Cl^-$, $Ca^{2+}$, $K^+$, $Mg^{2+}$, and $SO_4^{2-}$ are enriched relative to $Na^+$ in the atmosphere, whereas $Cl^-$ and $Br^-$ are impoverished. Many bubbles concentrate transition metals due to strong complexing with organics. The ratio of $Br^-/Cl^-$ is only slightly higher in air (both are neglected via bubble breaking (Duce, University of Rhode Island). The gaseous form of $Cl^-$ may be HCl, formed from the reaction of salt and sulfur trioxide or $NO_2$. Photochemical oxidation of $Br^-$ in the salt yields $Br_2$. The $I^-/Cl^-$ ratio in the atmosphere is 777 times higher than in the sea as $I_2$ is formed by photochemical oxidation of $I^-$. The rate of I lost from the ocean has been estimated to be $4 \times 10^{11}$ g/year. High concentrations of B in rain

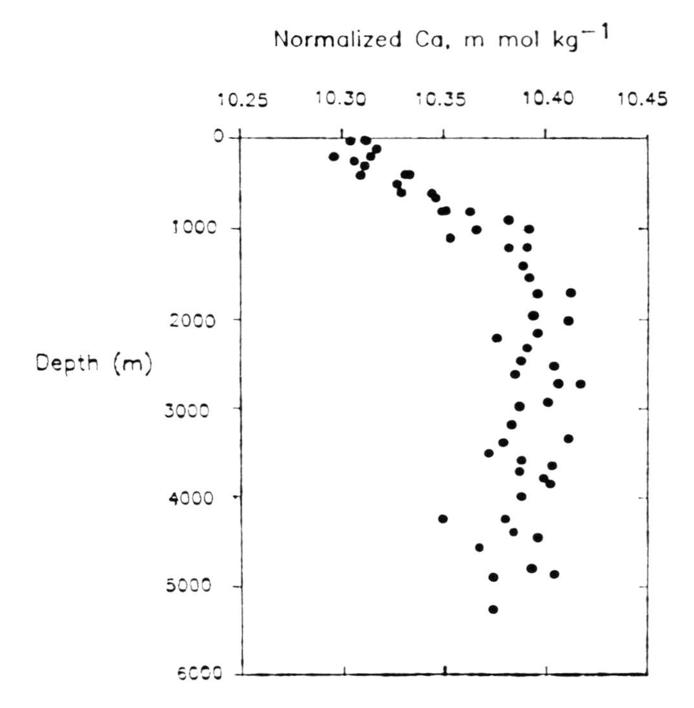


FIGURE 2.28. Depth profile of calcium in the North Pacific.

probably results from distillation from the sea surface of $B(OH)_3$ which has a high volatility.

## 6.7. Anoxic Basins

A number of anoxic basins exist (Black Sea, Cariaco Trench) in the world oceans (Richards, 1965). The $SO_4^{2-}/Cl^-$ ratios in these basins are quite low due to bacteria using $SO_4^{2-}$ as a source of $O_2$ giving off $H_2S$. The $HS^-$ is lost by precipitation of $FeS_2$ and other sulfides (ZnS, CuS, etc.). More will be said of these anoxic basins later.

## 6.8. Freezing

Sea ice has a higher $SO_4^{2-}/Cl^-$ ratio than seawater (due to $SO_4^{2-}$ incorporation into ice). Waters in the North Pacific in the minimum temperature layer (formed by melting ice) have been shown to be deficient in $SO_4^{2-}$. Differences may also occur in Ca/Cl due to precipitation of $CaCO_3$ in sea ice.

### 6.9. Interstitial Waters

Interstitial pore waters in sediments differ appreciably from seawater in some major components. The $Cl^-$ is within $\pm 1‰$ of overlying water with large variations from one locality to another. Changes in $Ca^{2+}$ can result from the dissolution of $CaCO_3$ due to the oxidation of plant material (giving $CO_2$); while changes in $SO_4^{2-}$ can occur due to the production of $H_2S$ by bacteria.

Changes in $K^+$ and other cations can occur by exchange with clay minerals. Mg may be depleted due to uptake by chlorite or reaction with $CaCO_3$ to form dolomite. In contrast $K^+$ is enriched due to the hydrolysis of feldspars. Changes in the composition of pore waters due to temperature effects have been shown to be important. Different ratios have been found for sediments when kept at *in situ* temperature. This is due to changes in the solution-solid equilibrium with temperature. If the samples are squeezed at *in situ* temperatures, the results are not affected.

## 7. ISOTOPIC VARIATIONS

Water is the principle constituent of seawater. In 1929 stable isotopes of oxygen were discovered and in 1932 deuterium was discovered. Thus, it became apparent that naturally occurring water is a mixture of several species differing in molecular weight. There are three known isotopes of hydrogen ($^1H$, $^2H$ or D (deuterium), and $^3H$ or T (tritium)) and six isotopes of oxygen ($^{14}O$, $^{15}O$, $^{16}O$, $^{17}O$, $^{18}O$, and $^{19}O$). Tritium is radioactive with a half-life of 12.5 years. The isotopes of $^{14}O$, $^{15}O$, and $^{19}O$ are also radioactive, but are short lived and do not occur significantly in natural waters. The precise isotopic content of natural water depends upon the origin of the sample; however, within the limits of variation the abundance is (Table 2.17) 99.73% is $^1H_2{}^{16}O$ (light water), 0.2% $^{18}O$ water, 0.04% $^{17}O$ water, and 0.032% $^1H^2H^{16}O$ water (HDO). It should be pointed out that the equilibrium

$$D_2O + H_2O = 2HDO \qquad (55)$$

is always maintained. The hydrogen in water is 0.032%/2 = 0.015% D. The presence of these isotopes can change some of the properties of water and, thus, must be considered a major component of seawater. Determination of D and $^{18}O$ by mass spectrophotometry is the best method of determining D and $^{18}O$. Prior to D determination, the water sample is converted to hydrogen

**TABLE 2.17**
**Composition of Water with Respect to the Different Forms of Water**

| Water molecule | Portion in total water | Portion in heavy water | Comparable concentration |
|---|---|---|---|
| $^{1}H\ ^{16}O$ | 99.73 | — | — |
| $^{1}H\ ^{18}O$ | 0.20 | 73.5 | Mg |
| $^{1}H\ ^{17}O$ | 0.04 | 14.7 | Ca |
| $^{1}H^{2}H^{16}O$ | 0.032 | 11.8 | K |
| $^{1}H^{2}H^{18}O$ | $6 \times 10^{-5}$ | 0.022 | N |
| $^{1}H^{2}H^{17}O$ | $1 \times 10^{-5}$ | 0.003 | Al |
| $^{2}H\ ^{16}O$ | $3 \times 10^{-6}$ | 0.001 | P |
| $^{2}H\ ^{18}O$ | $6 \times 10^{-9}$ | $2 \times 10^{-6}$ | Hg |
| $^{2}H\ ^{17}O$ | $1 \times 10^{-9}$ | $3 \times 10^{-7}$ | Au |

gas by reaction with hot zinc or uranium metal. In the determination of $^{18}O$, the water sample is equilibrated with $CO_2$ gas, which is then analyzed by mass spectrophotometry.

Absolute isotopic abundances cannot at present be determined with sufficient accuracy to be of use in studies of their natural variations. In earlier work, D analyses were reported relative to local tap water. This was unsatisfactory since differences occur in the D content of tap waters. Some workers, in their $^{18}O/^{16}O$ work on natural waters, used the average of their isotopic data for a number of deep water samples taken from the Atlantic, Pacific, and Indian Oceans as their standard. This, however, is not completely satisfactory. Craig (SCRIPPS) pointed out the desirability of a standard reference for both D and $^{18}O$ analyses. He determined the D/H ratio for a set of ocean waters similar to the ones chosen by earlier workers for their $^{18}O$ data. He suggested that both D and $^{18}O$ data could be specified in terms of the National Bureau of Standards isotopic reference samples No. 1 and No. 1-A. These are distilled water samples of large volumes, intended as a cross check calibration to mass spectrophometry labs. After consultation, Craig defined a standard mean ocean water (SMOW) in terms of the N.B.S. reference sample 1

$$D/H(SMOW) = 1.050\ D/H(NBS) - 1 \tag{56}$$

$$^{18}O/^{16}O(SMOW) = 1.008\ ^{18}O/^{16}O(NBS) - 1 \tag{57}$$

It was suggested that isotopic data for both isotopes be reported as parts per mille. Enrichment ($\delta$) relative to defined standard mean ocean water (SMOW) is given by

$$\delta = [(R_{sample}/R_{SMOW}) - 1]\ 1000 \tag{58}$$

where the values of R are isotopic ratios. SMOW is thus defined in terms of an actual water reference standard NBS-1 and provides a convenient and consistent zero reference level for reporting enrichment.

### 7.1. Deuterium

Ocean waters have widely different D/H ratios. Equatorial waters are richer in D than waters in high latitudes. D follows the same trend with depth as $^{18}O$, that is, surface waters contain relatively more than deep waters. D/H varies from 6410 to 6536 for the Atlantic, Pacific, and Indian Oceans. In summary, high latitude waters tend to contain less D than equatorial waters and deep waters contain less D than surface waters.

### 7.2. $^{18}O$ Results

The $^{18}O/^{16}O$ ratios of a number of freshwaters derived from snow, rain, lakes, and rivers as well as the oceans have been examined. The $^{18}O$ of freshwaters varies considerably, thus, $^{18}O$ in the oceans varies depending upon how much freshwater is in the source. Melted snow can have values of $^{18}O$ 3.5 times greater than the waters from the Mississippi. For waters collected in the Alaska and California currents, the low salinity and high $^{18}O$ content is due to the melting of snow and ice.

Deep waters contain less $^{18}O$ than surface samples. The differences are greater than would be expected due to evaporation. The deep water is diluted by sinking polar waters with a consequential lowering of $^{18}O$. Mediterranean waters (surface and deep) showed a much smaller difference because the high evaporation rate causes less fractionation of the isotopes, which in turn implies that freshwaters entering the ocean are less depleted in $^{18}O$ than in the other cases discussed. The deep waters of the Pacific have higher $^{18}O$ values compared to deep Atlantic values since the waters are older and oxidation enriches the waters in $^{18}O$.

The vapor pressure of HDO is lower than that of $H_2O$ which means that during evaporation $H_2O$ enters the vapor phase more readily than does HDO, with the results that D becomes preferentially concentrated in the residual water. This accounts for its higher concentrations in surface waters near the equator. In the precipitation process, the reverse is true. The residual phase vapor becomes depleted of HDO which is preferentially precipitated. As the raindrop falls, further isotopic enrichment can take place due to evaporation, especially in dry climates.

Deuterium analyses of co-existing ice and arctic waters showed that the ice contained 2% more D than waters from which it was formed. In general, $^{18}O$ follows the same pattern as D; however, since the relative masses between $^{18}O$ and $^{16}O$ are smaller, the effects are also smaller. The $^{18}O/^{16}O$ in snow is lower than in rain or seawater.

In general, the two isotopic ratios are quite variable in seawater. Both heavier isotopes are concentrated in surface water relative to deep waters, but for different reasons. Evaporation causes the effect for D/H while dilution by sinking polar waters produces a decrease of $^{18}O/^{16}O$ in deep water. Since organisms prefer $^{16}O$, older waters have a larger $^{18}O/^{16}O$ ratio than younger waters.

## 7.3. Isotopes Of Sulfur

Sulfur has four stable isotopes having atomic weights 32, 33, 34, and 36. The heaviest is the least abundantly occurring isotope. The $^{32}S/^{33}S$ ratio is 123.4 for seawater collected in the Pacific, Atlantic, Arctic, North, and Baltic Seas. This ratio is lower than that found in rain water, suggesting that the $SO_4^{2-}$ sulfur in precipitation is not derived from the salt spray. The most important agents in isotopic fractionation of S were $SO_4^{2-}$ reducing bacteria

$$^{34}S^{2-} + {}^{32}SO_4^{2-} = {}^{32}S^{2-} + {}^{34}SO_4^{2-} \tag{59}$$

When $SO_4^{2-}$ is reduced to $S^{-2}$, the equilibrium will produce $H_2S$ which is deficient in the heavier S isotope. This will occur only in sediments or anaerobic water and the effect will not alter the physical properties of seawater significantly. It is significant that $SO_4^{2-}$ minerals occurring in marine evaporite deposits have the same $^{32}S/^{33}S$ ratio as in seawater. The $SO_4^{2-}$ minerals associated with precipitation in anaerobic conditions have a higher ratio (as high as 23.2 for some shales).

## FURTHER READING

Aston, S. R., Estuarine Chemistry, chap. 41, *Chemical Oceanography,* Vol. 7, 2nd ed., J. P. Riley and R. Chester, Eds., Academic Press, New York, 361-440 (1978).

Borchert, H., Principles of Oceanic Salt Deposition and Metamorphism, chap. 19, *Chemical Oceanography,* Vol. 2, 1st ed., J. P. Riley and G. Skirrow, Eds., Academic Press, New York, 205-276 (1965).

Cox, R. A., The Physical Properties of Seawater, chap. 3, *Chemical Oceanography,* Vol. 1, 1st ed., J. P. Riley and G. Skirrow, Eds., Academic Press, New York, 73-120 (1965).

Culkin, F., The Major Constituents, chap. 4, *Chemical Oceanography,* Vol. 1, 1st ed., J. P. Riley and G. Skirrow, Eds., Academic Press, New York, 121-161 (1965).

Grasshoff, K., M. Ehrhart and K. Kremling, Eds., *Methods of Seawater Analysis* (1983).

Riley, J. P., Analytical Chemistry of Sea Water, chap. 21, *Chemical Oceanography,* Vol. 2, 1st ed., J. P. Riley and G. Skirrow, Eds., Academic Press, New York 295-424 (1965).

Riley, J. P., Analytical Chemistry of Sea Water, chap. 19, *Chemical Oceanography,* Vol. 3, 2nd ed., J. P. Riley and G. Skirrow, Eds., Academic Press, New York, 193-514 (1975).

Riley, J. P., Historical Introduction, chap. 1, *Chemical Oceanography,* Vol. 1, 2nd ed., J. P. Riley and G. Skirrow, Eds., Academic Press, New York 1-41 (1965).

Wallace, C., *Development of the Chlorinity-Salinity Concept,* Elsevier, New York (1974).

Whitfield, M., Electroanalytical Chemistry of Sea Water, chap. 20, *Chemical Oceanography,* Vol. 4, 2nd ed., J. P. Riley and G. Skirrow, Eds., Academic Press, New York, 1-154 (1975).

Wilson, T. R. S., Salinity and the Major Elements of Sea Water, chap. 6, *Chemical Oceanography,* Vol. 1, 2nd ed., J. P. Riley and G. Skirrow, Eds., Academic Press, New York 365-413 (1975).

# 3 Table of Contents

**Minor Elements in Seawater** .... 115
1. *Classification of Elements* .... 117
   1.1. *$d^0$ Cations* .... 118
   1.2. *$d^{10}$ Cations* .... 121
   1.3. *Transition Metals Between $d^0$ and $d^{10}$* .... 124
2. *Residence Times* .... 124
3. *Distribution of Trace Elements in the Oceans* .... 128
4. *Biological Interactions* .... 139
5. *Geochemical Balance* .... 147

*Further Reading* .... 155

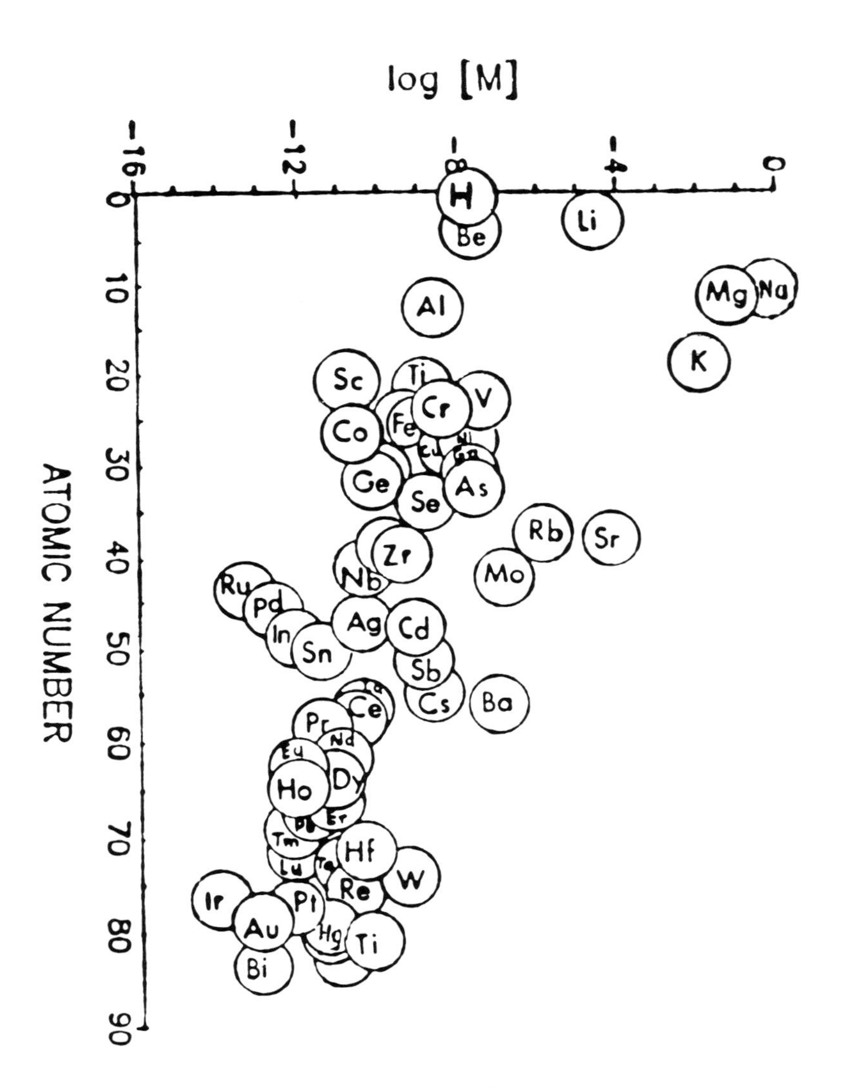


The range of concentrations of elements in seawater.

# 3 Minor Elements in Seawater

## 1. CLASSIFICATION OF ELEMENTS

The elements found in seawater (Figure 3.1) include most of those given in the Periodic Table. Of these, only 14 (O, H, Cl, Na, Mg, S, Ca, K, Br, C, Sr, B, Si, and F) have concentrations greater than 1 ppm. Most of these elements (with the exception of Si) are generally unreactive elements (both chemically and biologically). Many of the remaining elements, called minor, are involved in inorganic and biological reactions in the marine environment. The biolimited elements N, Si, and P will be discussed later. The inert gases will also be discussed separately in a later chapter.

Bruland (University of California, Santa Cruz) has tabulated the range and average concentration of a number of elements in seawater (S = 35). His results are tabulated in Table 3.1. He conveniently divides the elements into three classes based on concentration.

1. Major elements 0.05 to 750 m$M$
2. Minor elements 0.05 to 50 $\mu M$
3. Trace elements 0.05 to 50 n$M$

Trace Elements < 50 pmol kg$^{-1}$ · Trace Elements 0.05-50 nmol kg$^{-1}$ · Minor Elements 0.05-50 μmol kg$^{-1}$ · Major Elements 0.05-50 mmol kg$^{-1}$ · Major Elements > 50 mmol kg$^{-1}$

| Ia | IIa | IIIb | IVb | Vb | VIb | VIIb | VIIIb | | | Ib | IIb | IIIa | IVa | Va | VIa | VIIa |
|---|---|---|---|---|---|---|---|---|---|---|---|---|---|---|---|---|
| 3 Li I | 4 Be II | | | | | | | | | | | 5 B III | 6 C IV | 7 N V | 8 O 0 | 9 F -I |
| 11 Na I | 12 Mg II | | | | | | | | | | | 13 Al III | 14 Si IV | 15 P V | 16 S VI | 17 Cl -I |
| 19 K I | 20 Ca II | 21 Sc III | 22 Ti IV ? | 23 V V | 24 Cr VI | 25 Mn II | 26 Fe III | 27 Co II | 28 Ni II | 29 Cu II | 30 Zn II | 31 Ga III | 32 Ge IV | 33 As V | 34 Se VI | 35 Br -I |
| 37 Rb I | 38 Sr II | 39 Y III | 40 Zr IV | 41 Nb V | 42 Mo VI | 43 (Tc) VII | 44 Ru IV ? | 45 Rh III ? | 46 Pd II ? | 47 Ag I | 48 Cd II | 49 In III | 50 Sn IV | 51 Sb V | 52 Te VI ? | 53 I V |
| 55 Cs I | 55 Ba II | 57 La III | 72 Hf IV | 73 Ta V | 74 W VI | 75 Re VII | 76 Os (IV) ? | 77 Ir III ? | 78 Pt (IV,II) ? | 79 Au I | 80 Hg II | 81 Tl (III,I) | 82 Pb II | 83 Bi III | | |

| 58 Ce III | 59 Pr III | 60 Nd III | 61 (Pm) | 62 Sm III | 63 Eu III | 64 Gd III | 65 Tb III | 66 Dy III | 67 Ho III | 68 Er III | 69 Tm III | 70 Yb III | 71 Lu III |
|---|---|---|---|---|---|---|---|---|---|---|---|---|---|

FIGURE 3.1. Classification of elements in ocean waters.

The concentrations of elements in the ocean, because of their reactivity, have a wide range of concentrations (Table 3.1 and Figure 3.2).

Since many of these minor elements are metals, Goldberg (SCRIPPS) divided them into three classes based on their electronic structure (Table 3.2). This simple classification for metals is given below.

## 1.1. d$^0$ Cations

Ions of metallic elements with a rare gas configuration. These include the alkali metals ($Li^+$, $Na^+$, $K^+$, $Rb^+$, $Cs^+$, $Fr^+$), the alkaline earth metals ($Be^{2+}$, $Mg^{2+}$, $Ca^{2+}$, $Sr^{2+}$, $Ba^{2+}$, $Ra^{2+}$), and the lanthanide series ($La^{3+}$, $Ce^{3+}$, $Pr^{3+}$, $Nd^{3+}$, $Pm^{3+}$, $Sm^{3+}$, $Eu^{3+}$, $Gd^{3+}$, $Tb^{3+}$, $Dy^{3+}$, $Ho^{3+}$, $Er^{3+}$, $Tm^{3+}$, $Yb^{3+}$, $Lu^{3+}$) + $Al^{3+}$, $Sc^{3+}$, $Ti^{3+}$, and $Th^{4+}$.

This group is characterized by the fact that its members form few complexes, these mainly with $F^-$ and ligands where oxygen is the donor atom (e.g., $SO_4^{2-}$, $CO_3^{2-}$, and $PO_4^{3-}$). There is little or no evidence that these metal ions form complexes with the heavier halides. In a given series the stability of the complexes increase with increasing charge and for cations of the same charge with decreasing radius. This is shown in Table 3.3.

**TABLE 3.1**
**Speciations, Concentrations, and Distribution Types of Elements in Ocean Water**

| Element | Probable species | Range and av. concentration | Type of distribution |
|---|---|---|---|
| Li | $Li^{+}$ | 25 μ*M* | Conservative |
| Be | $BeOH^{+}$,$Be(OH)_2$ | 4—30 p*M*, 20 p*M* | Nutrient Type |
| B | $B(OH)_3$, $B(OH)_4$ | 0.416 m*M* | Conservative |
| C | $HCO_3^-$, $CO_3^{2-}$ | 2.0—2.5 m*M*, 2.3 m*M* | Nutrient Type |
| N | $NO_3^-$,($N_2$) | 0—45 μ*M* | Nutrient Type |
| F | $F^-$, $MgF^+$, $CaF^+$ | 68 μ*M* | Conservative |
| Na | $Na^+$ | 0.468 *M* | Conservative |
| Mg | $Mg^{2+}$ | 53.2 m*M* | Conservative |
| Al | $Al(OH)_4^-$, $Al(OH)_3$ | 5—40 n*M*, 2 n*M* | Mid-Depth-Min. |
| Si | $Si(OH)_4$ | 0—180 μ*M* | Nutrient Type |
| P | $HPO_4^{2-}$, $MgHPO_4$ | 0—3.2 μ*M* | Nutrient Type |
| S | $SO_4^{2-}$, $NaSO_4^-$, $MgSO_4$ | 28.2 m*M* | Conservative |
| Cl | $Cl^-$ | 0.546 *M* | Conservative |
| K | $K^+$ | 10.2 m*M* | Conservative |
| Ca | $Ca^{2+}$ | 10.3 m*M* | Conservative |
| Sc | $Sc(OH)_3$ | 8—20 p*M*, 15 p*M* | Surface Depletion |
| Ti | $Ti(OH)_4$ | few p*M* | ? |
| V | $HVO_4^{2-}$, $H_2VO_4^-$ | 20—35 n*M* | Surface Depletion |
| Cr | $CrO_4^{2-}$ | 2—5 n*M*, 4 n*M* | Nutrient Type |
| Mn | $Mn^{2+}$ | 0.2—3 n*M*, 0.5 n*M* | Depletion at Depth |
| Fe | $Fe(OH)_3$ | 0.1—2.5 n*M*, 1 n*M* | Surface and Depth Depletion |
| Co | $Co^{2+}$, $CoCO_3$ | 0.01—0.1 n*M*,0.02 n*M* | Surface and Depth Depletion |
| Ni | $NiCO_3$ | 2—12 n*M*, 8 n*M* | Nutrient Type |
| Cu | $CuCO_3$ | 0.5—6 n*M*, 4 n*M* | Nutrient Type, Scavenging |
| Zn | $Zn^{2+}$, $ZnOH^+$ | 0.05—9 n*M*, 6 n*M* | Nutrient Type |
| Ga | $Ga(OH)_4^-$ | 5—30 p*M* | ? |
| As | $HAsO_4^{2-}$ | 15—25 n*M*, 23 n*M* | Nutrient Type |
| Se | $SeO_4^{2-}$, $SeO_3^{2-}$ | 0.5—2.3 n*M*, 1.7 n*M* | Nutrient Type |
| Br | $Br^-$ | 0.84 n*M* | Conservative |
| Rb | $Rb^+$ | 1.4 μ*M* | Conservative |
| Sr | $Sr^{2+}$ | 90 μ*M* | Conservative |
| Y | $YCO_3$ | 0.15 n*M* | Nutrient Type |
| Zr | $Zr(OH)_4$ | 0.3 n*M* | ? |
| Nb | $NbCO_3$ | 50 p*M* | Nutrient Type(?) |
| Mo | $MoO_4^{2-}$ | 0.11 μ*M* | Conservative |
| Tc | $TcO_4^-$ | No stable isotope | ? |
| Ru | ? | <0.05 p*M* | ? |
| Rh | ? | ? | ? |

**TABLE 3.1 (continued)**
**Speciations, Concentrations, and Distribution Types of Elements in Ocean Water**

| Element | Probable species | Range and av. concentration | Type of distribution |
|---|---|---|---|
| Pd | ? | 0.2 p*M* | ? |
| Ag | $AgCl_2^-$ | 0.5—35 p*M*, 25 p*M* | Nutrient Type |
| Cd | $CdCl_2^-$ | 0.001—1.1 n*M*, 0.7 n*M* | Nutrient Type |
| In | $In(OH)_3$ | 1 p*M* | ? |
| Sn | $Sn(OH)_4$ | 1—12 p*M*, 4 p*M* | Surface Input |
| Sb | $Sb(OH)_6^-$ | 1.2 n*M* | ? |
| Te | $TeO_3^{2-}$, $HTeO_3^-$ | ? | ? |
| I | $IO_3^-$ | 0.2—0.5 μ*M*, 0.41 *M* | Nutrient Type |
| Cs | $Cs^+$ | 2.2 n*M* | Conservative |
| Ba | $Ba^{2+}$ | 32—150 n*M*,100 n*M* | Nutrient Type |
| La | $LaCO_3^+$ | 13—37 p*M*,30 p*M* | Surface Depletion |
| Ce | $CeCO_3^+$ | 16—26 p*M*,20 p*M* | Surface Depletion |
| Pr | $PrCO_3^+$ | 4 p*M* | Surface Depletion |
| Nd | $NdCO_3^+$ | 12—25 p*M*, 10 p*M* | Surface Depletion |
| Sm | $SmCO_3^+$ | 3—5 p*M*, 4 p*M* | Surface Depletion |
| Eu | $EuCO_3^+$ | 0.6—1 p*M*, 0.9 p*M* | Surface Depletion |
| Gd | $GdCO_3^+$ | 3—7 p*M*, 6 p*M* | Surface Depletion |
| Tb | $TbCO_3^+$ | 0.9 p*M* | Surface Depletion |
| Dy | $DyCO_3^+$ | 5—6 p*M*, 6 p*M* | Surface Depletion |
| Ho | $HoCO_3^+$ | 1.9 p*M* | Surface Depletion |
| Er | $ErCO_3^+$ | 4—5 p*M*, 5 p*M* | Surface Depletion |
| Tm | $TmCO_3^+$ | 0.8 p*M* | Surface Depletion |
| Yb | $YbCO_3^+$ | 3—5 p*M*, 5 p*M* | Surface Depletion |
| Lu | $LuCO_3^+$ | 0.9 p*M* | Surface Depletion |
| Hf | $Hf(OH)_4$ | <40 p*M* | ? |
| Ta | $Ta(OH)_5$ | <14 p*M* | ? |
| W | $WO_4^{2-}$ | 0.5 n*M* | Conservative |
| Re | $ReO_4^-$ | 14—30 p*M*, 20 p*M* | Conservative |
| Os | ? | ? | ? |
| Ir | ? | 0.01 p*M* | ? |
| Pt | $PtCl_4^{2+}$ | 0.5 p*M* | ? |
| Au | $AuCl_2^-$ | 0.1—0.2 p*M* | ? |
| Hg | $HgCl_4^{2-}$ | 2—10 p*M*, 5 p*M* | ? |
| Tl | $Tl^+$, TCl | 60 p*M* | Conservative |
| Pb | $PbCO_3$ | 5—175 p*M*, 10 p*M* | Surface Input, Depletion at Depth |
| Bi | $BiO^+$, $Bi(OH)_2^+$ | <0.015—.24 p*M* | Depletion at Depth |

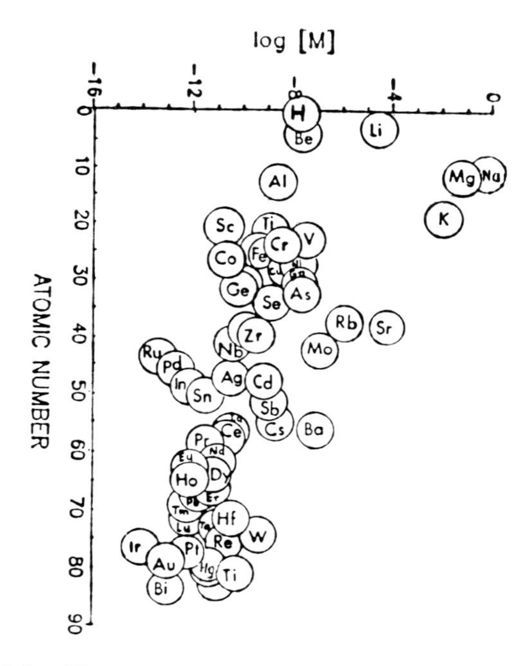


FIGURE 3.2. The range of concentrations of elements in seawater.

Later in the text we will discuss the various inorganic complexes of the $d^0$ metal ions that are the major components of seawater.

## 1.2. $d^{10}$ Cations

Cations with an outer shell of 18 electrons. These include $Ag^+$, $Zn^{2+}$, $Ga^{3+}$, and $Sc^{4+}$. The univalent $d^{10}$ metal ions behave in a different way from the $d^0$ metals. For these ions the halide complexes are quite strong and increase in stability with increasing atomic weight or size of the ligand.

| | AgF | AgCl | AgBr | AgI |
|---|---|---|---|---|
| log K | −0.3 | 3.0 | 4.3 | 8.1 |

This is due to the increase in polarizability of the $d^{10}$ metals and the large halide ligands. The complexes, thus, have more covalent character. Since the concentration of $Cl^-$ in seawater is much larger than that of the other halides, the Cl complexes will always dominate (the values of $K_{MX}$ for the heavier halides are not sufficiently large to compensate for the low concentration). It is possible however for $OH^-$ ions to compete successfully with $Cl^-$. For $Cl^-$ complexes to predominate $\log K_{MCl} - \log K_{MOH}$ must be greater than $-5.4$. An examination of the values of $\log K_{MCl}$ and $\log K_{MOH}$ for various metal ions is shown in Table 3.4.

**TABLE 3.2**
**Electronic Configuration of Atoms**

| | | | K | L | | M | | | N | | | |
|---|---|---|---|---|---|---|---|---|---|---|---|---|
| **Period** | **Z** | **Element** | s | s | p | s | p | d | s | p | d | f |
| 1 | 1 | H | 1 | | | | | | | | | |
| | 2 | He | 2 | | | | | | | | | |
| 2 | 3 | Li | 2 | 1 | | | | | | | | |
| | 4 | Be | 2 | 2 | | | | | | | | |
| | 5 | B | 2 | 2 | 1 | | | | | | | |
| | 6 | C | 2 | 2 | 2 | | | | | | | |
| | 7 | N | 2 | 2 | 3 | | | | | | | |
| | 8 | O | 2 | 2 | 4 | | | | | | | |
| | 9 | F | 2 | 2 | 5 | | | | | | | |
| | 10 | Ne | 2 | 2 | 6 | | | | | | | |
| 3 | 11 | Na | 2 | 2 | 6 | 1 | | | | | | |
| | 12 | Mg | 2 | 2 | 6 | 2 | | | | | | |
| | 13 | Al | 2 | 2 | 6 | 2 | 1 | | | | | |
| | 14 | Si | 2 | 2 | 6 | 2 | 2 | | | | | |
| | 15 | P | 2 | 2 | 6 | 2 | 3 | | | | | |
| | 16 | S | 2 | 2 | 6 | 2 | 4 | | | | | |
| | 17 | Cl | 2 | 2 | 6 | 2 | 5 | | | | | |
| | 18 | Ar | 2 | 2 | 6 | 2 | 6 | | | | | |
| 4 | 19 | K | 2 | 2 | 6 | 2 | 6 | | 1 | | | |
| | 20 | Ca | 2 | 2 | 6 | 2 | 6 | | 2 | | | |
| | 21 | Sc | 2 | 2 | 6 | 2 | 6 | 1 | 2 | | | |
| | 22 | Ti | 2 | 2 | 6 | 2 | 6 | 2 | 2 | | | |
| | 23 | V | 2 | 2 | 6 | 2 | 6 | 3 | 2 | | | |
| | 24 | Cr | 2 | 2 | 6 | 2 | 6 | 5 | 1 | | | |
| | 25 | Mn | 2 | 2 | 6 | 2 | 6 | 5 | 2 | | | |
| | 26 | Fe | 2 | 2 | 6 | 2 | 6 | 6 | 2 | | | |
| | 27 | Co | 2 | 2 | 6 | 2 | 6 | 7 | 2 | | | |
| | 28 | Ni | 2 | 2 | 6 | 2 | 6 | 8 | 2 | | | |
| | 29 | Cu | 2 | 2 | 6 | 2 | 6 | 10 | 1 | | | |
| | 30 | Zn | 2 | 2 | 6 | 2 | 6 | 10 | 2 | | | |
| | 31 | Ga | 2 | 2 | 6 | 2 | 6 | 10 | 2 | 1 | | |
| | 32 | Ge | 2 | 2 | 6 | 2 | 6 | 10 | 2 | 2 | | |
| | 33 | As | 2 | 2 | 6 | 2 | 6 | 10 | 2 | 3 | | |
| | 34 | Se | 2 | 2 | 6 | 2 | 6 | 10 | 2 | 4 | | |
| | 35 | Br | 2 | 2 | 6 | 2 | 6 | 10 | 2 | 5 | | |
| | 36 | Kr | 2 | 2 | 6 | 2 | 6 | 10 | 2 | 6 | | |

**TABLE 3.3**
**Stability Constants for the Formation of Fluoride and Hydroxide Complexes**

| Ion | Log $K_{MF}$ | Log $K_{MOH}$ | Radius (Å) |
|---|---|---|---|
| $Be^{2+}$ | 4.29 | 10.28 | 0.31 |
| $Mg^{2+}$ | 1.82 | 2.3 | 0.65 |
| $Ca^{2+}$ | 1.04 | 1.4 | 0.99 |
| $Sr^{2+}$ | — | 0.9 | 1.13 |
| $Ba^{2+}$ | 0.45 | 0.8 | 1.35 |

**TABLE 3.4**
**Stability Constants for the Formation of Chloride and Hydroxide Complexes**

| Ion | Log $K_{MCl}$ | Log $K_{MOH}$ | Log $K_{MCl}$ − Log $K_{MOH}$ |
|---|---|---|---|
| $Ag^{+}$ | 3.1 | 2.3 | 0.8 |
| $Cd^{2+}$ | 2.0 | 5.5 | −3.5 |
| $Hg^{2+}$ | 7.3 | 11.5 | −4.2 |
| $Zn^{2+}$ | −0.5 | 4.4 | −4.9 |
| $Cu^{2+}$ | 0.4 | 6.3 | −5.9 |
| $Pb^{2+}$ | 1.5 | 7.8 | −6.3 |

From Table 3.4 it is evident that $Cl^-$ complexes are greater than $OH^-$ complexes for $Ag^+$, $Cd^{2+}$, $Hg^{2+}$, and $Zn^{2+}$, but not for $Cu^{2+}$ and $Pb^{2+}$. The $SO_4^{2-}$ complexes cannot compete for the $d^{10}$ cations. It should be pointed out that the addition of further $Cl^-$ can occur for many of these complexes.

$$Hg^{2+} + Cl^- \rightarrow HgCl^+ \quad (1)$$

$$HgCl^+ + Cl^- \rightarrow HgCl_2^0 \quad (2)$$

$$HgCl_2^0 + Cl^- \rightarrow HgCl_3^- \quad (3)$$

$$HgCl_3^- + Cl^- \rightarrow HgCl_4^{2-} \quad (4)$$

$$HgCl_4^{2-} + Cl^- \rightarrow HgCl_5^{3-} \quad (5)$$

**TABLE 3.5**
**Stability Constants for the Formation of Organic Ligands with Metals**

| | Log K | | |
|---|---|---|---|
| Ion | EDTA | Ethylenediamine | Nitrilotriacetic acid |
| $Mn^{2+}$ | 14 | 2.7 | 7.4 |
| $Fe^{2+}$ | 14 | 4.3 | 8.3 |
| $Co^{2+}$ | 16 | 5.9 | 10.5 |
| $Ni^{2+}$ | 18 | 7.9 | 11.4 |
| $Cu^{2+}$ | 19 | 10.5 | 12.8 |
| $Zn^{2+}$ | 16 | 6.0 | 10.5 |

To evaluate the complete speciation of these heavy metals in seawater it is necessary to consider the competition of all the ligands ($Cl^-$, $Br^-$, $OH^-$, $HCO_3^-$, $CO_3^{2-}$, etc.). The speciation of the heavy metals ($Cd^{2+}$, $Hg^{2+}$, $Zn^{2+}$, $Cu^{2+}$, and $Pb^{2+}$) in seawater and river waters will be discussed later.

### 1.3. Transition Metals Between $d^0$ and $d^{10}$

The transition metal cations in which the d electrons are greater than zero and less than ten represent this group. These include $Mn^{2+}$, $Fe^{2+}$, $Co^{2+}$, and $Ni^{2+}$ ions. These metals have been extensively studied. From such work has evolved the so-called "Irving-Williams order" which states that for almost every ligand the stability of its complexes increases in the order

$$Mn^{2+} < Fe^{2+} < Co^{2+} < Ni^{2+} < Cu^{2+} > Zn^{2+}$$

An example of this order is shown in the following Table 3.5. The cause of this order is related to the stability of the electronic structure of the various metals.

## 2. RESIDENCE TIMES

An element could have a low concentration in seawater for two reasons: (1) it may be very reactive and thus be rapidly removed to the sediments, or

(2) it may occur in very low concentrations in the source — crystalline rocks or gaseous emanations from the interior of the earth. For example, $Al^{3+}$, although a minor element in seawater, is one of the most predominant constituents of igneous rocks; its high reactivity reduces its concentration. $Cs^{+}$, on the other hand, has a low value in seawater and in crystalline rocks. Insight into the comparative behavior of elements can thus be gained by considering the relative reactivity of the elements on the basis of the average time which they spend in seawater before removal to the sediments or their degree of undersaturation in seawater. These areas are discussed in the next sections.

Barth was the first to introduce the concept of the ocean being a simple reservoir for the elements introduced into it during the major sediment cycle. He assumed a steady state system in which the amount of an element entering/ per unit time is equal or compensated by the settling out of an equivalent amount. The residence time, $\tau$, can then be defined as the average time which a substance remains in seawater before removal by some precipitation or adsorption process

$$\tau = \frac{\text{Total mass of element in sea}}{\text{Mass supplied per year}}$$

It is further assumed that there is complete mixing of the element in a time which is short compared to the residence time. If one uses a steady rate model for the river input (Q) and sediment removal (R) of an element, A,

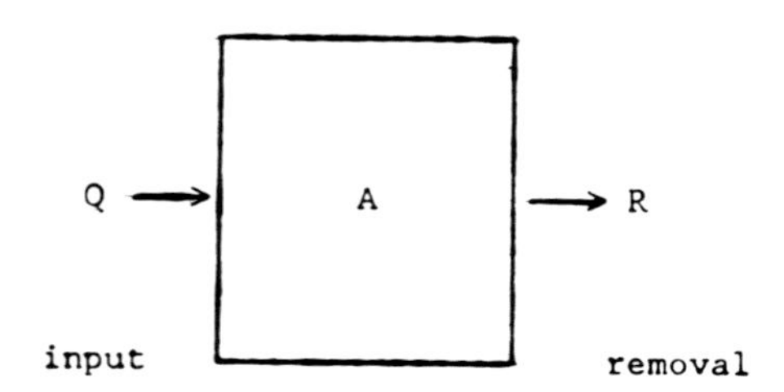


one has dA/dt = Q − R = 0. If the removal is proportioned to the concentration, we have R = k[A] and the residence time is given by

$$\tau = 1/k = [A]/R = [A]/Q \quad (6)$$

**TABLE 3.6**
**The Residence Times of Elements in Seawater**

| | Residence time (M years) | |
|---|---|---|
| **Element** | **River input** | **Sedimentation** |
| Na | 210 | 260 |
| Mg | 22 | 45 |
| Ca | 1 | 8 |
| K | 10 | 11 |
| Sr | 10 | 19 |
| Si | 0.935 | 0.01 |
| Li | 12 | 19 |
| Rb | 6.1 | 0.27 |
| Ba | 0.05 | 0.084 |
| Al | 0.0031 | 0.0001 |
| Mo | 2.15 | 0.5 |
| Cu | 0.043 | 0.05 |
| Ni | 0.015 | 0.018 |
| Ag | 0.25 | 2.1 |
| Pb | 0.00056 | 0.002 |

Elements are introduced into the ocean by three methods:

1. The fallout of substances from the atmosphere
2. Influx of river water
3. Interior of earth

Barth used estimates of the river input (Q) of various elements to estimate the residence time of elements. Since some elements come into the oceans from rivers in a solid phase (Si in clay minerals), care must be taken when estimating the river input. Care must also be taken to consider the recycling of elements from the sea to the land and back to the sea. The $Cl^-$, for example, coming into the oceans from rivers is largely recycled sea salt. Estimates of the residence times from river input are given in Table 3.6 and shown plotted vs. atomic number in Figure 3.3. Also given in this table are resident times determined from the sedimentation rates of elements (R).

The agreement between the two methods of calculating residence times is quite reasonable considering the simplicity of the model for the oceans. The values span six orders of magnitude — Na with $2.6 \times 10^8$ years to Al of 100 years. It should be noted that one further condition must be satisfied.

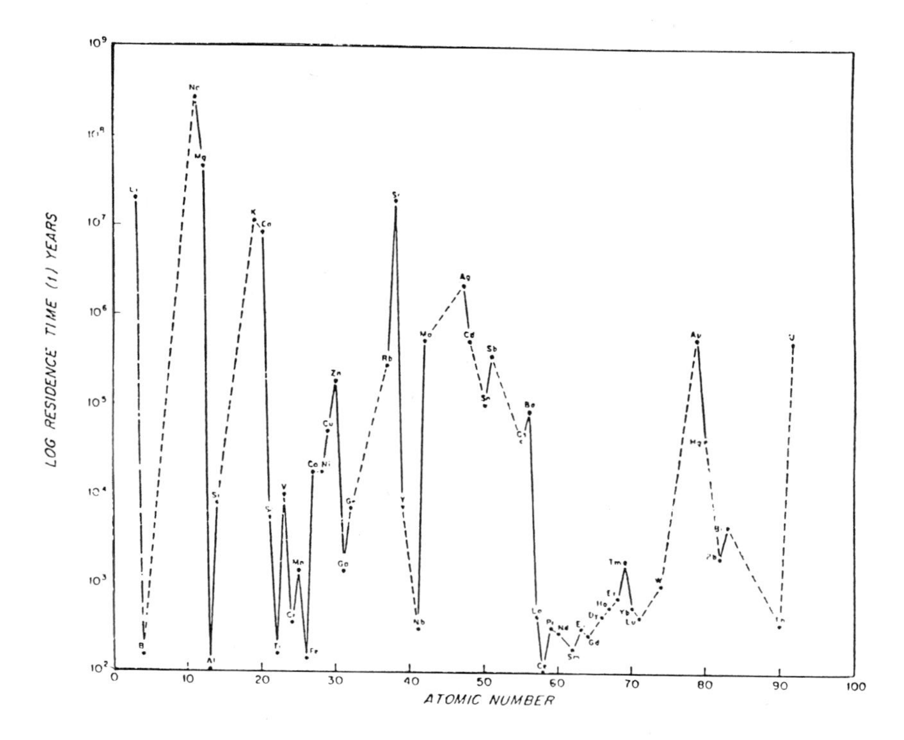


FIGURE 3.3. The residence time of elements in seawater vs. atomic number.

Both A and Q should not change during a period of the order of 3 to 4 times τ. Even for Na this assumption appears to be valid since $10^9$ years for the age of the oceans is in agreement with present day geological concepts.

The elements with long residence times are characterized by the lack of reactivity of their aqueous ions in the oceans. The decrease in residence times of the alkali metals in going from $Na^+$ to $Cs^+$ reflects variations in their ocean reactivities. The primary reactions regulating the concentrations of the alkaline metals may involve ion-exchange equilibria with clay minerals on the sea floor. For the cations, retention on clay surfaces increases with increasing ionic radius (i.e., decreasing hydrated radius). Thus, the decrease in residence times with increasing atomic numbers is in accord with the known behavior of the alkali metals.

Certain elements ($Be^{2+}$, $Al^{3+}$, $Ti^{3+}$, $Cr^{3+}$, $Fe^{2+}$, $Nb^{3+}$, and $Th^{4+}$) have residence times less than 1000 years, which is of the order of the mixing times for ocean waters. These elements enter the oceans as particulate particles from the continents or volcanic activity in the form of clay mineral, feldspars, etc. and, thus, they rapidly settle to the sediments. Some of these elements are also reactants with such substances as ferromanganese minerals and zeolites. Thus, their entry as solids and/or their high chemical reactivity can account for their low residence times. The absolute values of these residence times are somewhat tenuous due to the assumption that there is complete mixing of such elements in the oceans. Nevertheless, certain deductions can be made with regard to their geochemistry. Ocean to ocean and temporal variations of these elements are to be expected. The short residence times of $Th^{4+}$ in the oceans allows various water masses to maintain different concentrations.

Certain minor elements have intermediate values of residence times in years: Mn (7000), Zn (180,000), Co (18,000), and Cu (65,000 years). The reactivities of some of these elements are clearly connected with the formation of plant material.

Recently Whitfield and co-workers (Plymouth, England) have developed a semi-empirical correlation of residence times for elements

$$\log \tau = 2.6 \log [C_{SW}/C_{RW}] + a\,\Delta H_h + b \quad (7)$$

where $C_{SW}$ and $C_{RW}$ are the concentration of the element in seawater (SW) and river water (RW), $\Delta H_h$ is the heat of hydration of the element and a = 0.00452 and b = −0.6 are adjustable parameters. The reliability of this equation is shown in Figure 3.4.

A number of workers have examined the relative reactivity of elements in seawater on the basis of the degree of undersaturation (Table 3.7). The upper limit that an element might attain will be regulated by the solubility of its least soluble compound. Although this is generally true, one finds that many reactive elements have a short residence time and concentration well below the saturation limits.

## 3. DISTRIBUTION OF TRACE ELEMENTS IN THE OCEANS

In recent years there has been a rapid increase in our knowledge of the distribution of minor trace elements (mostly metals) in the oceans. This recent revolution is related to major advances in instrumentation and the elimination

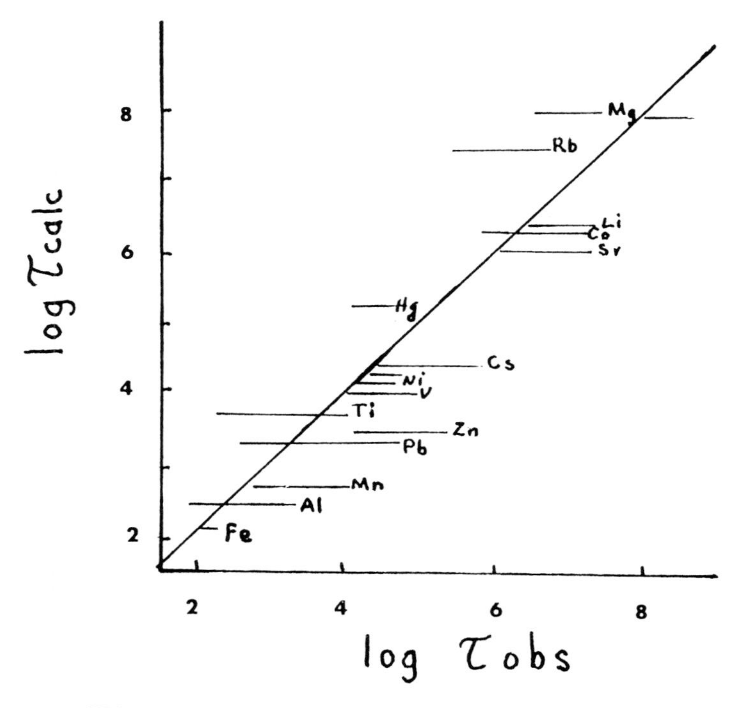


FIGURE 3.4. Calculated and observed residence times.

**TABLE 3.7**
**Comparison of Ratio of Saturated Concentration to Measured Values and Residence Times**

| Metal | R[a] | Years |
|---|---|---|
| $Pb^{2+}$ | 10,000—20,000 | 2,000 |
| $Ni^{2+}$ | 10,000—225,000 | 18,000 |
| $Co^{2+}$ | 50,000—400,000 | 18,000 |
| $Cu^{2+}$ | 133—266 | 50,000 |
| $Ba^{2+}$ | 3.7 | 84,000 |
| $Zn^{2+}$ | 120—250 | 180,000 |
| $Cd^{2+}$ | 40,000—10 million | 500,000 |
| $Ca^{2+}$ | 0.25—1.2 | 8 million |
| $Sr^{2+}$ | 2.75 | 190 million |
| $Ba^{2+}$ | 27 | 450 million |

[a] Measure of degree of undersaturation. R = saturation concentration/measured concentration.

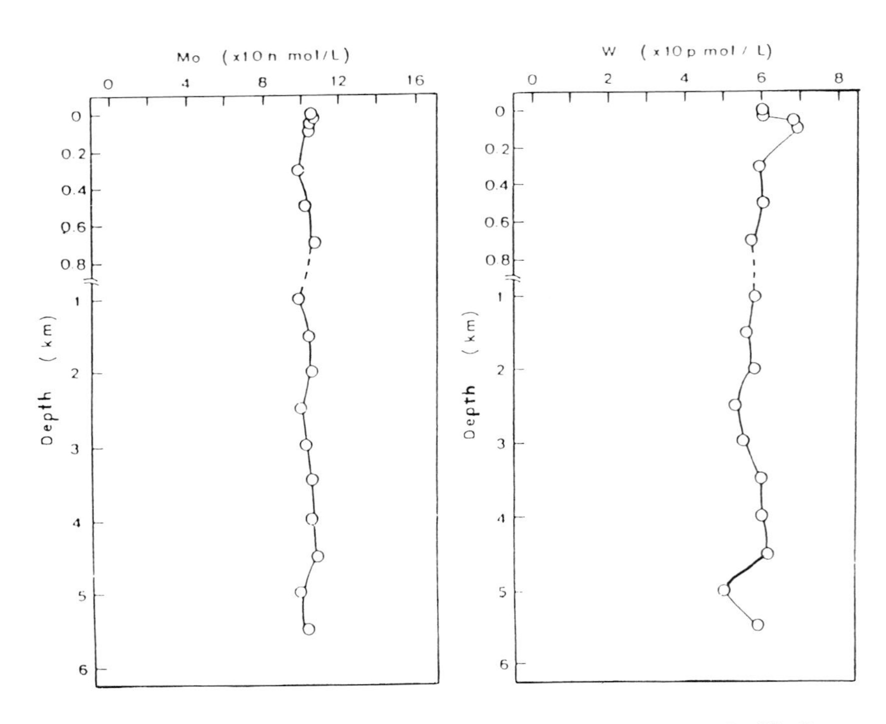


FIGURE 3.5. Profiles of molybdenum (Mo) and tungsten (W) in the Pacific Ocean.

or control of contamination during sampling, storage, and analysis. Bruland (1983) has reviewed these developments. Preconcentration techniques such as chelation on resins and coprecipitation have been used to separate the element from the major components of seawater, and measurements of trace elements at nmol and pmol $kg^{-1}$ levels have been achieved. The distribution of the elements were found to be consistent with known biological and physical behavior. Ultra-clean room techniques pioneered by Patterson (Cal Tech) in his studies of lead and the use of specially designed hydrowires (Kevlar) and sample bottles (Teflon coated) provided reliable data for trace metals such as Mn, Cu, Cd, Ni, Ba, and Fe. Surface values for Pb, Hg, Cu, Ni, and Zn were obtained using rafts and collecting the samples by hand. The types of profiles found for various elements can be divided into a number of general categories. Some are given below.

**Conservative profile** — A constant ratio of the concentration of the element to chlorinity or salinity is found for some elements due to low reactivity. Along with the major components of seawater, trace metals such as $Rb^+$ and $Cs^+$ and anions such as $MoO_4^{2-}$ exhibit this type of behavior (Figure 3.5). One might also expect elements such as Au and Ag to be conservative in seawater.

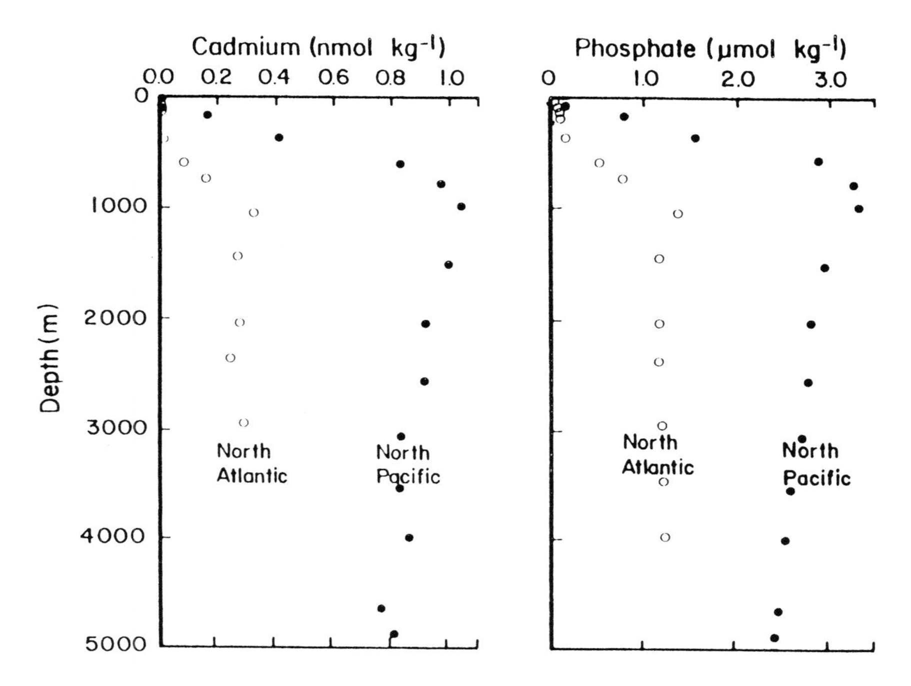


FIGURE 3.6. Profiles of cadmium (Cd) and phosphate ($PO_4$) in the Atlantic and Pacific Oceans.

**Nutrient type profile** — A depletion of an element in surface waters and an enrichment at depth is a nutrient type profile. The element is removed from the surface waters by plankton or biologically produced particulate matter. It is regenerated in deep waters when the biologically produced particulate matter is oxidized by bacteria. Three types of nutrient type profiles are found. First, the shallow water regeneration which leads to a maximum near 1 km, for example, $PO_4^{3-}$ and $NO^{3-}$. The metal Cd is also a good example of this type of nutrient behavior (Figure 3.6).* This behavior indicates that the element is associated with the soft parts of living and dead biological material. Second, a deep regeneration cycle leading to a deep maximum is observed for metals of this type, for example, $SiO_2$ and total alkalinity. Examples of this type also include Ba, Zn, and Ge (Figures 3.7, 3.8, and 3.9). Third, a combination shallow and deep generation is inferred from the nutrient type profiles of Ni and Se (Figures 3.10 and 3.11).

**Surface enrichment and depletion at depth** — Elements of this type have a surface input from the atmosphere, rivers, and land and are rapidly

* Figures 3.6, 3.8 to 3.12, and 3.14 from Bruland, 1983; Figures 3.13, 3.15, and 3.16 from Broecker and Peng, 1982.

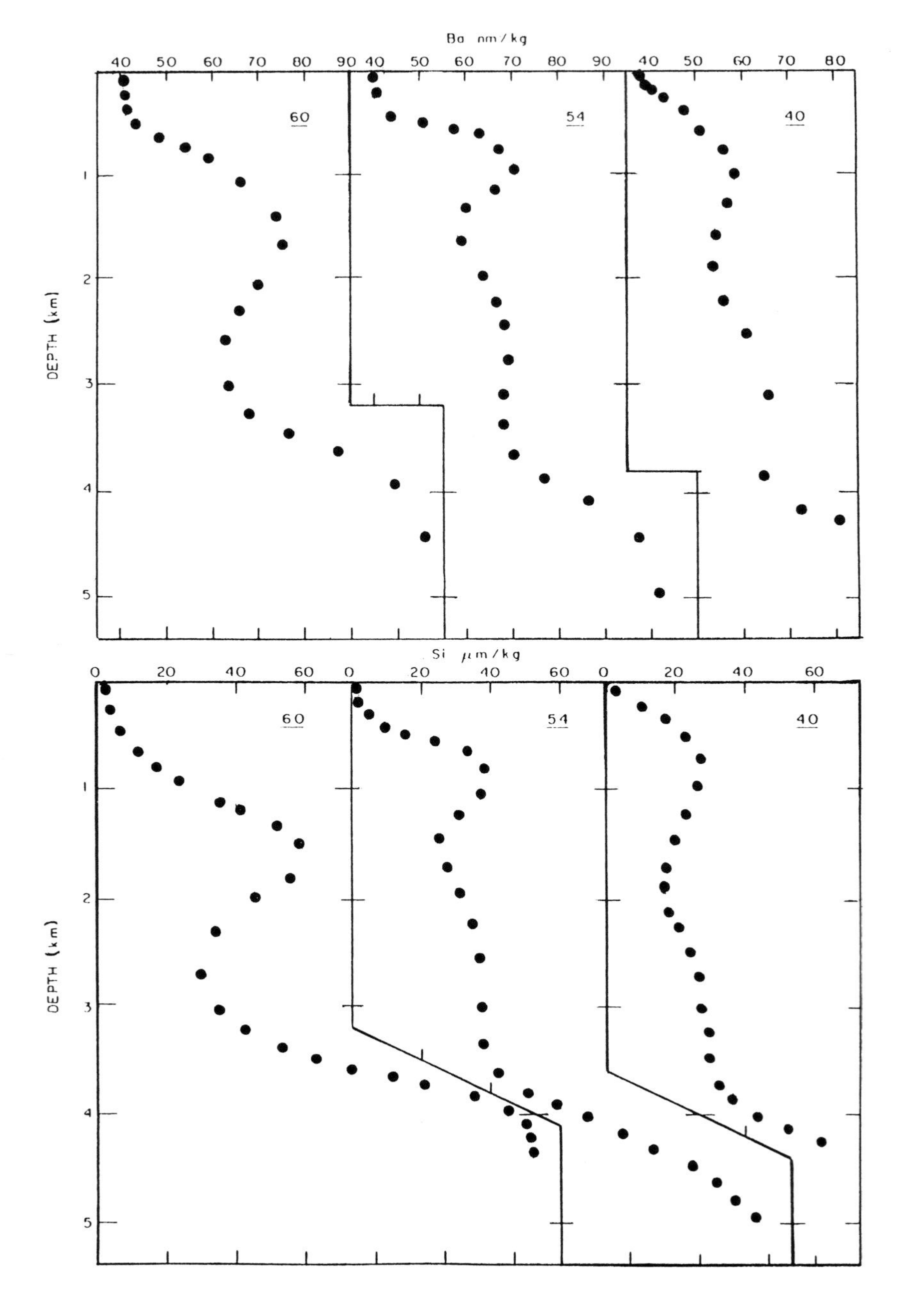


FIGURE 3.7. Comparison of barium (Ba) and silica (Si) profiles in the south Atlantic.

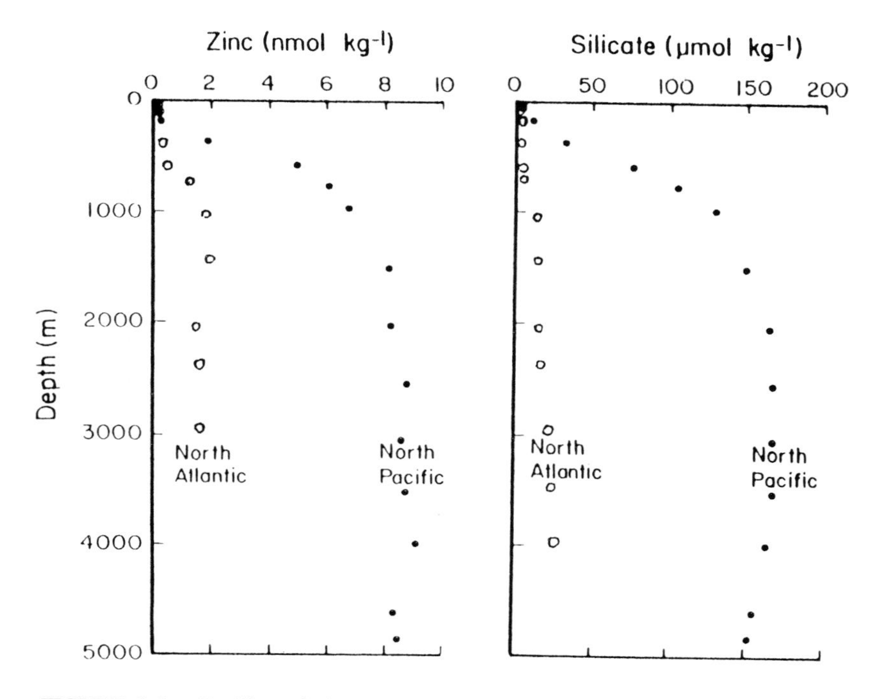


FIGURE 3.8. Profiles of zinc (Zn) and silicate ($Si(OH)_4$) in the Atlantic and Pacific Oceans.

removed from seawater. The residence times of these elements are very short. The metal Pb is a good example of an element entering the oceans via the atmosphere (Figure 3.12). The mechanism for the scavenging process for lead is not well defined at present. The metals Mn and $^{228}Ra$ are good examples of elements which enter surface waters via rivers or by release from shelf sediments (Figure 3.13). Elements of different oxidation states can also exhibit this type of profile. An *in situ* production in surface waters due to biological and photochemical processes can lead to a reduction process. The subsequent oxidation can lead to an oxidized form that is insoluble in seawater. The elements Cr(III), As(III), and I( − I) are examples that fall into this classification.

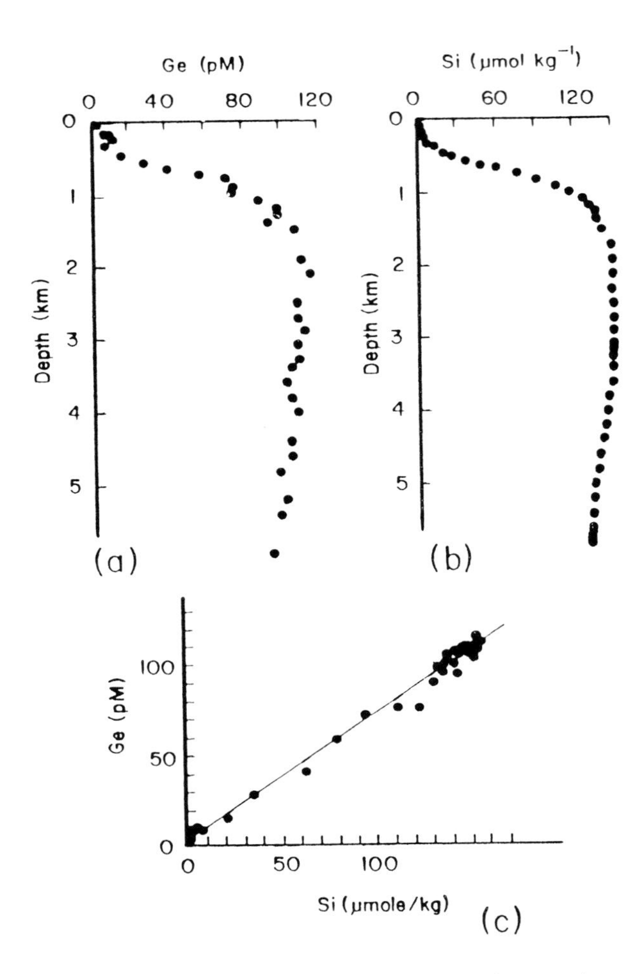


FIGURE 3.9. Comparison of the profiles of germanium (Ge) and silica (Si) in Pacific waters.

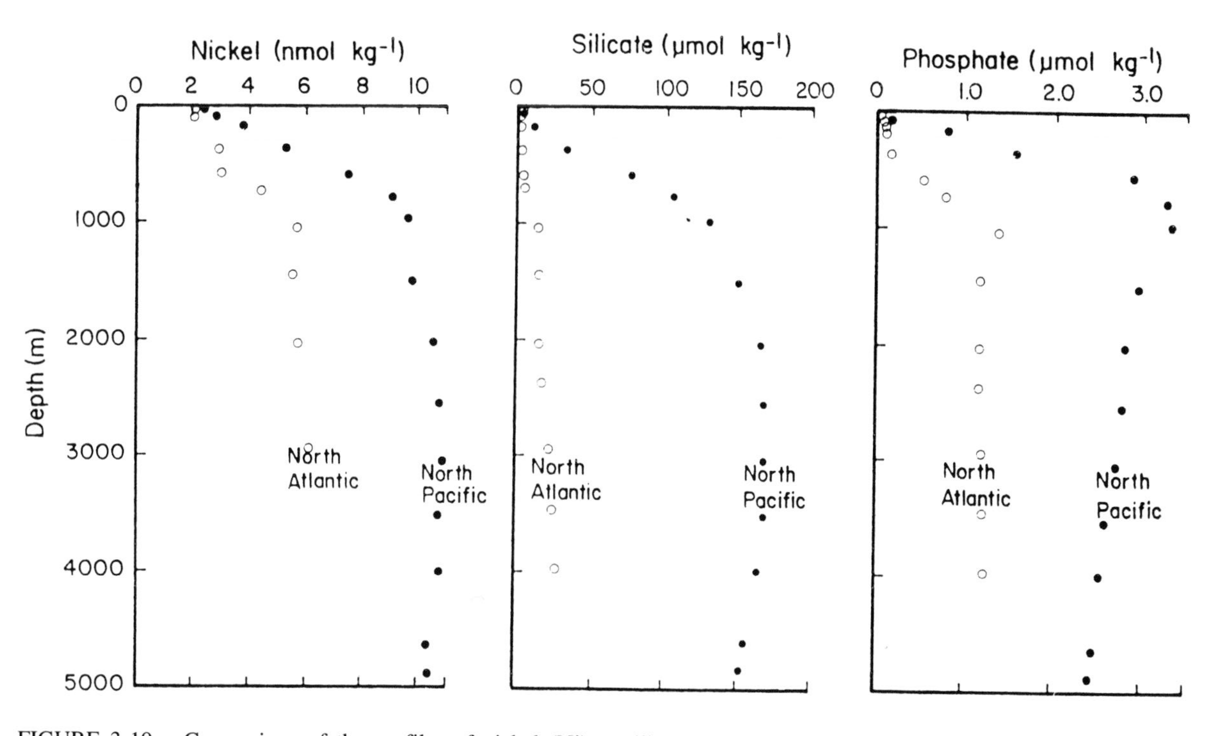


FIGURE 3.10. Comparison of the profiles of nickel (Ni) to silicate ($Si(OH)_4$) and phosphate ($PO_4$) in the Atlantic and Pacific Oceans.

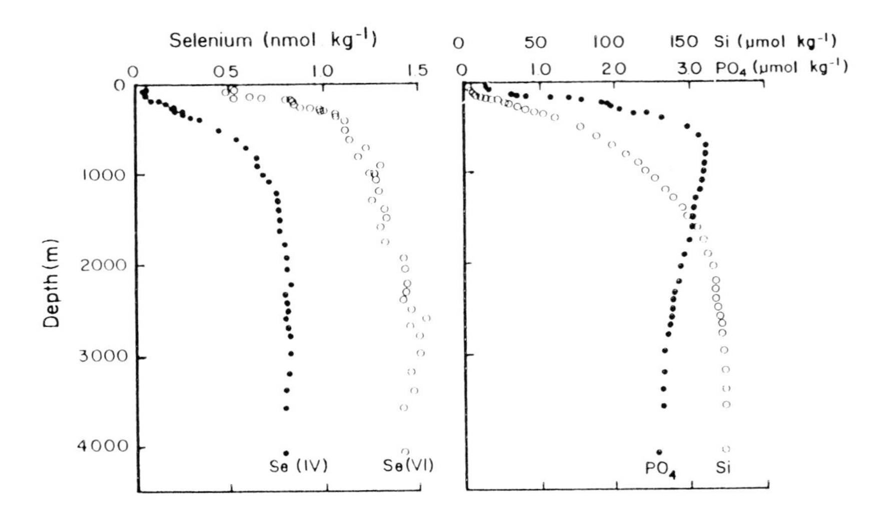


FIGURE 3.11. Comparison of the profiles of selenium (Se) to silica (Si) and phosphate ($PO_4$) in the Pacific Ocean.

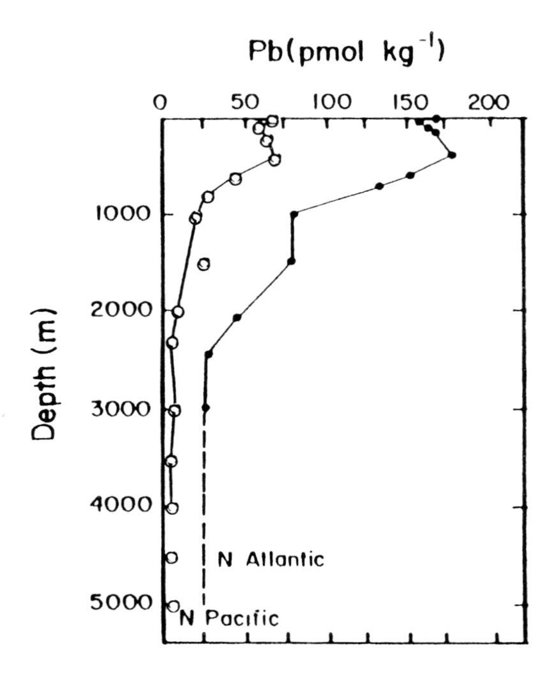


FIGURE 3.12. Lead (Pb) profiles in the Atlantic and Pacific Oceans.

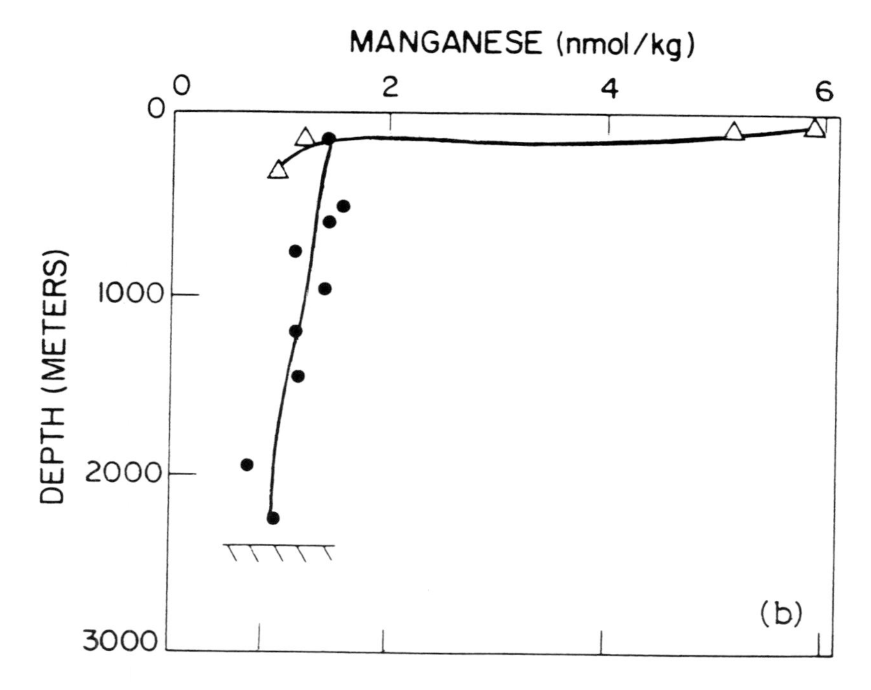


FIGURE 3.13. Profile of manganese (Mn) off the coast of the North Pacific.

**Mid-depth minima** — A mid-depth minimum can result from a surface input and regeneration at or near the bottom or scavenging throughout the water column. The metals Cu, Sn, and Al show this type of profile (Figure 3.14).

**Mid-depth maxima** — A profile of this type can result from a hydrothermal input from the mid-ocean ridge system. The elements Mn and $^{3}He$ are good examples of this type of profile (Figures 3.15 and 3.16).

**Mid-depth maxima or minima in the sub-oxic layer** (~1 km) — A large sub-oxic layer exists in some regions of the Pacific and Indian oceans. Reduction and oxidation processes in the water column or adjacent slope sediments can yield maxima of the reduced form (Mn(II) and Fe(II)) and minima of the reduced form if it is insoluble or scavenged by solid phases (Cr(III)).

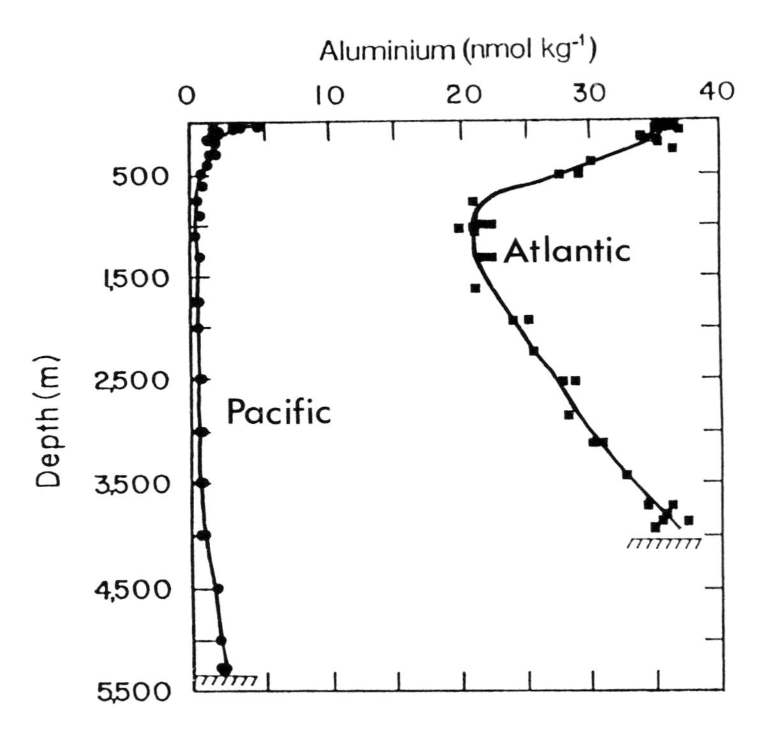


FIGURE 3.14. Profiles of aluminum (Al) in Atlantic and Pacific Ocean waters.

**Maxima and minima in anoxic waters** — In areas of restricted circulation such as the Black Sea, Cariaco Trench, and fjords, the water can become anoxic (devoid of $O_2$) with the production of $H_2S$. Near the interface between the two waters redox processes can occur that cause maxima and minima due to solubility changes of the various species. Mn(II) and Fe(II), for example, have a maxima due to the increased solubility of the reduced form (Figures 3.17 and 3.18) while a minima can occur for Cr(III). More will be said about anoxic water later. Many of the metals studied have higher concentrations in the Pacific than the Atlantic (Table 3.8). Exceptions are Pb and Al. The higher values in the deep Pacific are attributed to the older water accumulating more metals. For Pb and Al, the higher surface input in the Atlantic leads to higher values in deep waters. Since these metals are quickly scavenged from seawater, they do not accumulate in the deep waters of the Pacific. The higher concentration of most metals on shelves (Table 3.9) indicates a land source (rivers); the higher concentration (Table 3.10) in the Atlantic central gyre indicates an atmospheric input.

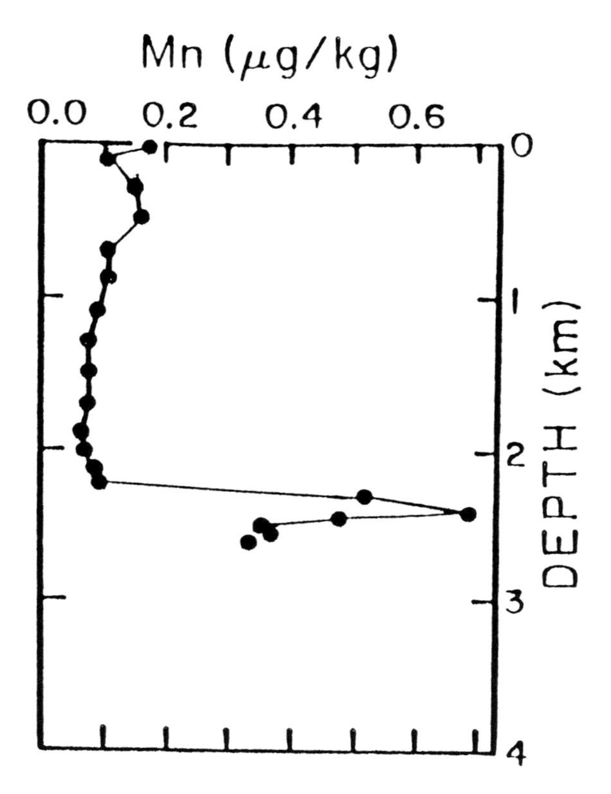


FIGURE 3.15. Profile of manganese (Mn) in the Pacific Ocean showing hydrothermal input from mid-Pacific Ridge.

## 4. BIOLOGICAL INTERACTIONS

The interactions of trace elements with marine organisms have been studied by many workers. Bowen has summarized these interactions which are outlined below in terms of concentration factors (concentration in the organism relative to a given volume of seawater).

1. $Cl^-$ is rejected by organism.
2. $Na^+$, $Mg^{2+}$, $Br^-$, $F^-$, and $SO_4^{2-}$ have concentrations in organisms similar to seawater (concentration factors of 1.0).
3. Most of the other elements, with the exception of the noble gases, are strongly concentrated in living tissue.

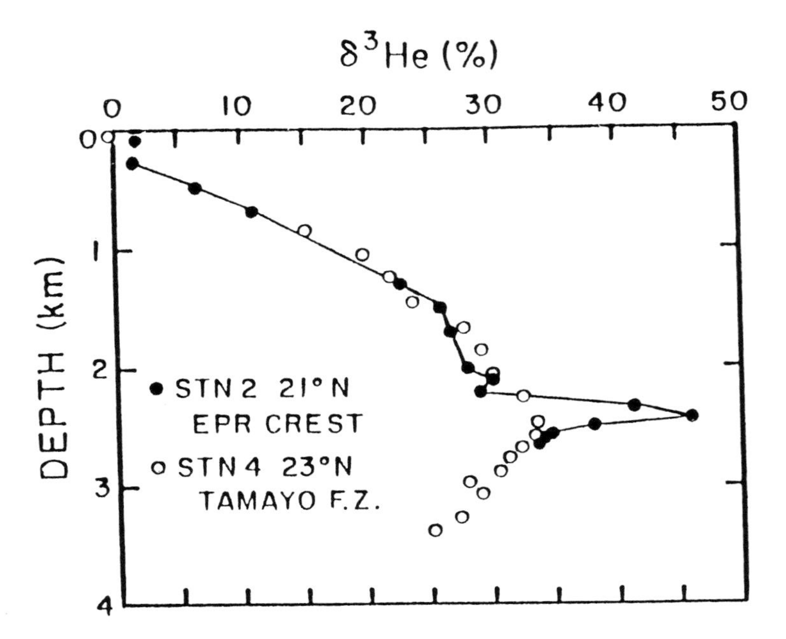


FIGURE 3.16. Profile of helium (He) in the Pacific Ocean showing hydrothermal input from mid-Pacific Ridge.

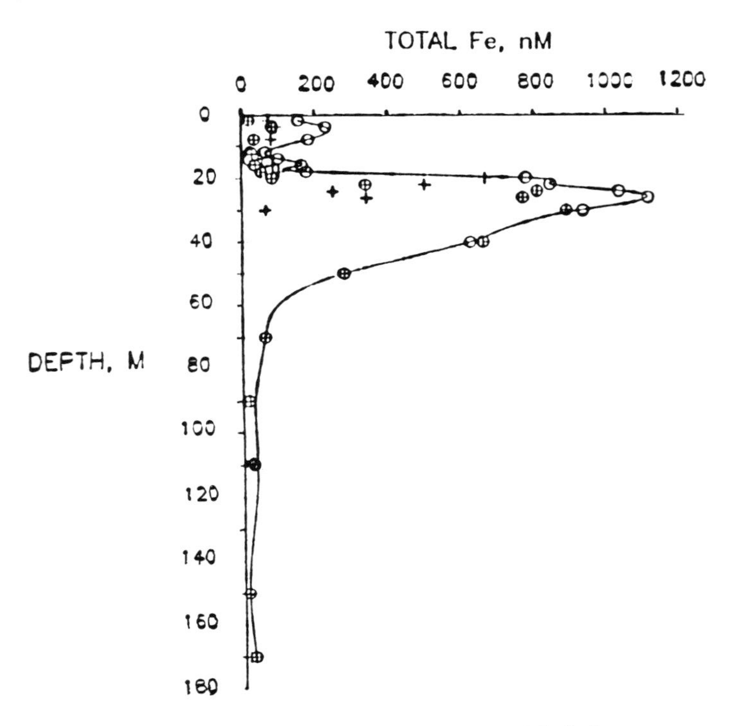


FIGURE 3.17. Profile of iron (Fe) in the Black Sea.

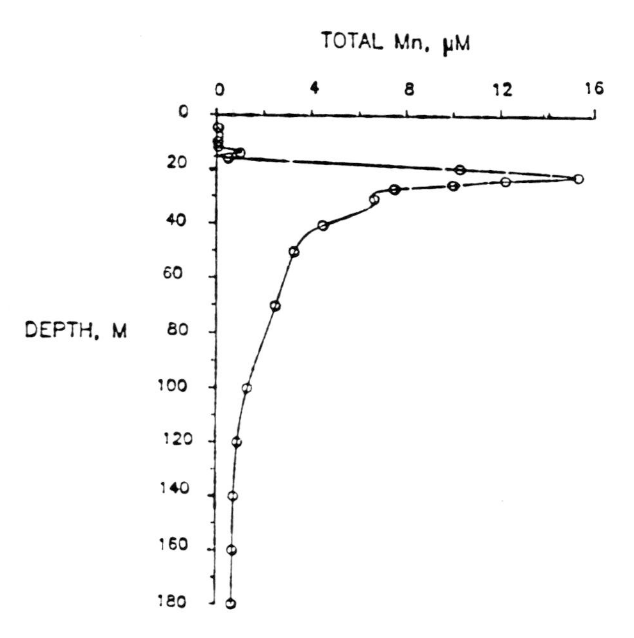


FIGURE 3.18. Profile of manganese (Mn) in the Black Sea.

**TABLE 3.8**
**Metals in Atlantic and Pacific Deep Waters (n*M*)**

| Metal | Atlantic | Pacific | P/A |
|---|---|---|---|
| Cd | 0.29 | 0.94 | 3.2 |
| Zn | 1.5 | 8.2 | 5.5 |
| Ni | 5.7 | 10.4 | 1.8 |
| Cu | 1.7 | 2.7 | 1.6 |
| Mn | 0.6 | — | — |

4. The order of affinity of organisms for cations is
$4+ > 3+ > 2+$ transition $> 2+$ Group IIA $> 1+$ Group I metals.
For plankton the order is
$Fe^{3+} > Al^{3+} > Ti^{3+} > Cr^{3+} > Ga^{3+} > Zn^{2+} > Pb^{2+} > Cu^{2+} > Mn^{2+} > Co^{2+} > Cd^{2+}$
This order does not agree with the Irving-Williams order.

**TABLE 3.9**
**Metals on Shelf vs. Open Sea Surface Waters**

| Metal | Shelf | Open |
|---|---|---|
| Mn | 21 n*M* | 2.4 n*M* |
| Ni | 5.9 n*M* | 2.3 n*M* |
| Cu | 4.0 n*M* | 1.2 n*M* |
| Zn | 2.4 n*M* | 0.06 n*M* |
| Cd | 200 p*M* | 2 p*M* |

**TABLE 3.10**
**Metals in Central Gyres**

| Metal | Atlantic | Pacific |
|---|---|---|
| Mn | 2.4 n*M* | 1.0 n*M* |
| Cu | 1.2 n*M* | 0.5 n*M* |
| Ni | 2.1 n*M* | 2.4 n*M* |
| Zn | 0.06 n*M* | 0.06 n*M* |
| Cd | 2 p*M* | 2 p*M* |

5. Heavy group elements of a particular class are taken up more strongly than lighter elements.
6. The affinity for anions increases with increasing ionic charge and, in a given group, with increasing weight of the central atom: $F^- > Cl^- > Br^- > I^-$; $SO_4^{2-} > MoO_4^{2-} > WO_4^{2-}$
7. The lower organisms concentrate elements more strongly than higher organisms.
8. Heavy metals are frequently concentrated in the digestive or renal organs.

The biosphere can affect minor elements by:

1. The regulation of dissolved and particulate organic material as a function of time and space.
2. The increased concentration of elements in living and nonliving organic material. These effects can account for the movement of elements from

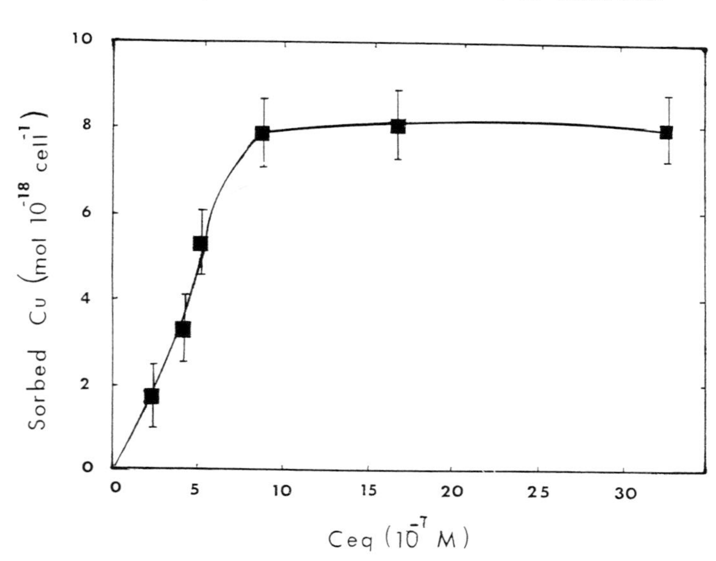


FIGURE 3.19. Copper adsorption isotherm to the surface of bacteria.

surface waters to the sediments. This movement can be caused by:

a. Active uptake by organism ($Fe^{2+}$, $Zn^{2+}$, $Mn^{2+}$)
b. Passive uptake by organisms (heavy metals)
c. Adsorption on particulate matter ($Pb^{2+}$, $Cu^{2+}$)
d. Remobilization from sediments by oxidation ($Mn^{2+}$)
e. Precipitation ($Fe^{3+}$)

The active uptake of metals may be due to use by living tissue in enzyme systems (V, Cr, Mn, Fe, Co, Ni, Cu, Zn, and Mo). Passive uptake may be due to the adsorption on organic particles due to interactions with surface groups (carboxylic, phenolic).

It is well known that a number of metals are concentrated in biological materials. This has led to the suggestion that their source in the deep ocean comes from the death of marine organisms with subsequent remineralization during oxidation. An alternate explanation may be that metals may be carried to the deep ocean by adsorption on organic matter both living and dead.

We have found, for example, that $Cu^{2+}$ is adsorbed on bacteria that are living or dead (Figure 3.19). The absorption is quite fast (Figure 3.20) and the $Cu^{2+}$ can be desorbed by acidifying the solutions. The adsorption increases with increasing pH which indicates that surface OH and COOH groups may

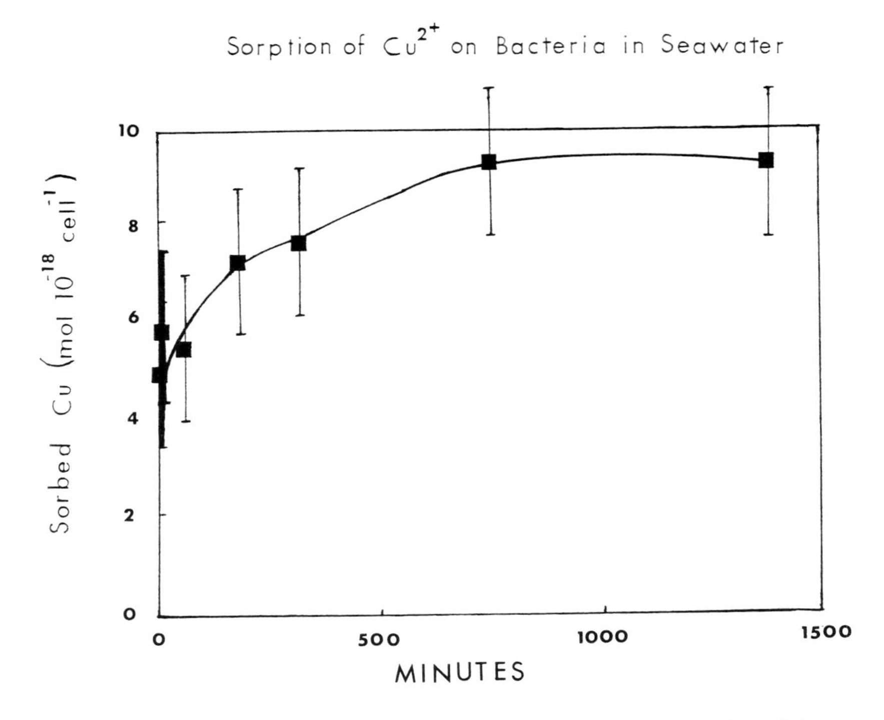


FIGURE 3.20. Sorbed copper to the surface of bacteria as a function of time.

be involved in the process. Recently, Fisher (Stony Brook) has shown that adsorption of a number of metals to living and dead phytoplankton was similar and proportional to the hydrolysis constants of the metals (Figure 3.21) and to the residence times (Figure 3.22). These results suggest that the scavenging of metals in the oceans may be the result of ionic interactions with surface groups on living and dead organic matter in the oceans. Fisher also showed (Figure 3.23) that toxicity of metals to plankton was related to hydrolysis constants.

The above discussion is not meant to rule out nonbiological transport, since it is well known that many metals can be coprecipitated by $SiO_2$ and $CaCO_3$ or adsorbed on minerals in ocean waters. In an oxidizing environment, for example, many metals ($Fe^{2+}$, $Cu^{2+}$, $Ni^{2+}$, $Co^{2+}$, etc.) are concentrated on Mn nodules. In a reducing environment a number of metals are coprecipitated or adsorbed by pyrite ($FeS_2$). The oxidation of $Fe^{2+}$ and $Mn^{2+}$ results in the formation of solids that are quite active in the adsorption of metals. In summary, each metal may be influenced by biological or nonbiological processes. Future studies are needed to elucidate that these processes actually are responsible for the distribution of elements in the oceans.

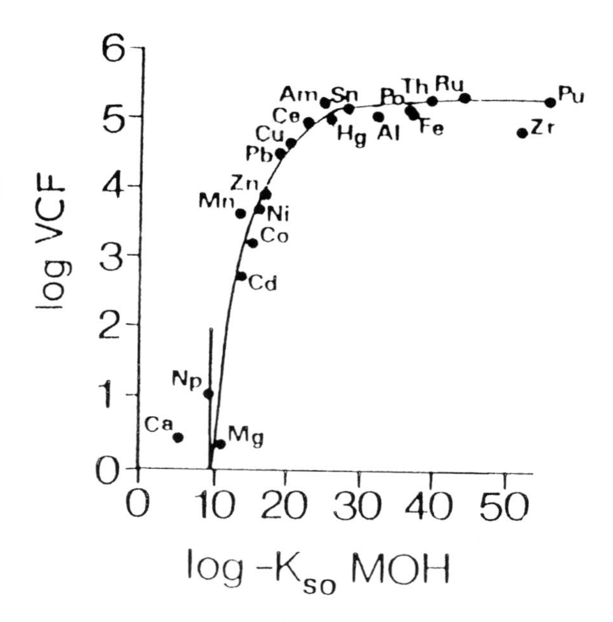


FIGURE 3.21. Values of the volume concentration factor (VCF) for metals to phytoplankton vs. the solubility of metal hydroxide. (From Fisher, N. S., *Limnol. Oceanogr.*, 31(2), 443, 1986.)

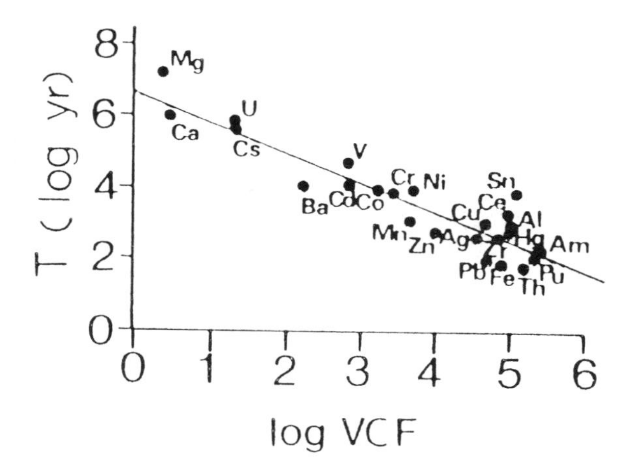


FIGURE 3.22. The residence time for metals vs. the volume concentration factor. (From Fisher, N. S., *Limnol. Oceanogr.*, 31(2), 443, 1986.)

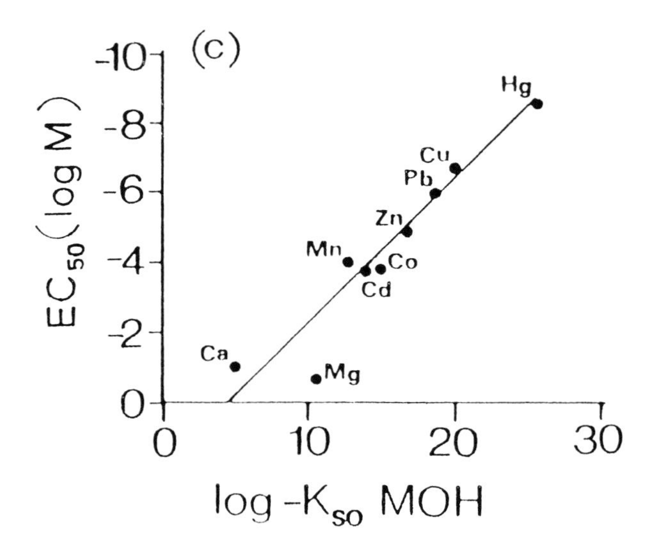


FIGURE 3.23. The effective concentration of metal that reduces phytoplankton growth rate by 50% ($EC_{50}$) vs. the solubility of metal hydroxide. (From Fisher, N. S., *Limnol. Oceanogr.*, 31(2), 443, 1986.)

Although most minor elements that enter the oceans eventually are removed to the sediments, many are recycled. The removal process is related to the interactions of elements with particles. The production, sinking, and decomposition of particulate matter are thus important in controlling the recycling of metals. Phytoplankton are the primary producers of particles in the surface oceans. These plants are grazed on by zooplankton, and packaged into fecal pellets, that have sinking rates of a few meters to thousands of meters per day. The oxidation and dissolution of these particles can recycle some of the elements. Recent work has shown that large particles are responsible for the bulk transport of materials to the deep sea. These particles have been studied by catching them in sediment traps or by using large volume pumping systems. This cycling of C, N, P, and S as well as trace metals is presently being studied in a large multi-institutional study called the Joint Global Ocean Flux Study (JGOFS).

**TABLE 3.11**
**Present Concentration of Major Components of Seawater with the Amount Added by Rivers**

| | Na | Mg | Ca | K | Cl | $SO_4$ | $CO_3$ | $NO_3$ |
|---|---|---|---|---|---|---|---|---|
| Present | 129 | 15 | 2.8 | 2.7 | 150 | 8 | 0.3 | 0.01 |
| Added by rivers | 196 | 122 | 268 | 42 | 157 | 84 | 342 | 11 |
| Excess added | 67 | 107 | 265 | 39 | 7 | 76 | 342 | 11 |

## 5. GEOCHEMICAL BALANCE

There is a continual interaction between the oceans and the land. The water from the oceans evaporates, comes down as rain, and attacks the rocks and soil. The rivers of the world carry the weathered products (dissolved and suspended) to the oceans. Sediments are formed on continental shelves and fine particles are carried throughout the oceans. A number of geochemists have been concerned with geochemical processes with an interest in the evolution of seawater, sedimentary cycling, and the controls of the composition of the oceans. By comparing the present composition of dissolved ions in the oceans (mol/cm$^2$ of surface of the earth) with the amount delivered, attempts have been made to prepare a geochemical balance. Sillen tabulated the amounts of some of the major components of seawater delivered by the rivers over the last 100 million years. These estimates are compared to the amounts in the present oceans in Table 3.11. Most of the elements show an excess amount added over this period of time. The $Na^+$ and $Cl^-$ delivered to the oceans is mostly sea spray that is being washed back to the oceans. The other ions, however, are the results of weathering of rocks and soils. In addition to the dissolved materials, 300 to 600 kg/cm$^3$ of solid material has been delivered to the oceans in the last 100 million years. These solids are mostly clay minerals which can participate in ion-exchange reactions with the components of seawater and go through phase transformations.

A number of geochemists have attempted to develop a geochemical balance between the material added to the rivers and the amount delivered to the sediments. These attempts were started in 1933 by Goldsmith. He considered the reaction

$$\text{igneous rocks} + \text{volatiles} \rightarrow \text{seawater} + \text{sediments} + \text{air} \qquad (8)$$

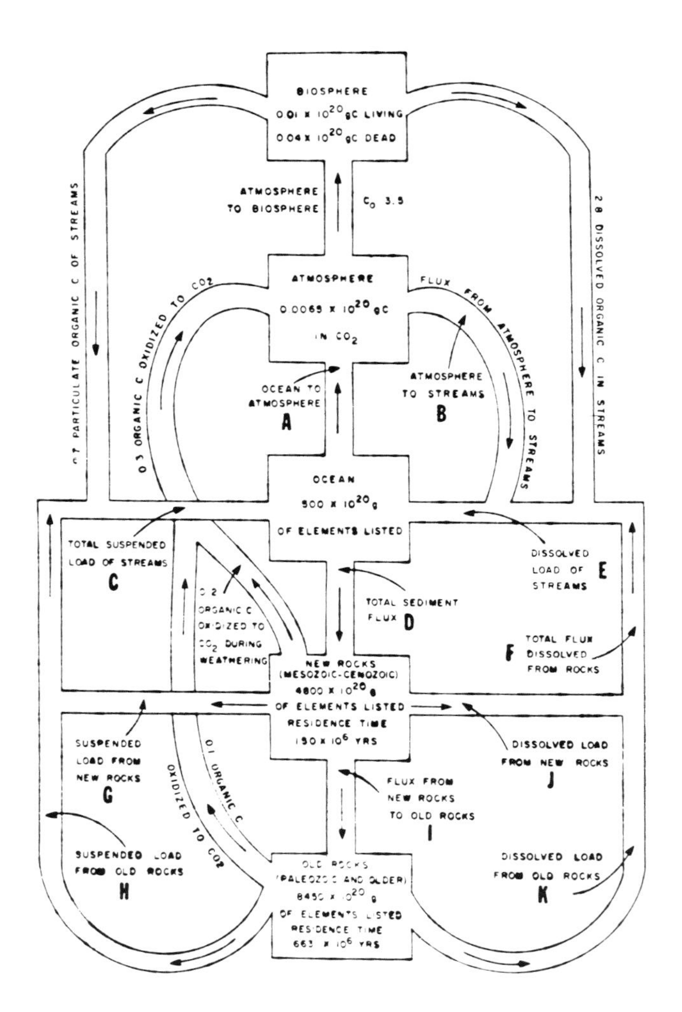


FIGURE 3.24. The circulation of elements in the sediment cycle (Mackenzie, 1975).

to control the composition of rocks and seawater and attempted to balance the reaction. He concluded that 600 g of rock would lead to 600 g of sediments. More recent attempts have been made by Garrels and Mackenzie (Mackenzie, 1975). The steady state model for the circulation of elements in the sedimentary cycle is shown in Figure 3.24. The fluxes are in units of $10^{14}$g/year. Unlike the earlier model of Goldsmith, recent work has the weathering reaction going both ways, that is, reverse weathering or igneous rocks are formed during sea floor spreading. The earlier models gave a reasonable balance between the rocks and the sediments for most elements (Table 3.12). For the

**TABLE 3.12**
**The Concentration of Various Components of the Geochemical Cycle of the Oceans**

| Component | Rock | Volatile | Air | SW | Sediments |
|---|---|---|---|---|---|
| $H_2O$ | — | 54.90 | — | 54.90 | — |
| Cl(HCl) | — | .94 | — | .55 | .40 |
| $Na(NaO_{.5}, NaOH)$ | 1.47 | — | — | .47 | 1.00 |
| $Ca(CaO, Ca(OH)_2)$ | 1.09 | — | — | .01 | 1.08 |
| $Mg(MgO, Mg(OH)_2)$ | .87 | — | — | .05 | .82 |
| $K(KO_{.5}, KOH)$ | .79 | — | — | .01 | .78 |
| $Si(SiO_2)$ | 12.25 | — | — | — | 12.25 |
| $Al(AlO_{1.5}, Al(OH)_3$ | 3.55 | — | — | — | 3.55 |
| $C(CO_2)$ | .03 | 1.05 | — | .002 | 1.08 |
| C(s) | — | 1.01 | — | — | 1.01 |
| $O_2$ | — | .022 | .022 | — | — |
| $Fe(FeO, Fe(OH)_2)$ | .52 | — | — | — | .53 |
| $(FeO_{1.5}, FeOH)$ | .38 | — | — | — | .38 |
| $Ti(TiO_2)$ | .12 | — | — | — | .12 |
| S | .02 | .06 | — | .03 | .05 |
| F(HF) | .05 | — | — | — | .05 |
| $P(PO_{2.5}, H_3PO_4)$ | .04 | — | — | — | .04 |
| $Mn(MnO_{1\ to\ 2})$ | .05 | — | — | — | .05 |
| $N_2$ | — | .082 | .082 | — | — |

elements B, S, Cl, As, Se, Br, and I, an excess was found presumably due to a volcanic origin. With the recent discovery of the flux of materials coming from hydrothermal vents, these values will need some modification for some elements.

Sillen suggested some imaginary experiments to represent the real system. He visualized the formation of the oceans being the results of mixing water with a number of minerals. This mixing, performed perhaps by a well known bartender, took place over a period of time to ensure equilibrium. The first model he considered is given below in Figure 3.25.

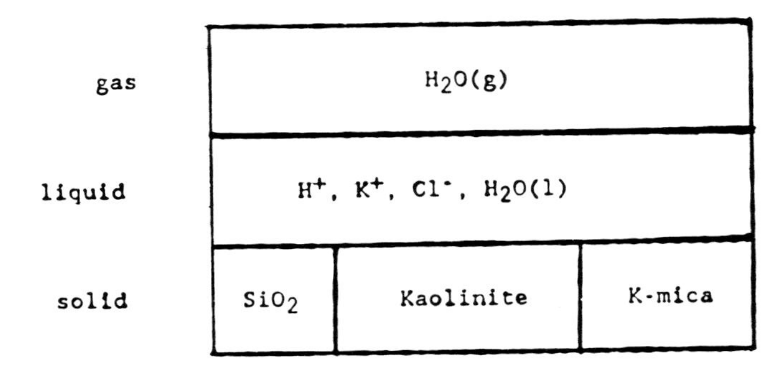


FIGURE 3.25. Sillon's ion exchange model.

The ion exchange reaction

$$\underset{\text{Kaolinite}}{1.5\ Al_2Si_2O_5(OH)_4(s)} + K^+ \rightarrow$$

$$\underset{\text{K-mica}}{KAl_3Si_3O_{10}(OH)_2(s)} + 1.5\ H_2O + H^+ \tag{9}$$

can control the ratio of $H^+$ in the oceans.

Applying equilibrium conditions gives for this reaction (at constant Cl and temperature)

$$K = [H^+]/[K^+] \tag{10}$$

$$[K^+] + [H^+] = [Cl^-] + K_w/[H^+] \tag{11}$$

The ratio of $[H^+]/[K^+]$ is fixed; thus, the addition of HCl or KOH will return it to its original value as long as $[Cl^-]$ and temperature are constant. The system is, thus, a pH stat rather than a buffer as long as Cl and temperature are constant and no phase will completely disappear. This simple result suggests that the pH and main ionic concentrations of seawater are controlled by clay exchange reactions. When Sillen made this suggestion in 1959, most oceanographers were shocked as they thought that the pH of the oceans was controlled by the carbonate system and the clays acted only as the walls of the container. Experimental studies yield $K = 10^{-6}$ to $10^{-6.5}$ (25°C) for the reaction, while the ratio is $10^{-6.2}$ in seawater. For the equilibrium between kaolinite and montmorillonite in a five component system with $Na^+$, we have

$$1.5\ Al_2Si_2O_5(OH)_4(s) + Na^+ \rightleftarrows NaAl_3Si_3O_{10}(OH)_2(s) + 1.5\ H_2O + H^+ \quad (12)$$

the ratio of $H^+/Na^+$ also comes near the ratio in seawater. Since clays (fresh) are transformed in a day or two, forming new phases in a couple of years, most workers have found the silicate theory to be reasonable. Recent workers have also demonstrated that $SiO_2$ dissolves more rapidly in seawater than was once thought. Sillen also considered a more complicated nine component system: (1) seawater; (2) $SiO_2$ (quartz); (3) $CaCO_3$ (calcite); (4) kaolinite; (5) illite; (6) chlorite; (7) montmorillonite; (8) phillipsite; and (9) atmosphere. Mg was found to be a problem as well as $SiO_2$. More recent results on hydrothermal vents can explain these differences.

Whitfield and Turner have shown a correlation between the residence time of an element in the oceans and the partitioning of that element between rocks and seawater. They define the ocean-rock partition coefficient $K_Y$ by

$$K_Y = \overline{Y}_C/\overline{Y}_A \quad (13)$$

where $Y_A$ is the rate of transport of material from crustal rock and $Y_C$ is the rate of transport from the oceanic reservoir. The value of $Y_A$ has been corrected for recycled sea spray. The logarithm of the mean residence time was found to be a linear fraction of the log of the partition coefficient

$$\log \tau = a \log K_Y + b \quad (14)$$

(Figure 3.26). The value of $\tau_R$ = 1000 revs per M years is the stirring revolution of the ocean which is assumed to be 2000 M years old and in a steady state.

They also showed that the values of the partition coefficients ($K_Y$) were related to the electronegativity of the elements, $Q_{YO}$, by

$$\log K_Y = a_1 Q_{YO} + b \quad (15)$$

where the electronegativities are given by (ev)

$$Q_{YO} = (X_Y - Y_O)^2 \quad (16)$$

where the value of $Q_{YO}$ is a measure of the electrostatic contribution to the

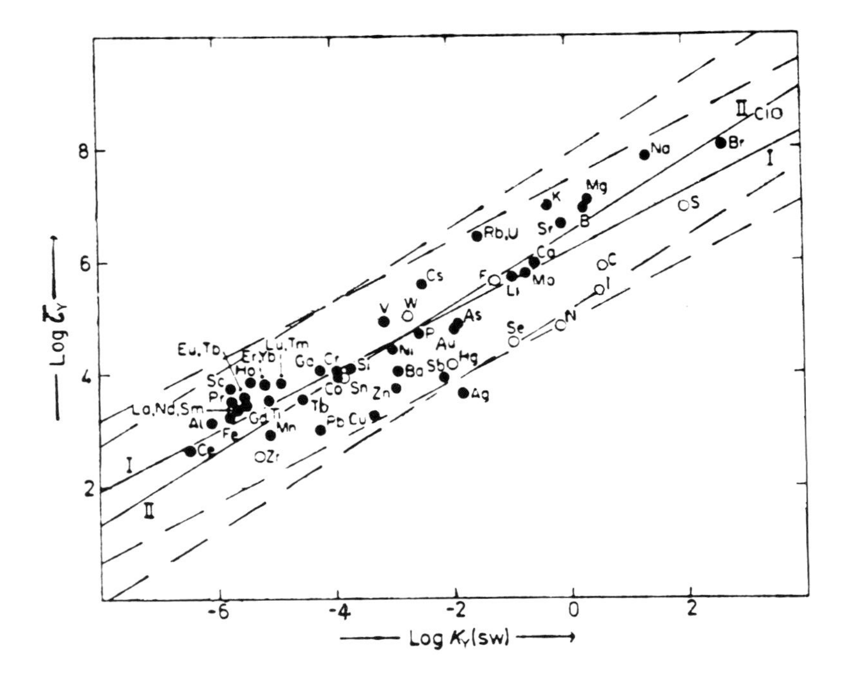

FIGURE 3.26. The log of the mean residence time (τ) vs. the log of the partition coefficient ($K_y$) (Turner, Dickson, and Whitfield, 1980).

Y–O bond energy or a measure of the attraction of an element (Y) for the oxide-based mineral. The values X are values of the electronegativities of various elements.

Meybeck has recently examined chemical weathering by considering the ions formed from various rock types. The abundance of the major rock types are given in Table 3.13. His estimates for the origins of the major components of the rock types released to the rivers are given in Table 3.14. The sources of the dissolved river salts carried to the oceans by various rock types are given in Table 3.15.

**TABLE 3.13**
**Outcrop Abundance of Major Rock Types on Land (Percent by Area)**

| | | | |
|---|---|---|---|
| Plutonic Rocks | 11% | Granite | 10.40% |
| | | Gabbro and periodtite | 0.60% |
| Metamorphic Rocks | 15% | Gneiss | 10.40% |
| | | Mica schist | 1.50% |
| | | Quartzite | 0.80% |
| | | Marble | 0.40% |
| | | Amphibolite | 1.90% |
| Volcanic Rocks | 7.9% | Basalt | 4.15% |
| | | Andesite | 3.00% |
| | | Rhyolite | 0.75% |
| Sandstones | 15.8% | Quartz sandstone | 12.60% |
| | | Arkose | 0.80% |
| | | Graywacke | 2.40% |
| Shales | 33.1% | | |
| Carbonate Sedimentary Rocks | 15.9% | Detrital carb. rock | 5.90% |
| | | Dolomite | 3.65% |
| | | Limestone | 6.35% |
| Evaporites | 1.3% | Gypsum | 0.75% |
| | | Halite | 0.55% |

**TABLE 3.14**
**Origins of Weathering Products by Rock Types (% of Total Amount Released)**

| | $SiO_2$ | $Ca^{2+}$ | $Mg^{2+}$ | $Na^+$ | $K^+$ | $Cl^-$ | $SO_4^{2-}$ | $HCO_3^-$ | $\Sigma^+$ (meq/l) |
|---|---|---|---|---|---|---|---|---|---|
| Granite | 10.9 | 0.60 | 1.2 | 5.9 | 4.9 | | 2.0 | 1.6 | 1.6 |
| Gabbro | 0.75 | 0.0 | 1.1 | 0 | 0 | | 0.3 | 0.3 | 0.3 |
| Gneiss and Micaschists | 11.7 | 1.10 | 2.8 | 6.6 | 8.2 | | 4.6 | 2.0 | 2.4 |
| Miscellaneous Metamorphic | 1.5 | 4.7 | 3.2 | 0.3 | 1.6 | | 1.7 | 4.7 | 3.7 |
| Volcanic rocks | 11.1 | 1.8 | 4.9 | 5.4 | 6.6 | | 0.5 | 4.2 | 3.1 |
| Sandstone | 16.6 | 2.1 | 3.8 | 5.2 | 19.6 | | 9.6 | 2.3 | 3.2 |
| Shale | 35.1 | 19.9 | 30.7 | 22.6 | 4.10 | 7.4 | 30.3 | 22.5 | 23.1 |
| Carbonate rocks | 11.3 | 60.4 | 39.3 | 3.5 | 13.1 | | 8.6 | 59.5 | 46.7 |
| Gypsum | 0.5 | 7.2 | 7.2 | 4.9 | 1.6 | 10.0 | 32.7 | 1.7 | 6.8 |
| Rock Salt | 0.4 | 2.2 | 5.8 | 45.6 | 3.3 | 82.6 | 9.6 | 1.2 | 9.1 |
| Total % | 100 | 100 | 100 | 100 | 100 | 100 | 100 | 100 | 100 |
| $10^{12}$ g/year | 320 | 504 | 118 | 132 | 24 | 120 | 280 | 1950 | |

**TABLE 3.15**
**Sources of River Dissolved Load Carried to the Oceans (in % of Total Amount Derived from Weathering)**

| Model | $SiO_2$ | $Ca^{2+}$ | $Mg^{2+}$ | $Na^+$ | $K^+$ | $Cl^-$ | $SO_4^{2-}$ | | $HCO_3^-$ | $\Sigma^+$ |
|---|---|---|---|---|---|---|---|---|---|---|
| | | | | | | | pyr.[a] | $SO_3$[b] | | |
| Atmosphere | | | | | | | | | 67 | |
| Silicates | 92.5 | 26 | 48 | 46 | 95 | 0 | 40 | 18 | | 35 |
| Carbonates | 0 | 67 | 42 | 0 | 0 | 0 | 0 | 0 | 33 | 51 |
| Evaporites | 0 | 7 | 10 | 54 | 5 | 100 | 0 | 42 | | 14 |
| Amorphous silica | 7.5 | | | | | | | | | |

$\Sigma^+$ = sum of cations.

[a] $SO_4^{2-}$ resulting from pyrite and organic sulfur oxidation.

[b] $SO_4^{2-}$ resulting from sulfate mineral dissolution.

## FURTHER READING

Brewer, P., Minor Elements in Sea water, chap. 7, *Chemical Oceanography,* Vol. 1, 2nd ed., J. P. Riley and G. Skirrow, Academic Press, Eds., New York, 416-496 (1975).

Bruland, K. W., Trace Elements in Sea-water, chap. 45, *Chemical Oceanography,* Vol. 8, 2nd ed., J. P. Riley and R. Chester, Eds., Academic Press, New York, 157-220 (1983).

Garrels, R. M. and C. L. Christ, *Solutions, Mineral and Equilibria,* Harper and Row, New York (1965).

Goldberg, E., Minor Elements in Sea Water, chap. 5, *Chemical Oceanography,* Vol. 1, 1st ed., J. P. Riley and G. Skirrow, Eds., Academic Press, New York, 163-196 (1965).

Mackenzie, F. T., Sedimentary Cycling and the Evolution of Sea Water, chap. 5, *Chemical Oceanography,* Vol. 1, 2nd ed., J. P. Riley and G. Skirrow, Eds., Academic Press, New York, 309-364 (1975).

Nicholls, G. D., The Geochemical History of the Oceans, chap. 20, *Chemical Oceanography,* Vol. 2, 1st ed., J. P. Riley and G. Skirrow, Eds., Academic Press, New York, 277-294 (1965).

# 4 Table of Contents

**Ionic Interactions** .......... 159
1. *Water — The Unique Solvent* .......... 161
2. *Review of the Structure of Water* .......... 165
    2.1. *Uniformist, Average Models* .......... 166
    2.2. *Mixture Models* .......... 168
        2.2.1. *Ice-Like Models* .......... 169
        2.2.2. *Cluster Theories* .......... 171
        2.2.3. *Clathrate Cage Models* .......... 172
        2.2.4. *Significant Structure Theory and Euken's Polymer Model* .......... 172
3. *Ion-Water Interactions* .......... 173
    3.1. *Electrostriction* .......... 179
4. *Ion-Ion Interactions* .......... 188
    4.1. *Ion-Pairing Model* .......... 198
    4.2. *Specific Interaction Model* .......... 206
5. *Physical Properties of Seawater* .......... 216
*Further Reading* .......... 224

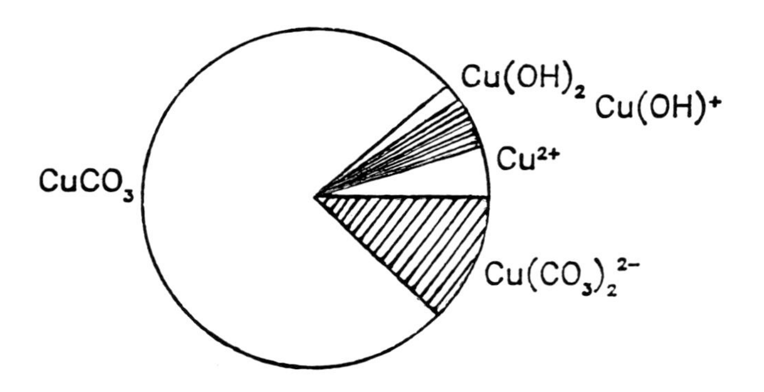


The various forms of $Cu^{2+}$ in seawater.

# 4 Ionic Interactions

In recent years there has been an increasing concern for the geochemical fate of trace metals in the marine environment. Many so-called ''base line'' studies have been made to determine the total concentration of these metals in the solution phase, the gas phase, and in the solid phase (i.e., on and in living and nonliving material). Although these studies have been useful in indicating the fate of trace metals in the marine environment, little progress has been made in determining the mechanisms involved (physical, chemical, and biological) in the transfer of metals between phases. Part of the reason for this lack of understanding of trace metals in the marine environment comes from the difficulties in detecting experimentally the true activity (both thermodynamic and biochemical) of the trace metals; due to their low concentrations conventional electrode methods of determining thermodynamic activity are difficult to apply. Although it is possible to detect the activity of some metals, we are at present forced to use models to estimate the thermodynamic activity of most trace metals in order to make comparisons with biological studies.

The importance of knowing the true biological activity (rather than total concentration) of metal ions in seawater has been suggested by a number of workers. For example, we have found that the growth depression by copper of the bacterium *Vibrio alginolyticus* was affected by complex formation.

This variation in copper toxicity is due to the presence of differing concentrations of free copper due to the formation of complexes with ligands such as fulvic acids.

Synthetic ligands (like NTA — nitriloacetic acid) and natural organic ligands can have similar effects in the marine environment. Three possible consequences may follow the addition of such a ligand: (1) metals may become solubilized and transported from regions of high concentration to uncontaminated regions, (2) due to over-complexation, necessary metals may not be available for the growth of marine organisms, and (3) by reducing the natural metal toxicity increased numbers of pathogenic organisms may occur.

The important role that free or complexed metal ions play in productivity has also been pointed out by a number of workers. For example, Barber and Ryther reported that phytoplankton blooms in newly upwelled waters off the Peru coast are conditioned by the organics in seawater. They suggested that the organic compounds are necessary to make the trace metals available for growth. However, Steeman-Neilson and Wium-Anderson suggested that the organic conditioning is due to the reduction in copper toxicity (due to complexation). Although it is not possible at present to state with certainty which of the suggested roles of organic ligands (or both) is true, we can state that it is the form of the metal (not the total amount) that is important in determining its biological and geochemical activity. The determination of the form of a given element is called speciation.

We shall now discuss the ionic interactions that can affect the state or structure of metal ions in seawater. Whether the thermodynamic activities can be directly related to biological activities remains to be seen. It must also be kept in mind that the most probable thermodynamic state may not be the state found in the marine environment due to the slow kinetics of the formation and degradation of metal complexes and various redox pairs.

In order to understand ionic interactions in aqueous electrolyte solutions, various chemical models have been developed. The development of a chemical model for natural waters has been strongly influenced by the coupling of physical chemical solution theory and marine chemical analytical and experimental data. The major trends of the development of a chemical model for seawater have resulted from the application of methods developed on the physical chemistry of electrolyte solutions. In the past it has taken a considerable amount of time for the physical chemical trends to influence marine chemistry. For example, it took 40 years before the Arrhenius theory was accepted in oceanography and 30 years before the quantitative aspects of the Debye-Huckel and Bjerrum theories were applied to seawater by Garrels and

Thompson. Although Wirth introduced the oceanographic community to the theoretically derived concentration behavior for densities in 1940, it has only been recently used to represent the equation of state of seawater.

In recent years, this trend has changed and, hopefully, marine chemistry shall accept and apply these new theories and models to geochemical and biochemical problems. The application of chemical models to examine ionic interactions in natural waters has been confined to two major areas: (1) those that are concerned with the bulk thermodynamic and transport properties of marine waters, and (2) those that are concerned with the effect of marine waters on the activity of dissolved solutes. More will be discussed later. To understand the state and structure of an ion in seawater there are a number of steps one must take. We will examine the state of an ion in seawater by considering two processes: (1) the ion-water interactions that occur when an ion is transferred from the ideal gas state to an infinitely large reservoir of water (i.e., where ions cannot interact with one another)

$$M^+ \text{ (ideal gas)} \rightarrow M^+ \text{ (infinite dilute solution)} \tag{1}$$

and (2) the ion-ion interactions that occur when an ion is transferred from infinite dilution to seawater (where interactions of all of the ions in the mixture affect the state (i.e., plus-plus, plus-minus, and minus-minus interactions)

$$M^+ \text{ (infinite dilute solution)} \rightarrow M^+ \text{ (seawater)} \tag{2}$$

Before these two processes are discussed, we will examine the unique properties of water as a solvent system.

## 1. WATER — THE UNIQUE SOLVENT

Although water makes up 96.5% of ocean waters and accounts for many of the unique physical and chemical properties of seawater, its importance has been neglected by many oceanographers. Horne (1969) attempted to stress the importance of understanding the nature of water and its interactions in natural waters. As chemical oceanography becomes a less descriptive science and moves to one that attempts to probe the chemical processes occurring in the oceans, this approach will become more popular. One must keep in mind that the ultimate aim of a marine chemist is to obtain a molecular level understanding of the chemical processes that occur in the marine environment. To do this one eventually must understand the role water plays in these molecular interactions. With this in mind we will discuss some of the unique properties of water and its structure in this section.

The unique position of water compared to other similar compounds in nature can be of great importance in various meterological, oceanographic, geochemical, and biochemical processes. This uniqueness can be demonstrated by comparing its physical and chemical properties to other liquids. Let us first examine the boiling point of water (the temperature at which the vapor pressure is equal to the atmospheric pressure) compared to other liquids of similar structure (Figure 4.1). It is obvious from this figure that water appears to have a much higher boiling point than expected. It should be kept in mind that the boiling point ($T_b$) of a given series usually increases with increasing molecular weight. This higher boiling point indicates that it takes more energy than expected to get the water molecule into the gas state. The melting point ($T_m$) for water is also higher than expected. This indicates that it takes more energy to melt ice than expected.

It is also possible to show the uniqueness of water by comparing its other physical and chemical properties with other liquids. This is done in Table 4.1. The high dielectric constant of water (80) is the result of the high dipole moment. This also results in a low dissociation ($K = 1 \times 10^{-14}$) of $H_2O \rightarrow H^+ + OH^-$. The great dissolving properties of $H_2O$ are also due to its high dielectric constant. Only a few inorganic compounds have higher dielectric constants: D ($NH_3$) = 23, D (HF) = 85, D (HCN) = 95, and D ($SO_2$) = 140). The hydrating properties of water are well known; however, the causes are not clearly known.

The unique bond angle and ability to form hydrogen bonds results in long range order in water (and ice) that is unique compared to most other liquids. Each water molecule has the opportunity to combine with up to four other water molecules through hydrogen bonds (Figure 4.2).

All of these unique properties of water can be traced to the structure of the individual water molecules and how they interact with one another. One might expect the atoms in water to be at a bond angle of 180°, but they actually form a bond angle of 105°.

As a result of the intermolecular forces in the water molecule not being completely balanced (105° bond angle), the water molecule has an electric dipole. The separation of the negative charge on the unpaired electrons and the positive charge on the protons results in a dipole moment of 1.84D

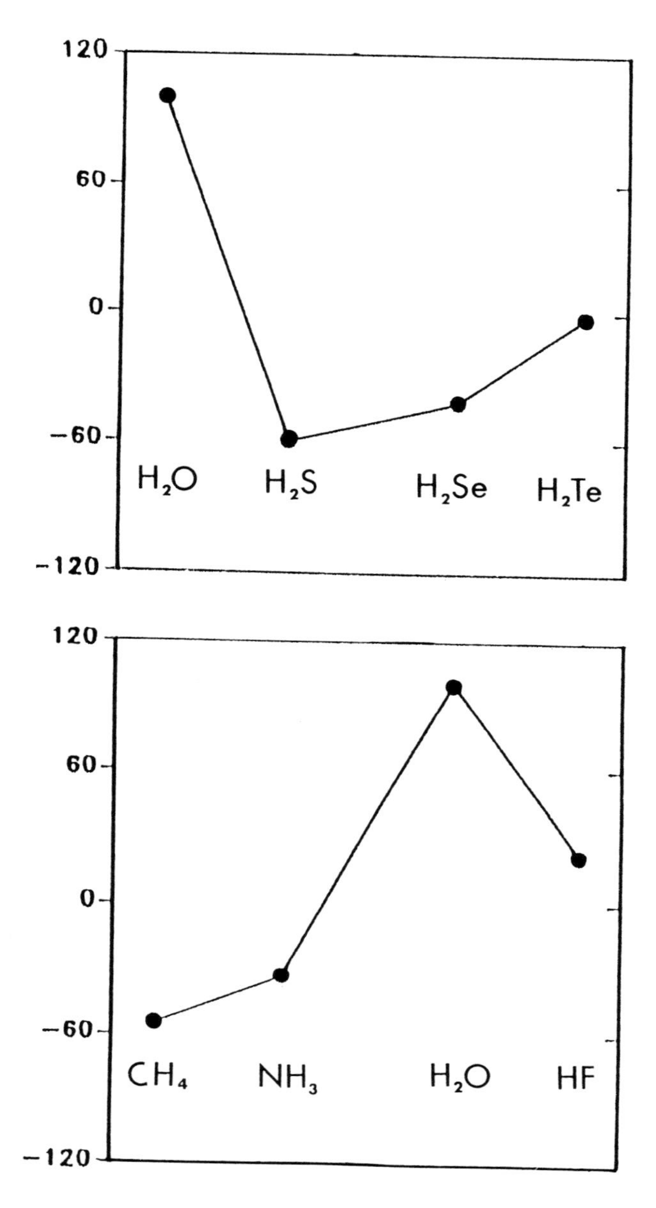


FIGURE 4.1. Boiling points of compounds similar to water.

**TABLE 4.1**
**Comparison of Physical Properties of $H_2O$, MeOH, and *n*-Heptane**

| | $H_2O$ | MeOH | *n*-Heptane |
|---|---|---|---|
| Mol wt | 18 | 32 | 100 |
| Dipole moment (Debye's) | 1.84 | 1.70 | >0.2 |
| Dielectric constant | 80 | 24 | 1.97 |
| Density (g/cm$^3$) | 1.0 | 0.79 | 0.73 |
| Bp (°C) | 100 | 65 | 98.4 |
| Mp (°C) | 0 | −98 | −97 |
| Sp ht (cal/g·degree) | 1.0 | 0.56 | 0.5 |
| H vap (cal/g) | 540 | 263 | 76 |
| H fus (cal/g) | 79 | 22 | 34 |
| Surface tension (dg/cm) | 73 | 23 | 25 |
| Viscosity 20°C (poise) | 0.01 | 0.006 | 0.005 |
| Compressibility (25°C) (atm$^{-1}$) | $4.57 \times 10^{-11}$ | $12.2 \times 10^{-11}$ | $14 \times 10^{-11}$ |

(D = Debye, one D is equivalent to the separation of two point charges of opposite sign a distance of an Å or $10^{-10}$ m). This is shown in Figure 4.3. Since there are actually two points of negative and two points of positive charge separation on a water molecule, there exists a quadrupole moment. This is shown in Figure 4.4. This quadrupole moment results from dipole-dipole interactions which are the electrical analogy to moments of inertia.

The dipole-dipole interactions of two water molecules cause the hydrogen bonding in water. A given water molecule has the ability to form four hydrogen bonds (Figure 4.5).

The energies involved in hydrogen bonding are in general a lot larger than most dipole-dipole interactions, thus, they should be classified separately. The ΔH for the formation of a hydrogen bond is small (1 to 10 kcal/mol) compared to most compounds that form chemical bonds of 100 kcal/mol. The hydrogen bond is not completely electrostatic in nature. It has some covalent character due to the sharing of the unpaired electrons on the oxygen of one water molecule with protons of another. This hydrogen bonding in water causes many of its physical chemical properties to be different. The effect of temperature and pressure on many of its properties is unique compared to many liquids. This is shown in Figures 4.6 and 4.7. The effect these anomalies have on natural waters is given in Table 4.2.

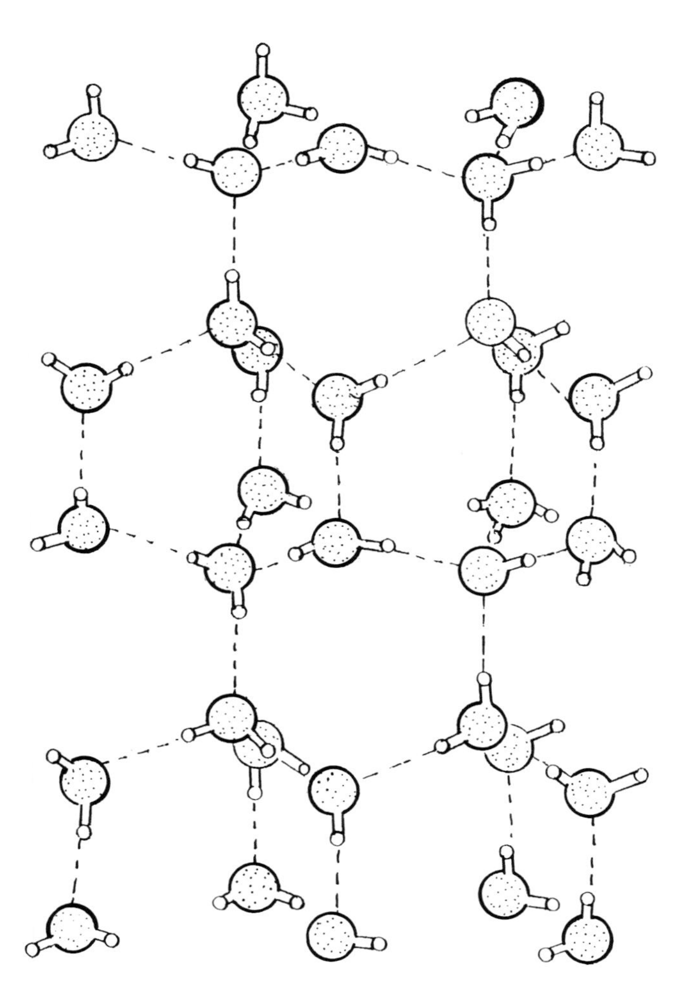

FIGURE 4.2. The structure of Ice I.

## 2. REVIEW OF THE STRUCTURE OF WATER

Several reviews and monographs have appeared summarizing the status of research on the structure of water. An excellent survey is given in the books edited by Franks. We can divide the water structure models into two major categories: the uniformist or average models and the mixture models. Both models recognize that water is a very structured liquid; the major difference between the two models is that in the mixture models the simultaneous existence of at least two different states of water is thought to be present.

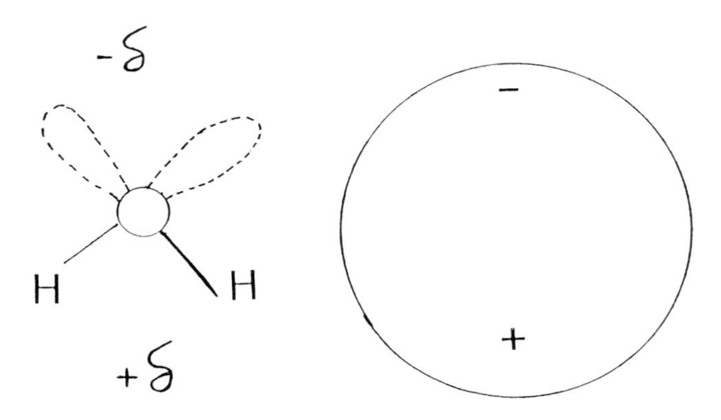


FIGURE 4.3. The water dipole.

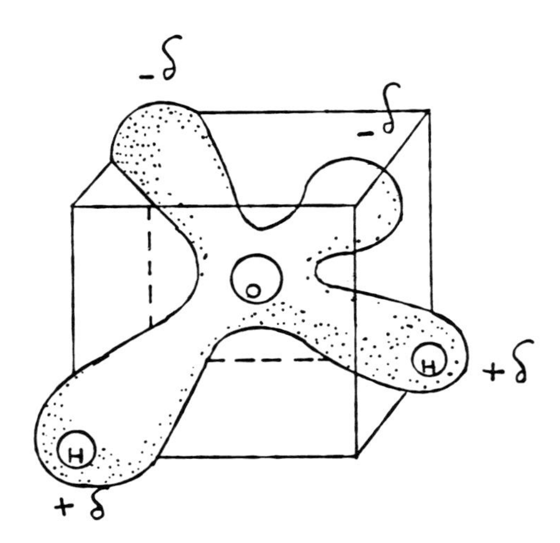


FIGURE 4.4. The three-dimensional structure of the water molecule.

## 2.1. Uniformist, Average Models

Bernal and Fowler, Pople, Wall and Horning, Falk, and Kell were proponents of this model. The basic element in the uniformist view is that there exists in water no local domains of structure different from that of any other arbitrarily chosen element of water. In the process of averaging, the individual water molecule behaves at any time much like any other water molecule is behaving. Bernal and Fowler's original model has been used and has served very well in many applications. Pople's treatment gave a far greater qualitative

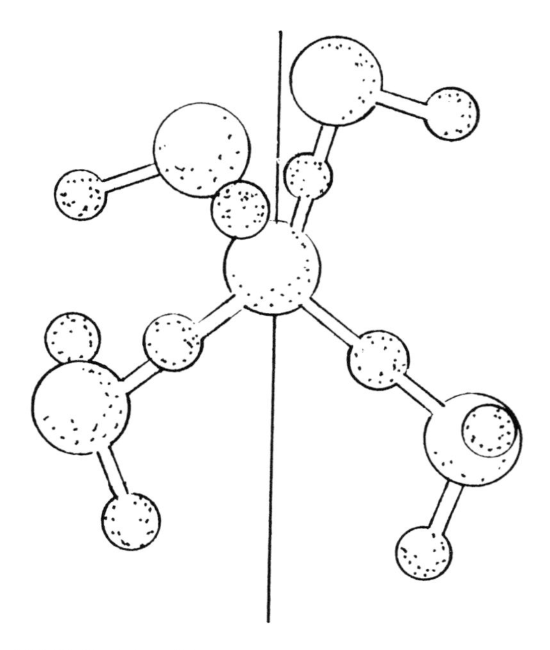

FIGURE 4.5. The hydrogen bonded structure of water molecules.

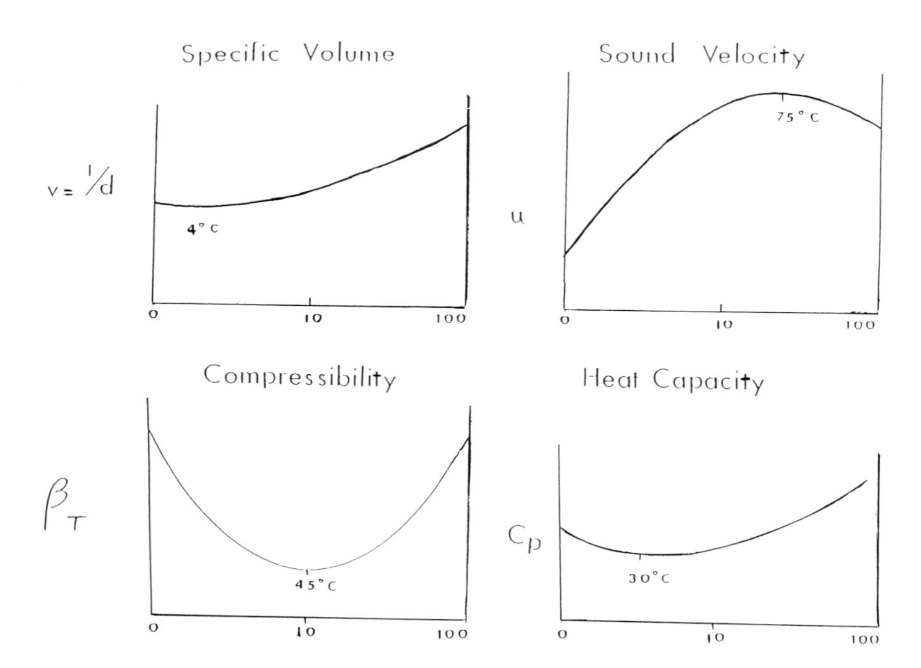


FIGURE 4.6. The effect of temperature on the specific volume (l/density), sound speed, compressibility, and heat capacity for water.

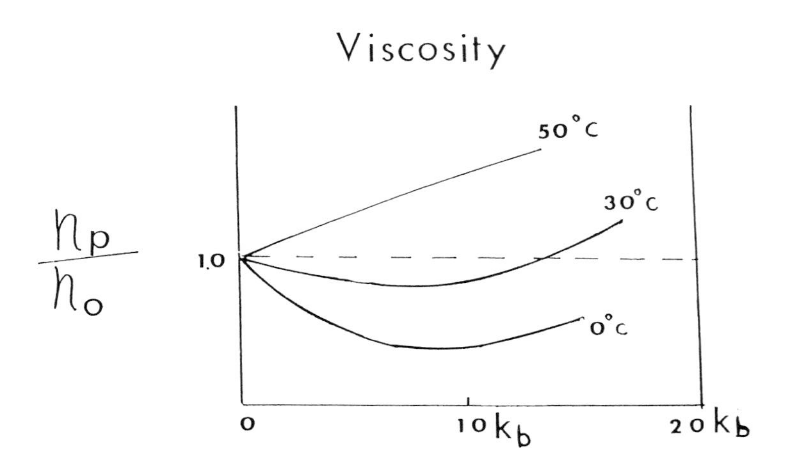


FIGURE 4.7. The effect of pressure on the viscosity of water at various temperatures.

insight into the structure of water. Pople explained the maximum density of liquid water by stating two opposing effects — increase in volume due to expansion of the lattice structure and a bending of the H-bonds. Thus, this model treats liquid water as an "ice like" lattice with differences due to the bending of bonds, not the breaking of bonds.

## 2.2. Mixture Models

These models have received more notice over the years. We can divide the mixture models into the following categories:

1. Broken-down ice lattice models — i.e., Ice-like units in equilibrium with monomers.
2. Cluster models — i.e., H-bonded clusters in equilibrium with monomers.
3. Clathrate models — Clathrate-like cages in equilibrium with monomers.
4. Significant structure model or Eucken's polymer model — Bulky species are not necessarily a monomer.

In each case at least two different species of water exist: a bulky species representing some type of structured units and a dense species such as a monomer (sketches of two popular mixture models are shown in Figures 4.8 and 4.9).

**TABLE 4.2**
**Summary of Anomalous Properties of $H_2O$**

| Property | Results |
|---|---|
| 1. High heat capacity | Prevents extreme ranges of temperature<br>Heat transfer by water movement is large<br>Maintains uniform body temperatures |
| 2. High heat of fusion | Thermostatic effect of freezing and melting |
| 3. High heat of evaporation | Important in transfer of heat from water to atmosphere |
| 4. Thermal expansion | Fresh $H_2O$ and dilute SW have their maximum density above Tm (the melting point); Controls the temperature density distribution and vertical circulation in lakes |
| 5. High surface tension | Important in cell physiology<br>Controls certain surface behavior and drop formation |
| 6. High dielectric constant | Important in causing salts to ionize and become electrolytes (dissolving power) |
| 7. Low dissociation | $H^+$ and $OH^-$ behavior very important in many geological and biological processes; Dissociation (pH sensitive) |
| 8. High transparency | Absorbs radiant energy in the IR and UV little is visible; Important to physical and biological process |
| 9. High conduction of heat | Important only on small scale as in living cells, eddy conductance is greater |
| 10. Compressibility B changes with temperature 0—50°C in inverse manner | Electrostriction changes with temperature; Causes changes in thermochemical properties — solubility, etc. |
| 11. Heat capacity (Cp) changes with temperature 0—30°C in inverse manner | Thermoproperties are strange compared to other solvents |
| 12. High viscosity | Important to physical behavior (waves, etc.) |
| 13. Density of solid ice is less than density of liquid at melting point | Important for many geochemical, atmospheric, and biological processes |

### 2.2.1. Ice-Like Models

Rowlands was apparently the first to suggest an equilibrium between Ice I and monomers. Samoilov proposed a model in which the monomeric $H_2O$ molecules are "tucked away" in the interstitial spaces of an "ice like" lattice. Tamann suggested that there should be as many types of structures in liquid water (Figure 4.10) as there are types of ice (e.g., Ice I, II, etc.). The very careful X-ray diffraction study by Danford and Levy was found to be in good agreement with a broken-down ice structure model. They suggest that the monomeric water molecules occupy positions in the interstices of the ice.

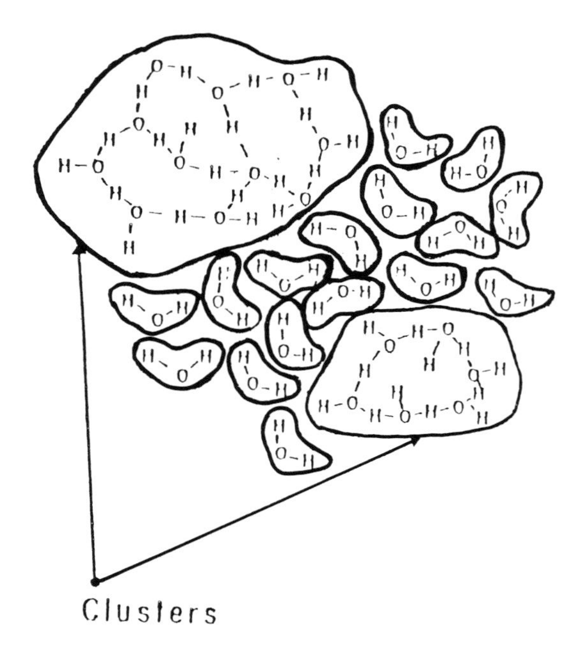


FIGURE 4.8. The Frank and Wen Flickering Cluster Model for the structure of water.

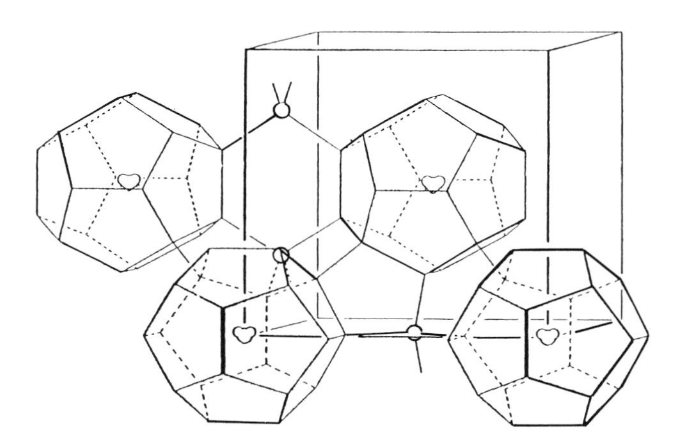

FIGURE 4.9. The Pauling Self-Clathrate Model for the structure of water.

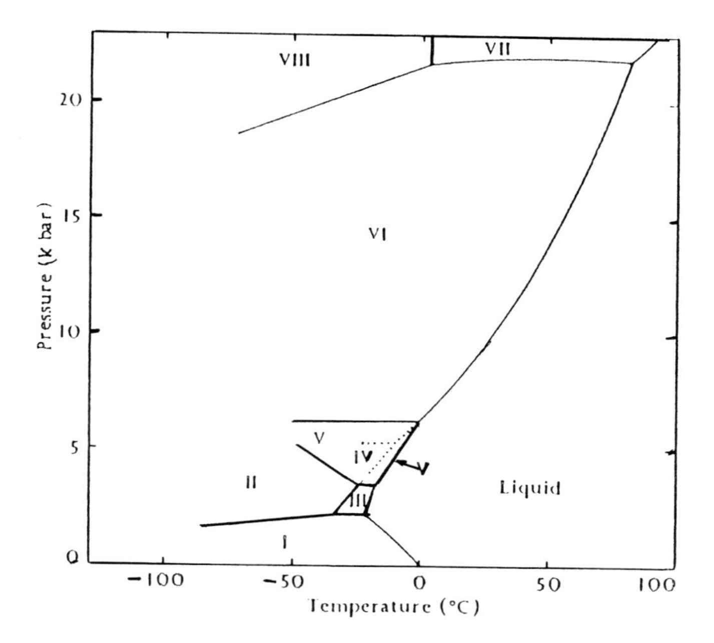


FIGURE 4.10. Phase diagram for the various forms of ice.

(This is similar to Samilov's model). Although the model has been used to successfully reproduce the X-ray data, many adjustable parameters were used, thus, there has been criticism of this model. Davis and Litovitz suggested a model involving puckered hexagonal rings similar to the ice rings formed in the basal plane of ice. Density variations are accounted for by assuming two rings exist — an open-packed, ice-like and a closed-packed, ring structure.

### 2.2.2. Cluster Theories

Stewart was the first to suggest the existence of clusters due to his X-ray work. The clusters contained 10,000 water molecules. Nemethy and Scheraga build on the Frank and Wen concept of flickering clusters (Figure 4.8) using statistical thermodynamics. They have used this model to calculate the thermodynamic properties of $H_2O$. The authors considered unbonded water molecules and water molecules with 1, 2, 3, and 4 H-bonds/molecule. The average cluster varies from 91 to 25 from 0 to 70°C. At 25°C the cluster size is about 50 molecules. Monomeric molecules range from fractions of 0.24 to 0.29 from 0 to 70°C (i.e., 24% bonds are broken at 0° upon melting). Others have made similar calculations by assigning energy bands instead of levels for different states of the H bonding. Only three species are considered; free $H_2O$, and one OH group bonded and two OH groups bonded molecules. They have been able to calculate the thermodynamic properties with great accuracy using these models. Luck has also formulated a model using clusters of $H_2O$ molecules (100 to 700 molecules).

### 2.2.3. Clathrate Cage Models

Pauling originally postulated the clathrate cage model (Figure 4.9). He noted the existence of clathrate hydrates of many inert gases and suggested that water is its own clathrate hydrate. Frank and Quist developed this model and successfully calculated thermodynamic properties of water in good agreement with experimental values. In the clathrate model, discrete sites exist in the host lattice (for other $H_2O$ molecules or solute molecules).

### 2.2.4. Significant Structure Theory and Eucken's Polymer Model

The bulky species is "ice-like" (not necessarily like Ice I). The dense species is not necessarily a monomer, but may be another ice-like species of higher density.

Manchi and Eyring used significant structure theory — a specific element exists with fluidized vacancies in addition to the individual monomers. Others have extended this treatment by recognizing that the amount of free monomeric water is small. Thus, the significant structures are cage-like clusters with 16 molecules (density of Ice I) with, in equilibrium, an Ice III-like structure. Fluidized vacancies are produced during melting (contraction occurs due to the packing of single $H_2O$ molecules into voids of Ice I-like clusters). This model has been used to successfully predict the minimum at 4° in density, the vapor pressure, and the specific heat (combined with Eyring's rate theory this model has been used to calculate the pressure dependence of the relative viscosity of $H_2O$).

Eucken treated water as a mixture of distinctly associated species of dimers, tetramers, and octamers. Although this method is probably not correct, Wicke has suggested that dimers may exist near the critical point. Eucken's theory is interesting in the fact that it is able to estimate thermodynamic properties that agree with the experimental results; however, the model is not correct. It is thus important to note that the fact that a model is able to calculate accurate properties of water does not prove the model is correct. A more critical test of a water theory is its ability to predict correctly, even if only quantitatively, a large variety of widely different properties of water.

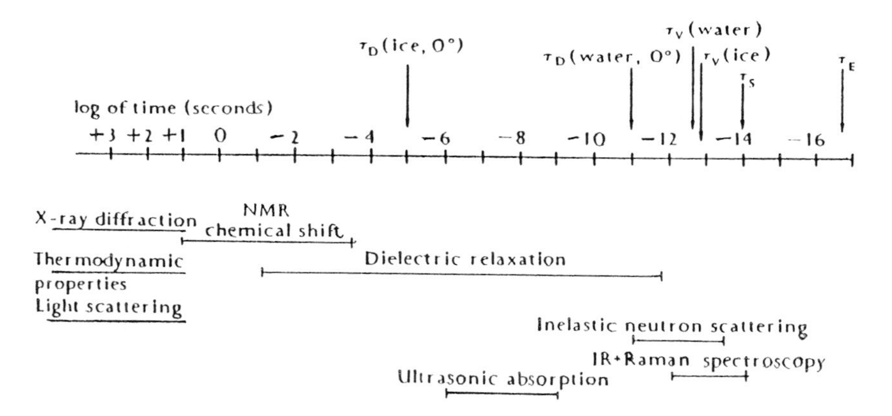


FIGURE 4.11. Comparison of the times of water movements to the times required to make various measurements.

One of the major difficulties of the mixture models that assume the dense species is monomeric water is in explaining how mere van der Waals forces can provide enough attraction for the single water molecule to avoid escaping into the vapor state. The water molecules in a clathrate cage must also possess some unique properties. It is difficult to see how $H_2O$ molecules with their large dipole moment can exist in a clathrate cage without strong interactions. Experimental studies have not proved that $H_2O$ molecules reside in clathrate cages.

It is important to understand that when we think of the structure in a liquid, it is necessary to consider the time scale used to measure that structure. Depending upon the experimental techniques used (Figure 4.11), one will take a picture that is related to the shutter speed of the camera. Thermodynamic measurements see an average structure. The time molecules move (i.e., the time for molecular displacements) is $10^{-5}$ s in ice and $10^{-11}$ s in the liquid phase.

## 3. ION-WATER INTERACTIONS

To understand the behavior of ions in seawater it is important to be able to understand the interactions of ions with water molecules. To study these ion-water interactions one must study the thermodynamic and transport properties of electrolytes at infinite dilution. In practice it is not possible to make

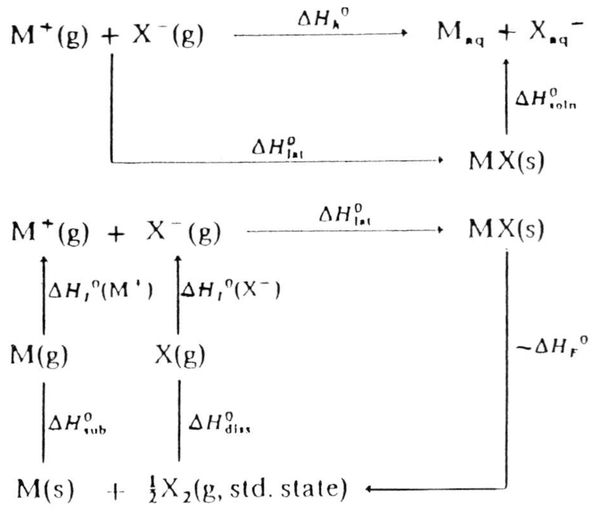

$$M^+(g) + X^-(g) \xrightarrow{\Delta H_h^0} M_{aq} + X_{aq}^-$$

$$\Delta H^0_{lat}: M^+(g) + X^-(g) \rightarrow MX(s) \qquad \Delta H^0_{soln}: MX(s) \rightarrow M_{aq} + X_{aq}^-$$

$$M^+(g) + X^-(g) \xrightarrow{\Delta H^0_{lat}} MX(s)$$

$$\Delta H_I^0(M^+): M(g) \rightarrow M^+(g) \qquad \Delta H_I^0(X^-): X(g) \rightarrow X^-(g)$$

$$\Delta H^0_{sub}: M(s) \rightarrow M(g) \qquad \Delta H^0_{diss}: \tfrac{1}{2}X_2 \rightarrow X(g)$$

$$-\Delta H_F^0: MX(s) \rightarrow M(s) + \tfrac{1}{2}X_2(g, \text{std. state})$$

$$\begin{aligned}\Delta H_H^0 &= \Delta H^0_{lat} + \Delta H^0_{soln}\\ &= -\Delta H_F^0 + \Delta H^0_{diss} + \Delta H^0_{sub} + \Delta H_I^0(M^+)\\ &\quad + \Delta H_I^0(X^-) + \Delta H^0_{soln}\end{aligned}$$

$\Delta H_h^0$ = heat of hydration
$\Delta H_I^0$ = heat of gas phase ionization
$\Delta H^0_{sub}$ = heat of sublimation
$\Delta H^0_{diss}$ = heat of dissociation
$\Delta H_F^0$ = heat of formation
$\Delta H^0_{lat}$ = heat of crystal lattice formation
$\Delta H^0_{soln}$ = heat of solution

FIGURE 4.12. The methods used to calculate the enthalpies of hydration for an electrolyte.

direct measurements at infinite dilution; thus, the infinite dilution thermodynamic properties are extrapolated from experimental results at finite low concentrations (with the aid of the Debye-Huckel equations for long range ion-ion interactions). Since one normally studies the ion-water interactions in solution where no ion-ion interactions occur, it is necessary to select an initial state devoid of ion-ion interactions. The initial state normally selected is that of ions in a vacuum at an infinitely low pressure (i.e., the ideal gas). One considers then the changes in properties such as free energy $\Delta G_h^\circ$, enthalpy $\Delta H_h^\circ$, and entropy $\Delta S_h^\circ$ for the process denoted by Equation 1 (called, respectively, the hydration free energies, enthalpies, and entropies). The methods used to calculate these thermodynamic hydration functions are discussed elsewhere. The methods used to calculate the $\Delta H_h^\circ$ are shown in Figure 4.12.

**TABLE 4.3**
**The Thermodynamics of Hydration of Ions at 25°C**

| Ion | r (Å) | $-\Delta G_h^\circ$ (kcal/mol) | $-\Delta H_h^\circ$ (kcal/mol) | $-\Delta S_h^\circ$ (cal/mol · degree) |
|---|---|---|---|---|
| $H^+$ | — | 260.5 | 269.8 | 31.3 |
| $Li^+$ | 0.60 | 122.1 | 132.1 | 33.7 |
| $Na^+$ | 0.95 | 98.2 | 106.0 | 26.2 |
| $Ag^+$ | 1.26 | 114.5 | 122.7 | 27.6 |
| $K^+$ | 1.33 | 80.6 | 85.8 | 17.7 |
| $Tl^+$ | 1.40 | 82.0 | 87.0 | 16.7 |
| $Rb^+$ | 1.48 | 75.5 | 79.8 | 14.8 |
| $NH_4^+$ | 1.60 | — | 84.8 | — |
| $Cs^+$ | 1.69 | 67.8 | 72.0 | 14.1 |
| $Cu^+$ | 0.96 | 136.2 | 151.1 | 50.0 |
| $Be^{2+}$ | 0.31 | — | 594.6 | — |
| $Mg^{2+}$ | 0.65 | 455.5 | 477.6 | 74.3 |
| $Ni^{2+}$ | 0.72 | 494.2 | 518.8 | 82.4 |
| $Co^{2+}$ | 0.74 | 479.5 | 503.3 | 80.0 |
| $Zn^{2+}$ | 0.74 | 484.6 | 506.8 | 74.5 |
| $Fe^{2+}$ | 0.76 | 456.4 | 480.2 | 79.8 |
| $Mn^{2+}$ | 0.80 | 437.8 | 459.2 | 72.1 |
| $Cu^{2+}$ | 0.96 | 498.7 | 519.7 | 73.9 |
| $Cd^{2+}$ | 0.97 | 430.5 | 449.8 | 65.2 |
| $Ca^{2+}$ | 0.99 | 380.8 | 398.8 | 60.8 |
| $Hg^{2+}$ | 1.10 | 436.3 | — | — |
| $Sr^{2+}$ | 1.13 | 345.9 | 363.5 | 59.2 |
| $Pb^{2+}$ | 1.20 | 357.8 | 371.9 | 47.4 |
| $Ba^{2+}$ | 1.35 | 315.1 | 329.5 | 48.5 |
| $Al^{3+}$ | 0.50 | 1103.3 | 1141.0 | 126.6 |
| $Fe^{3+}$ | 0.64 | 1035.5 | 1073.4 | 127.5 |
| $Cr^{3+}$ | 0.69 | — | 1079.4 | — |
| $Y^{3+}$ | 0.93 | 859.5 | 891.5 | 107.6 |
| $Sc^{3+}$ | 0.81 | 929.3 | 962.7 | 112.5 |
| $La^{3+}$ | 1.15 | — | 811.9 | — |

Since we are interested in the transfer of ions ($M^+$) rather than electrolytes (MX), it is necessary to make some nonthermodynamic assumptions concerning the differences between the properties of cations and anions. The details of such methods are discussed elsewhere. Once the selection is made for the absolute thermodynamic quantity of one ion (usually the proton), the values for the other ions can be easily determined by the additivity principle

$$\Delta H_h^\circ(MX) = \Delta H_h^\circ(M^+) + \Delta H_h^\circ(X^-) \tag{3}$$

Values of $\Delta G_h^\circ$, $\Delta H_h^\circ$, and $\Delta S_h^\circ$ for some metal ions are given in Table 4.3.

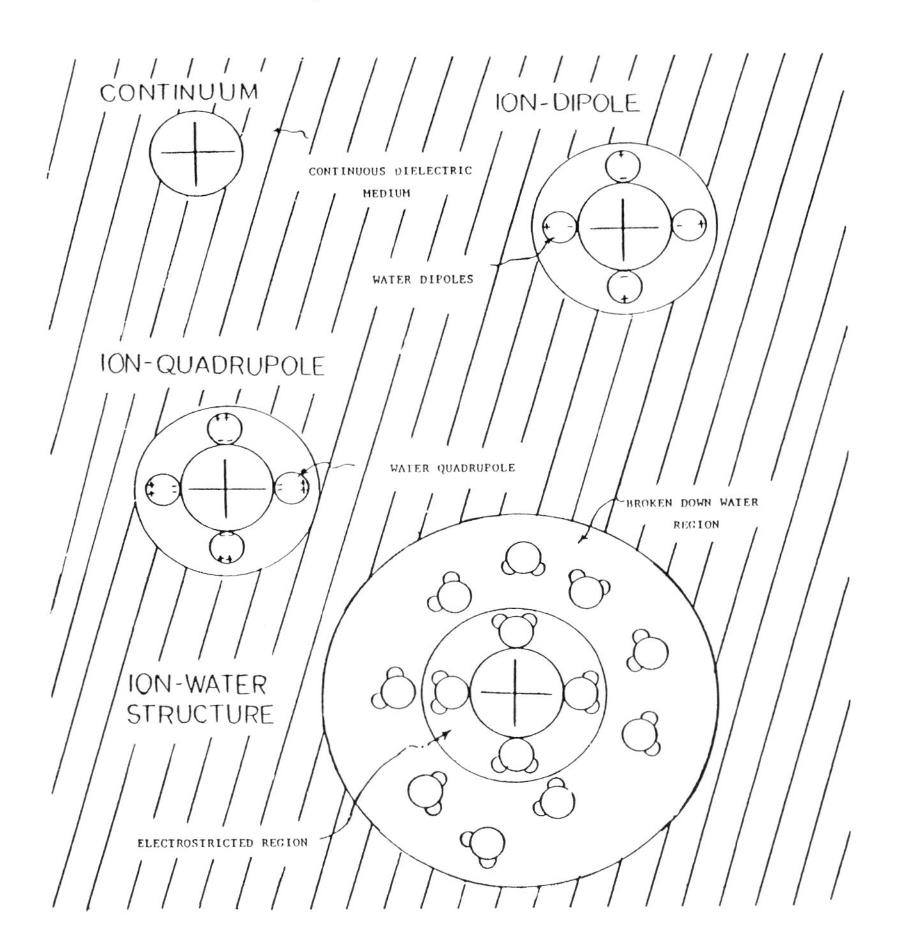


FIGURE 4.13. Models used to explain ion-water interactions.

To truly understand ion-water interactions one must know the structure of water. Since the structure of water is very complex, one must use simple models for the interaction between ions and water molecules. These models serve as mental pictures and reproduce approximately what occurs in the real system. The better they are able to predict the experimental properties of the real system, the better they serve as aids to understanding the real system. Before we discuss these models (Figure 4.13) we will first examine what happens when a salt is added to water.

The addition of salt to water complicates the structures and properties of water to a large degree. We will start by considering the effect of adding sufficient NaCl to water to make a synthetic seawater of 3.5% NaCl by weight.

The addition of 35 g of NaCl to 965 g of $H_2O$ gives the following molality

$$m = (35/965)(1000/58.48) = 0.62 \tag{4}$$

What happens when the solution is formed?

$$NaCl(c) \rightarrow Na^+(aq) + Cl^-(aq) \tag{5}$$

1. There is about 0.6 kcal of heat absorbed (an endothermic reaction). This absorption of heat indicates that the solution process involves the making and breaking of chemical bonds. Since the heat involved in forming the NaCl(s) from the gaseous ions ($M^+ + X^-$) is quite large

$$Na^+(g) + Cl^-(g) \rightarrow NaCl\ (crystal)$$

$$\Delta H^\circ = 181\ kcal/mol \tag{6}$$

The heat involved in hydrating the ions $Na^+$ and $Cl^-$ is also quite large

$$Na^+(g) + Cl^-(g) \rightarrow Na^+(aq) + Cl^-(aq) \tag{7}$$

This can be clearly demonstrated by examining the Born-Haber cycle for the hydration of NaCl ($\Delta H_h$ is the heat of hydration of the ions). This heat of hydration for the salt (MX) is given by the

$$M^+(g) + X^-(g) \rightarrow M^+(aq) + X^-(aq) \tag{8}$$

It is related to the heats of the reactions of the formation of MX in the solid state

$$M^+(g) + X^-(g) \rightarrow MX(s) \tag{9}$$

which is called the lattice heat and the heat of solution of the salt

$$MX(g) \rightarrow M^+(aq) + X^-(aq) \tag{10}$$

$$\Delta H_h = \Delta H_{LATTICE} + \Delta H_{SOLN} \tag{11}$$

For NaCl

$$\Delta H_h = 181 + 0.6 = 182 \text{ kcal mol}^{-1} \tag{12}$$

The fact that $\Delta H_h$ is the same order of magnitude as $\Delta H_{LATTICE}$ indicates that the energies involved in solvating the ions are the same order of magnitude as forming ionic bonds in a crystal.

2. The solution formed after the addition of NaCl no longer freezes at 0°C, but at approximately −2.3°C. The freezing point depression is given approximately by

$$\Delta T_f = -1.86\ \nu\ m \tag{13}$$

where $\nu$ is the number of ions formed when the salt completely dissociates. This equation gives $\Delta T_f = -2.3°C$, $T_f = -2.3°\ C$. This indicates that the interaction of $Na^+$ and $Cl^-$ with water breaks down the structure of water.

3. The solution also boils at a higher temperature. The boiling point elevation is given by

$$\Delta T_b = 0.52\ \nu\ m \tag{14}$$

This equation gives $\Delta T_b = 0.63$ or $T_b = 100.63°C$. This indicates that the hydration interactions tend to keep molecules in the liquid state.

4. The vapor pressure of the NaCl solution is lower than pure water. The ratio of the vapor pressure over the solution $P/P_{H_2O} = 0.98$. This 2% depression in vapor pressure also indicates that the hydration "ties up" the water molecules, making it harder for them to go into the gas state.

5. The electrical conductance of the solution is increased by 10,000 ($10^{-6}$ $\Omega^{-1}$ $cm^{-1}$ to $4.68 \times 10^{-2}\ \Omega^{-1}\ cm^{-1}$). This indicates that the ions are able to carry a charge through solution.

6. The temperature of the maximum density is lowered by 8°C to −4°C. This (like seawater) is below the freezing point. These results indicate that the hydration process breaks down the water structure.

7. The osmotic pressure (26 atm) is created. This pressure is related to the vapor pressure, freezing point, and boiling point effects (it is a colligative property). It can be determined from

$$\pi = -(RT/V_{H_2O}) \ln a_{H_2O} \tag{15}$$

where $a_{H_2O}$ is the activity of water in the solution ($P/P_{H_2O}$) and $V_{H_2O}$ is the molar volume of water in the solution ($V_{H_2O} = MW/\rho$). This osmotic pressure can act as a driving force for the diffusion of water through membranes. The $H_2O$ molecules interact very strongly with ions. Many electrolytes hold onto their water molecules with such tenacity that the solids crystalize with a definite number of water molecules.

An understanding of ion-water interactions and ion-ion interactions is thus a prerequisite for making any progress in understanding the chemistry of seawater. One might ask the question, "how many $H_2O$ molecules are hydrated to the $Na^+$ or $Cl^-$ ions"? Estimates range from 2 to 70 depending upon how the measurements were made. This large variety of numbers is due to the fact that the hydration volume around an ion does not have a definite boundary. Thus, some methods count the $H_2O$ molecules that are not firmly bound by the ion. Another question one might ask is, how many of the hydrated water molecules move when the ion moves? To answer this question we must know the exact time that an $H_2O$ molecule remains on an ion. This is normally not long and it is probably not realistic to try to distinguish between the stationary and the kinetic hydration atmosphere of an ion.

For most of the ions of importance in seawater chemistry the major factor determining the extent or strength of hydration is the charge density (the Z/r ratio). The higher the charge density the greater the hydration.

### 3.1. Electrostriction

Another unexpected phenomenon upon the formation of a NaCl solution is electrostriction. For example, the density of solid NaCl is 2.165 g/cm$^3$. The volume of 35 g of NaCl would thus occupy 16.2 cm$^3$/mol. The density of water at 25°C is 0.997 g/cm$^3$, thus the volume of 965 of water is 967.9 cm$^3$. If the volumes are conservative on mixing the solution would have a volume of 16.2 + 967.9 cm$^3$ = 984.1 cm$^3$. Since the density of the solution is 1.0232 g/cm$^3$, the actual volume is 977.3 cm$^3$. Thus, the volume of the solution is decreased by 984.1 − 977.3 = 6.8 cm$^3$. This decrease in volume is called electrostriction and is caused by the ion-water interactions. The ions draw the water molecules inward compressing the solvent. The water molecules near an ion have a higher density than the bulk water. This effect is important for two reasons: (1) this hydration changes the mobility of ions, and (2) the effect of pressure on ionic equilibria forces reactions to the smallest volume. Since ions in solution have a smaller effective volume, pressure will force solids to have a higher solubility.

The continuum model is an example of a crude model that can serve as an approximation for the real system. Drude and Nernst first used this model to explain the decrease in volume that occurs when an electrolyte is dissolved in water. Born popularized the model and his name is normally attached to its use. In the model an ion is pictured as a solid sphere of radius r bearing a charge Ze (where Z is the valence and e is the electrostatic charge) and the solvent is a structureless continuous dielectric medium. Using electrostatics the $\Delta G_h°$ (in units of kcal mol$^{-1}$) is given by (at 25°C)

$$\Delta G_h^o = -(Ne^2Z^2/2r)(1 - 1/D) = -1.63.89\ Z^2/r \qquad (16)$$

where N is Avogadro's number, r is the radius in Angstrom units (1Å = 1 × $10^{-8}$ cm), and D is the dielectric constant of water (78.36 at 25°C). By appropriate differentiation of Equation 16 with respect to temperature (T), it is possible to determine the other thermodynamic hydration functions $\Delta S_h°$ (in units of cal deg$^{-1}$ mol$^{-1}$) and $\Delta H_h°$ (in units of kcal mol$^{-1}$)

$$\Delta S_h^\circ = (Ne^2Z^2/2r)(\partial \ln D/\partial T)_P = -9.649\ Z^2/r \qquad (17)$$

$$\Delta H_h^\circ = (-Ne^2Z^2/2r)[1 - 1/D - (T/D)(\partial \ln D/\partial T)_P]$$

$$= -\ 166.78\ Z^2/r \qquad (18)$$

A comparison of the experimental values of $\Delta G_h°$, $\Delta H_h°$, and $\Delta S_h°$ plotted vs. $Z^2/r$ is shown in Figures 4.14, 4.15, and 4.16. It is quite apparent from these figures that the Born model offers a reasonable first approximation to the magnitude, radius, and charge dependence of $\Delta G_h°$, $\Delta H_h°$, and $\Delta S_h°$. A close examination of the data shows a number of significant deviations. For example, values of $\Delta H_h°$ for the transition metals (Ca to Zn) given in Figure 4.17 do not increase in magnitude with increasing atomic number (decreasing radius). This is due to the 3d orbitals not being spherically symmetrical (i.e., the hydrated water molecules do not have the same energy).

By further differentiation of Equations 16, 17, and 18 it is possible to obtain information about the size and structure of the hydration sphere. From the pressure dependence of $\Delta G_h°$, one obtains the volume change (electrostriction)

$$V° \text{ (elect)} = (-NZ^2e^2/2Dr)(\partial \ln D/\partial P)_T = -4.175\ Z^2/r \qquad (19)$$

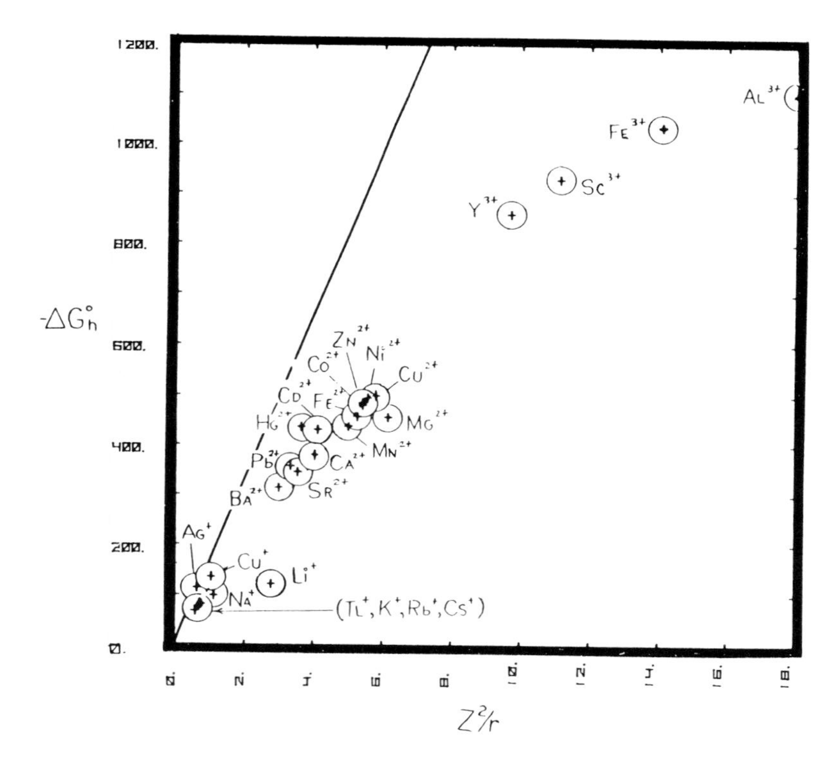


FIGURE 4.14. Values of the free energy of hydration for metals vs. the charge (Z) squared divided by the crystal radii (r).

Further differentiation of Equation 19 with respect to (T) and (P) gives the electrostriction partial molal expansibility ($E^\circ = \partial V^\circ/\partial T$) and compressibility ($K^\circ = -\ \partial V^\circ/\partial P$)

$$E^\circ\ (\text{elect}) = (-NZ^2e^2/2Dr)[(\partial^2 \ln D/\partial T\partial P) - (\partial \ln D/\partial T) \times (\partial \ln D/\partial P)_T] = -2.74 \times 10^{-2}\ Z^2/r \tag{20}$$

$$K^\circ\ (\text{elect}) = (NZ^2e^2/2Dr)[(\partial \ln D/\partial P^2)_T - (\partial \ln D/\partial P)_T^2] = -8.31 \times 10^{-4}\ Z^2/r \tag{21}$$

Similar differentiation of the solution component of Equation 16 with respect to temperature yields the electrostatic partial molal entropy — ($-\partial G^\circ/\partial T = S^\circ$) — and heat capacity — $\partial(S^\circ/\partial T)/\partial T = \partial H^\circ/\partial T = C^\circ_P$.

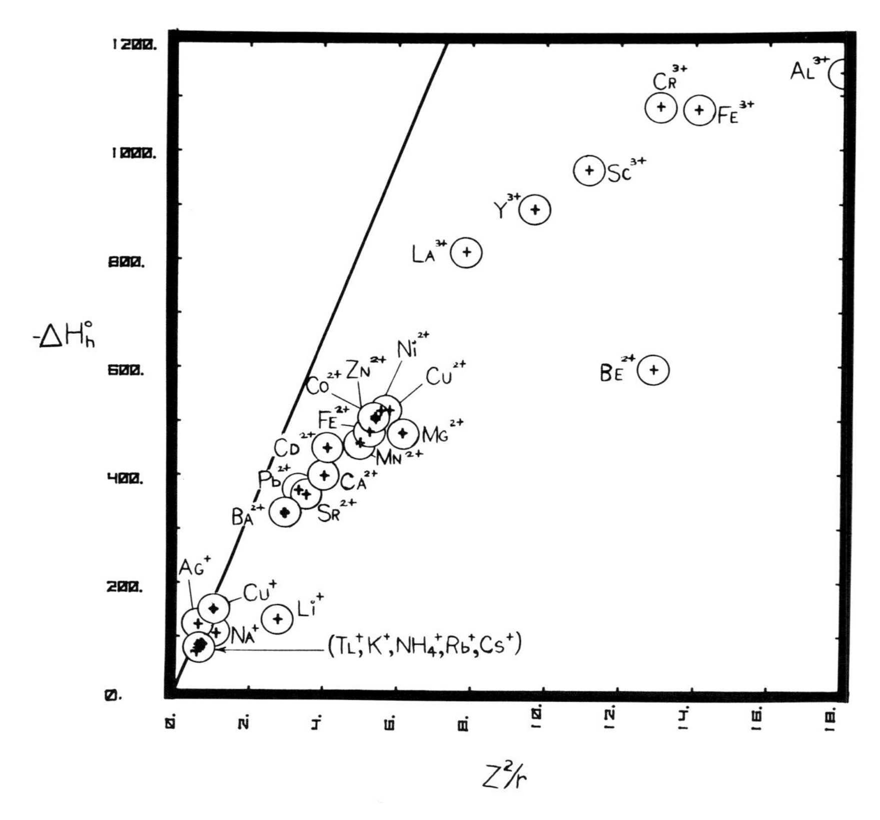


FIGURE 4.15. Values of the enthalpy of hydration for metals vs. the charge (Z) squared divided by the crystal radii (r).

$$S^\circ\ (\text{elect}) = (NZ^2e^2/2Dr)(\partial \ln D/\partial T)_P = -9/65\ Z^2/r \tag{22}$$

$$C_P^\circ\ (\text{elect}) = (NZ^2eT^2/2Dr)[(\partial^2 \ln D/\partial T^2)_P - (\partial \ln D/\partial T)_P^2]$$

$$= -12.96\ Z^2/r \tag{23}$$

The partial molal properties of ions in solution contain a minimum of two terms, an intrinsic contribution and an electrical contribution. For example, for the partial molal volume of an ion, we have

$$V^\circ\ (\text{ion}) = V^\circ\ (\text{int}) + V^\circ\ (\text{elect}) \tag{24}$$

where the intrinsic partial molal volume, $V^\circ$ (int), is equal to the size of the ion, $V^\circ$ (cryst) $= (4\pi\ N/3)r^3 = 2.52\ r^3$ (when r is expressed in Å units) plus

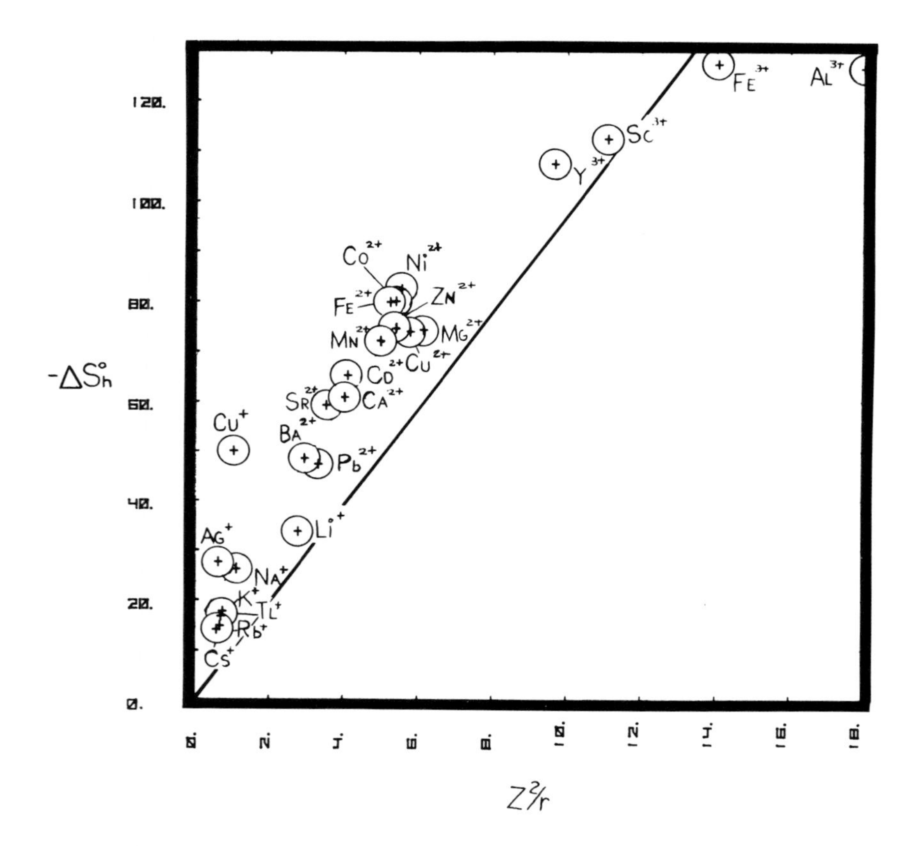


FIGURE 4.16. Values of the entropy of hydration for metals vs. the charge (Z) squared divided by the crystal radii (r).

the packing effects, and the electrostriction partial molal volume V° (elect) is the decrease in volume due to ion-water interactions. Thus, to plot the various partial molal properties vs. $Z^2/r$, one must estimate the intrinsic term. For V (int) one can use the semiempirical values from

$$V^\circ\ (\text{int}) = 4.48\ r^3 \tag{25}$$

while for S° (int) one can use

$$S^\circ\ (\text{int}) = 3/2\ \ln[\text{A.W.}] \tag{26}$$

where A.W. is the atomic weight. Plots of V° (elect) and S° (elect), calculated by using Equations 25 and 26, vs. $Z^2/r$ are shown in Figures 4.18 and 4.19. Although the general features of these figures agree with the Born model, a

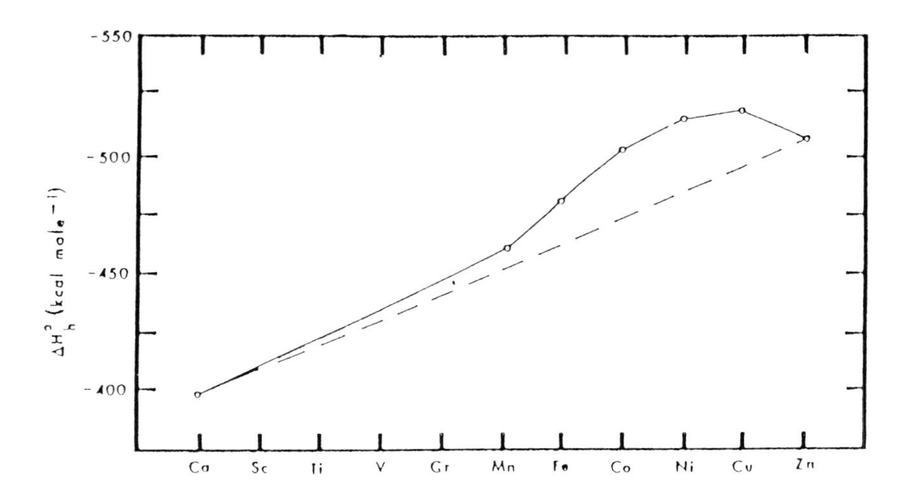


FIGURE 4.17. Values of the enthalpy of hydration for the transition metals.

look at the fine structure shows some discrepancies. For example, the V° (elect) of many divalent and trivalent ions appear to be nearly independent of r and the charge dependence of V° (elect) does not appear to be directly related to $Z^2$.

A number of workers have attempted to extend the Born model to account for these discrepancies by adjusting the size of the radii (e.g., Latimer et al. have added 0.85 Å to cations and 0.1 Å to anions to obtain linear plots) and adjusting the dielectric constant of the solvent as one approaches an ion (e.g., Laidler and Pegis have suggested the effective dielectric constant is 2 near an ion). These methods however, do not consider the structure of the water molecule. In recent years, workers have used structural hydration models to examine ion-water interactions (Figure 4.13). The structural models consider the ion-dipole interactions, ion-quadrupole interactions, and effects related to the structure of water. Although a full discussion of these models is beyond the scope of this chapter, they can be summarized as follows. By considering the interaction of an ion with a water dipole, one can account for the molecular structure of the water molecule, and by considering the interaction of an ion with a water quadrupole, one considers the difference in the thermodynamic properties of cations and anions of the same size. The water structure effects lead to a region between the orientated dipoles in the electrostricted region and bulk water; the water molecules are partially orientated by the ion and also affected by the bulk water structure. Many workers have divided ions into two classes: (1) structure-makers, which have a net effect of making

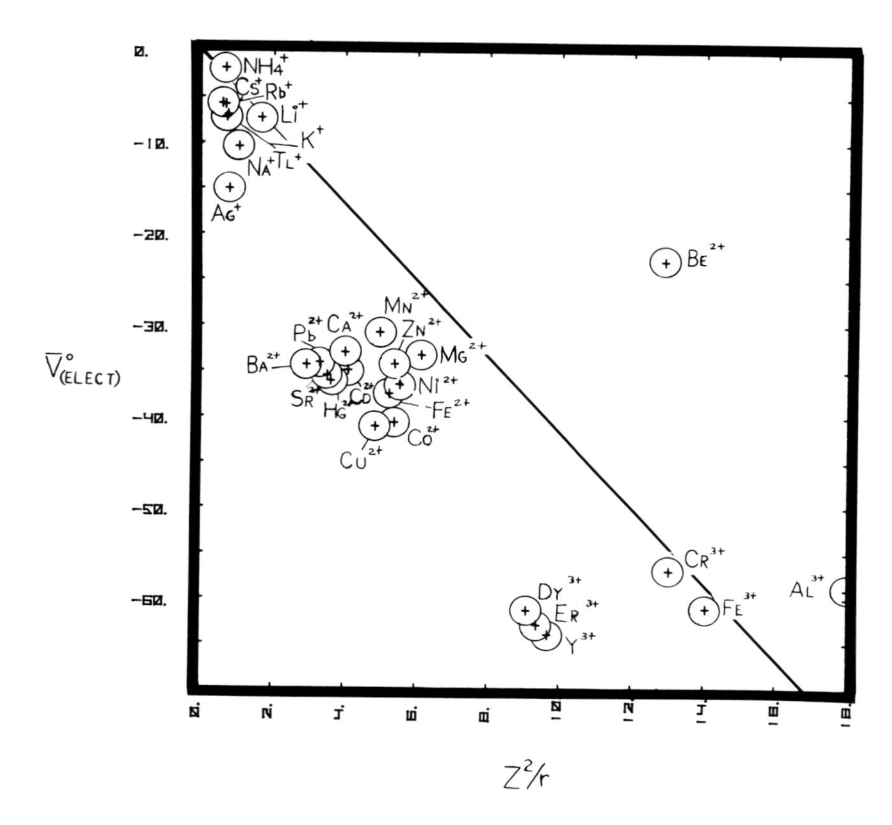


FIGURE 4.18. Values of the molar volume of electrostriction for metals vs. the charge (Z) squared divided by the crystal radii (r).

more structure around the ion, and (2) structure-breakers, which have a net effect of breaking down the structure of water. In general, the use of these terms is ambiguous as we know little about the structure being made or broken. By confining our arguments to hydration effects, V° (elect), it is possible to discuss ion-water interactions using Equation 24 or its equivalent for other thermodynamic properties.

If we use a hydration model for ion-water interactions, the V° (elect) can be related to the number of water molecules affected by the ion (i.e., the hydration number h)

$$V° \text{ (elect)} = V° \text{ (ion)} - V° \text{ (int)} = h(V_E^\circ - V_B^\circ) \tag{27}$$

where $V_E^\circ$ is the molal volume of water in the electrostricted region and $V_B^\circ$

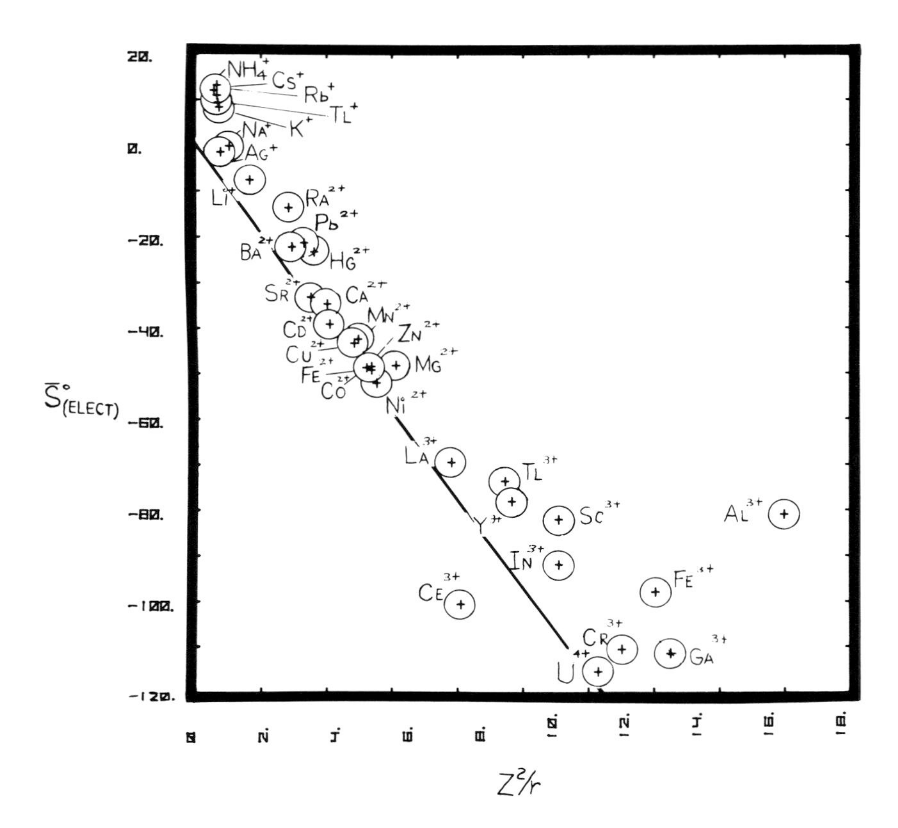


FIGURE 4.19. Values of the molar entropy of electrostriction for metals vs. the charge (Z) squared divided by the crystal radii (r).

is the molal volume of water in the bulk phase (18.0 $cm^3/mol$). Due to the difficulties of determining V° (int) it is not possible to unambiguously solve this equation. By differentiating Equation 24 with respect to pressure we have the partial molal compressibility

$$K^\circ \text{ (ion)} = K^\circ \text{ (int)} + K^\circ \text{ (elect)} \tag{28}$$

If we assume = K° (int) = 0, we can combine this equation with the differential of Equation 27 (assuming h and $V_E^\circ$ are not functions of pressure)

$$\begin{aligned} K^\circ \text{ (ion)} &= K^\circ \text{ (elect)} = -\partial V^\circ \text{ (elect)}/\partial P = h(\partial V_B^\circ/\partial P) \\ &= -hV_B^\circ \beta_B^\circ \end{aligned} \tag{29}$$

**TABLE 4.4**
**Hydration Numbers of Some Solutes at 25°C Determined from Compressibility Data**

| Ion | h | Ion | h |
|---|---|---|---|
| Cations | | Anions | |
| $Li^+$ | 2.8 | $F^-$ | 5.6 |
| $Na^+$ | 3.7 | $Cl^-$ | 2.0 |
| $K^+$ | 2.9 | $Br^-$ | 1.2 |
| $Rb^+$ | 2.9 | $I^-$ | 0.1 |
| $Cs^+$ | 2.5 | $OH^-$ | 6.4 |
| $Ag^+$ | 3.4 | $SO_4^{2-}$ | 8.6 |
| $NH_4^+$ | 0.4 | $CO_3^{2-}$ | 12.8 |
| $Mg^{2+}$ | 7.8 | | |
| $Zn^{2+}$ | 5.9 | | |
| $Cu^{2+}$ | 7.6 | Ion Pairs | |
| $Cd^{2+}$ | 8.5 | $MgSO_4^0$ | 15.3 |
| $Ca^{2+}$ | 6.5 | $MnSO_4^0$ | 14.2 |
| $Ba^{2+}$ | 9.2 | $LaSO_4^+$ | 17.3 |
| $La^{3+}$ | 14.7 | $LaFeCN_6^0$ | 18.3 |

where $\beta^\circ_B = -(1/V^\circ_B)(\partial V^\circ_B/\partial P)$ is the compressibility of bulk water ($45.25 \times 10^{-6}$ $bar^{-1}$ at 25°C). By rearrangement of Equation 29, we have for the hydration number

$$h = -K^\circ\,(\text{ion})/V^\circ_B\,\beta^\circ_B \tag{30}$$

Hydration numbers for some cations and anions calculated from Equation 30 are given in Table 4.4. By examining V° (elect) and K° (ion) = K° (elect) for various ions it is possible to calculate $(V^\circ_E - V^\circ_B)$. Combining equations, we have

$$V^\circ\,(\text{elect}) = -[(V^\circ_E - V^\circ_B)/V^\circ_B\,\beta^\circ_B]\,K^\circ\,(\text{elect}) = -\,k\,K^\circ\,(\text{elect}) \tag{31}$$

A plot of V° (elect) vs. K° (ion) is shown in Figure 4.20. From Figure 4.20, we obtain k = 4800 bars, which can be compared to

$$k = -(\partial \ln D/\partial P)/[(\partial^2 \ln D/\partial P) - (\partial \ln D/\partial P)^2] = 5000\text{ bars} \tag{32}$$

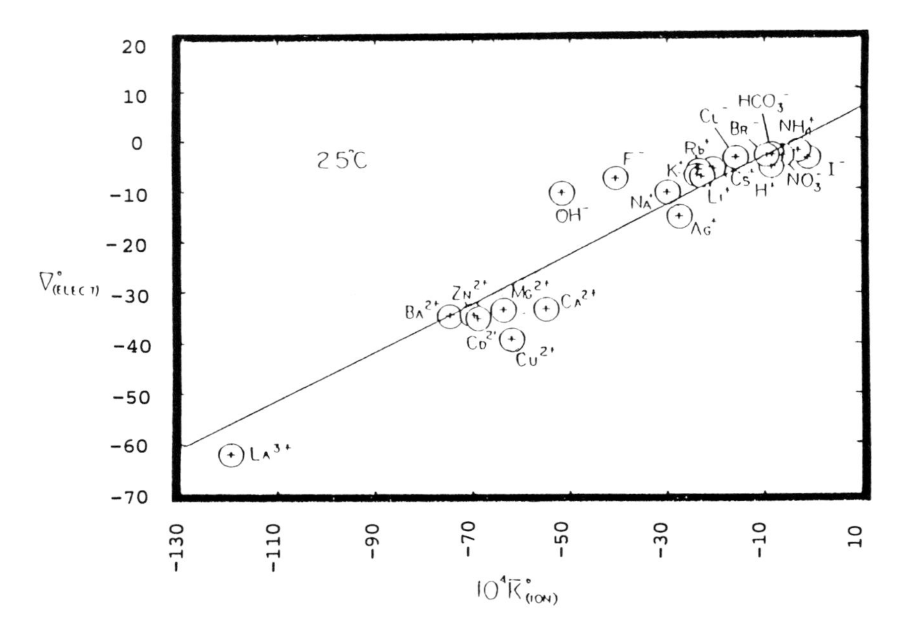


FIGURE 4.20. Correlation of the molar volumes and compressibilities of electrostriction for metals.

from the Born model. Using the value of k = 4800 bars, we obtain $(V°_E - V°_B) = -3.9\ cm^3\ mol^{-1}$. Combining this value with $V°_B = 18.0\ cm^3\ mol^{-1}$, we find $V°_E = 14.1\ cm^3\ mol^{-1}$ which is much larger than the crystal molal volume of water $V°\ (cryst) = 2.52 \times (1.38)^3 = 6.6\ cm^3\ mol^{-1}$ or the value corrected for packing effects $V°\ (int) = 4.48 \times (1.38)^3 = 11.8\ cm^3\ mol^{-1}$. Thus, the water molecules in the electrostricted region are not as tightly packed as one might expect; part of this difference may, however, be due to the water molecules in the so-called broken-down region. The solution properties of the other partial molal properties can also be treated by using the hydration model.

## 4. ION-ION INTERACTIONS

Now that we have a reasonable understanding of the structure of an ion in solution at infinite dilution, we can consider what happens as the concentration is increased. To obtain an understanding of these ion-ion interactions experimentally, the activity coefficient as well as its pressure $(V - V^0)$ and

DEBYE - HÜCKEL

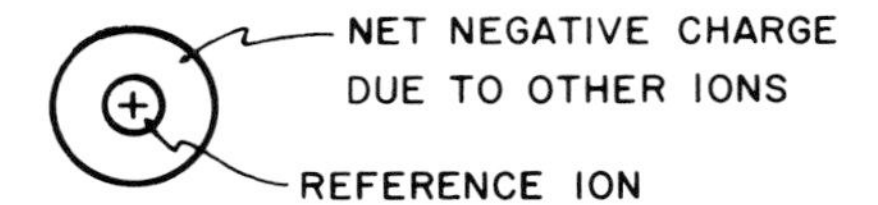


BJERRUM

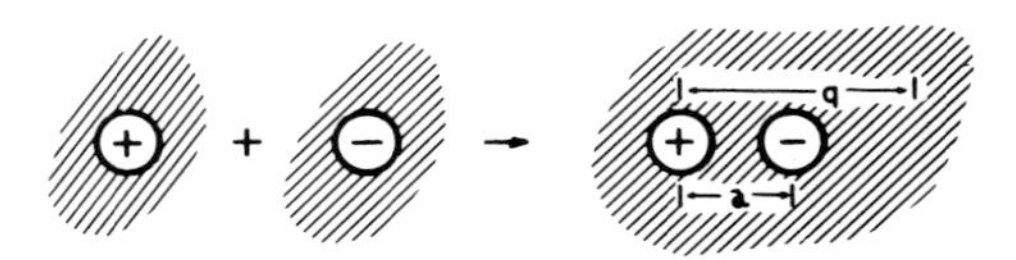


CLUSTER

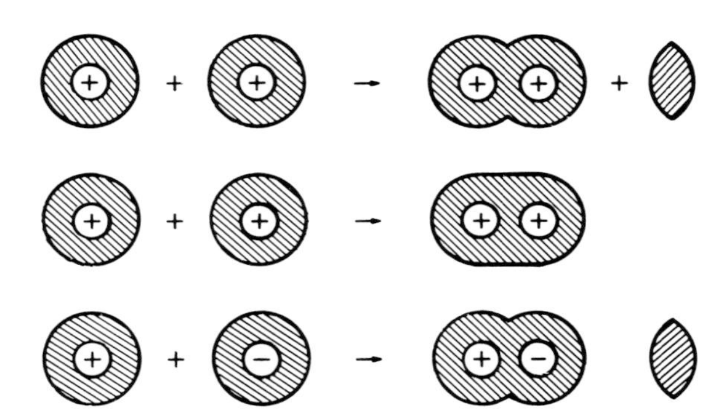

FIGURE 4.21. Models used to explain ion-ion interactions.

temperature (H − H$^0$) dependence are studied. By studying the thermodynamic properties of two ion systems ($M^+$, $X^-$), one can obtain an understanding of plus-minus interactions and by studying three ion systems ($M^+$, $N^+$, $X^-$ or $M^+$, $X^-$, $Y^-$) one can study plus-plus and minus-minus interactions. As in the case of ion-water interactions, it is useful to use models (Figure 4.21) to examine these interactions. A thorough discussion of the models used to treat ion-ion interactions is given elsewhere. Some of the models include the continuum models (the Debye-Huckel theory and Bjerrum ion pairing theory). These models assume that the nonideal behavior of an electrolyte is due entirely to electrical effects. The structural models attempt to account for hydration effects as well as specific interactions. The more recent cluster theories of Friedman make no attempt to separate the electrical and nonelectrical interactions (except in the limit). They also consider the importance of all the possible interactions in solution (plus-plus, plus-minus, and minus-minus).

The starting point for all discussions of ion-ion interactions is the Debye-Huckel theory. The theory predicts that the mean activity coefficient ($\gamma\pm$) of an electrolyte is given by

$$\ln \gamma \pm = -S_f I^{1/2}/(1 + A_f a I^{1/2}) \tag{33}$$

where $S_f$ and $A_f$ are constants related to absolute temperature (T) and the dielectric constant (D) of water (for a 1-1 electrolyte, $S_f = 0.5116$ and $A_f = 0.3292$ at 25°C), $I = {}^1/_2\Sigma\ \nu_i Z_i^2 m_i$ is the molal ionic strength ($\nu_i$ is the number, $Z_i$ is the charge, and $m_i$ is the molality of ionic species(i), and (a) is the ion size parameter in angstrom units. This equation serves as a limit in dilute solutions; however, it fails at the high ionic strength of seawater because of

1. Defects in some of the basic assumptions (e.g., treating ions as point charges in a continuous dielectric medium)
2. Deviations that occur due to noncoulombic effects such as hydration (the Debye-Huckel theory considers only electrical effects)

The classical method of examining the deviations from the Debye-Huckel theory in concentrated solutions is to use various extended forms involving one or more arbitrary constants. The difference between this form and the experimental data is attributed to noncoulombic effects. For example, Guggenheim used the equation

$$-\log \gamma \pm = 0.551\ Z_m Z_x I^{1/2}/(1 + I^{1/2}) + 2\nu\ B_{MX} m \tag{34}$$

where $\nu = 2\nu_M\nu_X/(\nu_M + \nu_X)$, I is the molal ionic strength, and m is the molality. By differentiating Equation 34 with respect to temperature and pressure, it is possible to examine the specific interaction model as a function of temperature and pressure.

The most popular method of treating the deviations from the Debye-Huckel theory in concentrated solutions is the ion pairing method of Bjerrum. This method assumes that short-range interactions can be represented by the formation of ion pairs

$$M^+ + A^- \rightarrow MA^\circ \tag{35}$$

A characteristic association constant is assigned to this formation

$$K_A = a_{MA}/a_M a_A = ([MA]/[M^+][A^-])\ \gamma_{MA}/\gamma_M \gamma_A \quad (36)$$

where $a_i$, [i], and $\gamma_i$ are, respectively, the activity, molal concentration, and activity coefficient of species i. There are four classes of ion pairs (Figure 4.22):

1. Complexes — When the ions are held in contact by covalent bonds
2. Contact ion pairs — When the ions are in contact and linked electrostatically (with no covalent bonding)
3. Solvent-shared ion pairs — Pairs of ions linked electrostatically, separated by a single water molecule
4. Solvent-separated ion pairs — Pairs of ions linked electrostatically but separated by more than one water molecule.

Bjerrum defined the distance between oppositely charged ions which can be classified as being associated by $q = Z_+Z_-e^2/2DkT$, where $Z_i$ is the charge on the ion i; e is the electrostatic charge; D is the dielectric constant; k is the Boltzmann constant; and T is the absolute temperature. In this treatment, two ions of opposite charge are considered to form an ion pair when they are between a°, the ion size parameter, and q. This can include ion pairs of classes 2, 3, and 4. The Bjerrum theory predicts greater ion pair formation, the higher the valencies and the smaller the dielectric constant of the solvent, which is in agreement with experimental results.

Many workers have criticized the theory because of the arbitrary cutoff distance. It has now been superseded by other theories. For example, the model of Fuoss considers only anions on the surface of a cation in volume, $V = 2.52a^3$, to be ion pairs. Fuoss obtained

$$K_A = (4\ \pi\ Na^3/3000)\ \exp(Z_+Z_-e^2/DakT) \quad (37)$$

where the first term is the excluded volume around the cation. Others have made further elaborations on these methods and discussed the shortcomings of the model.

More recent studies of electrolyte solutions have been made by using the cluster expansion method of Friedman. This method in simple terms considers all the interactions in a solution and makes no attempt to separate coulombic

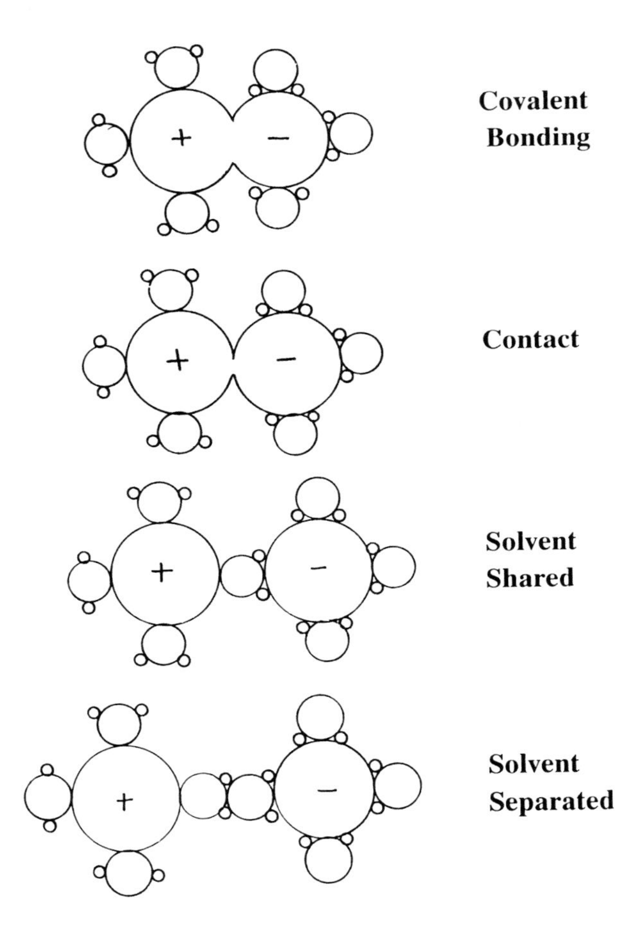


FIGURE 4.22. Types of ion pairs.

and noncoulombic terms. For example, for the major sea salts (NaCl + $MgSO_4$) there are a number of possible interactions to consider:

| Interactions | Possible Types |
|---|---|
| Plus-Plus | Na-Na, Mg-Mg, Na-Mg |
| Minus-Minus | Cl-Cl, $SO_4$-$SO_4$, Cl-$SO_4$ |
| Plus-Minus | Na-Cl, Mg-$SO_4$, Mg-Cl, Na-$SO_4$ |

These interactions can be represented by the following cross-square diagram:

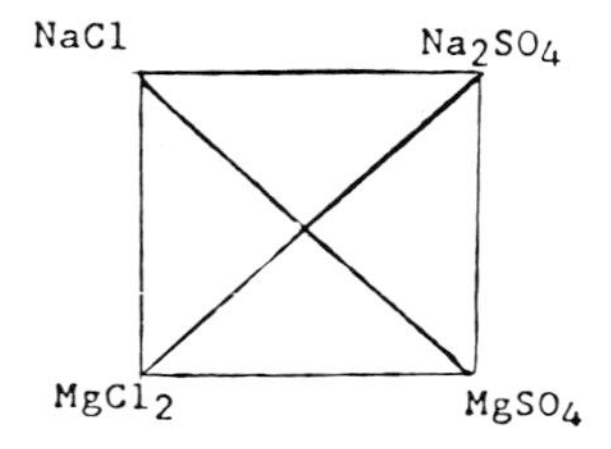


By studying the mixtures along the side of this diagram one can obtain some information about plus-plus and minus-minus interactions; by studying individual salts and the sum around the sides ($MgSO_4 = MgCl_2 + Na_2SO_4 - 2NaCl$) plus-minus interactions can be studied. The cross terms represent the mixtures (or simple seawater). Since the plus-plus and minus-minus terms are small, the total activity coefficients can be estimated from

$$\log \gamma \pm {}^{T}(MX) = \log \gamma \pm {}^{0}(MX) + \text{plus-plus terms} + \text{minus-minus terms} \quad (38)$$

where $\log \gamma \pm {}^{0}(MX)$ is the value for MX in itself at the ionic strength of the mixture and the other terms are related to interactions due to mixing. For example, for NaCl in seawater we have

$$\text{plus-plus} = (\text{Na-Mg}) + (\text{Na-K}) + (\text{Na-Ca}) + \ldots \quad (39)$$

$$\text{minus-minus} = (\text{Cl-SO}_4) + (\text{Cl-HCO}_3) + (\text{Cl-Br}) + \ldots \quad (40)$$

where the terms in parentheses are weighted according to the composition of the mixture. More will be said about this later.

Before we examine the use of the methods described in the last section to determine the activity of a metal ion in seawater

$$a_M = [M]_T \gamma_T(M) \quad (41)$$

we will consider some of the factors that control the state of an ion in seawater:

1. Eh
2. pH
3. Inorganic ligands
4. Organic ligands

The Eh of seawater may control the oxidation state of a metal ion. For a metal ion that can exist in two oxidation states we have

$$\mathrm{Ox} + \mathrm{ne^-} = \mathrm{Red} \tag{42}$$

where Ox is the oxidized form, Red is the reduced form, and n is the number of electrons ($e^-$) transferred. The equilibrium constant is given by

$$\log K = \log a_{Red} - \log a_{Ox} + n\ pE \tag{43}$$

where $a_i$ is the activity of i and pE $= -\log a_e$, the log of the activity of an electron. Rearranging this equation we have

$$pE = pE^0 + (1/n)\log a_{Ox}/a_{Red} \tag{44}$$

where $pE^0 = (1/n)\log K$. Since pE = Eh/(2.303RT/F), we have the more familiar form

$$Eh = Eh^0 + (2.303RT/nF)\log a_{Ox}/a_{Red} \tag{45}$$

where $Eh^0 = (2.303RT/nF)\log K = (0.0591/n)\log K$ at 25°C.

The upper theoretical limit of the pE or Eh of oxygenated water is controlled by the reaction

$${}^1/_2O_2(g) + 2H^+ + 2e = H_2O(aq) \tag{46}$$

Using log K = 41.6, log $aH_2O$ = −0.01, and pH = 8.1 at 25°C, the pE is given by

$$pE = (\log P + 50.7)/4 \tag{47}$$

at log P = −0.69, pE = 12.5, or Eh = 0.73 V. The experimentally measured values (about 0.5 to 0.6 V) for the Eh of open surface seawater are lower than this theoretical value. By using the reaction

$$O_2 + 2H^+ + 2e^- = H_2O_2 \tag{48}$$

one obtains a lower theoretical pE = 6.3 or Eh = 0.4 V (taking $[H_2O_2] = 10^{-7}$), which is closer to the experimentally determined values.

For anoxic conditions the negative pE or Eh are thought to be controlled by the reactions

$$SO_4^{2-} + 9H^+ + 8e^- = HS^- + 4H_2O \quad (49)$$

$$SO_4^{2-} + 8H^+ + 6e^- = S^0(s) + 4H_2O \quad (50)$$

Using log K = 34.0 and 36.6, respectively, for Reactions 49 and 50, the pE is given by

$$pE = (-\log [HS^-] - 41.4)/8 \quad (51)$$

$$pE = (-\log (S_4^0) - 30.4)/6 \quad (52)$$

at $(HS^-) = 10^{-3}$ to $10^{-6}$, pE = −4.8 to −4.4, or Eh = −0.28 to −0.26.

The pH and Eh environments that one encounters in marine waters are shown in Figure 4.23. The upper and lower limits are determined by the properties of water. The cross-textured and dotted bands represent the stability band for oxygen (4 to 260 *M*) and sulfide ($10^{-3}$ to $10^{-6}$ *M*) concentrations as given by Reactions 47 and 51. Most ocean waters have pH values between 7.6 and 8.3 and Eh values greater than 0.2 V. A sample calculation of the oxidation state of a metal in seawater is made for the iron system examined by Kester and Byrne. The two oxidation states of iron are related by

$$Fe^{3+} + e^- = Fe^{2+} \quad (53)$$

Using Equation 45 the oxidation state can be determined from

$$Eh = Eh^0 + (2.3RT/nF) \log [a_{FeIII}/a_{FeII}] \quad (54)$$

Using $Eh^0 = 0.771$ V at 25°C, we obtain $a_{Fe3+}/a_{Fe2+} = 2.43 \times 10^{-18}$ when Eh = −0.27 V. It should be pointed out that these ratios are at infinite dilution and do not include the effects of complex formation (which usually stabilizes a system against reduction). To determine the concentration ratios at higher ionic strengths, one must estimate the stoichiometric activity coefficients of the ions

$$[Fe^{3+}]_T/[Fe^{2+}]_T = [a_{Fe3+}/a_{Fe2+}] \; \gamma_{TFe2+}/\gamma_{TFe3+} \quad (55)$$

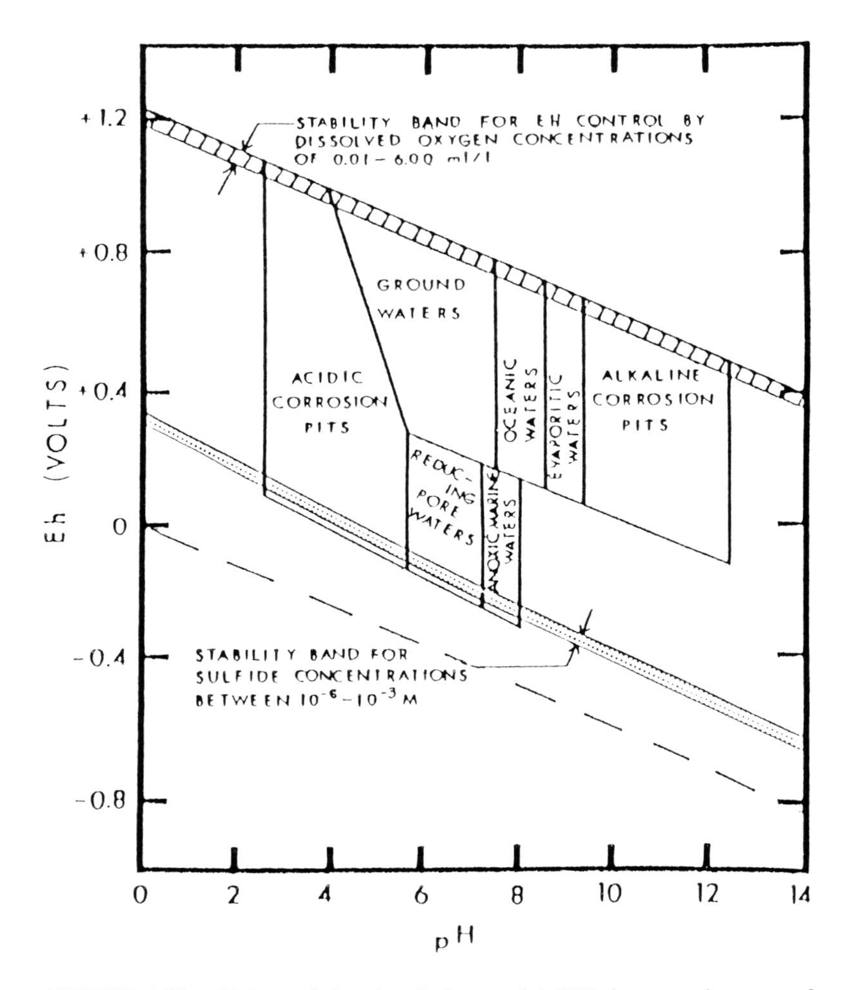


FIGURE 4.23. Values of the electrical potential (Eh) for natural waters of various pH.

Kester and Byrne have also examined the ion pairing of $Fe^{3+}$ and $Fe^{2+}$ in seawater which can be used to estimate the activity coefficient ratio. Although redox calculations are quite simple, some of the problems inherent in these calculations are as follows:

1. Equilibrium is not reached (i.e., the process is kinetically controlled).
2. Biological activity may change the oxidation state.
3. Photochemical processes may control the state.
4. Other important species may be neglected (e.g., organic complexes).
5. Unreliable analytical and thermodynamic data are available for the actual system.

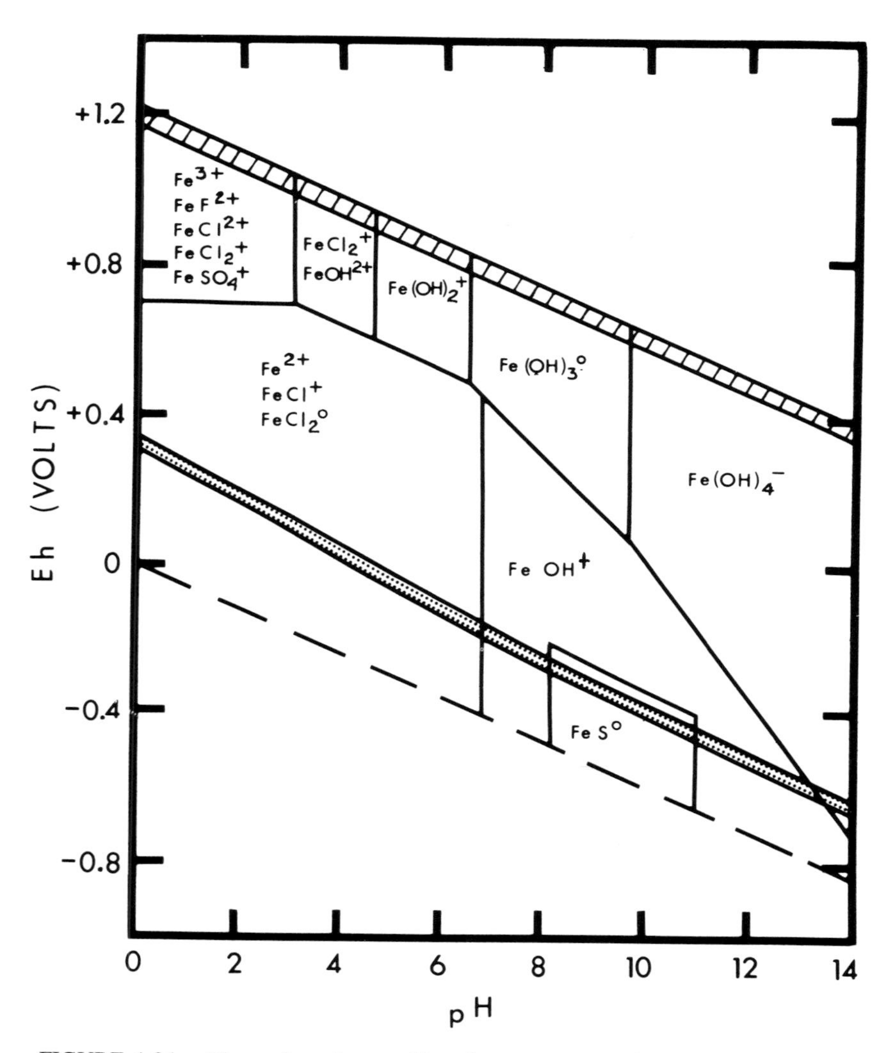


FIGURE 4.24. The various forms of iron in seawater as a function of $E_h$ and pH.

The pH can affect the metals directly by hydrolysis equilibrium

$$M^{2+} + H_2O = M(OH)^+ + H^+ \tag{56}$$

and by affecting the form of the ligands

$$HCO_3^- = H^+ + CO_3^{2-} \tag{57}$$

The effect of pH on the state of iron in marine waters has been examined by Kester and co-workers. These results are shown in Figure 4.24. The major form of iron at the pH and Eh of seawater is $Fe(OH)_3^0$. Since the $OH^-$ concentration is a function of pH, the relative forms of $Fe^{3+}$ are quite pH dependent. Near the pH limits of 7 and 10, there is a transition between $Fe(OH)_2^+ \rightarrow Fe(OH)_3^0$ and $Fe(OH)_3^0 \rightarrow Fe(OH)_4^-$. These results suggest that iron can alternate between a cationic and anionic species, which would strongly influence the ion-exchange characteristics of colloidal iron as well as its transport properties.

The inorganic ligands affecting metals include the major anionic components of seawater ($Cl^-$, $SO_4^{2-}$, $HCO_3^{2-}$, $Br^-$, $B(OH)_4^-$, and $F^-$) and some of the minor anionic components ($OH^-$, $H_2PO_4^-$, and $NO_3^-$ to mention a few). It is difficult to classify the types of organic ligands one must consider, since little is known about the composition of the organics in seawater. EDTA is frequently used as a model; however, the more important organic ligands may be made up of humic and fulvic acids and their derivatives. For the present, it is not possible to consider these organic ligands, although we realize they may be of great importance. It should be pointed out that all the model calculations made by using various organic ligands indicate that organic ligands have little effect on the speciation of most trace metals (i.e., compared to the major inorganic ligands).

We now examine the use of the ion pairing model and the specific interaction model to estimate the total activity coefficient of the major ions of seawater. Owing to the large amount of thermodynamic data available for the major sea salts it is possible to apply both methods. As we see later, for the minor or trace metals, we can use at present only the ion-pairing methods.

## 4.1. Ion-Pairing Model

The speciation forms that various metals can exist in are shown in Figure 4.25. Most models confine the speciation to the dissolved species. To examine the colloidal forms of metals, one must use filters of various sizes or use such techniques as dialysis to separate the various forms. Some of the methods used to determine individual species include:

1. Physical separation based on size
2. Equilibrium distributions using adsorption, ion exchange, redox equilibria, etc.
3. Potentiometric methods using ion selective electrodes
4. Electrode kinetics such as polarography
5. Detection of structure using optical methods
6. Bioassay techniques where metals may act as a catalyst or inhibitor

filterable; membrane filterable; dialysable; in true solution

| Free metal ions | Inorganic ion pairs; inorganic complexes | Organic complexes, chelates | Metal species bound to high molecular wt. org. material | Metal species in the form of highly dispersed colloids | Metal species sorbed on colloids | Precipitates organic particles remains of living organisms |
|---|---|---|---|---|---|---|
| Diameter range: | ← 10 Å | | 100 Å | | 1000 Å | |
| *Examples:* | | | | | | |
| $Cu^{2+}$ aq. | $Cu_2(OH)_2^{2+}$ | Me-SR | Me-lipids | FeOOH | $Me_x(OH)_y$ | |
| $Fe^{3+}$ aq. | $Pb(CO_3)^o$ | Me-OOCR | Me-humic-acid polymers | $Fe(OH)_3$ | $MeCO_3$, MeS etc. on clays, | |
| $Pb^{2+}$ aq. | $CuCO_3$<br>AgSH<br>$CdCl^+$<br>$CoOH^+$<br>$Zn(OH)_3^-$<br>$Ag_2S_3H_2^{2-}$ | 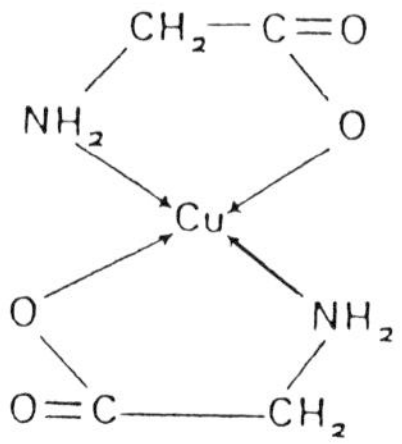 Cu chelate with $NH_2$–$CH_2$–C=O / O ligands (glycinate) | "lakes"<br>"Gelbstoffe"<br>Me-polysaccharides | Mn(IV) oxides<br>$Mn_7O_{13} \cdot 5H_2O$<br>$Na_4Mn_{14}O_{27}$<br>$Ag_2S$ | FeOOH or Mn(IV) on oxides | |

FIGURE 4.25. The various forms of elements in natural waters.

**TABLE 4.5**
**Division of Solutes into Hard and Soft Acids**

| Hard acids | Borderline | Soft acids |
|---|---|---|
| All A-metal cations plus $Cr^{3+}$, $Mn^{3+}$, $Fe^{3+}$, $Co^{3+}$, $Bi^{3+}$, $UO^{2+}$, $VO^{2+}$ | All bivalent transition metal cations plus $Zn^{2+}$, $Pb^{2+}$, $Bi^{3+}$ | All B-metal cations minus $Zn^{2+}$, $Pb^{2+}$ |
| As well as species like $BF_3$, $BCl_3$, $SO_3$, $RSO_2^+$, $RPO_2^+$, $CO_2$, $RCO^+$, $R_3C^+$ | $SO_2$, $NO^+$, $B(CH_3)_3$ | All metal atoms, bulk metals, $I_2$, $Br_2$, ICN, $I^+$, $Br^+$ |

The classification of metals made by Goldberg, which was discussed earlier, can be expanded using the Pearson division of solutes into hard and soft acids (Table 4.5).

When using the ion-pairing model, the activity of the free ion *i* is given up

$$a_i = [i]_F \gamma_F(i) \tag{58}$$

where $[i]_F$ is the molal concentration and $\gamma_F(i)$ is the activity coefficient of the free or uncomplexed ion *i*. This value of $\gamma_F$ is assumed to be only a function of the ionic strength and independent of the relative composition. Since the activity coefficient of the ion is also related to the total concentration and activity coefficient by

$$a_i = [i]_T \gamma_T(i) \tag{59}$$

the total or stoichiometric activity coefficient is given by

$$\gamma_T(i) = ([i]_F/[i]_T)\gamma_F(i) \tag{60}$$

This total activity coefficient is the one desired to obtain activities from total concentrations (i.e., using Equation 59). The term $\alpha_F = [i]_F/[i]_T$ is the fraction of free ions in a solution of fixed composition and ionic strength. If a series of one-to-one complexes are formed

$$M_i^+ + X_i^- \rightarrow M_iX_i \tag{61}$$

the ion-pairing constant for the formation of $M_iX_i^\circ$ is given by

$$K^*_{MX} = [M_iX_i^\circ]/[M_i^+][X_i^-] \quad (62)$$

$$K^*_{MX} = K_{MX}[\gamma_F^{(M)}\gamma_F^{(X)}/\gamma_F^{MX}] \quad (63)$$

where $K_{MX}$ is the thermodynamic constant in pure water; $K^*_{MX}$ is the stoichiometric constant, and $\gamma_F(i)$ are the activity coefficients of species *i*. The total concentration of $M_i^+$ and $X_i^-$ is given by

$$[M_i]_T = [M_i]_F + \Sigma[M_iX_i^\circ] \quad (64)$$

$$[X_i]_T = [X_i]_F + \Sigma[M_iX_i^\circ] \quad (65)$$

where $\Sigma[M_iX_i^\circ]$ is the sum of all the various ion pairs in the solution. By combining these equations with Equation 62 we have

$$\alpha_M = [M]_F/[M]_T = (1 + \Sigma\, K^*_{MX}[X_i]_F)^{-1} \quad (66)$$

$$\alpha_X = [X]_F/[X]_T = (1 + \Sigma\, K^*_{MX}[M_i]_F)^{-1} \quad (67)$$

These equations can be solved by a series of iterations if $K^*_{MX}$ is known. Several computer programs are available to aid in these iterations. The results, however, are dependent upon the quality of the values of $K^*_{MX}$ which are functions of ionic strength and, to a degree, the composition of the solutions. The fraction of a given ion pair can be obtained from

$$[MX_i]/[M]_T = K^*_{MX}[X_i]_F\, \alpha_M \quad (68)$$

$$[M_iX]/[X]_T = K^*_{MX}\,[M_i]_F\, \alpha_X \quad (69)$$

It should be pointed out that the form of the given complex does not affect the thermodynamic activity of M or X.

$$\gamma_T^{(M)} = \alpha_M\, \gamma_F^{(M)} \quad (70)$$

$$\gamma_T^{(X)} = \alpha_X\, \gamma_F^{(X)} \quad (71)$$

**TABLE 4.6**
**Ion-Pairing Constants for the Major Ion Pairs in Seawater at S = 35 and t = 25°C**

| Ion | $K^*_{HX}$ | $K^*_{NaX}$ | $K^*_{KX}$ | $K^*_{MgX}$ | $K^*_{CaX}$ | $K^*_{SrX}$ |
|---|---|---|---|---|---|---|
| $Cl^-$ | — | — | — | — | — | — |
| $SO_4^{2-}$ | 31.5 | 2.2 | 1.6 | 10.3 | 10.9 | 7.1 |
| $HCO_3^-$ | — | 0.28 | — | 2.1 | 2.2 | 2.2 |
| $Br^-$ | — | — | — | — | — | — |
| $CO_3^{2-}$ | — | 2.4 | — | 78 | 141 | 141 |
| $B(OH)_4^-$ | — | 0.7 | — | 9.6 | 13.0 | 7.1 |
| $F^-$ | 887 | 0.24 | — | 16 | 2.4 | 2.4 |
| $OH^-$ | — | 2.6 | — | 49 | 5.5 | 1.8 |

For the major ionic components of seawater, Millero and Schreiber have given ionic strength functions for $K^*_{MX}$ and $\gamma_F(i)$ for a number of ions. These equations can be used to calculate the speciation and activity coefficients of the major components of natural waters using a personal computer.

To understand how these equations are used, it is useful to do a sample calculation for the major components of seawater. The desired ion-pairing constants for $K^*_{MX}$ are given in Table 4.6. The appropriate equations are

$$\alpha_{SO4} = (1 + K^*_{NaSO_4}[Na]_F + K^*_{KSO4}[K]_F + K^*_{MgSO_4}[Mg]_F + K^*_{CaSO4}[Ca]_F + K^*_{SrSO_4}[Sr]_F)^{-1} \quad (72)$$

$$\alpha_{HCO_3} = (1 + K^*_{NaHCO_3}[Na]_F + K^*_{MgHCO_3}[Mg]_F + K^*_{CaHCO3}[Ca]_F + K^*_{SrHCO_3}[Sr]_F)^{-1} \quad (73)$$

$$\alpha_{CO_3} = (1 + K^*_{NaHCO_3}[Na]_F + K^*_{MgCO_3}[Mg]_F + K^*_{CaCO_3}[Ca]_F + K^*_{SrCO_3}[Sr]_F)^{-1} \quad (74)$$

$$\alpha_{B(OH)4} = (1 + K^*_{NaB}[Na]_F + K^*_{MgB}[Mg]_F + K_{CaB}[Ca]_F + K_{SrF}[Sr]_F)^{-1} \quad (75)$$

$$\alpha_{OH} = (1 + K^*_{NaOH}[Na]_F + K^*_{MgOH}[Mg]_F +$$

$$K^*_{CaOH}\,[Ca]_F + K^*_{SrOH}\,[Sr]_F)^{-1} \tag{76}$$

$$\alpha_F = (1 + K^*_{NaF}\,[Na]_F + K^*_{Mg}\,[Mg]_F + K^*_{CaF}\,[Ca]_F + K^*_{SrF}\,[Sr]_F)^{-1} \tag{77}$$

$$\alpha_H = (1 + K^*_{HSO4}\,[SO_4]_F + K^*_{HF}\,[F]_F)^{-1} \tag{78}$$

$$\alpha_{Na} = (1 + K^*_{NaSO4}\,[SO_4]_F + K^*_{NaHCO3}\,[HCO_3]_F + K^*_{NaCO3}\,[CO_3]_F + K^*_{NaB}\,[B(OH)_4]_F + K^*_{NaF}\,[F]_F + K^*_{NaOH}\,[OH]_F)^{-1} \tag{79}$$

$$\alpha_K = (1 + K^*_{KSO4}\,[SO_4]_F)^{-1} \tag{80}$$

$$\alpha_{Mg} = (1 + K^*_{MgSO4}\,[SO_4]_F + K^*_{MgHCO3}\,[HCO_3]_F + K^*_{MgCO3}\,[CO_3]_F + K^*_{MgB}\,[B(OH)_4]_F + K^*_{MgF}\,[F]_F + K^*_{MgOH}\,[OH]_F)^{-1} \tag{81}$$

$$\alpha_{Ca} = (1 + K^*_{CaSO4}\,[SO_4]_F + K^*_{CaHCO3}\,[HCO_3]_F + K^*_{CaCO3}\,[CO_3]_F + K^*_{CaB}\,[B(OH)_4]_F + K_{CaF}\,[F]_F + K^*_{CaOH}\,[OH]_F)^{-1} \tag{82}$$

$$\alpha_{Sr} = (1 + K^*_{SrSO4}\,[SO_4]_F + K^*_{SrHCO3}\,[HCO_3]_F + K^*_{SrCO3}\,[CO_3]_F + K^*_{SrB}\,[B(OH)_4]_F + K^*_{SrF}\,[F]_F + K^*_{SrOH}\,[OH]_F)^{-1} \tag{83}$$

To start the iteration one can assume that $\alpha_M = 1.0$ and calculate the values of $\alpha_X$. These values of $\alpha_X$ are used to make an estimate of $[X]_F = \alpha_X\,[X]_T$ and the values of $\alpha_M$ are estimated. The new $\alpha_M$ is used to determine $[M]_F = \alpha_M\,[M]_T$ and the process is repeated until a self-consistent set of $\alpha_i$ are found. A sample of the values obtained by various iterations is given in Table 4.7. It is clear from this table that after three iterations, a self-consistent set of $\alpha_i$ is obtained. These values can be used to determine the total activity

**TABLE 4.7**
**Calculations of the Fraction of Free Cations and Anions After Various Iterations**

| | Iteration | | | |
|---|---|---|---|---|
| **Ion** | **1st** | **2nd** | **3rd** | **4th** |
| $H^+$ | 1.0000 | 0.7331 | 0.7003 | 0.7003 |
| $Na^+$ | 1.0000 | 0.9764 | 0.9758 | 0.9757 |
| $K^+$ | 1.0000 | 0.9834 | 0.9827 | 0.9827 |
| $Mg^{2+}$ | 1.0000 | 0.8882 | 0.8875 | 0.8875 |
| $Ca^{2+}$ | 1.0000 | 0.8757 | 0.8750 | 0.8750 |
| $Sr^{2+}$ | 1.0000 | 0.9078 | 0.9075 | 0.9075 |
| $OH^-$ | 0.1998 | 0.2143 | 0.2143 | 0.2143 |
| $HCO_3^-$ | 0.7845 | 0.7964 | 0.7965 | 0.7966 |
| $F^-$ | 0.4955 | 0.5225 | 0.5226 | 0.5227 |
| $B(OH)_4^-$ | 0.4988 | 0.5208 | 0.5209 | 0.5209 |
| $SO_4^{2-}$ | 0.3614 | 0.3755 | 0.3756 | 0.3757 |
| $CO_3^{2-}$ | 0.1258 | 0.1378 | 0.1379 | 0.1380 |

coefficients of the various ions (Table 4.8). It is also possible to determine the fraction of the various forms of each metal and anion as shown in Figures 4.26 and 4.27.

The validity of these total activity coefficients can be demonstrated by calculating the dissociation constant for boric acid in seawater. The $K^*_{HB}$ is given by

$$K^*_{HB} = K_{HB}\ \gamma_{HB}/\gamma_H\gamma_B = 10^{-9.24} \times 1.09/(0.688 \times 0.351)$$

$$= 10^{-8.54} \qquad (84)$$

This calculated value can be compared to the measured value of 8.60 measured by Hansson. Similar calculations for other acids are shown in Table 4.9. The calculated values are in good agreement with the measured values and demonstrate the reliability of the model.

The speciation of minor components of seawater can be made using the concentrations of free cations and anions given in Table 4.8. The calculations can be made without making any iterations, since minor components do not

**TABLE 4.8**
**Values for $[i]_T$, $\alpha_i$, $[i]_F$, $\gamma_F^{(i)}$, and $\gamma_T^{(i)}$ for the Major Ionic Components of Seawater at S = 35 and t = 25°C**

| Ion | $[i]_T$ | $\alpha_i$ | $[i]_F$ | $\gamma_F^{(i)}$ | $\gamma_T^{(i)}$ |
|---|---|---|---|---|---|
| $H^+$ | — | 0.7263 | — | 0.967 | 0.688 |
| $Na^+$ | 0.48610 | 0.9763 | 0.47458 | 0.707 | 0.690 |
| $K^+$ | 0.01058 | 0.9801 | 0.01037 | 0.623 | 0.615 |
| $Mg^{2+}$ | 0.05474 | 0.8912 | 0.04878 | 0.285 | 0.255 |
| $Ca^{2+}$ | 0.01066 | 0.8844 | 0.00943 | 0.256 | 0.228 |
| $Sr^{2+}$ | 0.00009 | 0.9201 | 0.00008 | 0.218 | 0.231 |
| $Cl^-$ | 0.56577 | 1.0000 | 0.56577 | 0.623 | 0.623 |
| $OH^-$ | — | 0.2769 | — | 0.863 | 0.236 |
| $F^-$ | 0.00007 | 0.4884 | 0.00003 | 0.681 | 0.333 |
| $Br^-$ | 0.00087 | 1.0000 | — | — | — |
| $HCO_3^-$ | 0.00193 | 0.7924 | 0.00153 | 0.673 | 0.536 |
| $B(OH)_4^-$ | 0.00009 | 0.5321 | 0.00005 | 0.633 | 0.351 |
| $SO_4^{2-}$ | 0.02927 | 0.3747 | 0.01097 | 0.219 | 0.085 |
| $CO_3^{2-}$ | 0.00020 | 0.1604 | 0.00003 | 0.205 | 0.029 |
| $H_2PO_4^-$ | — | 0.7859 | — | 0.483 | 0.395 |
| $HPO_4^{2-}$ | — | 0.2968 | — | 0.157 | 0.050 |
| $PO_4^{3-}$ | — | 0.0016 | — | 0.0252 | 0.000041 |

affect the speciation of the major components. A sample calculation for the $PO_4^{3-}$ ion is shown below

$$\alpha_{PO_4} = (1 + K^*_{NaPO_4}[Na]_F + K^*_{MgPO_4}[Mg]_F + K^*_{CaPO_4}[Ca]_F)^{-1}$$

$$= (1 + 3.3 \times 0.4746 + 6{,}918 \times 0.0488 + 31{,}623 \times 0.0094)^{-1} = 0.0016 \quad (85)$$

Since the value of $\gamma_F = 0.0252$ for $PO_4^{3-}$, the value of $\gamma_T = 4 \times 10^{-5}$

For the calculation of the fraction of free trace metals in seawater, one needs reliable values of $K^*_{MX}$ for the metal with the major anions at an ionic strength of 0.7. Reliable values of $K^*_{MX}$ are not normally available. Whitfield and co-workers have made these calculations using the best available data Their results are given in Table 4.10. Since reliable values of $K^*_{MX}$ are available for $Cu^{2+}$ (Table 4.11), it is possible to demonstrate how the method is used.

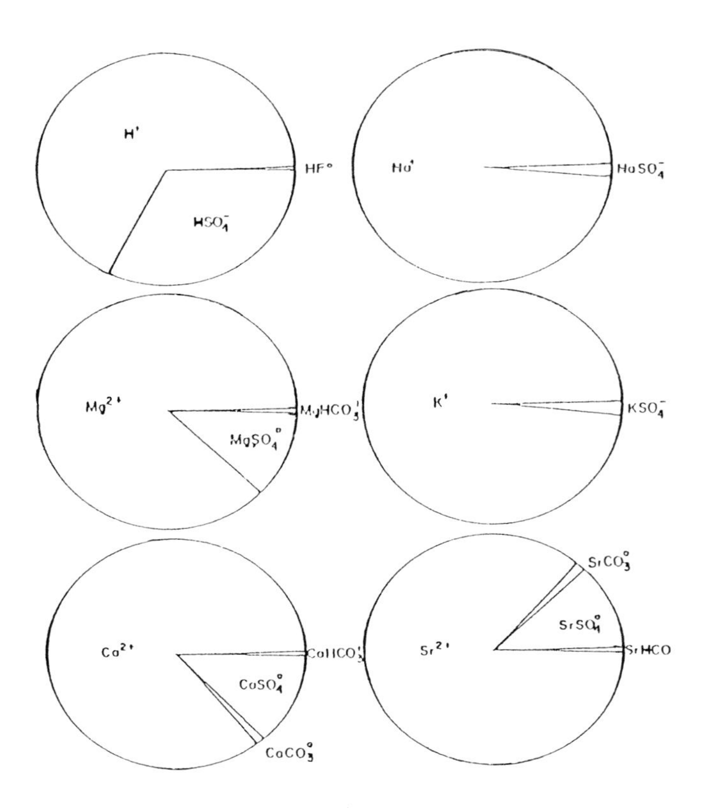

FIGURE 4.26. The speciation of cations in seawater.

The fraction of free $Cu^{2+}$ is given by

$$\alpha_{Cu} = (1 + K^*_{CuSO4}[SO_4]_F + K_{CuHCO3}[HCO_3]_F + K_{CuCO3}[CO_3]_F + K^*_{Cu(CO3)2}[CO_3]F^2 + K^*_{Cu(OH)}[OH]_F + K^*_{Cu(OH2)}[OH]^2_F)^{-1} \quad (86)$$

The various forms of $Cu^{2+}$ calculated from Equation 86 are given in Table 4.12 and shown in Figure 4.28. At a pH = 8.1 and 25°C, the dominant forms are $CuCO_3°$ (74%) and $Cu(CO_3)_2^{2-}$ (14%).

## 4.2. Specific Interaction Model

Although the ion-pairing model yields reliable estimates of activity coefficients in dilute solutions (below 1 m), it is not as reliable at higher ionic strengths. This is largely due to the fact that it does not consider the interactions of ions of the same sign (plus-plus and minus-minus interactions). Ionic

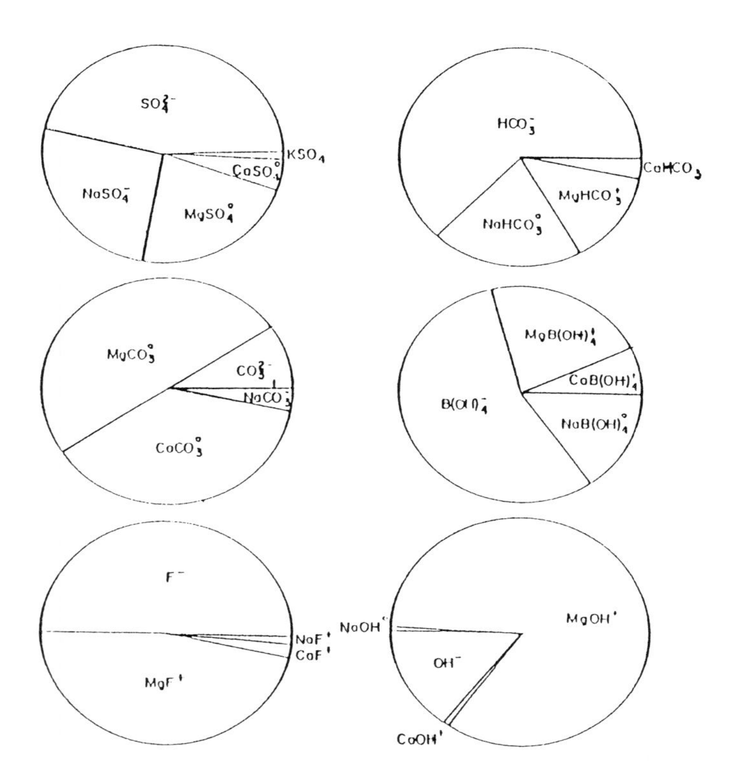


FIGURE 4.27. The speciation of anions in seawater.

**TABLE 4.9**
**Comparison of the Measured and Calculated $pK^*_{HA}$ for Acids in Seawater at 25°C**

| Acid | Calculated | $pK^*_{HA}$ Measured | Δ |
|---|---|---|---|
| $H_2O$ | 13.21 | 13.18 | 0.03 |
| $B(OH)_3$ | 8.59 | 8.60 | −0.01 |
| $H_2CO_3$ | 5.86 | 5.84 | 0.02 |
| $HCO_3^-$ | 8.96 | 8.93 | 0.03 |
| $NH_4^+$ | 9.345 | 9.351 | −0.006 |
| $H_3PO_4$ | 1.55 | 1.57 | −0.02 |
| $H_2PO_4^-$ | 6.14 | 5.94 | 0.20 |
| $HPO_4^{2-}$ | 9.11 | 8.93 | 0.18 |

**TABLE 4.10**
**The Fraction of Free Metals in Seawater and the Dominant Forms**

Calculated Speciation of Cations in Model Fresh Water at pH = 6

| Cation | Free | OH | F | Cl | $SO_4$ | $CO_3$ | Log α |
|---|---|---|---|---|---|---|---|
| $Ag^+$ | 72 | * | * | 28 | * | * | 0.15 |
| $Al^{3+}$ | * | 90 | 10 | — | — | * | 3.14 |
| $Au^+$ | * | — | — | 100 | — | — | 6.08 |
| $Au^{3+}$ [a] | * | 100 | — | * | — | — | 21.98 |
| $Ba^{2+}$ | 99 | * | * | * | 1 | * | 0.01 |
| $Be^{2+}$ | 15 | 57 | 28 | * | * | * | 0.82 |
| $Bi^{3+}$ | * | 100 | * | * | * | — | 9.08 |
| $Cd^{2+}$ | 96 | * | * | 2 | 2 | * | 0.02 |
| $Ce^{3+}$ | 72 | * | 3 | * | 22 | 3 | 0.14 |
| $Co^{2+}$ | 98 | * | * | * | 2 | * | 0.01 |
| $Cr^{3+}$ | * | 98 | * | * | 1 | — | 2.41 |
| $Ca^+$ | 100 | — | 0 | * | — | — | 0.00 |
| $Cu^+$ | 95 | 0 | 0 | 5 | 0 | 0 | 0.02 |
| $Cu^{2+}$ | 93 | 1 | * | * | 2 | 4 | 0.03 |
| $Dy^{3+}$ | 65 | 1 | 6 | * | 21 | 7 | 0.19 |
| $Er^{3+}$ | 63 | 1 | 7 | * | 19 | 10 | 0.20 |
| $Eu^{3+}$ | 71 | 1 | 3 | * | 21 | 4 | 0.15 |
| $Fe^{2+}$ | 99 | * | * | * | 1 | * | 0.01 |
| $Fe^{3+}$ | * | 100 | * | * | * | * | 6.41 |
| $Ga^{3+}$ | * | 100 | * | * | — | * | 7.80 |
| $Gd^{3+}$ | 63 | 1 | 5 | * | 22 | 9 | 0.20 |
| $Hf^{4+}$ | * | 100 | * | * | * | * | 13.28 |
| $Hg^{2+}$ | * | 8 | * | 92 | * | * | 6.88 |
| $Ho^{3+}$ | 65 | 1 | 7 | * | 19 | 8 | 0.19 |
| $In^{3+}$ | * | 100 | * | * | * | * | 5.53 |
| $La^{3+}$ | 73 | * | 1 | * | 25 | 1 | 0.14 |
| $Li^{3+}$ | 100 | * | — | 0 | * | 0 | 0.00 |
| $Lu^{3+}$ | 59 | 1 | 8 | * | 15 | 17 | 0.23 |
| $Mn^{2+}$ | 98 | * | * | * | 2 | * | 0.01 |
| $Nd^{3+}$ | 70 | 1 | 3 | * | 24 | 3 | 0.15 |
| $Ni^{2+}$ | 98 | * | * | * | 2 | * | 0.01 |
| $Pb^{2+}$ | 86 | 2 | * | 1 | 4 | 7 | 0.06 |
| $Pr^{3+}$ | 72 | 1 | 2 | * | 23 | 2 | 0.14 |
| $Rb^+$ | 100 | — | — | * | — | — | 0.00 |
| $Sc^{3+}$ | * | 43 | 41 | * | * | 15 | 2.80 |
| $Sm^{3+}$ | 68 | 1 | 3 | * | 25 | 4 | 0.17 |
| $Sn^{4+}$ [a] | — | 100 | — | — | — | — | 24.35 |
| $Tb^{3+}$ | 67 | 1 | 6 | * | 22 | 4 | 0.18 |
| $Th^{4+}$ | * | 100 | * | * | * | * | 7.94 |

**TABLE 4.10 (continued)**
**The Fraction of Free Metals in Seawater and the Dominant Forms**

| Cation | Free | OH | F | Cl | $SO_4$ | $CO_3$ | Log α |
|---|---|---|---|---|---|---|---|
| $TiO^{2+}$ [a] | * | 100 | — | — | * | — | 7.17 |
| $Tl^{+}$ | 100 | * | * | * | * | — | 0.00 |
| $Tl^{3+}$ | * | 100 | — | * | * | — | 14.62 |
| $Tm^{3+}$ | 66 | 1 | 8 | * | 20 | 6 | 0.18 |
| $U^{4+}$ | * | 100 | * | * | * | — | 14.02 |
| $UO_2^{2+}$ [a] | 12 | 18 | 8 | * | 1 | 60 | 0.91 |
| $Y^{3+}$ | 63 | 1 | 17 | * | 14 | 4 | 0.20 |
| $Yb^{3+}$ | 58 | 1 | 7 | * | 17 | 17 | 0.24 |
| $Zn^{2+}$ | 98 | * | * | * | 2 | * | 0.01 |
| $Zr^{4+}$ | * | 100 | * | * | * | — | 14.33 |

Calculated Speciation of Cations in Model Fresh Water at pH = 8.2

| Cation | Free | OH | F | Cl | $SO_4$ | $CO_3$ | Log α |
|---|---|---|---|---|---|---|---|
| $Ag^{+}$ | * | * | * | 100 | * | * | 5.26 |
| $Al^{3+}$ | * | 100 | * | — | — | * | 9.22 |
| $Au^{+}$ | * | — | — | 100 | — | — | 12.86 |
| $Au^{3+}$ [a] | * | 100 | * | 9 | 5 | * | 27.30 |
| $Ba^{2+}$ | 86 | * | * | 9 | 5 | * | 0.07 |
| $Be^{2+}$ | * | 99 | 2 | * | * | * | 2.74 |
| $Bi^{3+}$ | * | 100 | * | * | * | — | 14.79 |
| $Cd^{2+}$ | 3 | * | * | 97 | * | * | 1.57 |
| $Ce^{3+}$ | 21 | 5 | 1 | 12 | 10 | 51 | 0.68 |
| $Co^{2+}$ | 58 | 1 | * | 30 | 5 | 6 | 0.24 |
| $Cr^{3+}$ | * | 100 | * | * | * | — | 5.82 |
| $Ca^{+}$ | 93 | — | — | 7 | — | — | 0.03 |
| $Cu^{+}$ | * | — | — | 100 | — | — | 5.18 |
| $Cu^{2+}$ | 9 | 8 | * | 3 | 1 | 79 | 1.03 |
| $Dy^{3+}$ | 11 | 8 | 1 | 5 | 6 | 68 | 0.94 |
| $Er^{3+}$ | 8 | 12 | 1 | 4 | 4 | 70 | 1.08 |
| $Eu^{3+}$ | 18 | 13 | 1 | 10 | 9 | 50 | 0.74 |
| $Fe^{2+}$ | 69 | 2 | * | 20 | 4 | 5 | 0.16 |
| $Fe^{3+}$ | * | 100 | * | * | * | * | 11.98 |
| $Ga^{3+}$ | * | 100 | * | * | — | * | 15.35 |
| $Gd^{3+}$ | 9 | 5 | 1 | 4 | 6 | 74 | 1.02 |
| $Hf^{4+}$ | * | 100 | * | * | * | — | 22.77 |
| $Hg^{2+}$ | * | * | * | 100 | * | * | 14.24 |
| $Ho^{3+}$ | 10 | 8 | 1 | 5 | 5 | 70 | 0.99 |
| $In^{3+}$ | * | 100 | * | * | * | * | 11.48 |
| $La^{3+}$ | 38 | 5 | 1 | 18 | 16 | 22 | 0.42 |
| $Li^{3+}$ | 99 | * | — | — | 1 | — | 0.00 |

**TABLE 4.10 (continued)**
**The Fraction of Free Metals in Seawater and the Dominant Forms**

| Cation | Free | OH | F | Cl | $SO_4$ | $CO_3$ | Log α |
|---|---|---|---|---|---|---|---|
| $Lu^{3+}$ | 5 | 21 | 1 | 1 | 1 | 71 | 1.32 |
| $Mn^{2+}$ | 58 | * | * | 37 | 4 | 1 | 0.23 |
| $Nd^{3+}$ | 22 | 8 | 1 | 19 | 12 | 45 | 0.66 |
| $Ni^{2+}$ | 47 | 1 | * | 34 | 4 | 14 | 0.33 |
| $Pb^{2+}$ | 3 | 9 | * | 47 | 1 | 41 | 1.51 |
| $Pr^{3+}$ | 25 | 8 | 1 | 12 | 13 | 41 | 0.61 |
| $Rb^{+}$ | 95 | — | — | 5 | — | — | 0.02 |
| $Sc^{3+}$ | * | 100 | * | * | * | * | 7.41 |
| $Sm^{3+}$ | 18 | 10 | 1 | 8 | 11 | 52 | 0.75 |
| $Sn^{4+}$ [a] | * | 100 | — | — | — | — | 32.05 |
| $Tb^{3+}$ | 16 | 11 | 1 | 8 | 9 | 55 | 0.80 |
| $Th^{4+}$ | * | 100 | * | * | * | * | 0.80 |
| $TiO^{2+}$ [a] | * | 100 | — | — | * | — | 11.14 |
| $Tl^{+}$ | 53 | * | * | 45 | 2 | — | 0.28 |
| $Tl^{3+}$ | * | 100 | — | * | * | — | 20.49 |
| $Tm^{3+}$ | 11 | 21 | 1 | 5 | 6 | 55 | 0.94 |
| $U^{4+}$ | * | 100 | * | * | * | — | 23.65 |
| $UO_2^{2+}$ [a] | * | * | * | * | * | 100 | 6.83 |
| $Y^{3+}$ | 15 | 14 | 3 | 7 | 6 | 54 | 0.81 |
| $Yb^{3+}$ | 5 | 9 | 1 | 2 | 3 | 81 | 1.30 |
| $Zn^{2+}$ | 46 | 12 | * | 35 | 4 | 3 | 0.34 |
| $Zr^{4+}$ | * | 100 | * | * | * | — | 23.96 |

Calculated Speciation of Cations in Model Fresh Water at pH = 9

| Cation | Free | OH | F | Cl | $SO_4$ | $CO_3$ | Log α |
|---|---|---|---|---|---|---|---|
| $Ag^{+}$ | 65 | * | * | 25 | * | 9 | 0.18 |
| $Al^{3+}$ | * | 100 | * | — | — | * | 12.95 |
| $Au^{+}$ | * | — | — | 100 | — | — | 6.07 |
| $Au^{3+}$ [a] | * | 100 | — | * | — | — | 30.93 |
| $Ba^{2+}$ | 96 | * | * | * | 1 | 3 | 0.02 |
| $Be^{2+}$ | * | 100 | * | * | * | * | 4.47 |
| $Bi^{3+}$ | * | 100 | * | * | * | — | 18.01 |
| $Cd^{2+}$ | 47 | 4 | * | 1 | 1 | 47 | 0.33 |
| $Ce^{3+}$ | * | 5 | * | * | * | 95 | 2.37 |
| $Co^{2+}$ | 20 | 7 | * | * | * | 73 | 0.70 |
| $Cr^{3+}$ | * | 100 | * | * | * | — | 9.08 |
| $Ca^{+}$ | 100 | — | — | * | — | — | 0.00 |
| $Cu^{+}$ | 95 | — | — | 5 | — | — | 0.02 |
| $Cu^{2+}$ | * | 3 | * | * | * | 96 | 2.62 |

**TABLE 4.10 (continued)**
**The Fraction of Free Metals in Seawater and the Dominant Forms**

| Cation | Free | OH | F | Cl | $SO_4$ | $CO_3$ | Log α |
|---|---|---|---|---|---|---|---|
| $Dy^{3+}$ | * | 46 | * | * | * | 54 | 3.01 |
| $Er^{3+}$ | * | 78 | * | * | * | 22 | 3.54 |
| $Eu^{3+}$ | * | 18 | * | * | * | 82 | 2.48 |
| $Fe^{2+}$ | 27 | 8 | * | * | * | 65 | 0.57 |
| $Fe^{3+}$ | * | 100 | * | * | * | * | 14.99 |
| $Ga^{3+}$ | * | 100 | * | * | — | * | 19.31 |
| $Gd^{3+}$ | * | 14 | * | * | * | 86 | 2.92 |
| $Hf^{4+}$ | * | 100 | * | * | * | — | 27.61 |
| $Hg^{2+}$ | * | 100 | * | * | * | * | 11.79 |
| $Ho^{3+}$ | * | 48 | * | * | * | 52 | 3.08 |
| $In^{3+}$ | * | 100 | * | * | * | * | 14.57 |
| $La^{3+}$ | 2 | 10 | * | * | * | 88 | 1.78 |
| $Li^{+}$ | 100 | * | — | — | * | — | 0.00 |
| $Lu^{3+}$ | * | 92 | * | * | * | 8 | 4.21 |
| $Mn^{2+}$ | 62 | 1 | * | * | 1 | 35 | 0.20 |
| $Nd^{3+}$ | * | 9 | * | * | * | 90 | 2.33 |
| $Ni^{2+}$ | 9 | 2 | * | * | * | 90 | 1.07 |
| $Pb^{2+}$ | * | 5 | * | * | * | 95 | 2.73 |
| $Pr^{3+}$ | 1 | 9 | * | * | * | 90 | 2.23 |
| $Rb^{+}$ | 100 | — | — | * | — | — | 0.00 |
| $Sc^{3+}$ | * | 100 | * | * | * | * | 10.82 |
| $Sm^{3+}$ | * | 14 | * | * | * | 86 | 2.49 |
| $Sn^{4+}$ [a] | * | 100 | — | — | — | — | 36.27 |
| $Tb^{3+}$ | * | 32 | * | * | * | 68 | 2.67 |
| $Th^{4+}$ | * | 100 | * | * | * | * | 19.86 |
| $TiO^{2+}$ [a] | * | 100 | — | — | * | — | 13.15 |
| $Tl^{+}$ | 100 | * | * | * | * | — | 0.00 |
| $Tl^{3+}$ | * | 100 | — | * | * | — | 23.57 |
| $Tm^{3+}$ | * | 86 | 8 | * | * | 14 | 3.51 |
| $U^{4+}$ | * | 100 | * | * | * | — | 28.77 |
| $UO_2^{2+}$ [a] | * | * | * | * | * | 100 | 8.61 |
| $Y^{3+}$ | * | 14 | * | * | * | 85 | 2.57 |
| $Yb^{3+}$ | * | 62 | * | * | * | 38 | 3.59 |
| $Zn^{2+}$ | 6 | 78 | * | * | * | 16 | 1.20 |
| $Zr^{4+}$ | * | 100 | * | * | * | — | 28.81 |

*Note:* —Indicates ligand not considered. *: Indicates calculated abundance < 1%.

[a] Classified as fully hydrolyzed oxidation states.

**TABLE 4.11**
**Stability Constants of $Cu^{2+}$ Complexes in Seawater at 25°C**

| Complex | Log $K^{*}_{MX}$ |
|---|---|
| $CuOH^{+}$ | −8.14 |
| $Cu(OH)_2$ | −16.73 |
| $CuSO_4$ | 1.37 |
| $CuCO_3$ | 5.67 |
| $Cu(CO_3)_2^{2-}$ | 9.34 |
| $CuHCO_3^{+}$ | 1.06 |

**TABLE 4.12**
**Speciation of $Cu^{2+}$ in Seawater (pH = 8.1, S = 35, and t = 25°C)**

| Species | Fraction | % |
|---|---|---|
| $Cu^{2+}$ | 0.0390 | 3.9 |
| $CuOH^{+}$ | 0.0486 | 4.9 |
| $Cu(OH)_2$ | 0.0220 | 2.2 |
| $CuSO_4$ | 0.0101 | 1.0 |
| $CuCO_3$ | 0.7379 | 73.8 |
| $Cu(CO_3)_2^{2-}$ | 0.1417 | 14.2 |
| $CuHCO_3^{+}$ | 0.0006 | 0.1 |

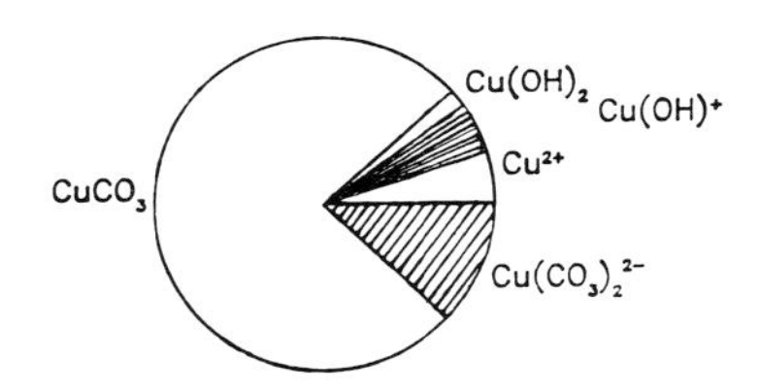


FIGURE 4.28. The various forms of $Cu^{2+}$ in seawater.

solution theory suggests that activity coefficients in electrolyte mixtures should be of the form

$$\ln \gamma = \text{D.H.} + \Sigma B_{ij} m_j + \Sigma C_{ijk} m_j m_k \tag{87}$$

where D.H. is a Debye-Huckel term, $B_{ij}$ is a term related to binary interactions of all the components (plus-minus, plus-plus, and minus-minus) and $C_{ijk}$ is related to ternary interactions of all the components (plus-minus-plus, plus-minus-minus, and minus-plus-plus). Pitzer has given some general equations that can be used to estimate the activity coefficient of ions that account for all the ionic interactions that occur in a mixed electrolyte solution. In their simplest form they are

$$\ln_{\gamma M} = Z_M^2 f_\gamma + 2\Sigma_a m_a (B_{Ma} + EC_{Ma}) + Z_M^2 \Sigma_c \Sigma_a\, m_c m_a B'_{ca} + Z_M \Sigma_c \Sigma_a\, m_c m_a C_{ca} \tag{88}$$

$$\ln_{\gamma X} = ZX^2 f^{\gamma} + 2\Sigma_c m_c (B_{cX} + EC_{cX}) + Z_X \Sigma_c \Sigma_a m_c m_a B'_{ca} +$$

$$Z_X \Sigma_c \Sigma_a m_c m_a C_{ca} \quad (89)$$

where $Z_i$ is the charge on the ion, a and c are anions and cations of the media, $m_i$ are the molalities, $E = {}^1/_2 \Sigma m_i Z_i$ is the equivalent molality, $f^{\gamma}$ is a Debye-Huckel term

$$f^{\gamma} = -0.392[I^{1/2}/(1 + 1.2I^{1/2}) + (2 + 1.(1 + 1.2I^{1/2})] \quad (90)$$

and the interaction terms $B_{MX}$, $B_{MX}{}^1$, and $C_{MX}$ are given for one-to-one electrolytes by

$$B_{MX} = \beta^{(0)}_{MX} + (\beta'_{MX}/2I)[1 - (1 + 2I^{1/2})\exp(-2I^{1/2}) \quad (91)$$

$$B'_{MX} = \beta^{(1)}_{MX}/2I^2)[-1 + (1 + 2I^{1/2} + {}^2I)\exp(-2I^{1/2}) \quad (92)$$

$$C_{MX} = C^{\phi}/(2|Z_M Z_X|^{1/2}) \quad (93)$$

Values of $\beta^0$, $\beta^1$, and $C\phi$ for various electrolytes are tabulated elsewhere (Pitzer, 1979).

Although the equations may appear to be quite complicated, they are easy to use if the calculations are made on a personal computer. These equations account for all the binary interactions. To account for triplet interactions, it is necessary to consider two more terms

$$\ln_{\gamma M} = \text{Equation 88} + \Sigma_c m_c (2\Theta m_c + \Sigma_a m_a \Psi m_{ca}) + \Sigma\Sigma m_a m_{a'} \Psi'_{aa} M \quad (94)$$

$$\ln_{\gamma X} = \text{Equation 89} + \Sigma_a m_a (2\Theta_{Xa} + \Sigma_c m_c \Psi'_{cc} X) + \Sigma\Sigma m'_{cc} \Psi'_{cc} X \quad (95)$$

The values of $\Theta_{ij}$ are related to the interaction of like charged ions (e.g., $Na^+ - K^+$) and the values of $\Psi_{ijk}$ are related to the triplet interactions (e.g., $Na^+ - K^+ - Cl^-$). If the values of $B_{MX}$, $B'_{MX}$, and $C_{MX}$ are calculated at a given ionic strength (as well as $f^{\gamma}$ and E), the calculation of $\gamma$ is straightforward. This can be demonstrated by giving the equations for M and X in a simple seawater media made up of $Na^+$, $Mg^{2+}$, $Cl^-$, and $SO_4^{2-}$. The equations are

$$\ln \gamma_M = Z_M^2 f^\gamma + 2m_{Cl}(B_{MCl} + EC_{MCl}) + 2m_{SO}(B_{MSO_4} + EC_{MSO_4}) + m_{Na}m_{Cl}(Z_M^2 B'_{NaCl} + Z_M C_{NaCl}) + m_{Mg}m_{Cl}(Z_M^2 B'_{MgCl} + Z_M C_{MgCl}) + m_{Na}m_{SO_4}(Z_M^2 B'_{NaSO_4} + Z_M C_{NaSO_4}) + m_{Mg}m_{SO_4}(Z_M^2 B^1_{MgSO_4} + Z_M C_{MgSO_4}) \quad (96)$$

$$\ln \gamma_X = Z_X^2 f^\gamma + 2m_{Na}(B_{NaX} + EC_{NaX}) + 2m_{Mg}(B_{MgX} + EC_{MgX}) + m_{Na}m_{Cl}(Z_M^2 B'_{NaCl} + Z_X C_{NaCl}) + m_{Mg}m_{Cl}(Z_X^2 B'_{MgCl} + Z_X C_{MgCl}) + m_{Na}m_{SO_4}(Z_X^2 B'_{NaSO_4}) + Z_X C_{NaSO_4}) + m_{Mg}m_{SO_4}(Z_X^2 B_{MgSO_4} + Z_X C_{MgSO_4}) \quad (97)$$

It first should be noticed that the double sum terms, $m_{Na}m_{Cl}$ etc., are the same in both equations and only functions of the media. It is thus possible to calculate and sum all the $B'$ and C terms into

$$Z_M^2 \Sigma\Sigma + Z_M \Sigma \quad (98)$$

where

$$\Sigma\Sigma = B'_{NaCl} + B'_{MgCl_2} + B'_{NaSO_4} + B'_{MgSO_4} \quad (99)$$

$$\Sigma = C_{NaCl} + C_{MgCl_2} + C_{NaSO_4} + C_{MgSO_4} \quad (100)$$

For seawater, $\Sigma = 0.03503$ and $\Sigma\Sigma = 0.00027$ which can be compared to $\Sigma = 0.03234$ and $\Sigma\Sigma = 0.00033$ in NaCl and $\Sigma = 0.03543$ and $\Sigma\Sigma = 0.00026$ in NaCl + $MgSO_4$. Since the values of $\Sigma$ and $\Sigma\Sigma$ do not vary too much, it is not a strong function of the media composition. If the Pitzer $B_{MX}$ and $C_{MX}$ are known for $Na^+$ and $Cl^-$ salts, it is possible to make a reliable estimate of $\gamma_T$ for many ions for a seawater media. If the appropriate terms are available for $Mg^+$ and $SO_4^{2-}$ salts, the estimates are more reliable. Some estimates for some simple ions are given in Table 4.13.

To make reliable estimates for some of the anions it is necessary to consider the strong interactions that occur between $Mg^{2+}$ and $Ca^{2+}$. For $H^+$

**TABLE 4.13**
**Estimated Activity Coefficients of Ions in Various Media**

| Ion | NaCl | NaCl + $MgSO_4$ | Seawater |
|---|---|---|---|
| $H^+$ | 0.779 | 0.739 | 0.546 |
| $Na^+$ | 0.664 | 0.668 | 0.667 |
| $K^+$ | 0.619 | 0.629 | 0.628 |
| $NH_4^+$ | 0.616 | 0.625 | 0.624 |
| $Mg^{2+}$ | 0.283 | 0.240 | 0.240 |
| $Ca^{2+}$ | 0.259 | 0.215 | 0.215 |
| $Sr^{2+}$ | 0.254 | 0.212 | 0.212 |
| $Ba^{2+}$ | 0.224 | 0.192 | — |
| $Mn^{2+}$ | 0.252 | 0.217 | — |
| $Fe^{2+}$ | 0.255 | 0.218 | — |
| $Co^{2+}$ | 0.257 | 0.220 | — |
| $Ni^{2+}$ | 0.266 | 0.225 | — |
| $Cu^{2+}$ | 0.223 | 0.193 | — |
| $Zn^{2+}$ | 0.235 | 0.206 | — |
| $F^-$ | 0.595 | 0.620 | 0.299 |
| $Cl^-$ | 0.664 | 0.668 | 0.667 |
| $Br^-$ | 0.688 | 0.694 | 0.692 |
| $OH^-$ | 0.670 | 0.672 | 0.216 |
| $HCO_3^-$ | 0.552 | 0.597 | 0.556 |
| $B(OH)_4^-$ | 0.513 | 0.559 | 0.418 |
| $CO_3^{2-}$ | 0.164 | 0.134 | 0.039 |
| $SO_4^{2-}$ | 0.131 | 0.115 | 0.113 |

it is also necessary to consider the strong interactions with $SO_4^{2-}$. The values for $H^+$, $F^-$, $OH^-$, $HCO_3^-$, $B(OH)_4^-$, and $CO_3^{2-}$ in seawater in Table 4.13 have been corrected for these effects. Presently, reliable Pitzer parameters and the necessary ion-pairing constants are available for all the major components of seawater at 250°C. More recent work has extended the parameters to 250°C for $Na^+$, $Mg^{2+}$, $Ca^{2+}$, $Cl^-$, and $SO_4^{2-}$. Future work will include reliable data for more ions over a wide range of temperatures.

The correction for ion pairs is simple to make. For example, for $H^+$ the desired $\gamma_T$ is given by

$$\gamma_T = \gamma_F (1 + \beta_{HSO_4} [SO_4])^{-1} \tag{101}$$

where $\gamma_F$ is the value for all the anions except $SO_4^{2-}$. Using $\beta_{HSO_4} = 12.0$ one obtains

$$\gamma_T = 0.739\,(1 + 12.0 \times 0.02947) = 0.546 \qquad (102)$$

The reliability of the Pitzer generated activity coefficients can be demonstrated by calculating the $K_2^*$ for carbonic acid in seawater. The value is given by

$$K_2^* = K_2\, \gamma_{HCO_3}/\gamma_H \gamma_{CO_3} = 10^{-10.33} \times 0.556/0.546 \times 0.039 = 10^{-8.91} \qquad (103)$$

This calculated value can be compared to the measured value of $pK_2^* = 8.93$ measured by Hansson.

The major advantage of the Pitzer equations is that they can be used over a wide range of compositions and ionic strengths without iterations. The equations can be used to estimate the free activity coefficients of minor cations and anions needed to use the ion-pairing model that must be considered for strong interactions (such as metal-$OH^-$ interactions).

## 5. PHYSICAL PROPERTIES OF SEAWATER

Ionic interactions in seawater can also affect the physical properties of ocean waters. Since the composition of natural waters can be quite different (Figures 4.29 and 4.30), it would be useful to have models that can be used to describe how the ionic components affect the physical properties of marine waters. In recent years, a great deal of progress has been made in interpreting and modeling the physical-chemical properties of mixed electrolyte solutions. These models have been used to estimate the properties of seawater using known properties for sea salts. This is done by using the apparent molal properties ($\phi$) of the solution (Figure 4.31). The apparent molal property is related to the change that occurs when a salt is added to water. The apparent molal property is defined by

$$\phi = \Delta\rho/n = (P\text{-}P^\circ)/n \qquad (104)$$

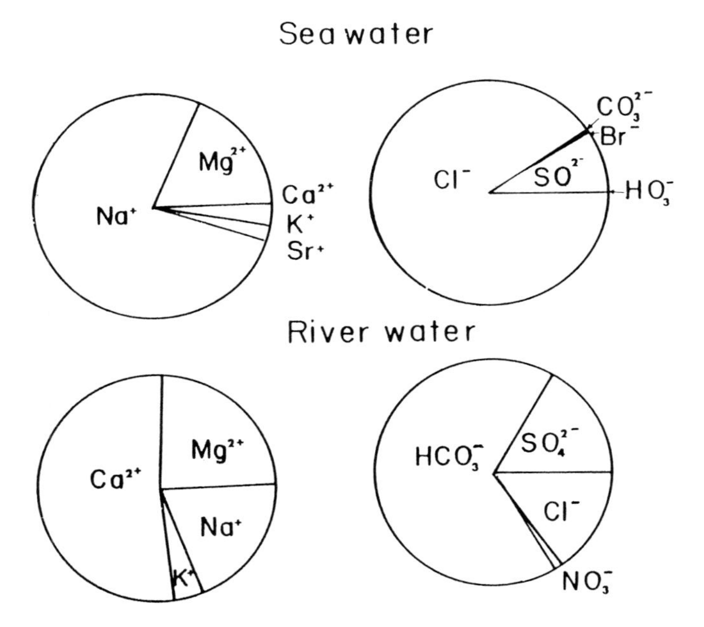


FIGURE 4.29. Comparison of the composition of the major components in river and sea waters.

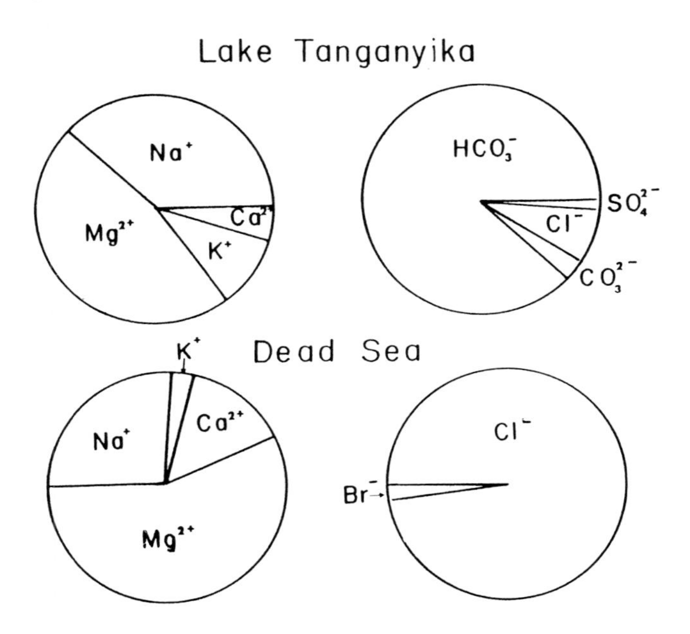


FIGURE 4.30. Comparison of the composition of Lake Tanganyika and Dead Sea waters.

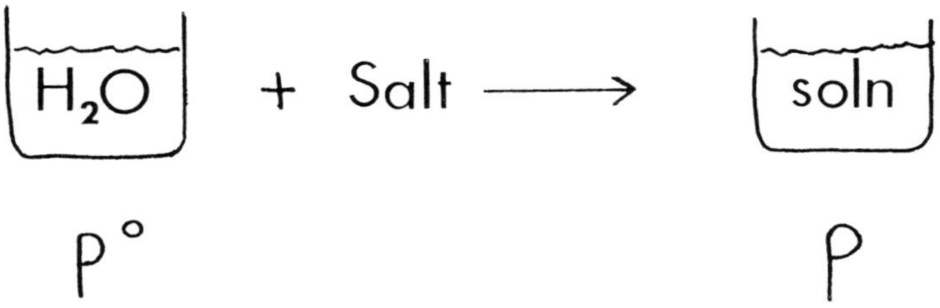

FIGURE 4.31. Formation of a solution from the addition of a salt to water.

where n is the number of moles of salt added, P is the property of the solution, and $P^\circ$ is the property of water. The major reason why the apparent molal property is useful is the fact that it is nearly additive for a mixture. The additivity, called Young's rule, is given by

$$\Phi = \Sigma N_i \phi_i \tag{105}$$

where $N_i = n_i/n_T$ is the mole fraction of electrolyte *i* in the mixture and $\phi_i$ is the molal property of *i* at the ionic strength of the mixture. In terms of the ionic components of a mixed electrolyte solution, the equation becomes

$$\Phi = \Sigma_M \Sigma_X E_M E_X \ \phi(MX) \tag{106}$$

where $E_M$ and $E_X$ are the equivalent fractions of cations (M) and anions (X), and $\phi(MX)$ is the apparent molal property of electrolyte MX at the ionic strength of the mixture. For the major components of seawater, this sum can be made three ways:

1. $\Phi(SW) = E_{Na}E_{Cl} \ \phi(NaCl) + E_{Mg}E_{SO4} \ \phi(MgSO_4)$

2. $\Phi(SW) = E_{Na}E_{SO4} \ \phi(Na_2SO_4) + E_{Mg}E_{Cl} \ \phi(MgCl_2)$

3. $\Phi(SW) = E_{Na}E_{Cl} \ \phi(NaCl) + E_{Na}E_{SO4} \ \phi(MgSO_4) +$

$$E_{Mg}E_{Cl} \ \phi(MgCl_2) + E_{Mg}E_{SO4} \ \phi(MgSO_4)$$

Experimentally, it is found that the third summation works best because it considers the weighted sum of all the possible cation-anion interactions. These

**TABLE 4.14**
**Comparison of the Measured and Calculated Densities (g $cm^{-3}$) of Seawater at 25°C (S = Salinity)**

| I | S | δ d, $10^6$ |
|---|---|---|
| 0.11 | 5 | −4 |
| 0.21 | 10 | −7 |
| 0.31 | 15 | −7 |
| 0.41 | 20 | −5 |
| 0.51 | 25 | 1 |
| 0.61 | 30 | 11 |
| 0.72 | 35 | 24 |

plus-minus interactions represent the major ionic interactions that occur in the mixture. Once the Φ for the mixture is estimated, a given physical property can be determined from

$$P = P^{\circ} + \Phi\, e_T \tag{107}$$

where $e_T$ is the total equivalents of the ionic components of the mixture.

Comparisons of the measured and calculated densities of seawater using this simple additivity method are shown in Table 4.14. The calculated values are in good agreement with the measured values. At higher ionic strengths, e.g., for brines, the estimates are not as reliable (Table 4.15). These larger errors at higher ionic strengths are related to excess mixing parameters. These excess mixing properties can be studied by mixing two electrolyte solutions at a constant ionic strength. For the mixing of the major sea salts, there are six possible mixtures that can be given by the diagram

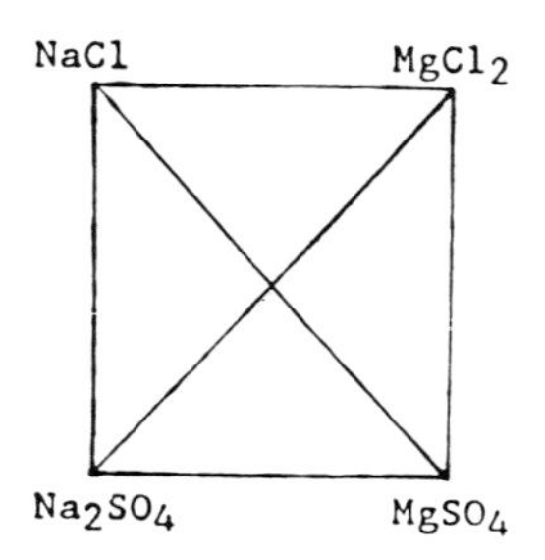

**TABLE 4.15**
**Comparison of the Measured and Calculated Relative Densities (g cm$^{-3}$) of Red Sea Brines Diluted with Water at 25°C**

| | Measured | |
|---|---|---|
| I | $(d - d_o)10^3$ | $\delta d, 10^6$ |
| 0.470 | 18.360 | −18 |
| 1.218 | 46.136 | 81 |
| 1.752 | 64.986 | 134 |
| 2.614 | 93.984 | 199 |
| 3.561 | 123.974 | 211 |
| 4.234 | 144.207 | 235 |
| 5.333 | 175.610 | 297 |
| 6.124 | 197.281 | 444 |

The mixing of the salts around the sides of this diagram have either a common cation or anion during mixing. A number of studies by Young and co-workers have shown the excess mixing properties $\Delta P_{EX}$ follow some simple rules. They are

1. The values of $\Delta P_{EX}$ in dilute solutions are not very large and can be assumed to be zero. This leads to the additivity of $\Phi$ or Young's first rule.
2. The values of $\Delta P_{EX}$ for mixtures with a common anion or cation are not strongly affected by the common ion. For example, the $\Delta P_{EX}$ for mixing NaCl and KCl is nearly the same as mixing NaBr and KBr.

$$\Delta P_{EX}(\text{NaCl-KCl}) = \Delta P_{EX}(\text{NaBr-KBr}) \tag{108}$$

Since this mixing process is largely related to cation-cation and anion-anion interactions, this means that as a first approximation, plus-plus and minus-minus interactions, are independent of the other ions in the solutions.

3. The third rule is called the cross square rule and is given by

$$\Sigma \square = \Sigma X \tag{109}$$

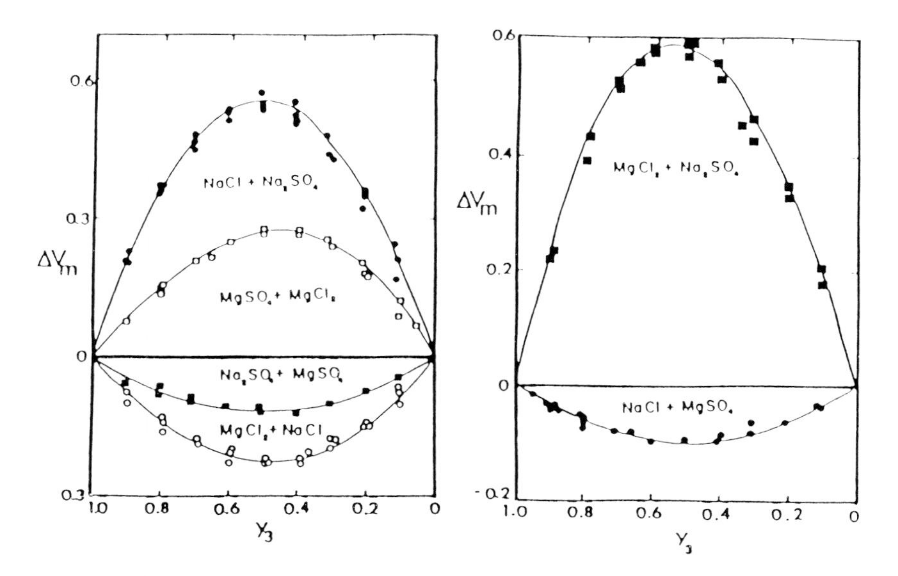


FIGURE 4.32. Values for the volumes of mixing for the major sea salts.

which states the sum of the excess mixing properties around the sides of the diagram given above is equal to the sum of the excess properties of the cross mixtures. For the major sea salts this gives

$$\Delta P_{EX}(NaCl + Na_2SO_4) + \Delta P_{EX}(MgCl_2 + MgSO_4) +$$

$$\Delta P_{EX}(NaCl + MgCl_2) + \Delta P_{EX}(MgSO_4 + Na_2SO_4) =$$

$$\Delta P_{EX}(NaCl + MgSO_4) + \Delta P_{EX}(Na_2SO_4 + MgCl_2) \qquad (110)$$

This simplifies the estimation of the $\Delta P_{EX}$ for a complicated mixture like seawater. The addition to Young's simple rule gives

$$\Phi = \Sigma_M \Sigma_X E_M E_X \, \phi(MX) + \Delta P_M/e_T \qquad (111)$$

where $\Delta P_M/e_T$ is given by

$$\Delta P_M/e_T = (e_T/4)\,[\Sigma\, E_M E_N E_X\, (Z_M + Z_X)(Z_N + Z_X)\Delta P(M,N)^X +$$

$$\Sigma E_X E_N E_M\, (Z_M + Z_X)(Z_M + Z_Y)\Delta P(X,Y)^M] \qquad (112)$$

where $Z_i$ is the absolute charge or ion *i*, $\Delta P(M,N)^X$ is the excess mixing properties for MX + NX, and $\Delta P(X,Y)^M$ is the excess mixing properties for NX + NY. This equation attempts to account for cation-cation and anion-anion interactions in the mixture by neglecting triplicate interaction (Young's third rule).

Since the values of $\Delta P_{EX}$ are symmetrical around the ionic strength fraction (Figure 4.32), the values of $\Delta P$ needed to estimate $\Delta P_M$ for the mixture can be the value near y = 0.5.

For a seawater solution, the value of $\Delta P_M/e_T$ is given by

$$\Delta P_M/e_T = (e_T/4)[E_{Na}E_{Mg}E_{Cl}(Z_{Na} + Z_{Cl})(Z_{Mg} + Z_{Cl}) \Delta P(NaCl + MgCl) + E_{Na}E_{Mg}E_{SO4}(Z_{Na} + Z_{SO4})(Z_{Mg} + Z_{SO4}) \Delta P(Na_2SO_4 + MgSO_4) + E_{Cl}E_{SO4}E_{Na}(Z_{Na} + Z_{Cl}) (Z_{Na} + Z_{SO4}) \Delta P(NaCl + Na_2SO_4) + (Z_{Mg} + Z_{Cl})(Z_{Mg} + Z_{SO4}) \Delta P(MgCl_2 + MgSO_4) \qquad (113)$$

The importance of using these mixing terms to estimate the physical properties of a mixture is demonstrated in Table 4.16. The estimated densities are in better agreement with the measured values when the excess mixing terms are considered ($\Delta V_{EX}$ is this case, the volume of mixing the major sea salts).

For most physical properties in dilute solutions, the estimates can be made without the $\Delta P_{EX}$ term. The apparent molal property, like the volume, can be considered to be made up of two major components: an ion-water interaction term and an ion-ion interaction term. The physical property of seawater is given by

$$P = P^\circ + \Sigma \text{ ion-water} + \Sigma \text{ ion-ion} \qquad (114)$$

where the term Σ ion-water is the weighted sum of the ion-water interactions of the major ions of the solution and Σ ion-ion is the weighted sum of the ion-ion interactions of the major ions of the solution. The first term is completely additive and is related to ion-water interactions in pure water. The second term is related to all the possible ion-ion interactions in the solutions.

**TABLE 4.16**
**Comparison of the Measured and Calculated Densities for Mixtures of NaCl + $MgSO_4$ (g $cm^{-3}$) at I = 3.0 and t = 25°C as a Function of the Ionic Strength Fraction**

| $y_{MgSO_4}$ | a | b |
|---|---|---|
| 0 | 0 | 0 |
| 0.1 | −113 | −12 |
| 0.2 | −214 | −34 |
| 0.3 | −286 | −50 |
| 0.4 | −340 | −79 |
| 0.5 | −367 | −99 |
| 0.6 | −366 | −115 |
| 0.7 | −344 | −123 |
| 0.8 | −273 | −107 |
| 0.9 | −161 | −71 |
| 1.0 | 0 | 0 |

*Note:* a: Without excess terms;
b: with excess terms.

These interactions can be estimated using the Pitzer approach and is related to

$$\Sigma \text{ ion-ion} = \text{D.H.} + \Sigma \text{ Binary} + \Sigma \text{ Ternary} \tag{115}$$

where the binary interactions are related to $\beta^0$, $\beta^1$, and $\Theta$ and the ternary interactions are related to $C^\phi$ and $\psi$. Thus, the use of Young's rule is embodied in the formulation of the Pitzer equations. The $\beta^0$, $\beta^1$, and $C^\phi$ terms are for all the binary components (NaCl, $Na_2SO_4$, $MgCl_2$, etc.) and the $\Theta$ and $\Psi$ terms are for all the ternary mixtures (NaCl + $Na_2SO_4$, NaCl + $MgCl_2$, etc.). This general approach, although somewhat complicated, can account for all the possible interactions in a stepwise manner.

## FURTHER READING


Franks, F., *Water,* Vols. 1—6, Plenum Press, New York, 1972.

Horne, R. A., *Marine Chemistry,* Wiley-Interscience, New York, 1969.

Kramer, C.J.M. and Duinker, J.C., *Complexation of Trace Metals in Natural Waters,* Martinius Nijhoff/W. Junk, The Hague (1984).

Morel, F.M.M., *Principles of Aquatic Chemistry,* Wiley and Sons, New York, 446 pp. (1983).

Millero, F.J., Influence of Pressure on Chemical Processes in the Sea, chap. 43, *Chemical Oceanography,* Vol. 8, 2nd ed., J.P. Riley and R. Chester, Eds., Academic Press, New York, 1-88 (1983).

Stumm, W., and Brauner, P.A., Chemical Speciation, chap. 3, *Chemical Oceanography,* Vol. 1, 2nd ed., J.P. Riley and G. Skirrow, Eds., Academic Press, New York, 173-239 (1975).

Stumm, W. and Morgan, J.J., *Aquatic Chemistry,* Wiley and Sons, (1970).

van den Berg, C.M.G., Electroanalytical Chemistry of Sea-Water, chap. 51, *Chemical Oceanography,* Vol. 9, 2nd ed., J.P. Riley, Ed., Academic Press, New York, 198-246 (1989).

Whitfield, M., Sea Water as an Electrolyte Solution, chap. 2, *Chemical Oceanography,* Vol. 1, 2nd ed., J.P. Riley and G. Skirrow, Eds., Academic Press, New York, 44-171 (1975).

Wong, C.S. et al., Eds., *Metals in Sea Water,* NATO Conf. Ser. IV., Plenum Press, New York (1983).

# 5 Table of Contents

**Dissolved Gases Other than $CO_2$** .............................. 227
1. *Composition of the Atmosphere* .............................. 228
2. *Dissolution of Gases in Seawater* .............................. 232
3. *Air-Sea Exchange* .............................. 234
4. *Nonreactive Gases* .............................. 242
5. *Dissolved Oxygen in Seawater* .............................. 247
6. *Other Nonconservative Gases* .............................. 260
7. *Structural Aspects of the Solubility of Gases* .............................. 263

*Further Reading* .............................. 266

## 1. COMPOSITION OF THE ATMOSPHERE

The atmosphere is made up of the major gases ($N_2$, $O_2$, and Ar) and the minor unreactive gases (Ne, He, Kr, and Xe). Water vapor is the most varied component of the atmosphere. Unstable minor gases (CO, $NO_2$, and $CH_4$) are produced by biological processes and man's activities. These gases will vary from place to place because of the different sources and sinks.

Dalton's law of partial pressures can be used to represent the composition of the atmosphere. This law simply states that the total partial pressure, $P_t$, of a mixture of gases in a fixed volume, V, is equal to the sum of the partial pressures of the components of the mixture. For the atmosphere this gives

$$P_t = \Sigma P_i = P_{N2} + P_{O2} + P_{Ar} + P_{H2O} \quad (1)$$

where the values of $P_i$ are the partial pressures of the major gaseous components i. If the gases are assumed to obey the ideal gas law, the partial pressure of each gas is given by

$$P_i = n_iRT/V \quad (2)$$

where $n_i$ is the number of moles of gas i (the number of molecules are equal to $n_iN$, where Avogadro's number $N = 6.024 \times 10^{23}$ molecules per mol), $R = 0.082057$ ($dm^3$ atm $mol^{-1}$ $K^{-1}$) and T is the absolute temperature ($T = t°C + 273.15$).

The composition of the gases in dry air, expressed as the mole fraction, is

$$X_i = n_i/n_T = P_i/P_t \quad (3)$$

where $n_T = \Sigma n_i$, the sum of the moles of gases in the atmosphere that are given in Table 5.1 (Kester, 1975). The errors (±) of the mole fractions reflect the constancy and precision of the composition. The mole fraction of the various gases are equal to the partial pressure ($P_i$) for ideal gases ($P_t = 1$ atm). The non-ideal gas behavior can be estimated from the van der Waals equation of state

$$(P_i + n_i^2 a/V^2)(V - n_ib) = n_iRT \quad (4)$$

**TABLE 5.1**
**Abundance and Properties of Atmospheric Gases[a]**

| Gas | Mole fraction in dry air coefficients ($X_i$) | Van der Waals at STP | | Molar volume ($dm^3\ mol^{-1}$) |
|---|---|---|---|---|
| | | a | b | |
| $N_2$ | $0.78084 \pm 0.00004$ | 1.390 | 0.03913 | 22.391 |
| $O_2$ | $0.20946 \pm 0.00002$ | 1.360 | 0.03183 | 22.385 |
| Ar | $(9.34 \pm 0.01) \times 10^{-3}$ | 1.345 | 0.03219 | 22.386 |
| $CO_2$ | $(3.3 \pm 0.1) \times 10^{-4}$ | 3.592 | 0.04267 | 22.296 |
| Ne | $(1.818 \pm 0.004) \times 10^{-5}$ | 0.2107 | 0.01709 | 22.421 |
| He | $(5.24 \pm 0.004) \times 10^{-6}$ | 0.03412 | 0.02370 | 22.436 |
| Kr | $(1.14 \pm 0.01) \times 10^{-6}$ | 2.318 | 0.03978 | 22.350 |
| Xe | $(8.7 \pm 0.1) \times 10^{-8}$ | 4.194 | 0.05105 | 22.277 |

[a] Kester (1975).

where the a-term is related to intermolecular attraction and b is related to the finite volume and compressibility of the gases. The coefficients a and b for Equation 4 are given in Table 5.1 (Kester, 1975). The molar volume of the gases at 0°C and 1 atm (standard temperature and pressure, STP) calculated from Equation 4 are also given in Table 5.1. The values deviated from ideal gas behavior (22.414 $dm^3\ mol^{-1}$) by +0.1% for He to −0.6% for Xe. For accurate calculations, Equation 4 should be used. More exact equations of state for these gases are available; however, near STP, Equation 4 is adequate to ± 0.05%.

The mole fractions of the gases given in Table 5.1 do not vary geographically or with altitude (to 95 km). The water vapor, however, does vary significantly. These variations are accounted for by making corrections for the humidity (%) of the air at a given temperature. The partial pressure of water vapor is given by

$$P_{H2O} = (h/100)\ P_O \tag{5}$$

when $P_O$ is the vapor pressure of dry water (atm) at a given temperature

$$\ln P_O = -0.63536311 \times 10^4\ T^{-1} + 0.3404926034 \times 10^2 - 0.19509874 \times 10^{-1}\ T + 0.12811805 \times 10^{-4}\ T^2 \tag{6}$$

**TABLE 5.2**
**The Composition of Minor Gases in the Atmosphere**

| Species | $X_i$ actual | Reliability | Source | Sink |
|---|---|---|---|---|
| $CH_4$ | $1.6 \times 10^{-6}$ | High | Biog. | PhotoChem. |
| CO | $0.5\text{-}2 \times 10^{-7}$ | Fair | Photo., Anthr. | PhotoChem. |
| $O_3$ | $10^{-7}\text{-}10^{-8}$ | Fair | Photo. | PhotoChem. |
| $NO + NO_2$ | $10^{-8}\text{-}10^{-12}$ | Low | Lightn., Anthr., Photo. | PhotoChem. |
| $HNO_3$ | $10^{-9}\text{-}10^{-11}$ | Low | Photo. | Rainout |
| $NH_3$ | $10^{-9}\text{-}10^{-10}$ | Low | Biog. | Photo., rainout |
| $N_2O$ | $3 \times 10^{-7}$ | High | Biog. | Photo. |
| $H_2$ | $5 \times 10^{-7}$ | High | Biog., Photo | Photo. |
| OH | $10^{-15}\text{-}10^{-12}$ | Very low | Photo. | Photo. |
| $HO_2$ | $10^{-11}\text{-}10^{-13}$ | Very low | Photo. | Photo. |
| $H_2O_2$ | $10^{-10}\text{-}10^{-18}$ | Very low | Photo. | Rainout |
| $H_2CO$ | $10^{-10}\text{-}10^{-9}$ | Low | Photo. | Photo. |
| $SO_2$ | $10^{-11}\text{-}10^{-10}$ | Fair | Anthr., Photo., Volcanic | Photo. |
| $CS_2$ | $10^{-11}\text{-}10^{-10}$ | Low | Anthr., Biol., | Photo. |
| OCS | $10^{-10}$ | Fair | Anthr., Biol., Photo. | Photo. |
| $CH_3CCl_3$ | $0.7^{-2} \times 10^{-10}$ | Fair | Anthr. | Photo. |

At 15°C and h = 80% the vapor pressure of water is 0.01342 atm. Thus, the water will contribute 1.3% to the total pressure (similar to Ar). The partial pressure of the other gases can be converted to dry air using

$$P_i = [P_t - (h/100)\,P_{H_2O}]\,X_i \tag{7}$$

The composition of some minor gases in the atmosphere are given in Table 5.2. The source of these gases is the result of biological, industrial, and photochemical processes. Earlier work on these gases is quite sparse. More recent studies have been concerned with these minor gases. The concentration of these gases in the atmosphere will be variable due to the different industrial and oceanic sources.

The stable isotopic composition of atmospheric gases is summarized in Table 5.3. (Kester, 1975). The values of the mole percent were calculated using $5 \times 10^{21}$g or $1.71 \times 10^{20}$ moles for the total atmosphere.

**TABLE 5.3**
**Isotopic Abundance of Atmospheric Gases[a]**

| Element | Mass number | Mol % |
|---|---|---|
| H (in $H_2O$) | 1 | 99.98 |
| H (in $H_2O$) | 2 | 0.02 |
| He | 3 | $1.1 \times 10^{-4}$ |
| He | 4 | 100.0 |
| C (in $CO_2$) | 12 | 98.9 |
| C (in $CO_2$) | 13 | 1.1 |
| C (in $CO_2$) | 14 | $9.5 \times 10^{-13}$ |
| N | 14 | 99.62 |
| N | 15 | 0.38 |
| O | 16 | 99.757 |
| O | 17 | 0.039 |
| O | 18 | 0.204 |
| Ne | 20 | 90.92 |
| Ne | 21 | 0.257 |
| Ne | 22 | 8.82 |
| Ar | 36 | 0.337 |
| Ar | 38 | 0.063 |
| Ar | 40 | 99.600 |
| Kr | 78 | 0.354 |
| Kr | 80 | 2.27 |
| Kr | 82 | 11.56 |
| Kr | 83 | 11.55 |
| Kr | 84 | 56.90 |
| Kr | 86 | 17.37 |
| Xe | 124 | 0.096 |
| Xe | 126 | 0.090 |
| Xe | 128 | 1.919 |
| Xe | 129 | 26.44 |
| Xe | 130 | 4.08 |
| Xe | 131 | 21.18 |
| Xe | 132 | 26.89 |
| Xe | 134 | 10.44 |
| Xe | 136 | 8.87 |

[a] Kester (1975).

## 2. DISSOLUTION OF GASES IN SEAWATER

The concentration of a gas in solution is related to the partial pressure by Henry's law

$$P_i = k_i\,[i] \tag{8}$$

where [i] is the concentration of the dissolved gas (expressed in mol $kg^{-1}$ of solution) and $k_i$ is the Henry's law constant. The parameter $k_i$ will depend upon the particular gas, the salinity or ionic strength of the solution, the temperature and total pressure. Equilibria is obtained when the partial pressure of the gas in solution is equal to the value in the gas phase

$$P_i\,(\text{soln}) = P_i\,(\text{gas}) \tag{9}$$

The concentration of a gas at equilibrium is given by

$$[i] = P_i\,(\text{gas})/k_i \tag{10}$$

The concentration of a gas in solution can be expressed by a variety of scales. Most physical chemists express the concentration as molality (mol/kg $H_2O$) or mole fraction.These scales are useful since the magnitude is not affected by the composition of the gas phase and is independent of temperature. Much of the earlier solubility of gases in seawater were expressed in terms of the Bunsen coefficient which is the value of [i] expressed in $cm^3$ of a gas at STP per $cm^3$ of solution at the temperature in question when the $P_i$ = 1.0 atm. Since the volume of each gas is different at STP, this has led to confusion. For practical reasons the most convenient scale to use is mol $kg^{-1}$ of seawater when $P_i$ is the partial pressure and the total pressure is one standard atmosphere (1.013 bar). The correction for $P_i$ for different total pressures (in atm) and % humidities (h) from standard values is given by

$$P_i' = P_i\,(P_t - P_S\,h/100)/(1 - P_S) \tag{11}$$

where $P_i'$ is the corrected partial pressure of the gas and $P_S$ is the vapor pressure of water in seawater at a given t(°C) and salinity(S).

$$P_S = P_O + AS + BS^{3/2} + CS^2 \tag{12}$$

where

$$A = 0.70249 + 2.3938 \times 10^{-3}t - 3.7170 \times 10^{-6}t^2 \quad (13)$$

$$B = -2.1601 \times 10^{-2} + 4.8460 \times 10^{-6}t - 1.0492 \times 10^{-6}t^2 \quad (14)$$

$$C = 2.7984 \times 10^{-3} + 1.5520 \times 10^{-5}t - 2.7048 \times 10^{-8}t^2 \quad (15)$$

The appropriate h and $P_S$ can be obtained from the conditions of the surface waters. Care must be taken, however, since solubility is slow to respond to changes in $P_t$ and the h near the water-air interface may be different than observed on a ship (5 m above the interface). For a water parcel that has left the surface, it is reasonable to assume $P_t = 1$ atm and h = 100%.

In recent years more accurate data has become available for the solubility of gases in seawater. This more accurate data allows one to examine the departure of waters from equilibria. The supersaturation of $N_2$ and Ar, for example, can be related to entrapped bubbles. The super or undersaturation of $O_2$ can be related to photosynthesis and respiration.

The solubilities of gases (C is in mol $kg^{-1}$) in seawater have been fit by Weiss (Scripps) to equations of the form

$$\ln C = B_1 + B_2S \quad (16)$$

which conforms to the Setchenow salting out equation and

$$\ln C = A_1 + A_2/T + A_3 \ln T + A_4 T \quad (17)$$

the integrated van't Hoff equation. The final form used by Weiss (Scripps) is

$$\ln C = A_1 + A_2(100/T) + A_3 \ln(T/100) + A_4(T/100) + S[B_1 + B_2(T/100) + B_3(T/100)^2] \quad (18)$$

The coefficients $A_i$ are related to the solubility of the gas in water at various temperatures, while the coefficients $B_i$ are related to the solubility of the gas in seawater. It should be noted that the effect of temperature on the salting coefficient ($B_2$) is not of the proper van't Hoff form (Equation 17). The $B_2$ and $B_3$ terms should be functions of l/T and lnT. The values of $A_i$

**TABLE 5.4**
**Solubility of Gases in Seawater with the Constants for Equation 18 in mol kg$^{-1}$ Relative to Air at 1 atm at 100% Relative Humidity[a]**

| Gas | $A_1$ | $A_2$ | $A_3$ | $A_4$ |
|---|---|---|---|---|
| $N_2$ | −173.2221 | 254.6078 | 146.3611 | −22.0933 |
| $O_2$ | −173.9894 | 255.5907 | 146.4813 | −22.2040 |
| Ar | −174.3732 | 251.8139 | 145.2337 | −22.2046 |
| Ne | −166.8040 | 255.1946 | 140.8863 | −22.6290 |
| He | −163.4207 | 216.3442 | 139.2032 | −22.6202 |
| | $B_1$ | $B_2$ | $B_3$ | |
| $N_2$ | −0.054052 | 0.027266 | −0.0038430 | |
| $O_2$ | −0.037362 | 0.016504 | −0.0020564 | |
| Ar | −0.038729 | 0.017171 | −0.0021281 | |
| Ne | −0.127113 | 0.079277 | −0.0129095 | |
| He | −0.44781 | 0.023541 | −0.0034266 | |

[a] Kester (1975).

and $B_i$ for various gases when [i] is in units of mol kg$^{-1}$ are given in Table 5.4. Values of [i] expressed in these units for S = 35 seawater are given in Table 5.5. It is important to point out that the use of Equation 18 smooths out the variability of gas solubilities with respect to temperature and salinity (Kester, 1975). More recently the solubilities for $O_2$ have been replaced by the new measurements of Benson and Krause (Swarthmore).

$$\ln C = -135.29996 + 1.572288 \times 10^{-5}/T - 6.637149 \times 10^{7}/T^{2} + 1.243678 \times 10^{10}/T^{3} - 8.621061 \times 10^{11}/T^{4} - S(0.020573 - 12.142/T + 2363.1/T^{2}) \quad (19)$$

## 3. AIR-SEA EXCHANGE

Most of the gases in seawater originate from three sources: (1) the earth's atmosphere, (2) volcanic activity beneath the sea, and (3) chemical processes occurring in the sea (biological — photosynthesis and decomposition of organic matter, physical — radioactive decay). One of the most useful as-

**TABLE 5.5**
**Solubilities of $N_2$, $O_2$, Ar, Ne, and He in Seawater (S = 35 Equilibrated with the Atmosphere, $P_t$ = 1 atm at 100% Humidity)[a]**

| | mol kg$^{-1}$ | | | nmol kg$^{-1}$ | | | |
|---|---|---|---|---|---|---|---|
| °C | $N_2$ | $O_2$ | Ar | Ne | He | Kr | Xe |
| 0 | 616.4 | 349.5 | 16.98 | 7.88 | 1.77 | 4.1 | 0.66 |
| 5 | 549.6 | 308.1 | 15.01 | 7.55 | 1.73 | 3.6 | 0.56 |
| 10 | 495.6 | 274.8 | 13.42 | 7.26 | 1.70 | 3.1 | 0.46 |
| 15 | 451.3 | 247.7 | 12.11 | 7.00 | 1.68 | 2.7 | 0.39 |
| 20 | 414.4 | 225.2 | 11.03 | 6.77 | 1.66 | 2.4 | 0.33 |
| 25 | 383.4 | 206.3 | 10.11 | 6.56 | 1.65 | 2.2 | 0.29 |
| 30 | 356.8 | 190.3 | 9.33 | 6.36 | 1.64 | 2.0 | 0.25 |

[a] Kester (1975).

sumptions concerning the solubility of gases is that at one time every parcel water was at the sea surface and at that time became equilibrated (or nearly so) with the atmospheric gases. In the course of oceanic circulation, the unreactive gases are distributed throughout the water column by advection and diffusion.

The exchange of gases across the air-sea interface has been examined by a number of workers using various models. The simplest model, which is a stagnant film model, is shown in Figure 5.1.* The model considers three regions: (1) a turbulent atmospheric phase in which the partial pressure of each gas is uniform; (2) a turbulent liquid phase which has a uniform partial pressure; and (3) a laminar layer that separates the two turbulent regions. The motion of the liquid in the laminar layer is parallel to the air-sea interface. It is assumed that the gas is transfered through the laminar layer by molecular diffusion and that the layer or film represents the major resistance to the gas transfer. The laminar or thin film is regarded as permanent and has a thickness of $\tau$. The flux of gas across the interface can be examined using Fick's first law

$$dC_i/dt = D_i\ [dC_i/dz] \tag{20}$$

* Figures 5.1, 5.4, 5.5, 5.17, 5.18 from Kester, 1975; Figures 5.2, 5.6, 5.7, 5.10, 5.12 to 5.16 from Broecker and Peng, 1982.

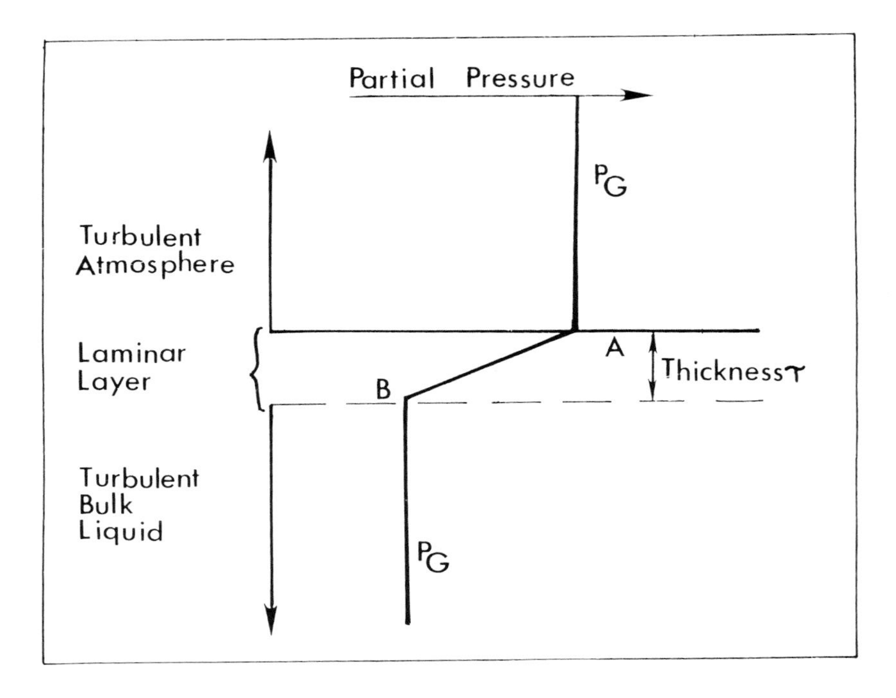


FIGURE 5.1. The laminar layer for the transport of gases across the air-sea interface.

where $C_i$ is the concentration of species *i*, t is the time, $D_i$ is the diffusion coefficient or diffusivity of species *i* and dC/dz is the gradient (z is the vertical distance). If this equation is combined with Henry's law ($P_i = k_iC_i$) we have

$$dC_i/dt = (A\ D_i/\tau\ k_i)\ [P_i\ (gas) - P_i\ (soln)] \tag{21}$$

where A is the interfacial area and $k_i$ is the Henry's law constant. This model assumes that the flux of the gas across the air-sea interface is directly proportional to the diffusion coefficient and inversely proportional to the Henry's law constant. The driving force of the gas exchanger is proportional to the partial pressure difference of the gas between the liquid and the atmosphere. The molecular diffusion coefficients for various gases are given in Table 5.6. At a given temperature the diffusion coefficients decrease with increasing molecular weight or size. An increase in temperature causes the diffusion coefficient to increase. The laminar layer model for various gases has led to boundary layers from 0.002 to 0.02 cm. The smaller thickness occurs with an increase in turbulence caused by an increase in the wind speed. Broecker

**TABLE 5.6**
**The Rates of Molecular Diffusion of Various Gases in Seawater**

| Gas | Mol wt | $D_i \times 10^{-5}$ $cm^2\ sec^{-1}$ 0°C | 24°C |
|---|---|---|---|
| He | 4 | 2.0 | 4.0 |
| Ne | 20 | 1.4 | 2.8 |
| $N_2$ | 28 | 1.1 | 2.1 |
| $O_2$ | 32 | 1.2 | 2.3 |
| Ar | 40 | 0.8 | 1.5 |
| Kr | 84 | 0.7 | 1.4 |
| Xe | 131 | 0.7 | 1.4 |
| Rn | 222 | 0.7 | 1.4 |
| $CO_2$ | 44 | 1.0 | 1.9 |
| $N_2O$ | 44 | 1.0 | 2.0 |

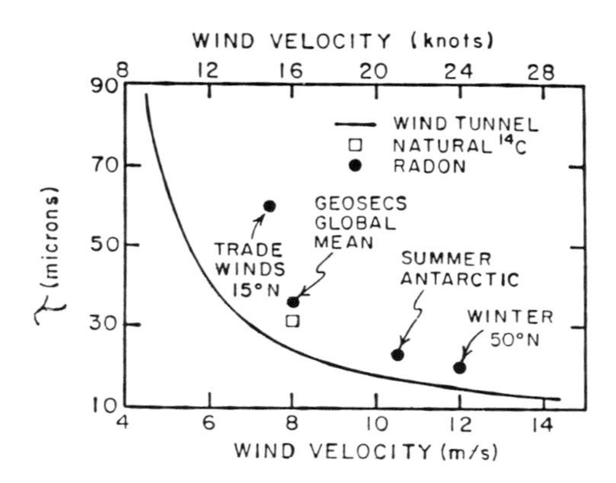


FIGURE 5.2. The thickness of the laminar layer as a function of the wind speed.

and co-workers (Lamont) have made a comparison of wind tunnel measurements of $\tau$ to values obtained using $^{14}C$ and Rn measurements. The results shown in Figure 5.2 are in reasonable agreement. The values of $\tau$ range from 10 to 90 microns ($10^{-6}$ m) depending on the wind speed. The solid line is from wind tunnel measurements. The solid circles are based on radon measurements in the ocean and the open square represents global averages based on $^{14}C$ measurements. The laboratory results give lower film thicknesses than the direct measurements. This may be due to larger waves in the oceans than in the wind tunnel or problems in estimating the appropriate wind speed above the water (10 cm above water in the wind tunnel vs. 15 m above the water in the sea). Both studies demonstrate the decrease in film thickness at high wind speeds.

A more general equation for the rate of transfer of a gas across the air-sea interface is given by

$$dC_i/dt = A\,(f_i/k_i)(P_i\,(\text{gas}) - P_i\,(\text{soln})) \tag{22}$$

where $f_i = D_i/\tau$ is called the exit coefficient or the transfer velocity. The laminar model equates $f_i$ to $D_i/\tau$. If one assumes that the rate determining factor is the time, $\theta$, which a volume element is at the air-sea interface then the transfer coefficient $f_i$ is given by

$$f_i = 2\,(D_i/\pi\Theta)^{1/2} \tag{23}$$

The existence of a laminar layer at the air-sea interface has been supported by observations.

Fick's first law can also be expressed as

$$F = dC/dt = k\,\Delta C \tag{24}$$

where F is the flux (mol $cm^{-2}$ $S^{-1}$), $\Delta C$ is the change in concentration across the air-sea interface (mol $cm^{-3}$), and k is called the transfer velocity (cm $s^{-1}$), permeability coefficient, mass transfer coefficient, absorption coefficient, exit coefficient, or piston velocity. If the thin film layer model is used, k is proportional to $D/\tau$ where D is the diffusion coefficient and $\tau$ is the boundary thickness. If the surface renewal model is used, k is proportional to $D^{1/2}$, and if a boundary layer model is used, k is proportional to $D^{2/3}$. The most popular model presently being used in oceanography to describe the flux across the air-sea interface is the model of Liss (East Anglia, U.K.). This model is a stagnant film model with a two-layer boundary between the interface (Figure 5.3).

As with the thin film model, the air above and the liquid below the gas and liquid films are assumed to be well mixed. Gas transfer occurs by molecular diffusion and the resistance to gas transport is diffusion across the gas-liquid interfacial layer. The value k depends on the turbulence in the gas

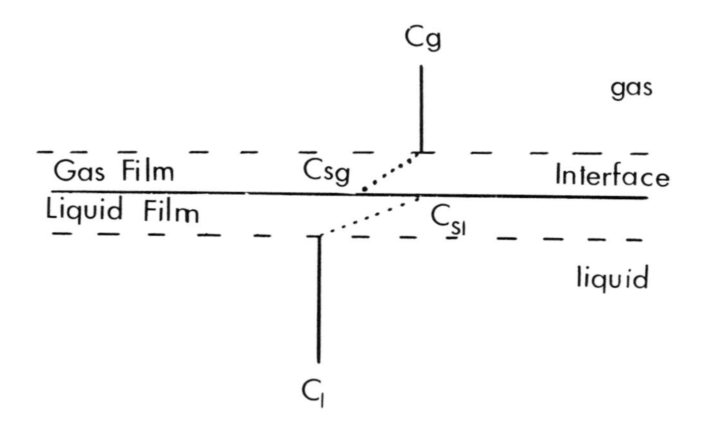


FIGURE 5.3. The gas film model for the transport of gases across the air sea interface.

and liquid phases and the chemical reactivity of the gas. The reciprocal of the exchange coefficient is a measure of the resistance to transfer across the interface. This resistance is divided into the resistance of both the air and water phases

$$1/k = R = R_g = R_l \tag{25}$$

The total flux is equal to the values through the two boundaries

$$F = k_g(C_g - C_{sg}) = k_l(C_{sl} - C_l) \tag{26}$$

where $C_g$ = concentration of the gas in the gas phase, $C_{sg}$ = concentration of gas in the surface film, $C_l$ = concentration of the gas in the liquid phase, and $C_{sl}$ = concentration of the gas in the surface film. In the surface film

$$C_{sg} = H\ C_{sl} \tag{27}$$

where H is the Henry's law constant. The substitution into Equation 26 gives

$$F = (C_g = H\ C_l)/(1/k_g + H/k_l) = (C_g/H - C_l)(1/k_l + 1/H\ k_g) \tag{28}$$

If we define

$$1/K_g = 1/k_g + H/k_l \tag{29}$$

$$1/K_l = 1/k_l + 1/H\ k_g \tag{30}$$

Equation 28 can be rewritten as

$$F = K_g\ (C_g - H\ C_l) = K_l\ (C_g/H - Cl) \tag{31}$$

The total resistance to transport across the air-sea interface (R) is given by

$$R = 1/K_g + 1/K_l \tag{32}$$

where $R_g = 1/K_g$ is the resistance of the gas phase and $R_l = 1/K_l$ is the resistance of the liquid phase. The value of R depends on the exchange constants of the gas and liquid phases and the Henry's law constant for the gas. For most gases the resistance of one phase dominates and controls the total resistance.

For gases that can react chemically with water ($CO_2$ and $SO_2$), the transport is more complex. This is due to the existence of not only a gas gradient, but also a gradient in the chemical species formed ($HCO_3^-$ and $HSO_3^-$). To account for the increase in the flux due to chemical reactions, workers have defined a term $\alpha$. This gives

$$1/K_l = 1/k_l\alpha + 1/H\ k_g = C_g/k_l\alpha H - C_l/k_l\alpha \tag{33}$$

For unreactive gases $\alpha = 1.0$ while for a gas like $SO_2$ $\alpha \simeq 2000$. For gases like $H_2O$, HCl, $SO_2$, and $HNO_3$ that partition strongly into water (low H) or react rapidly, the $Rg >> R_l$. For gases like $O_2$, $N_2$, $CO_2$, inert gases, $SF_6$, and freons that have high values of H and are nonreactive $R_l >> Rg$. The $\alpha$ for $CO_2$, for example, is about 1.02 to 1.03. Thus for most gases the $R_l$ or $k_l$ term will dominate.

Values of $k_l$ at the air-sea interface have been estimated by a number of workers. Broecker and Peng, using $^{14}C$ data, found a value of $k_l = 20$ cm $h^{-1}$ with a probable error of 20% or 5 cm $h^{-1}$. A number of workers have used $O_2$ measurements to estimate $k_l$. They found $k_l = 5$ to 15 cm $h^{-1}$ in the summer and $k_l = 40$ to 50 cm $h^{-1}$ in the winter. The higher values are probably related to the higher wind speeds in the winter. Broecker has also used the concentrations of radon $^{222}Rn$ near the interface to estimate values of $k_l$. These results give $k_l = 12$ to 15 cm $h^{-1}$ which corresponds to $k_l =$ 15 cm $h^{-1}$ for $CO_2$. The transfer velocity obviously depends upon the wind velocity; however, as seen in Figures 5.2 and 5.4, the relation is not straightforward.

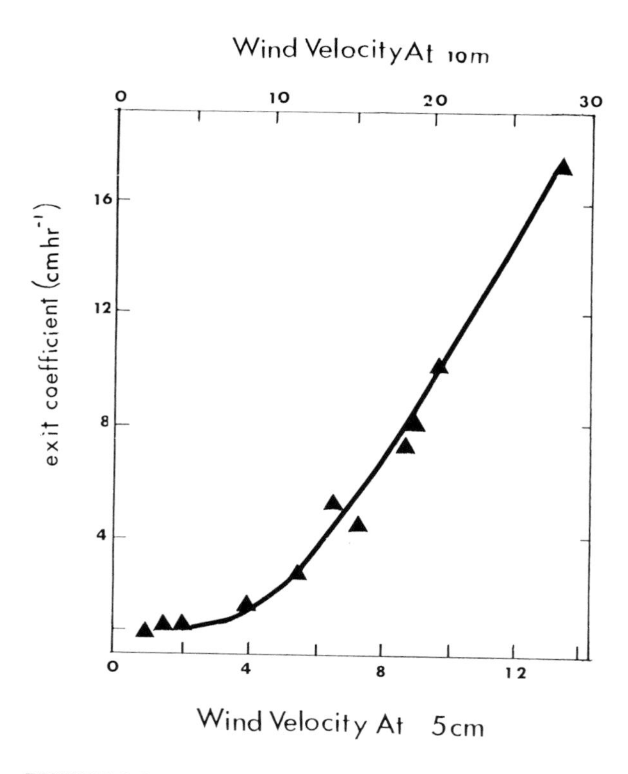


FIGURE 5.4. The exit coefficient (cm $h^{-1}$) across the air-sea interface as a function of wind speed.

Redfield used oxygen data in the Gulf of Maine to evaluate atmospheric exchange under natural conditions. In the spring he found $O_2$ was transferred from the sea to the atmosphere due to photosynthesis. In the summer the warming of the waters produced further evolution of $O_2$. In the winter $O_2$ was regained by the colder waters (40% to oxidize organic matter and 60% adsorbed). He determined the exchange coefficient, E = D/k = f/k to be 3 $\times$ $10^6$ $cm^3$ $month^{-1}$ $atm^{-1}$ from the oceans in the spring and 13 $\times$ $10^6$ $cm^3$ $month^{-1}$ $atm^{-1}$ to the ocean in the winter.

The effect of wind speed on the exchange of $O_2$ across the air-water interface is shown in Figure 5.4. The exit coefficient is constant below a wind speed of 3 m $s^{-1}$ (5 cm above the liquid). The coefficient is greatly increased above 3 cm $s^{-1}$, suggesting turbulent flow. The wind speeds of 5 cm are about $^1/_2$ the value at the standard meteorological height of 10 cm. Although these early studies have been improved upon, they do provide an insight into some of the factors controlling the gas transfer in natural waters. Broecker and co-workers have made a comparison of wind tunnel measurements with calculated values using $^{14}C$ and radon measurements. The results (Figure 5.2) are in good agreement.

Air bubbles can also affect the exchange of gases across the air-sea interface. Hydrostatic pressure can affect the rate of solution and composition of air bubbles. The radius of the dissolving bubbles decreases linearly with time and depth. The composition of the air bubble becomes enriched with $N_2$ because of the larger Henry's law coefficient for $O_2$ compared to $N_2$. Free bubbles exhibited a diffusion coefficient which is twice as large as stationary bubbles. This difference is probably related to differences in the interfacial boundary between stationary and free bubbles and the liquid.

Air bubbles are carried to 20 m below the sea surface during a storm. The partial pressure of the gases in bubbles at this depth would be three times the values at the sea surface. Gas exchange between a bubble and seawater is larger than across the air-sea interface because the laminar layer is thinner and the partial pressures are increased. Two populations of bubbles have been found: (1) those less than 40 m in radius associated with atmospheric particles deposited on the sea surface, and (2) those about 100 m in radius formed at the sea surface by collapsing waves. Air bubbles can change the magnitude of the flux of a gas as much as a barometric pressure change of 2% in 12 h (the passage of a storm). The rate of exchange by bubbles could be several orders of magnitude greater than seasonal fluxes; thus, bubbles are an important mechanism for gas exchange.

At low film thickness (high wind speeds) the chemical effects are negligible, but waves can influence the gas transfer. The transfer rate is proportional to the mean square slope of the wave. For capillary waves the enhancement was predicted to be a factor of nine. Waves produced in wave tunnels only show an enhancement of 10%. The wind tunnel results showed that the enhancement of the transfer rate varied as a function of the mean square slope of the waves divided by the friction velocity in the water. More work is needed to elucidate the many factors that control the transfer of gases across the air-sea interface.

## 4. NONREACTIVE GASES

Nitrogen and the noble gases are regarded as nonreactive. The distribution of nonreactive gases are affected by physical processes and the effects of temperature and salinity on the solubility. Studies of the distribution of the nonreactive gases can be useful in separating the effects of physical and biological processes on the distribution of reactive gases such as $O_2$ and $CO_2$. With the advent of better analytical techniques recent advances have been and are being made in this area of research.

The variations of nonreactive gases are studied by examining the solubility at 1 atm total pressure and 100% relative humidity. The degree of saturation $\sigma_i$ is given by

$$\sigma_i = [i]/[i]^* \times 100 \tag{34}$$

where [i]* is the solubility concentration at a given potential temperature and S and [i] is the measured gas concentration. The saturation anomaly $\Delta_i$ is also used

$$\Delta_i = ([i] - [i])/[i]^*) \times 100 \tag{35}$$

The examination of $\sigma_i$ and $\Delta_i$ permits one to follow the distribution without the bias of temperature and salinity. The physical processes which can cause nonreactive gases to depart from the expected concentration are

1. Departures from the standard pressure
2. Partial dissolution of air bubbles
3. Air injection
4. Differential heat and gas exchange
5. Mixing of waters of different temperature
6. Radiogenic or primordial addition (He)

The radiogenic production of $^{40}Ar$ from $^{40}K$ in seawater is not significant. It would take $10^{11}$ years of $^{40}K$ decay to produce the amount of $^{40}Ar$ in the oceans.

The effect of changes in barometric pressure and humidity affect all the gases by an equal percentage. At 30°C the $\Delta_i$ is increased by 0.9% when the humidity is decreased to 80%. To account for departures from the standard pressure and humidity one must study at least two different gases.

When a bubble of air is submerged below the surface, it will dissolve due to hydrostatic pressure. If the bubble partially dissolves before it goes back to the surface, all the gases will partially dissolve and increase $\Delta_i$. For partial dissolution the composition of the gases in the bubble will remain

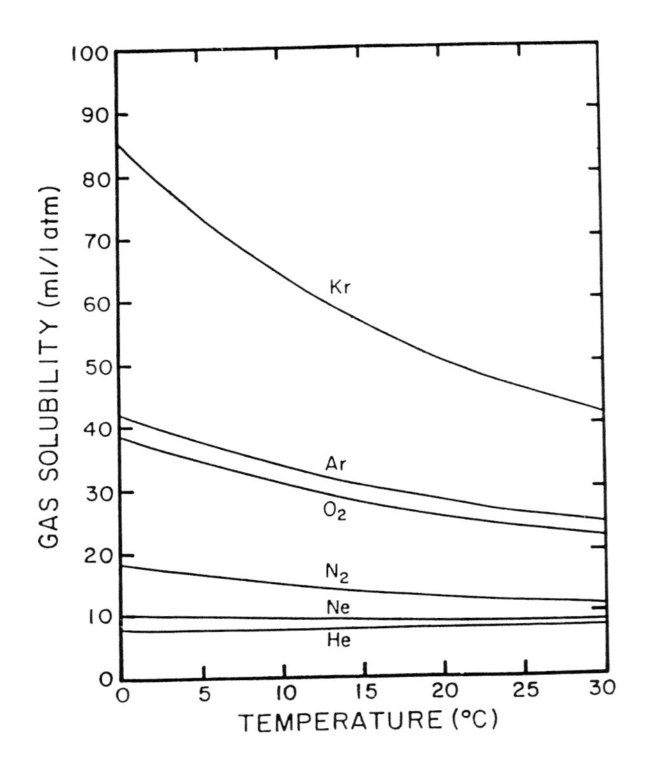


FIGURE 5.5. The effect of temperature on the solubility of gases in seawater.

unchanged. At a depth of 1 m the $\Delta_i$ is increased by +10% for all the gases. If the air bubble completely dissolves, it is called air injection. Each gas is affected differently for air injection due to the differences in solubility. The differences in the variation of the solubility of gases with temperature are shown in Figure 5.5. This difference in composition produces a large variation in $\Delta_i$ for different gases when air is injected or totally dissolved in seawater.

Since the solubility of gases vary differently with temperature, changes in temperature without an exchange with the atmosphere can cause $\Delta_i$ to vary. The mole fraction of each gas at equilibrium in solution is different from that of air (Table 5.7). For example, at 15°C a 1°C decrease in temperature will cause $\Delta_i$ to vary from −0.24% for He to −2.5% for Xe (Kester, 1975). When the heat exchange is more rapid than gas exchange (for example, in upwelling areas) saturation anomalies will occur. The mixing of waters of different temperatures will also produce saturation anomalies due to the non-linear behavior of solubility with temperature. For example, the mixing of waters at 0 and 30°C will produce saturation anomalies of 1.3% for He and 18% for Xe.

**TABLE 5.7**
**Comparison of the Mole Fraction ($X_i$) of Gases Dissolved in Saturated Seawater and in Dry Air and the Saturation Anomaly, Air, Resulting from the Injection of 1 cm³ of Air at STP per Kilogram of Seawater (All Seawater Values are Based on 15°C and S = 35)**

| | $N_2$ | $O_2$ | Ar | $Co_2$ | Ne |
|---|---|---|---|---|---|
| $X_{air}$ | 0.780 | 0.209 | 0.009 | 0.0003 | $18.2 \times 10^{-6}$ |
| $X_{SW}$ | 0.626 | 0.343 | 0.016 | 0.014 | $9.7 \times 10^{-6}$ |
| Air | +7.7% | +3.8% | +3.5% | 0.1% | +11.6% |

| | He | Kr | Xe |
|---|---|---|---|
| $X_{air}$ | $5.2 \times 10^{-6}$ | $1.1 \times 10^{-6}$ | $0.09 \times 10^{-6}$ |
| $X_{SW}$ | $2.3 \times 10^{-6}$ | $3.8 \times 10^{-6}$ | $0.54 \times 10^{-6}$ |
| Air | +13.8% | +1.8% | +1.0% |

**TABLE 5.8**
**Saturation Anomalies for Nitrogen[a]**

| Water mass | $N_2$ | |
|---|---|---|
| | Mean | SD |
| All surface waters (z < 10 m) | −0.17% | 3.4% |
| South Atlantic Central Water | −0.48% | 1.3% |
| Antarctic Intermediate Water | −1.42% | 3.0% |
| North Atlantic Deep Water | +0.32% | 1.9% |
| Antarctic Bottom Water | −0.04% | 2.2% |

[a] Kester (1975).

By studying several gases it is possible to determine the magnitudes of these processes. Xe and Kr are not sensitive to air injection, thus, they can be used to examine temperature effects. He is sensitive to air injection, but not affected by temperature effects. Since Ar and $O_2$ are affected by nearly the same amounts, Ar is the best nonreactive gas for separating the biological and physical processes for $O_2$.

The saturation anomalies for $N_2$ have been studied by a number of workers. The results (Kester, 1975) are shown in Table 5.8. All the waters are within ± 1.4% of the saturated values (which is within the standard error of 3%). The high standard errors could be related to real variations due to oceanic processes (air injection, etc.).

Since biochemical processes are known that convert $N_2$ to organically bound nitrogen and $N_2$ can be produced from organic nitrogen under anoxic conditions, one may question the degree to which $N_2$ is conservative. Benson and Parker's measurements of the ratio of $N_2$ to Ar indicate that for most ocean waters $N_2$ can be considered to be conservative to ± 1%. Richards and Benson's (1961) measurements of $N_2$/Ar in anoxic basins yielded $\Delta_{N2}$ = +2.3% for the Cariaco trench and +3.6% for the Damsfjord.

Due to the different solubility data used and the method of reporting the results, it is difficult to make a detailed comparison of various $N_2$ studies at the 1% level. Most ocean waters are generally within 1 to 2% of the saturation values except in anoxic waters which may be 2 to 4% supersaturated. As the analytical precision for $N_2$ has improved, the spread of $\Delta_{N2}$ has decreased.

The saturation anomalies of a gas can be attributed to three major components

$$\Delta_i = \delta_p + \delta_T + \delta_a \qquad (36)$$

The factor $\delta_p$ is due to differences in the effective and standard atmospheric pressure

$$\delta_p = \frac{[P_{atm} - (h/100)\,P_s]\,100}{1 - P_s} + \frac{Z}{10} \qquad (37)$$

where Z is depth in cm at which the air bubble is equated. Since $\delta_p$ does not depend upon a specific property of a gas, it is not possible to determine the various terms separately. The factor $\delta_T$ is due to changes in temperature, $\delta_T$, without gas exchange

$$\delta_T = [-100\ d\ \ln[i]/d\ T]\ \Delta T \qquad (38)$$

where $\Delta_T$ is the observed potential temperature minus the temperature at the time of equilibration. This term applies to changes in temperature due to radiative heating or cooling and geothermal heating (not mixing of waters of different temperatures). The factor $\delta_a$ is due to air injection and is given by

$$\delta_a = X_i\ 100\ a/[i]\ 22{,}390 \qquad (39)$$

where a is the $cm^3$ of air at STP injected/kg SW and 22,390 is the molar volume of air in $cm^3$ at STP.

These relations have been used by Kester (1975) for Atlantic waters ($t = 26°C$ and $S = 36$)

$$\Delta_{Ne} = 4.5 = \delta_p + 0.626\ \delta_T + 12.47\ \delta_a \tag{40}$$

$$\Delta_{Ar} = 2.0 = \delta_p + 1.62\ \delta_T + 4.22\ \delta_a \tag{41}$$

$$\Delta_{He} = 4.6 = \delta_p + 0.180\ \delta_T + 14.16\ \delta_a \tag{42}$$

The solution gives $\delta_p = -2.8\%$, $\delta_T = +1.7°C$, and $\delta_a = 0.50$ $cm^3$ air at STP/kg. The negative value of $\delta_p$ indicates that $P_{atm} < 1$ atm. A 1% increase in $\delta_p$ can be caused by an increase of 8 mm of Hg or an equilibration of air at 10 cm below the surface or a $h = 65\%$ at 25°C.

Deep waters of the oceans have been found to be injected by 0.5 to 1.0 $cm^3$ of air at STP per kg of seawater. He can also be injected into the deep oceans from the sea floor. The helium is produced by the radioactive decay of uranium and thorium in sediments and rocks. Direct injection of He has also been found to come from hydrothermal fluids. This excess helium injected into the deep oceans has been used by Craig and co-workers to study the movement of waters in the deep sea (Figure 5.6).

Saturation anomalies for He, Ne, and Ar in North Atlantic deep waters can be attributed to air injection. In the Pacific, however, the $\Delta_{He}$ of 7 to 12% are partially (50%) due to excess helium. The He anomaly can be studied by examining the $^3He/^4He$ ratio. Observed values (Pacific) of this ratio greater than equilibrium values (Figure 5.7) can be attributed to helium coming from the interior of the earth from active ridges. Recent measurements on hydrothermal waters (Figure 5.8) gives ratios of $^3He/^4He = 1.08 \times 10^{-5}$ compared to expected air injection values of $1.38 \times 10^{-6}$.

## 5. DISSOLVED OXYGEN IN SEAWATER

By far the most studied gas (other than $CO_2$) in ocean waters is oxygen. This is largely due to the fact that it is easy to measure and is of interest to oceanographers studying advective processes and biologists studying photo-

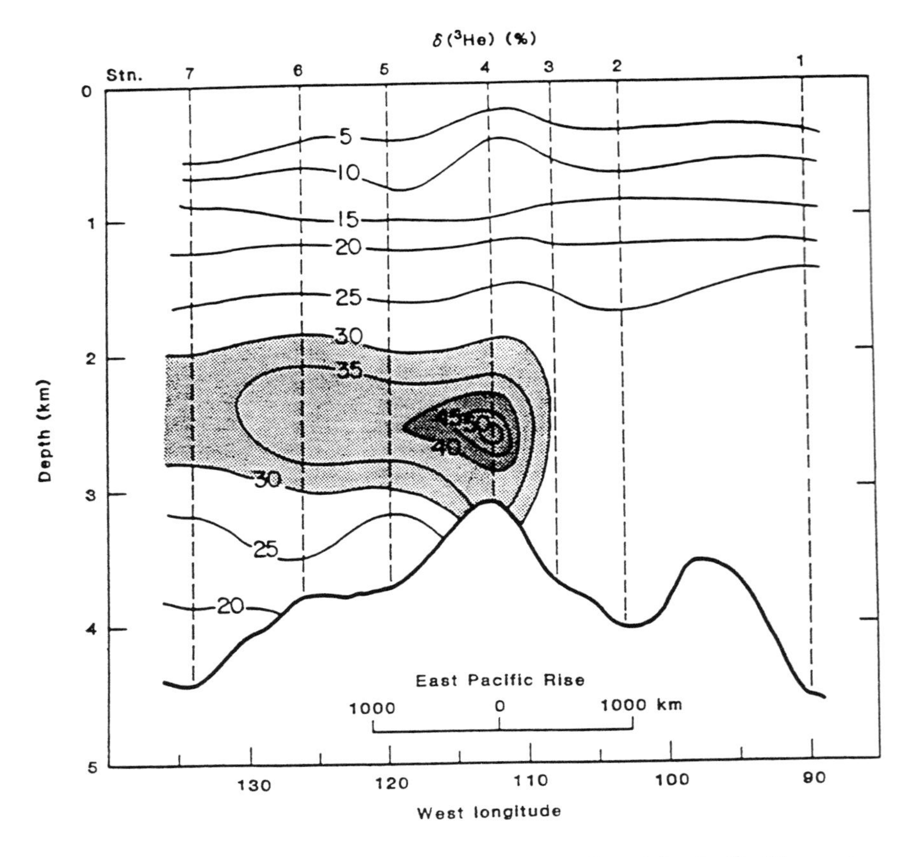


FIGURE 5.6. The concentration of Helium-3 to Helium-4 in the Pacific coming from active ridges.

synthesis and the oxidation of plant material. The analysis of $O_2$ in natural waters is still made by using the technique developed by Winkler in 1888 and modified by Carpenter in 1965. The collected sample is quickly fixed with $MnSO_4$ and NaOH. The reaction between NaOH and $MnSO_4$ forms manganous hydroxide

$$Mn^{2+} + 2OH^- \rightarrow Mn(OH)_2 \tag{43}$$

The $Mn(OH)_2$ reacts with the $O_2$ to form a tetravalent manganous compound

$$2Mn(OH)_2 + O_2 \rightarrow 2MnO(OH)_2 \tag{44}$$

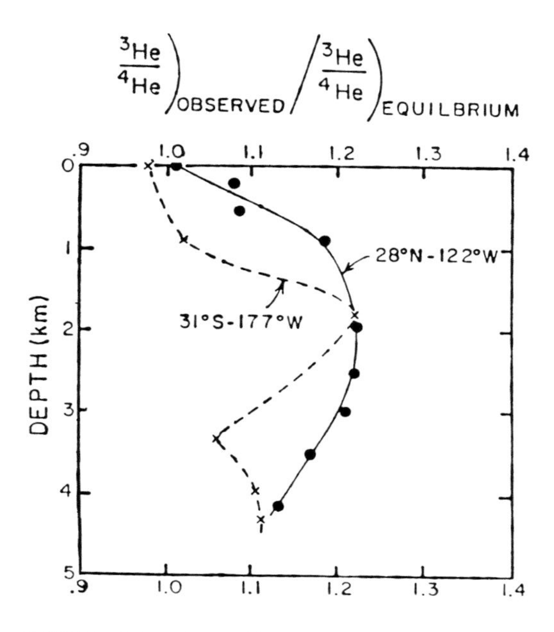


FIGURE 5.7. A profile of Helium-3 to Helium-4 in the Pacific Ocean showing the higher observed levels compared to equilibrium with the atmosphere.

This fixes the $O_2$. The solution is analyzed for $O_2$ after acidification in the presence of excess NaI. The manganous compound oxidizes the $I^-$ to $I_2$

$$MnO(OH)_2 + 4H^+ + 2I^- \rightarrow Mn^{2+} + I_2 + 3H_2O \tag{45}$$

The free $I_2$ formed is stabilized by the formation of $I_3^-$ with the excess $I^-$

$$I_2 + I^- \rightarrow I_3^- \tag{46}$$

The amount of $I_3^-$ formed is equivalent to the $O_2$ in the solution and is determined by titration with sodium thiosulfate

$$I_3^- + 2S_2O_3^{2-} \rightarrow 3I^- + S_4O_6^{2-} \tag{47}$$

A starch solution which forms a blue colored complex with $I_2$ is used to determine the endpoint. The concentration of $O_2$ can also be measured by using $O_2$ electrodes or by gas chromatography. The electrode system has been attached to CTD systems to obtain continuous profiles of $O_2$ in ocean waters.

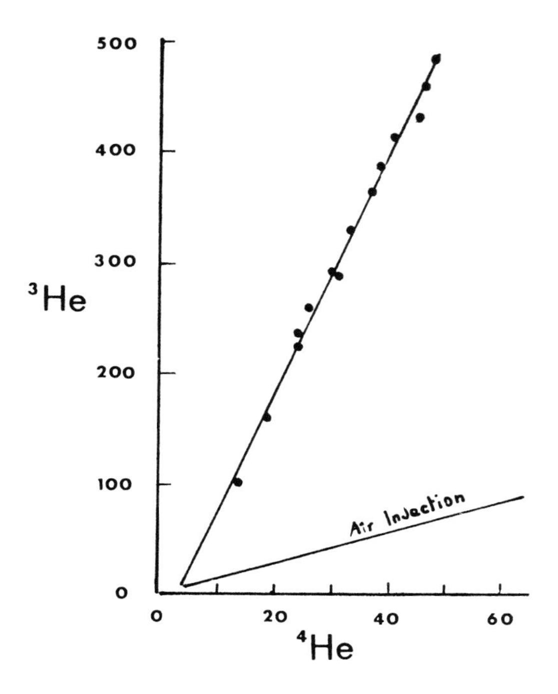


FIGURE 5.8. Values of Helium-3 vs. Helium-4 for Pacific Ocean waters compared to expected values from air injection.

The vertical distribution of $O_2$ in the major oceans is shown in Figure 5.9. In surface waters the $O_2$ concentrations in seawater are close to the expected values for the temperature and salinity of the waters. This is shown in Figure 5.10 for the measurements made during the GEOSECS program. The solid line represents the saturation value. The surface values are 7 μmol $kg^{-1}$ or about 3% supersaturated. This supersaturation can be attributed to bubble injection and photosynthesis. In the photosynthetic zone the $O_2$ will go through a maximum due to photosynthesis (Figure 5.11).

The most remarkable features of the $O_2$ profile in the major oceans are the minimum layer and the relatively high $O_2$ in deep waters. The minimum is the result of a balance between the biological oxidation of plant material and the advection of cold waters rich in $O_2$. The advection of oxygen rich waters in the oceans is more clearly demonstrated in the sections of $O_2$ in the major oceans given in Plates 12 to 14.* The intrusion of Antarctic and Arctic intermediate waters is evident in the Atlantic, Pacific, and Indian oceans from

* Plates 12 to 14 follow page 48.

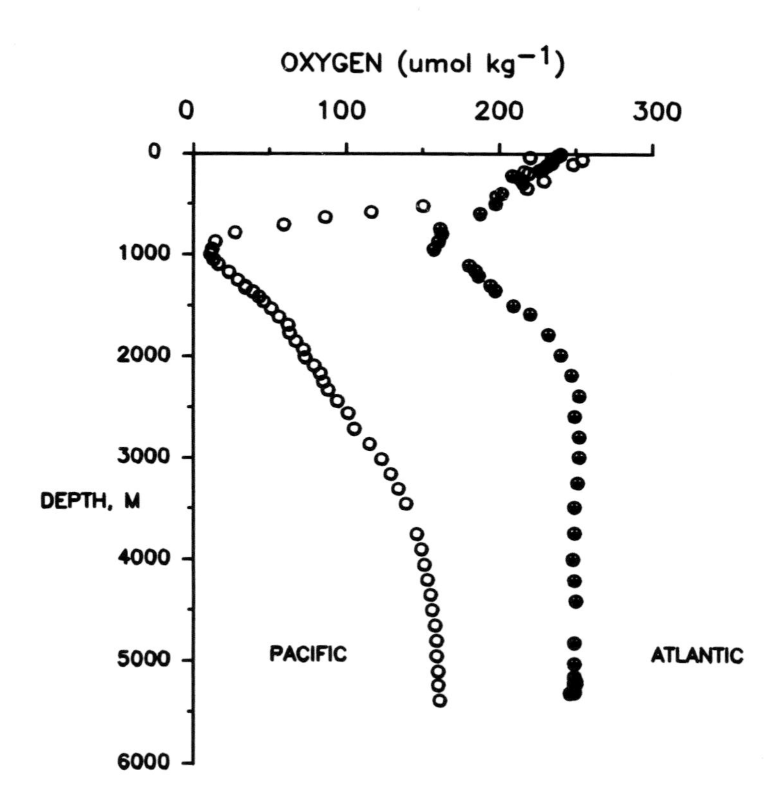


FIGURE 5.9. Profiles of oxygen in the Atlantic and Pacific Oceans.

the surface to 900 m. The North Atlantic Deep Water (NADW) is a region of high $O_2$ from 0 to 2000 m at 60N to a maximum at 3000 m in the South Atlantic. This NADW gradually loses $O_2$ as it flows northward in the deep Pacific and Indian Oceans from the circumpolar waters. The contributions of Antarctic bottom waters are also seen in the high $O_2$ in the southern oceans. The intermediate waters of the North Pacific are much lower in $O_2$ than in the North Atlantic. The intense minimum off the coast of South America is related to the high productivity occurring in the upwelled surface waters.

The distribution of $O_2$ in the oceans is the net result of:

1. Near equilibration of atmospheric oxygen in the surface mixed layer
2. Biological production in subsurface waters due to photosynthesis
3. Biological use of $O_2$ in respiration in all waters and oxidation of plant materials in intermediate waters
4. In deep waters the $O_2$ increases are due to the sinking of cold water rich in $O_2$

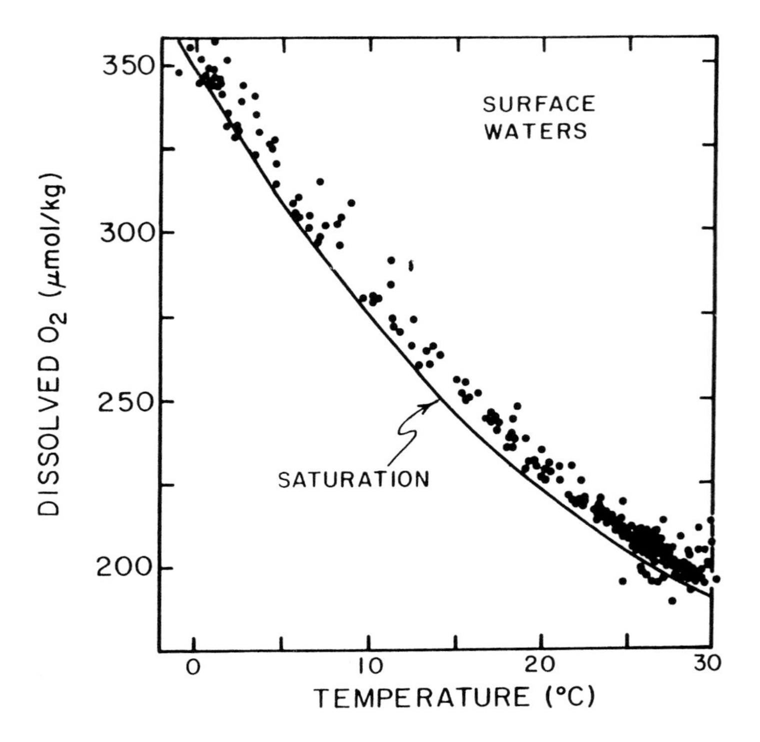


FIGURE 5.10. Comparison of the measured and calculated dissolved oxygen in surface seawaters as a function of temperature.

By examining the amount of $O_2$ that is utilized rather than the total concentration, it is easier to interpret the physical and biological effects. This will eliminate the differences due to conditions when the waters were in contact with the atmosphere. The contribution due to processes such as air injection, thermal effects, and atmospheric pressure can be accounted for by the methods described earlier. To do this properly, accompanying information is needed for other inert gases. By using Ar as a reference gas, one can make a reasonable correction for the deviations due to non-ideal equilibria. The concentration of oxygen consumed since the waters left the surface is given by

$$[O_2]_{surf} - [O_2]_{meas} = [O_2]^* \times [Ar]_{meas}/[Ar]^* - [O_2]_{meas} \qquad (48)$$

where $[i]^*$ are the predicted values at the given temperature and salinity and $[i]_{meas}$ are the measured values.

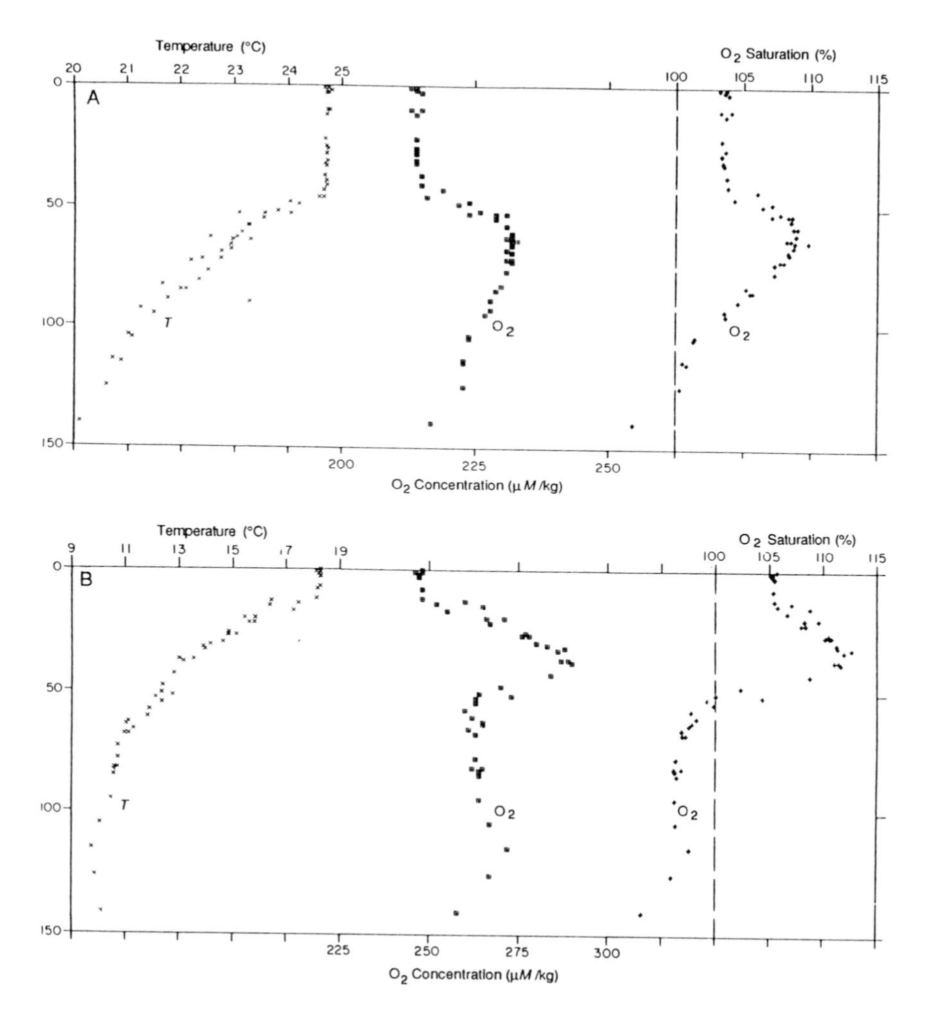


FIGURE 5.11. Profiles of temperature and oxygen in Pacific waters showing increases due to photosynthesis.

In the absence of Ar measurements, the apparent oxygen utilization (AOU) is determined from

$$\text{AOU} = [O_2]^* - [O_2]_{\text{meas}} \tag{49}$$

The apparent is used to recognize that the $[O_2]^*$ may depart from ideal values due to air injection, etc. As shown earlier, surface waters are 3% supersaturated. At upwelling areas the surface $O_2$ concentrations can be 20% less than saturation. This is shown in Figure 5.12 for Pacific waters. The surface

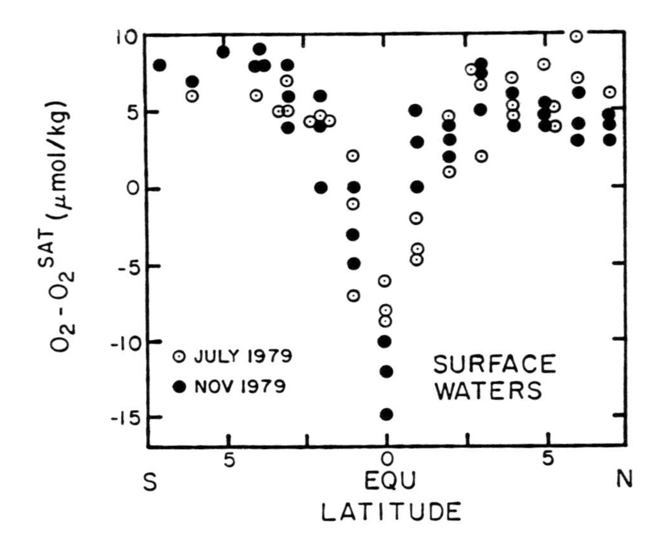


FIGURE 5.12. The differences in the measured and calculated dissolved oxygen concentrations in surface waters.

AOU values are 15 μmol/kg near the equator due to upwelling. This can be compared to values around −7 μmol/kg for other surface waters.

Peng and Broecker have determined values of AOU for various waters using the GEOSECS data. The distribution of AOU at a depth of 4000 m is shown in Figure 5.13. In the North Atlantic the AOU values are at 50 μ mol $kg^{-1}$ and they increase to 190 μ mol $kg^{-1}$ in the North Pacific ocean. Vertical sections of AOU in the Atlantic and Pacific are shown in Figure 5.14 and 5.15. The top sections to a depth of 1.5 km show a ''blow up'' of the surface waters. The high values of AOU near the equator are clearly shown in these blow ups. The dotted lines in these figures are $\sigma_0 = 27.0$ and $\sigma_0 = 27.5$ isopycnal horizons, respectively, for the upper and bottom lines in the blow ups. The values of AOU in the deep waters of the Atlantic, Pacific, and Indian oceans (Figure 5.16) clearly demonstrate the increase in AOU as the waters move from the North Atlantic to the North Pacific and Indian oceans. The dotted line on these deep water profiles represents the $\sigma_{4000} = 45.2$ isopynal horizon.

By examining the AOU at different locations with the distribution of nutrients, it is possible to determine the causes of the increase. The slow increase in deep waters follows the increase in nutrient concentrations (discussed later) and results from the oxidation of particulate organic material from the surface waters. The linear correlation of $O_2$ and salinity in inter-

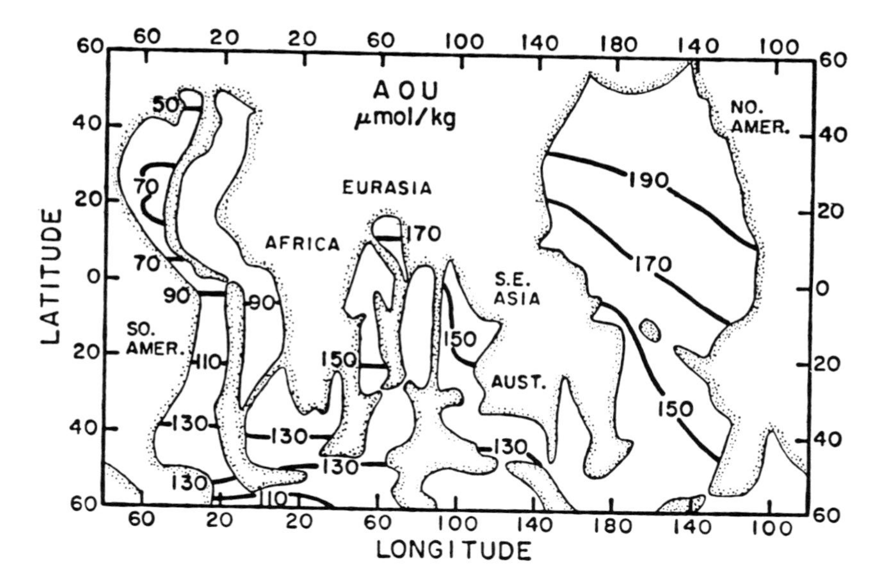


FIGURE 5.13. Apparent oxygen utilization in deep waters of the world oceans.

mediate waters may indicate that the observed distribution may result largely from mixing processes rather than *in situ* oxidation.

A number of workers have used advective diffusion models to analyze the $O_2$ minimum layer. The rate of the *in situ* oxygen consumption, R, is assumed to decrease exponentially with depth

$$R = R_0 e^{-\alpha 2} \tag{50}$$

where $R_0$ is the rate of $O_2$ utilization at the upper boundary of the model (500 m), and $\alpha$ accounts for the rate at which R decreases with depth (z). Horizontal mixing is assumed to be zero and the boundary values of $O_2$ are maintained by atmospheric exchange and horizontal transport. Wyriki (University of Hawaii) applied this model to $O_2$ in the Indian Ocean and found the results shown in Figure 5.17. The rate (R) is given by

$$R = A_z (\partial^2[O_2]/\partial z^2) - w (\partial[O_2]/\partial z) \tag{51}$$

where $A_z$ is the vertical eddy diffusion coefficient (independent of depth) and w is the vertical advection velocity. His results indicate that *in situ* consumption of $O_2$ is required to account for the $O_2$ minimum layer except for areas (off the west coast of South America and Africa) where the layer is

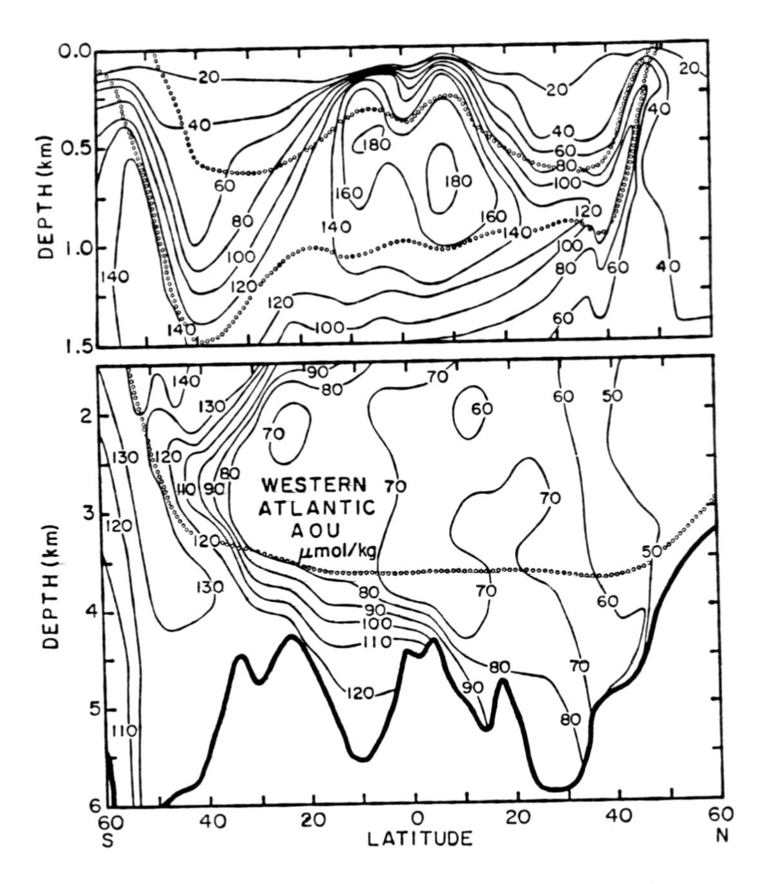


FIGURE 5.14. A section of apparent oxygen utilization of Atlantic Ocean waters.

near the surface where horizontal mixing is important. Craig (University of California, San Diego) and others have further developed this simple model. Culberson (U. Del.), for example, showed that the derived values of R were in reasonable agreement with estimates of the oxidation rate of Packard (University of Maine) obtained by using an enzyme analysis (Figure 5.18).

It should be pointed out that this vertical model is an over-simplification of the real oceans. The ultimate description of the distribution must use three dimensional models. The early work by Riley indicates that 90% of the oxidation of the organic matter produced by phytoplankton occurs in surface waters (<200 m). The remaining 10% is consumed in the deep oceans.

By examining the AOU distribution relative to other chemical parameters such as $PO_4$, $NO_3$, $CO_2$, and pH, a stoichiometric model was developed by

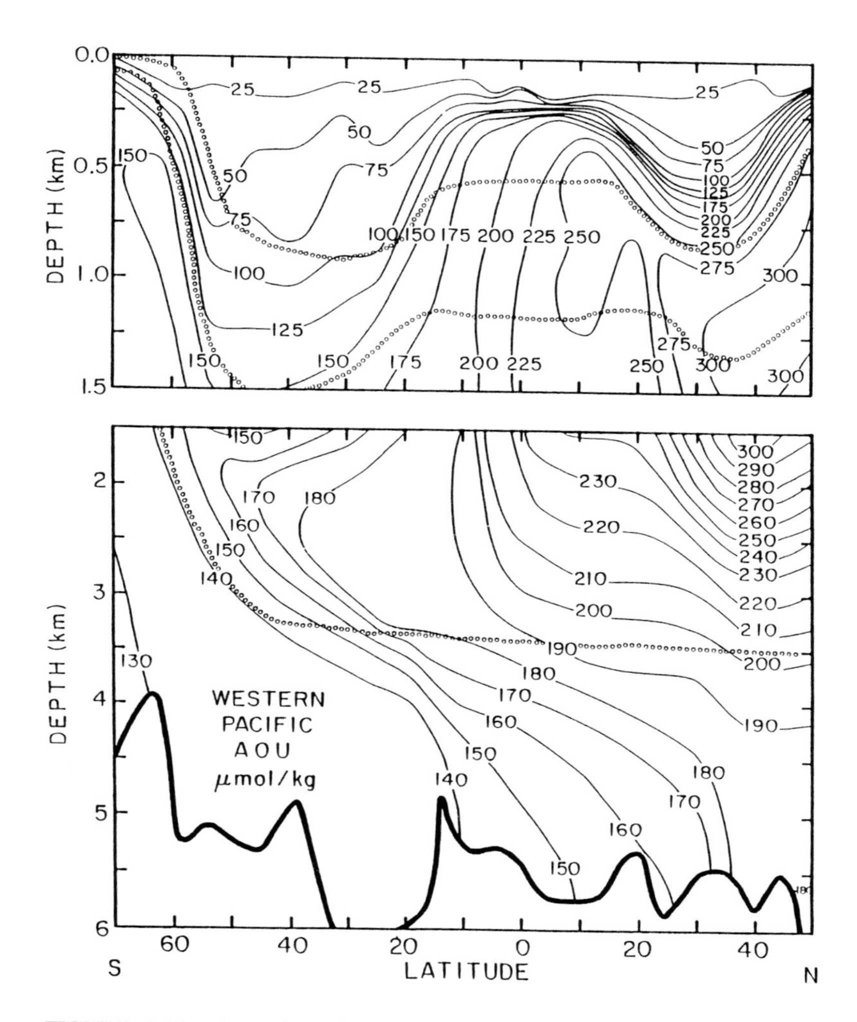


FIGURE 5.15. A section of apparent oxygen utilization of Pacific Ocean waters.

Redfield. The oxidation of organic matter by $O_2$ can be represented by

$$(CH_2O)_{106}\,(NH_3)_{16}\,H_3PO_4 + 138\,O_2 \underset{\text{photosynthesis}}{\overset{\text{oxidation}}{\rightleftarrows}} 106\,CO_2 +$$

$$16\,HNO_3 + H_3PO_4 + 122\,H_2O \qquad (52)$$

Details of using this simple model will be discussed later.

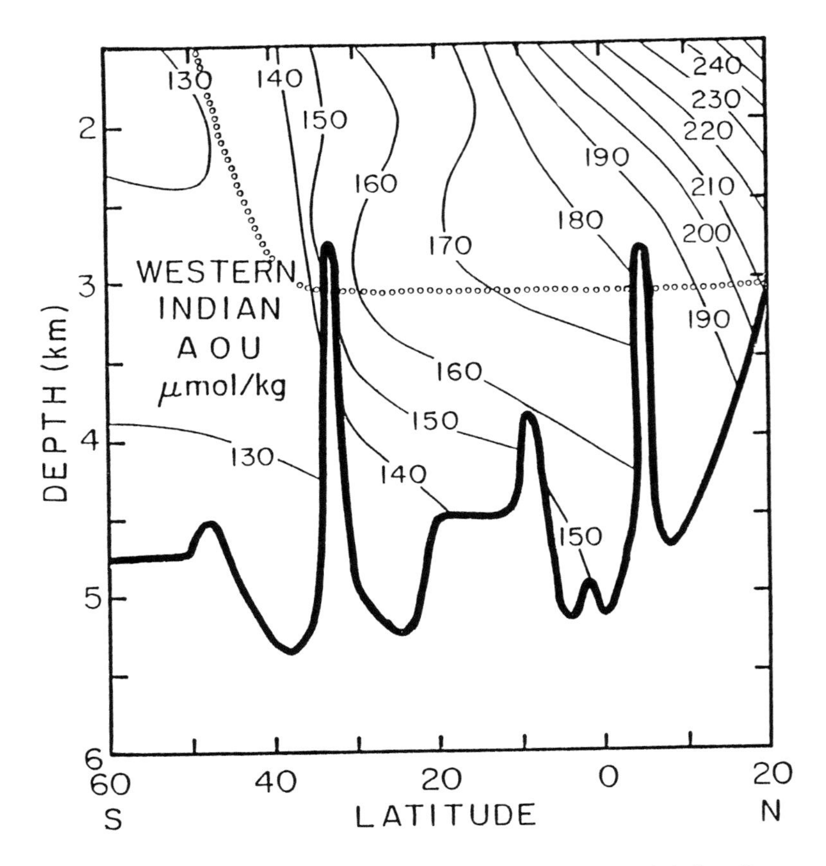


FIGURE 5.16. A section of apparent oxygen utilization of Indian Ocean waters.

With the recent development of *in situ* $O_2$ instruments, one might expect future work to provide useful information on the details of the oxidation of organic material in the oceans. Kester (1975) has demonstrated how these instruments can be used to look at small changes of $O_2$ as a function of depth due to mixing processes.

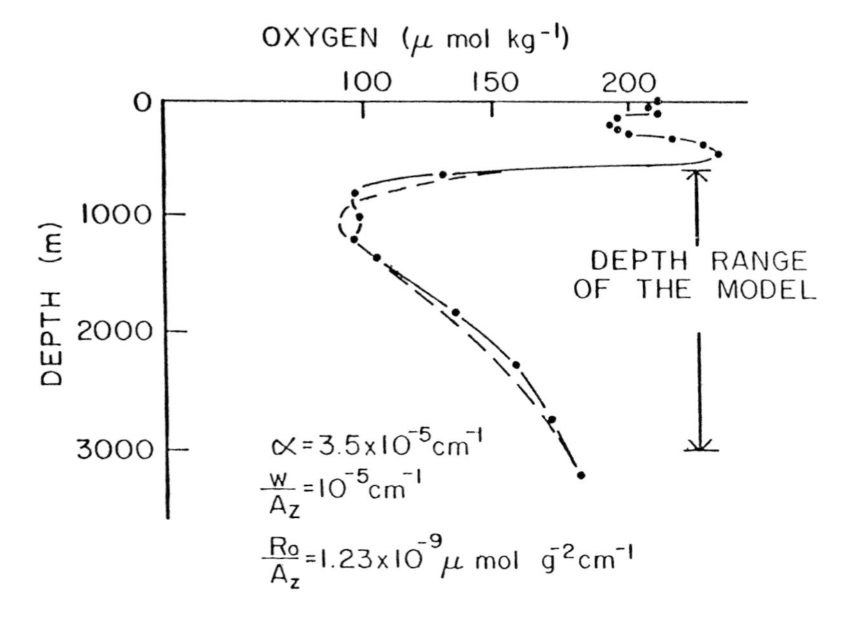


FIGURE 5.17. A comparison of the measured and calculated concentrations of oxygen using a one-dimensional mixing model.

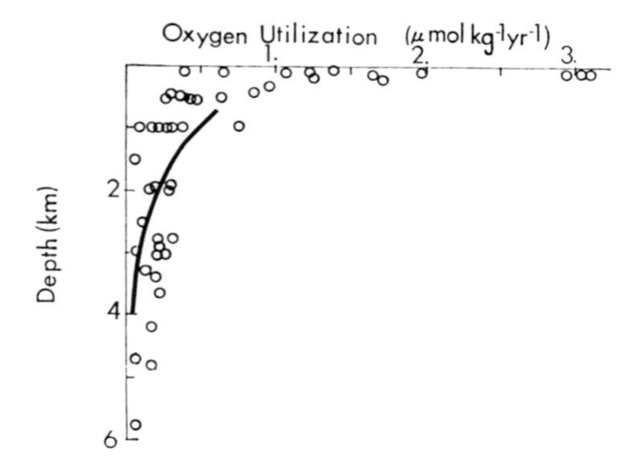


FIGURE 5.18. A profile of the measured and calculated oxygen utilization rates in the Atlantic Ocean.

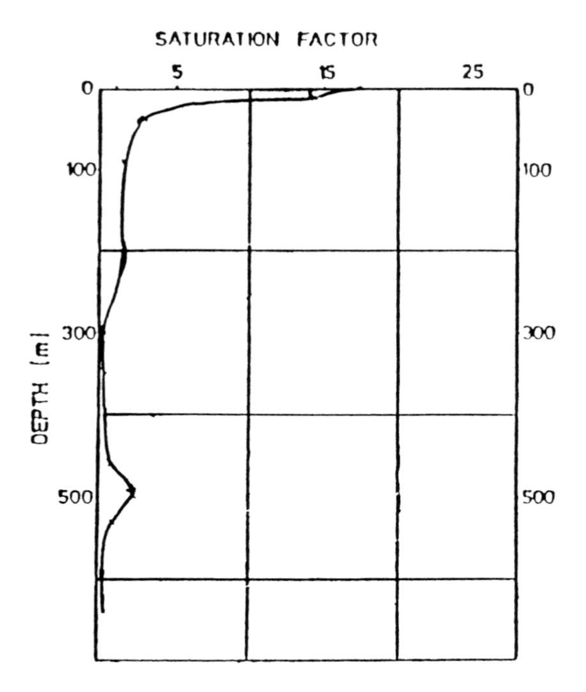


FIGURE 5.19. A profile of the saturation of carbon monoxide in the Atlantic Ocean.

## 6. OTHER NONCONSERVATIVE GASES

In recent years other nonconservative gases such as CO, $H_2$, $CH_4$, and $N_2O$ have been measured in seawater. A depth profile for CO in the eastern Atlantic is shown in Figure 5.19. The surface waters are supersaturated apparently due to bacterial activity or photochemical processes. The increase at 500 m is the location of water flowing out of the Mediterranean with its increase in microorganisms. The global cycle of CO is shown in Figure 5.20 ($10^{14}$ g/year). The oceans are a source of CO to the atmosphere.

A depth profile of $H_2$ in the Atlantic is shown in Figure 5.21. The values go through a maximum in the photosynthetic zone near the pynocline apparently due to biological activity. The depth profile of $CH_4$ in seawater (Figure 5.22) also goes through a maximum in surface waters. As shown in the section of $CH_4$ (Figure 5.23) the source may be from the shelf sediments.

As will be discussed later, the profiles of $N_2O$ in ocean waters (Figure 5.24) go through a maximum in the oxygen minimum layer.

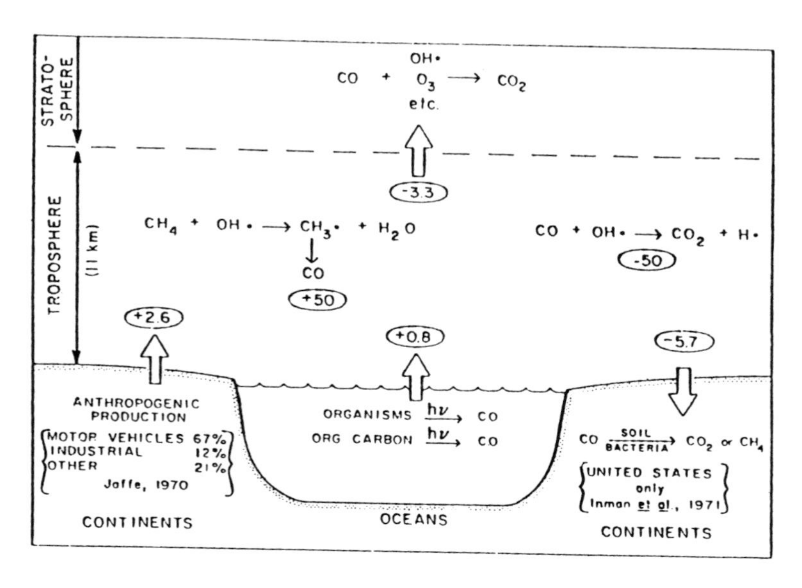


FIGURE 5.20. A sketch of the global cycle of carbon monoxide in the world.

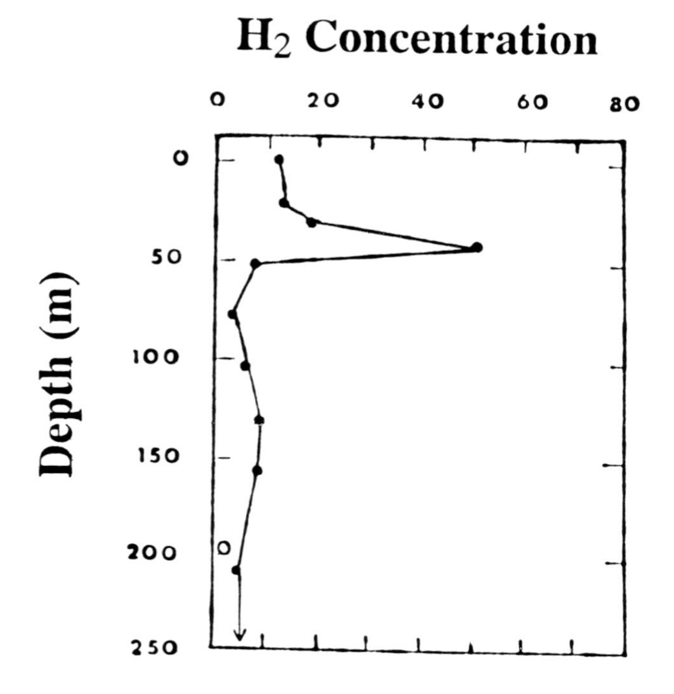


FIGURE 5.21. A profile of hydrogen in the Gulf of Mexico.

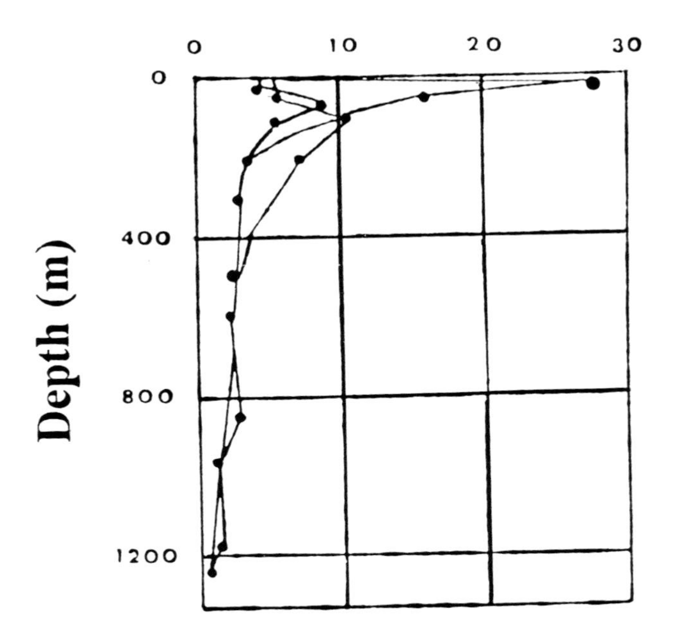


FIGURE 5.22. A profile of methane in the Gulf of Mexico.

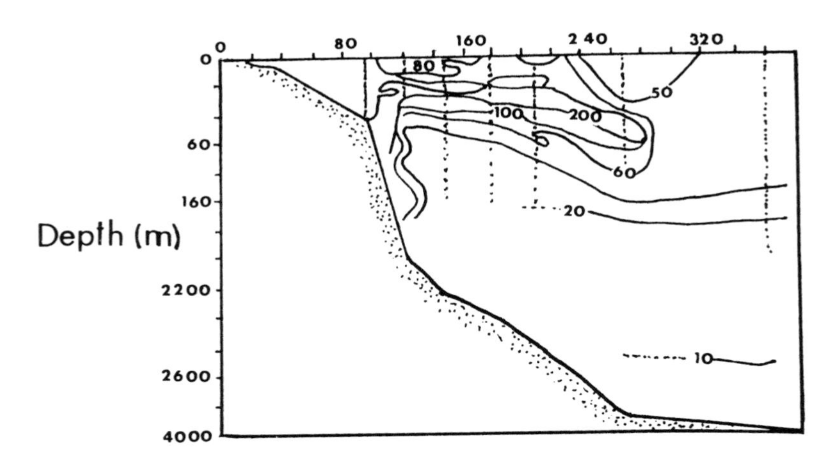


FIGURE 5.23. A section of methane off the Gulf coast.

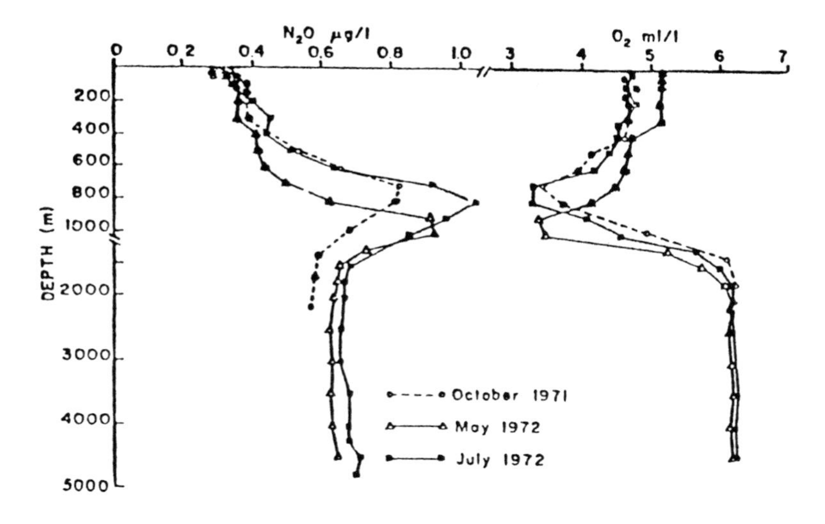


FIGURE 5.24. Profiles of nitrous oxide and oxygen in the Pacific Ocean.

## 7. STRUCTURAL ASPECTS OF THE SOLUBILITY OF GASES

Earlier we discussed the process of dissolving a charged ion in water and the energies involved. One might expect the dissolution of a gas molecule to involve very small energies and to be related entirely to packing or structural effects. Experimental studies of inert gases, however, indicate that $\Delta H_S$ and $\Delta S_S$ for the solution process are quite anomalous (Table 5.9). The values of $\Delta H_S$ and $\Delta S_S$ are both negative. This is in contrast to the solution of these gases into other liquids where the loosening of the solvent intermolecular forces results in an increase in entropy. These differences are normally interpreted in terms of changes in the structure of water in the vicinity of the dissolved gases. The dissolution process can be considered to be a two step process: (1) the creation of a hole or cavity in the solvent and (2) the introduction of the gas molecule into the cavity. This accounts for the large negative $\Delta H_S$ that may be an indication of a chemical bond between the gas and water. Frank and Evans, however, concluded that the addition of an inert gas or nonpolar molecule causes the water to appear to be more "ice-like"; that is, the water appears to be highly structured around an inert gas molecule. Others have suggested that the water structure does not build-up around the solute, but the structure is there to begin with. The introduction of the solute just causes the structure to shift locally. This local structure has been compared

**TABLE 5.9**
**Values of ΔH and ΔS for the Solution of Gases in Water (25°C)**

| Gas | $-\Delta H_S$ (kcal/mol) | $-\Delta S_S$ (cal deg/mol) |
|---|---|---|
| $H_2$ | 1.3 | 26 |
| $N_2$ | 2.1 | 29 |
| He | 0.8 | 27 |
| Ne | 1.9 | 29 |
| Ar | 2.7 | 30 |
| Kr | 3.6 | 32 |
| Xe | 4.5 | 34 |
| Rn | 5.1 | 34 |
| CO | 3.9 | 30 |
| $O_2$ | 3.0 | 31 |
| NO | 2.7 | 29 |
| $CO_2$ | 4.7 | 31 |
| COS | 5.8 | 35 |
| $N_2O$ | 4.8 | 32 |
| $CH_4$ | 3.2 | 32 |

to crystalline hydrates or by Pauling as some sort of clathrate. Many organic solutes have similar thermodynamic properties and the interaction is frequently called hydrophobic bonding.

The addition of an electrolyte to water interferes with the gas dissolution and the organization of water around the gas. This frequently results in a decrease in the solubility or a "salting out". This salting out is frequently a linear function of the molar ionic strength

$$\log C = \log C^\circ + k\, I_V \tag{53}$$

When k is negative the result is a "salting out", and when k is positive the result is a "salting in". Values of k for various gases in seawater are given in Table 5.10. The ratio of the solubility in water to the value in seawater is equal to the activity coefficient of the gas

$$\gamma_g = C^\circ/C \tag{54}$$

Values of $\gamma_g$ for various gases are given in Table 5.11.

**TABLE 5.10**
**Comparison of the Measured and Calculated Salting Coefficients of Gases in Seawater and NaCl at t = 25°C**

| | Seawater | | NaCl | |
|---|---|---|---|---|
| Gas | Meas. | Calc. | Meas. | Calc. |
| He | 0.092 | 0.099 | 0.090 | 0.100 |
| Ne | 0.102 | 0.099 | 0.106 | 0.100 |
| Ar | 0.122 | 0.123 | 0.131 | 0.122 |
| $O_2$ | 0.122 | 0.135 | 0.142 | 0.134 |
| $N_2$ | 0.132 | 0.144 | 0.141 | 0.143 |

**TABLE 5.11**
**Activity Coefficients of Gases in Seawater (S = 35, t = 25°C)**

| Gas | $\gamma_g$ |
|---|---|
| $N_2$ | 1.24 |
| $O_2$ | 1.22 |
| Ar | 1.22 |
| Ne | 1.18 |
| He | 1.16 |
| Kr | 1.23 |
| $CO_2$ | 1.17 |
| CO | 1.23 |
| $CH_4$ | 1.24 |
| $H_2S$ | 1.03 |

Masterton has shown that the salting out of gases in seawater can be accounted for by using

$$k = k_{\alpha} + k_{\beta} \tag{55}$$

This equation is derived from the scale particle theory. The $k_{\alpha}$ term results from the free energy of cavity formation and $k_{\beta}$ results from the free energy of interaction between the gas molecule and the surrounding water molecules and ions. The first term is calculated from the diameter of the gas and water molecules, the number of water and ions per milliliter, and the diameter of the ion. The $k_{\alpha}$ term is always positive and leads to salting out. It increases as the diameter of the gas molecule increases and becomes a smaller positive number as the temperature rises.

The $k_{\beta}$ is related to the polarizability of the gas molecules and the ions, the total number of electrons in the ion, and the dipole movement of water. The $k_{\beta}$ is negative for all gases between 0 to 40°C and thus leads to salting in. The magnitude of $k_{\beta}$ decreases with increasing temperature. A comparison of the measured and calculated values of k in NaCl and seawater are shown in Table 5.10. The agreement is quite good.

## FURTHER READING

Broecker, W.S. and Peng, T.H., *Tracers in the Sea,* Eldigo Press, New York, 1982.

Kester, D.R., Dissolved Gases Other than $CO_2$, chap. 8, *Chemical Oceanography,* Vol. 1, 2nd ed., J.P. Riley and G. Skirrow, Eds., Academic Press, New York, 498-556 (1975).

Liss, P.S., Chemistry of the Sea Surface Microlayer, chap. 10, *Chemical Oceanography,* Vol. 2, 2nd ed., J.P. Riley and G. Skirrow, Eds., Academic Press, New York, 193-243 (1975).

Richards, F.A., Dissolved Gases Other than Carbon Dioxide, *Chemical Oceanography,* Vol. 1, 1st ed., J.P. Riley and G. Skirrow, Eds., Academic Press, New York, 197-225 (1965).

# 6 Table of Contents

**The Carbonate System** .......... 269
1. *Acid-Base Equilibria in Seawater* .......... 272
2. *Equilibria of Carbonate Species* .......... 277
3. *Parameters of the $CO_2$ System in Seawater* .......... 282
4. *Distribution of Carbonate Species* .......... 291
   4.1. $P_{CO2}$ .......... 291
   4.2. *pH* .......... 294
   4.3. $A_T$ .......... 298
   4.4. $\Sigma CO_2$ .......... 305
5. *$CaCO_3$ Dissolution in Seawater* .......... 308

*Further Reading* .......... 319

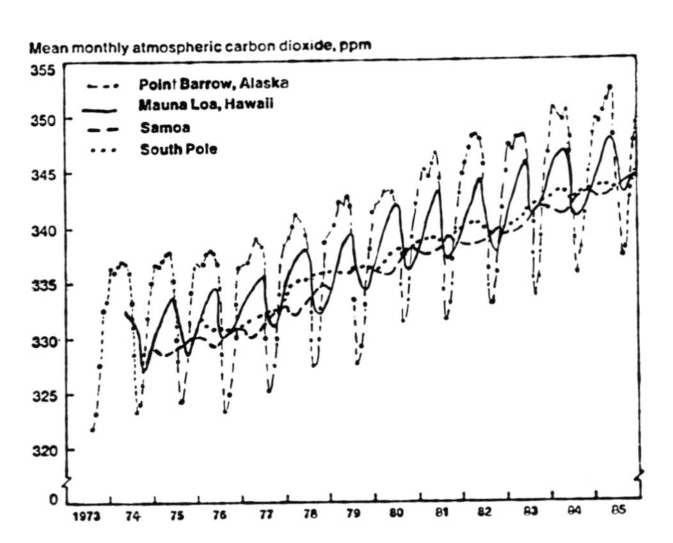


The difference in the annual cycling of carbon dioxide in the atmosphere over the last 24 years.

# 6 The Carbonate System

The major portion of carbon in the oceans occurs in the carbonate system. This system involves the following equilibria

$$CO_2(g) \rightarrow CO_2(aq) \quad (1)$$

$$CO_2(aq) + H_2O \rightarrow H^+ + HCO_3^- \quad (2)$$

$$HCO_3^- \rightarrow H^+ + CO_3^{2-} \quad (3)$$

$$Ca^{2+} + CO_3^{2-} \rightarrow CaCO_3(s) \quad (4)$$

The carbonate system is very important since it regulates the pH of seawater and controls the circulation of $CO_2$ between the biosphere, lithosphere, atmosphere, and oceans. Recent interest in the carbonate system has resulted from the "greenhouse effect" of $CO_2$. Atmospheric $CO_2$ has increased in the twentieth century (Figure 6.1) and since $CO_2$ can absorb infrared (IR) energy, the temperature of the earth will increase and could eventually melt the polar ice caps. The increase in $CO_2$ is related to the burning of fossil fuels (coal, petroleum, and natural gas) and the production of cement (Figure 6.2). Once the $CO_2$ is in the atmosphere, it is available for primary productivity and

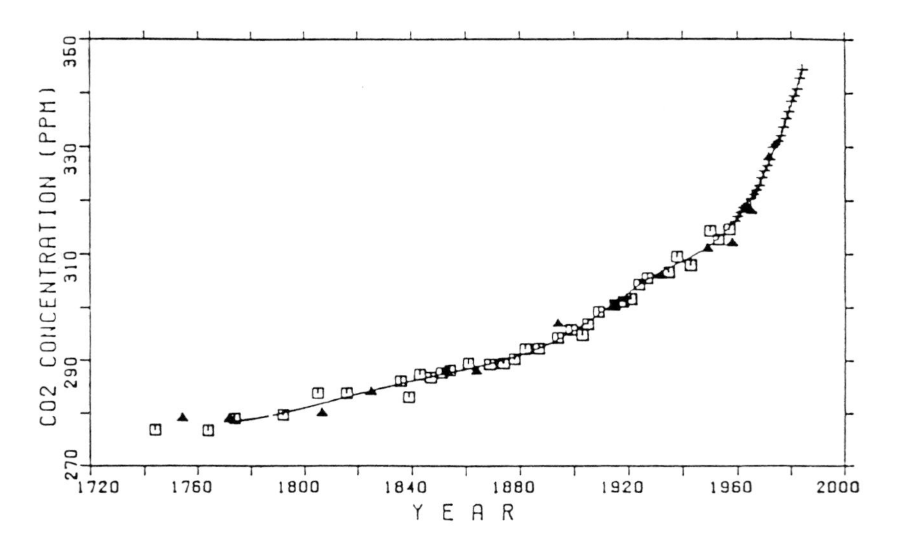


FIGURE 6.1. The partial pressure of carbon dioxide in the atmosphere over the last 3 centuries.

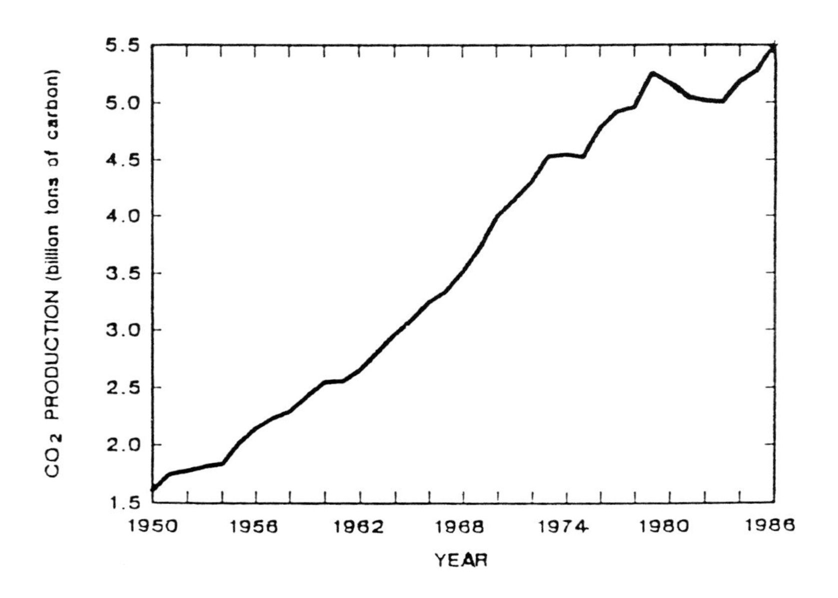


FIGURE 6.2. The carbon dioxide production over the last 30 years.

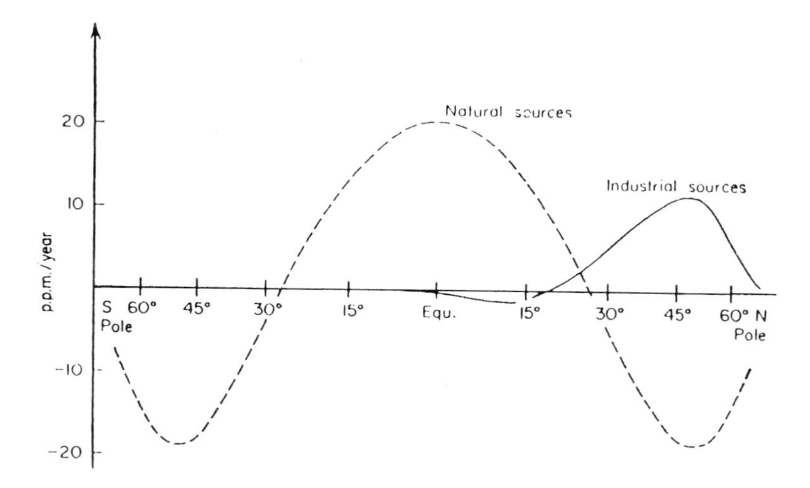


FIGURE 6.3. The anthropogenic inputs of carbon dioxide as a function of latitude.

weathering processes. It enters the oceans across the air-sea interface and participates in the equilibrium processes outlined by Equations 1 to 4. It also can be used by plants in primary productivity

$$CO_2 + H_2O \rightarrow CH_2O + O_2 \tag{5}$$

These processes are far from being simple since the rates of movement of $CO_2$ across the interface and from surface to deep waters varies with latitude, time, season, and biological processes. Diurnal and seasonal variations in the carbonate system are caused by the removal of $CO_2$ by photosynthesis and solar heating (Figure 6.3).* If the oceans were well mixed and in equilibrium with the atmosphere, most of the increased $CO_2$ would be absorbed. This, however, is not the case and the ocean response to increases in $CO_2$ is slow due to physical and chemical factors. The exchange involves the hydration of $CO_2$ which is a slow process relative to ionization. Approximate time scales for the mixing process can be determined using radioactive tracers to gain some idea of the mixing times. To use these estimates it is necessary to have some idea of the total carbon in various reservoirs. The most recent estimates are shown in Figure 6.4. The inorganic carbon estimates are reasonably accurate, but estimates for the carbon in the marine biosphere and humus are

* Figures 6.3, 6.12 to 6.14, 6.17, 6.18, and 6.32 from Skirrow, 1975; Figures 6.15, 6.24, 6.25, 6.33 to 6.35, and 6.38 from Broecker and Peng, 1982.

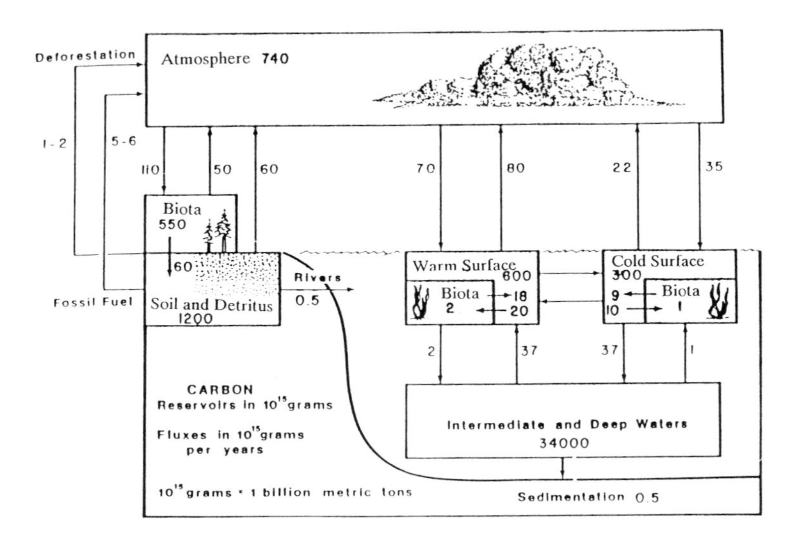


FIGURE 6.4. The reservoirs of carbon dioxide on the earth.

less precisely known. Most of the carbon in the oceans resides below the thermocline. The amount of carbon stored in carbonate rocks and sediments (not shown in Figure 6.4) is a lot larger than the cycled $CO_2$, but is not important in short time scales (years). Before we examine the distribution of $CO_2$ in the oceans, we will discuss the concepts needed to understand the carbonate system in the oceans.

## 1. ACID-BASE EQUILIBRIA IN SEAWATER

Before we discuss the equilibria of $CO_2$ in seawater, it is important to understand the concept of pH and understand how it is measured in seawater. Water is a weak electrolyte and dissociates according to Equation 6

$$H_2O \rightarrow H^+ + OH^- \tag{6}$$

This dissociation is defined by the equilibrium constant

$$K_w = a_H\, a_{OH}/a_{H2O} = [H^+][OH^-]\, \gamma_H\, \gamma_{OH}/a_{H2O} \tag{7}$$

where a = activity and $\gamma$ = activity coefficient. For dilute solutions $a_{H_2O}$ = 1.0 = $\gamma_H$ and $\gamma_{OH}$ and $K_w = 1 \times 10^{-14} = [H^+][OH^-]$ at 25°C. The pH is defined by

$$pH = -\log [H^+] = 7.0 \tag{8}$$

The thermodynamic dissociation constant varies as a function of temperature

$$\ln K_w = 149.9802 - 13847.26/T - 23.6521 \ln T \tag{9}$$

$K_w$ is also a function of pressure

$$\ln(K_W^P/K_w^0) = -(\Delta V^0/RT)P + (0.5\Delta K^0/RT)P^2 \tag{10}$$

The volume, $\Delta V^0$, and compressibility, $\Delta K^0$, changes are given by

$$\Delta V^0 = -25.60 + 0.2324\ t - 3.6246 \times 10^{-3}\ t^2 \tag{11}$$

$$10^3 \Delta K^0 = -7.33 + 0.1368\ t - 1.233 \times 10^{-3}\ t^2 \tag{12}$$

The concentration of $H^+ + OH^-$ is also affected by the major components in seawater. The stoichiometric product for the dissociation

$$K_w^* = K_w\ a_{H_2O}/\gamma_H\ \gamma_{OH} = [H^+]_T[OH^-]_T \tag{13}$$

The values of $K_w^*$ in seawater can be determined from

$$\ln K_w^* = \ln K_w + AS^{1/2} + BS \tag{14}$$

where

$$A = -97.9429 + 4149.915/T + 14.8269 \ln T \tag{15}$$

$$B = -2.3694 \times 10^{-2} \tag{16}$$

At 25°C and S = 35, $K_w^*$ is equal to $10^{-13.19}$ or $pK_w^* = 13.19$. This increase in the $pK_w^*$ from the value in fresh water is related to the interactions of $H^+$ with $SO_4^{2-}$ and $OH^-$ with $Mg^{2+}$

$$H^+ + SO_4^{2-} \rightarrow HSO_4^- \tag{17}$$

$$OH^- + Mg^{2+} \rightarrow MgOH^+ \tag{18}$$

As discussed earlier, this causes the free $H^+$ and $OH^-$ concentrations to decrease by 84 and 33%, respectively. The value of $K_w^*$ is thus given by

$$K_w^* = K_w\, a_{H_2O}/\gamma_H\, \gamma_{OH} = 10^{-14} \cdot 0.981/(0.71 \cdot 0.22) = 10^{-13.20} \tag{19}$$

From this short discussion of the values of $K_w$, it is possible to understand the various definitions of pH used in oceanography. The original definition of pH was made by Sorensen in 1909. He defined pH as measured for the cell

$$H_2(Pt)|Soln(X)||Salt\ Bridge|Reference\ Electrode \tag{20}$$

The Nernst equation gives

$$E = E^\circ + (2.30RT/F)pH(X) \tag{21}$$

The value of E° was determined by measuring the emf (E) in NaCl-HCl solutions of known $[H^+]$ determined using conductivity. Due to liquid junction differences between the solution-salt bridge reference solution, this method was not satisfactory. Bates developed a practical pH scale at the National Bureau of Standards. The NBS pH is defined by

$$pH_{NBS} = -\log a_H \tag{22}$$

Since it is not possible to determine individual activities of ions, this scale is based on a conventional definition of activity coefficients. A number of buffers were developed that had fixed values of pH on this scale at a given temperature. The scale is generally used by measuring the emf in a cell of the type

$$pH\ Glass\ Electrode|Solution(X)|KCl(aq)|Reference\ Electrode \tag{23}$$

where the reference electrode is normally a calomel electrode. The $pH_{NBS}$ of a solution is determined from emf measurements in (X) and the buffer. The values are determined from

$$pH(X) = pH(S) + [E(X) - E(S)]/(2.303RT/F) \tag{24}$$

In solutions of high ionic strength, this scale does not give reliable values due to differences in liquid junction potentials in the dilute buffers and the ionic media.

This difficulty has led various workers to develop pH scales that are defined using buffers solutions similar to those to be measured. The work in seawater is based on the total proton scale

$$pH_T = -\log [H^+]_T \tag{25}$$

where

$$[H^+]_T = [H^+]_F + [HSO_4^-] \tag{26}$$

(some workers have also considered the small amount of $H^+$ tied up as $HF^\circ$). The two scales are related by

$$10^{-pH_{NBS}} = f_H[H^+]_T = f_H[H^+]_F (1 + \beta_{HSO_4} [SO_4]_T + \beta_{HF}[F]_T) \tag{27}$$

where $f_H$ is the apparent total proton activity coefficient, $\beta_i$ values are the association constants, and $[H^+]_F$ is the free proton concentration. Buffers have also been developed that can be used to determine pH on the free scale

$$pH_F = -\log [H^+]_F \tag{28}$$

The free and total pH scales are related by

$$pH_T = pH_F - \log (1 + \beta_{HSO_4} [SO_4^{3-}]_T) \tag{29}$$

Since the values of $\beta_{HSO_4}$ are well known

$$\log \beta_{HSO_4} = A/T + B \tag{30}$$

where

$$A = 1226.966 + 65.6\ S^{1/2} \quad (31)$$

$$B = 6.09405 - 0.4502\ S^{1/2} + 1.3525 \times 10^{-2}\ S \quad (32)$$

the two scales are easy to convert from one to the other.

Since variations in the liquid junction potentials of various reference electrodes are different, it is better to use $pH_T$ or $pH_F$ scales. TRIS (Tris (hydroxymethyl) amino methane) buffers are available for seawater media that can be used to calibrate electrodes on these scales at a given temperature and salinity.

Although emf measurements are normally used to measure pH, it is also possible to use indicators that absorb light to measure pH. An indicator is an acid or base whose ionized and unionized species absorb light differently. If the stoichiometric constant is known for the indicator, the pH is given by

$$pH = pK^*_{HA} + \log C_A/C_{HA} \quad (33)$$

where $C_A$ and $C_{HA}$ are the concentrations of the acid anion and unionized acid. The values of $C_A$ and $C_{HA}$ are directly proportional to the absorbance of light at a given wavelength. This method of measuring pH may prove useful for monitoring changes over long periods of time.

When a solution contains a weak acid and its salt (e.g., acetic acid and sodium acetate), the addition of $H^+$ or $OH^-$ causes only a small change in the pH. The pH of this solution is given by Equation 33. Since $C_A = C^o_A - \Delta H^+$ and $C_{HA} = C^o_{HA} + \Delta H^+$, the ratio of $C_A/C_{HA}$ does not change

$$C_A/C^o_{HA} = (C^o_A - \Delta H^+)/(C^o_{HA} + \Delta H^+) \quad (34)$$

This buffering effect works best when $C^o_A = C^o_{HA}$ or the desired pH = $pK^*_{HA}$. The buffer capacity of an acid or base is defined by

$$\beta = \Delta C_B/\Delta pH \quad (35)$$

For a dibasic acid, $H_2A$, the buffer capacity is

$$\beta = 2.303 \left[ \frac{K^*_1 C_T C_H}{(K^*_1 + C_H)} + \frac{K^*_2 C_T C_H + C_H + C_{OH}}{(K^*_2 + C_H)^2} \right] \quad (36)$$

The maximum buffer capacity occurs where $C_A = C_{HA}$ and $C_{HA} = C_{H_2A}$.

To represent the concentrations of the components of an acid as a function of pH, one frequently uses a Bjerrum diagram. This diagram is simply a plot of the various forms of the acid (usually in % of the total) as a function of pH. For the ionization of a dibasic acid

$$H_2A \rightarrow H^+ + HA^- \tag{37}$$

$$HA^- \rightarrow H^+ + A^= \tag{38}$$

one has the following equations

$$K_1^* = [H^+][HA^-]/[H_2A] \tag{39}$$

$$K_2^* = [H^+][A^=]/[HA^-] \tag{40}$$

$$[H_2A]_T = [H_2A] + [HA^-] + [A^=] \tag{41}$$

The fraction of the various forms can be obtained by solving these equations. The solution gives

$$\alpha H_2A = (1 + K_1^*/[H^+] + K_1^*K_2^*/[H^+]^2)^{-1} \tag{42}$$

$$\alpha_{HA^-} = (1 + [H^+]/K_1^* + K_2^*/[H^+])^{-1} \tag{43}$$

$$\alpha_{A^=} = (1 + [H^+]/K_2^* + [H^+]^2/K_1^*K_2^*)^{-1} \tag{44}$$

From Equations 42 to 44, the values of $[H_2A] = [HA^-]$ at $pH = pK_1^*$ and $[HA^-] = [A^=]$ at $pH = pK_2^*$. Thus, if the values of $pK_1^*$ and $pK_2^*$ are known, it is possible to sketch the Bjerrum diagram (Figure 6.5).

## 2. EQUILIBRIA OF CARBONATE SPECIES

When $CO_2$ is in contact with water, equilibria as defined by Equations 1 to 4 will be established. Kinetics can affect the features of these reactions. Reaction 2 is first order with respect to $CO_2$ and has a first order rate constant

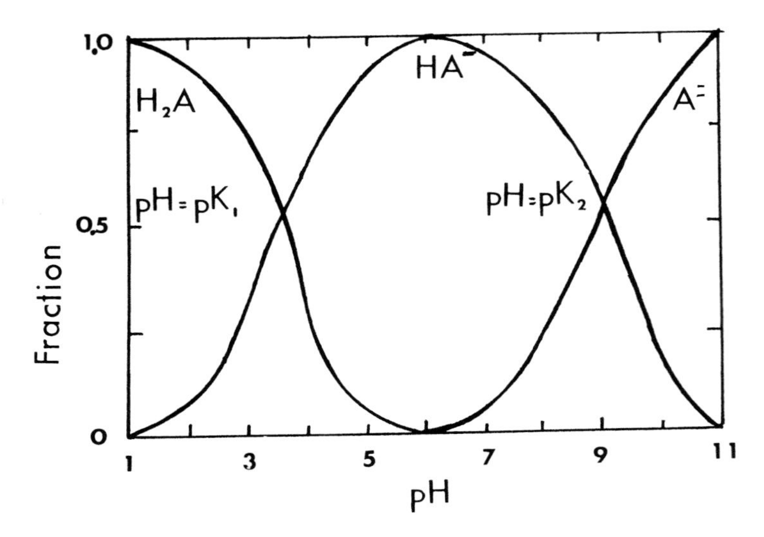


FIGURE 6.5. The fractions of a dibasic acid as a function of pH.

$k_1 = 0.03$ s or a half-time $t_{1/2} = \ln 2/k_1 = 23$ s. The reaction of $OH^- + CO_2 \rightarrow HCO_3^-$ is second order with respect to $[CO_2]$ and $[OH^-]$

$$-d[CO_2]/dt = k_2[CO_2][OH^-] \tag{45}$$

where $k_2 = 8500\ M^{-1}\ s^{-1}$. This process is important at high values of pH. The dehydration reaction, $H_2CO_3 \rightarrow CO_2 + H_2O$, is first order with respect to $[H_2CO_3]$ with rate constant $k_{-1} = 20\ s^{-1}$ and $t_{1/2} = 0.03$ s. The values for the forward and backward reactions

$$CO_2 + H_2O \underset{k_{-1}}{\overset{k_1}{\rightleftarrows}} H_2CO_3 \tag{46}$$

can be used to determine the equilibrium ratio

$$K = \frac{k_1}{k_{-1}} = \frac{0.03}{20} = \frac{1}{670} \tag{47}$$

This indicates that at equilibrium the concentration of $CO_2$ is about 670 times higher than $H_2CO_3$. This has led workers to use the so-called hydration convention to define the first ionization of carbonic acid (Equation 2). The stoichiometric association constant for the first ionization is defined by

$$K_1^* = [H^+]_T\ [HCO_3^-]_T/[CO_2^*] \tag{48}$$

where $[CO_2^*] = [CO_2] + [H_2CO_3]$ and the subscript T is used to denote total concentrations. The concentration of dissolved $CO_2$ is related to its partial pressure by

$$[CO_2^*] = P_{CO_2}\,\alpha_{CO_2} \tag{49}$$

where $\alpha_{CO_2}$ is the reciprocal of the Henry's law constant similar to the values described for other gases in Chapter 5. The values of $\alpha_{CO_2}$ can be obtained from

$$\ln \alpha_{CO_2} = -60.2409 + 93.4517(100/T) + 23.3585\ \ln(T/100) +$$
$$S[0.023517 - 0.023656(T/100) + 0.0047036(T/100)] \tag{50}$$

where $T = t°C + 273.15$. The solubility of $CO_2$ decreases with increasing temperature and salinity as for other gases. The solubility of $CO_2$ is greater than that of $O_2$ or $N_2$. The air ratios are $N_2{:}O_2{:}CO_2 = 240{:}630{:}1$ while the solution ratios are 28:19:1. Henry's law is not obeyed at high pH due to the formation of $HCO_3^-$ and $CO_3^{2-}$.

The stoichiometric value of $K_1^*$ is related to the thermodynamic value by

$$K_1 = a_H\, a_{HCO_3}/a_{CO_2}\, a_{H_2O} = K_1^*\, \gamma_H\, \gamma_{HCO_3}/\gamma_{CO_2}\, a_{H_2O} \tag{51}$$

$\gamma_{CO_2} = [CO_2]°/[CO_2]$ where the superscript zero denotes the solubility in pure water. The activity coefficients given above are the total values and include the effects due to the formation of ion pairs. The values of $pK_1^*$ in seawater can be calculated from

$$pK_1^* = pK_1 + (-840.39/T + 19.894\ \ln T)S^{1/2} + 0.00668S \tag{52}$$

The thermodynamic values of $pK_1$ in water are given by

$$pK_1 = 6320.81/T - 126.3405 + 19.568\ \ln T \tag{53}$$

The stoichiometric association constant for the second ionization of carbonic acid is defined by

$$K_2^* = [H^+]_T[CO_3^{2-}]_T/[HCO_3^-]_T \tag{54}$$

which is related to the thermodynamic value, $K_2$, by

$$K_2 = K_2^* \gamma_H \gamma_{CO_3} / \gamma_{HCO_3} \quad (55)$$

where the activity coefficients are total values which include the effect due to ionic interactions. The values of $K_2^*$ in seawater can be determined from

$$pK_2^* = pK_2 + (-690.59/T + 17.176 - 2.6719 \ln T)S^{1/2} + 1.0217S \quad (56)$$

where the thermodynamic value, $pK_2$, is given by

$$pK_2 = 5143.69/T - 90.1833 + 14.613 \ln T \quad (57)$$

The effect of pressure on $K_1^*$ and $K_2^*$ can be estimated from

$$\ln(K_i^P/K_i^0) = -(\Delta V_i/RT)P + (0.5\ \Delta K^0/RT)P^2 \quad (58)$$

where

$$-\Delta V_1 = 25.50 + 0.151(S - 34.8) - 0.1271t \quad (59)$$

$$-10^3 \Delta K_1 = 3.08 + 0.578(S - 34.8) - 0.0877t \quad (60)$$

$$-\Delta V_2 = 15.82 - 0.321(S - 34.8) + 0.0219t \quad (61)$$

$$-10^3 \Delta K_2 = -1.13 + 0.314(S - 34.8) + 0.1475t \quad (62)$$

Since boric acid makes up one of the major constituents of seawater, it is necessary to consider its ionization.

$$HB \rightarrow H^+ + B^- \quad (63)$$

where $HB = B(OH)_3$ and $B^- = B(OH)_4^-$. The dissociation constant is defined by

$$K_{HB}^* = [H^+]_T[B^-]_T/[HB]_T \quad (64)$$

The values of $pK^*_{KB}$ in seawater can be calculated from

$$\ln K^*_{HB} = \ln K_{HB} + (0.5998 - 75.25/T)S^{1/2} - 1.767S \qquad (65)$$

where the thermodynamic value is given by

$$\ln K_{HB} = 148.0248 - 8966.90/T - 24.4344 \ln T \qquad (66)$$

The effect of pressure on $K^*_{HB}$ can be determined from Equation 58 where

$$-\Delta V_{HB} = 29.48 - 0.295(S - 34.8) - 0.1622t \qquad (67)$$

$$-10^3\Delta K_{HB} = 2.84 - 0.354(S - 34.8) \qquad (68)$$

The solubility of $CaCO_3$ in its two major forms, calcite and aragonite, is also needed when studying the carbonate system. The stoichiometric solubility product is given by

$$K^*_{sp} = [Ca^{2+}][CO_3^{2-}] \qquad (69)$$

and is related to the thermodynamic value by

$$K^*_{sp} = K_{sp}/\gamma^T Ca \gamma^T CO_3 \qquad (70)$$

The values in seawater are given by

$$\ln K^*_{sp} = \ln K_{sp}(i) + (1.6233 - 118.64/T)S^{1/2} - 6.999S \qquad (71)$$

where $\ln K_{sp}$ is given by

$$\ln K_{sp}\ (\text{calcite}) = 303.1308 = 13348.09/T - 48.7537 \ln T \qquad (72)$$

$$\ln K_{sp}\ (\text{aragonite}) = 303.5363 - 13348.09/T - 48.7537 \ln T \qquad (73)$$

The effect of pressure on the solubility of calcite and aragonite is calculated from Equation 58 where

$$-\Delta V_c = 48.76 - 0.5304t \qquad (74)$$

$$-\Delta V_a = 46.0 - 0.5304t \quad (75)$$

$$-10^3\Delta K_c = 11.76 - 0.3692t \quad (76)$$

$$-10^3\Delta K_a = 11.76 - 0.3692t \quad (77)$$

## 3. PARAMETERS OF THE $CO_2$ SYSTEM IN SEAWATER

To characterize the various components of the carbonate system one must measure at least two of the four measurable parameters

1. pH
2. Alkalinity, $(A_T)$
3. Total $CO_2$, $(\Sigma CO_2)$
4. Partial pressure of $CO_2$, $(P_{CO_2})$

The pH can be measured using electrodes or indicators. If the electrodes are calibrated using seawater buffers, the accuracy is $\pm 0.01$ and the precision can be $\pm 0.002$ pH units. If the measurements are made at 25°C and 1 atm, it is necessary to determine the *in situ* values at a given depth in the ocean. The effect of temperature can be determined from

$$pH_t = pH_{25} + A(t - 25) + B(t - 25)^2 \quad (78)$$

$$10^3A = -9.296 + 32.505(pH_{25} - 8) + 63.9806(pH_{25} - 8)^2 \quad (79)$$

$$10^4B = 3.916 + 23.000(pH_{25} - 8) + 41.637\,(pH_{25} - 8)^2 \quad (80)$$

The effect of pressure can be estimated from

$$pH_t^P = pH_t^0 + AP$$

$$-10^3A = 0.424 - 0.0048(S - 35) - 0.00282t - 0.0816(pH_t^\circ - 8) \quad (81)$$

**TABLE 6.1**
**Contribution of Various Components to $A_T$ in Seawater**

| Species | % $A_T$ |
|---|---|
| $HCO_3^-$ | 89.8 |
| $CO_3^{2-}$ | 6.7 |
| $B(OH)_4$ | 2.9 |
| $SiO(OH)_3^-$ | 0.2 |
| $MgOH^+$ | 0.1 |
| $OH^-$ | 0.1 |
| $HPO_4^-$ | 0.1 |

The total alkalinity of seawater is defined as the concentration of all the bases that can accept $H^+$ when a titration is made with HCl to the carbonic acid endpoint. The value of $A_T$ is given by

$$A_T = [HCO_3^-] + 2[CO_3^{2-}] + [B(OH)_4^-] + [OH^-] - [H^+] + [SiO(OH)_3^-] + [MgOH^+] + 2[HPO_4^{2-}] + 3[PO_4^{3-}] \quad (82)$$

The percent of $A_T$ due to the various bases is shown in Table 6.1 for deep ocean waters with pH = 8, $SiO(OH)_3^- = 10^{-5.25}$, and $HPO_4^{2-} = 10^{-5.52}$. For most waters, $HCO_3^-$, $CO_3^{2-}$, and $B(OH)_4^-$ are the most important bases. For anoxic waters, $HS^-$ and $NH_3$ can also contribute to the total alkalinity.

The carbonate alkalinity, $A_C$, is defined by

$$A_C = [HCO_3^-] + 2[CO_3^{2-}] \quad (83)$$

and is calculated from

$$A_C = A_T - \Sigma B_i \quad (84)$$

where $\Sigma B_i = [B(OH)_4^-] + \ldots$, the sum of all the bases other than $HCO_3^-$ and $CO_3^{2-}$.

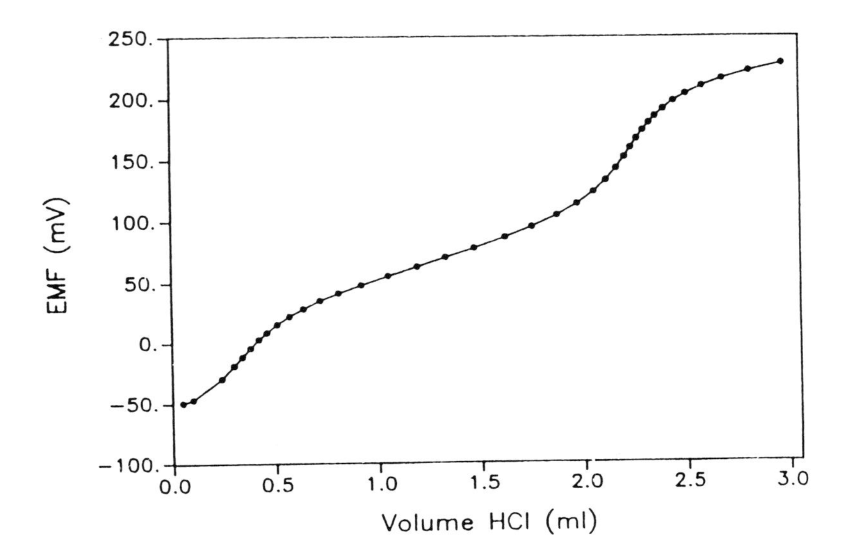


FIGURE 6.6. The emf of a pH electrode during the titration of seawater with hydrochloric acid.

The $[B(OH)_4^-]$ concentration, which is the largest source of $\Sigma B_i$, can be calculated from

$$[B(OH)_4^-] = K^*_{HB}[B]_T/(K^*_{HB} + [H^+]_T) \tag{85}$$

where $[B]_T = 1.212 \times 10^{-5}S$.

The $A_T$ for seawater is determined by titrating a given amount of seawater to the carbonic acid endpoint. The titration is followed by measuring the emf of a glass pH and reference electrode. A typical titration of 234 $cm^3$ of S = 35 seawater with 0.25 *N* HCl is shown in Figure 6.6. The titration shows two endpoints, $V_1$ and $V_2$. The value of $A_T$ is determined from

$$A_T = V_2\, N_{HCl}/W \tag{86}$$

where $N_{HCl}$ is the normality of the HCl solution and W is the weight of the seawater titrated (W = vol × density). The difference between the first and second endpoint can be used to determine the $\Sigma CO_2$

$$\Sigma CO_2 = (V_2 - V_1)\, N_{HCl}/W \tag{87}$$

The $pH_T$ of the initial solution before the addition of HCl can be determined from

$$pH_T = -(E - E^*)/(2.303RT/F) \quad (88)$$

The quantities $V_1$, $V_2$, and $E^*$ are determined from the measured emf (E) and the volume of added HCl (V) using an iteration technique. A reasonable approximation of $V_1$, $V_2$, and $E^*$ can be determined by calculating the $[H^+]_T$ from

$$[H^+]_T = 10^{[(E - 400)/k]} \quad (89)$$

where k = 2.303 RT/F = 59.16 mv at 25°C.

Using the emf data from 150 to 210 mv, the function $F_2$ is calculated from

$$F_2 = (V_0 + V)[H^+]_T \quad (90)$$

where $V_0$ is the initial volume of seawater. The values of $F_2$ are then fit to the linear equation

$$V = a + bF_2 \quad (91)$$

(Figure 6.7). At $F_2 = 0$, $V = V_2$. The value of $V_1$ is determined from the initial emf data ($\simeq -15$ to 50 mv) using the function

$$F_1 = (V_2 - V)[H^+]_T \quad (92)$$

The values of $F_1$ are fit to the equation

$$V = a + bF_1 \quad (93)$$

At $F_1 = 0$, $V = V_1$ (Figure 6.8). The value of $E^*$ is determined from the values of E past the second endpoint ($V > V_2$) using the equation

$$E^* = E - k \log \left[\frac{(V - V_2)}{(V_0 + V)} N_{HCl}\right] \quad (94)$$

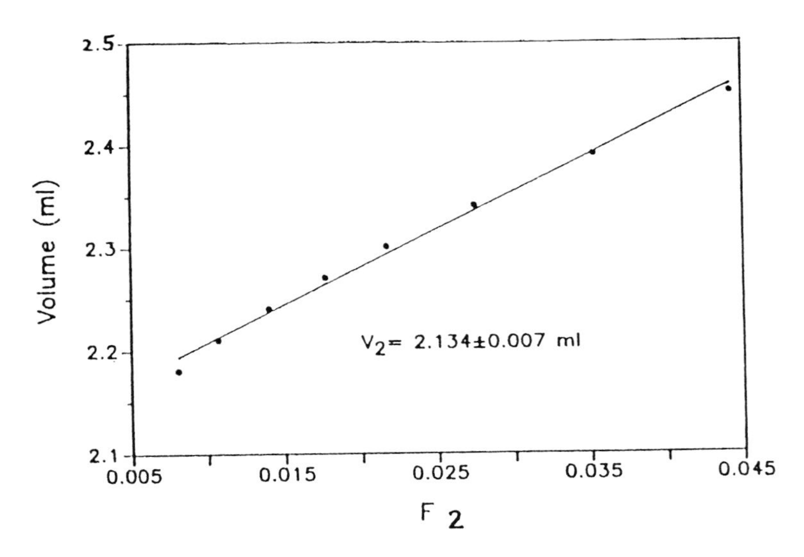


FIGURE 6.7. The Gran Function $F_2$ as a function of the volume of HCl added during the titration of seawater.

The average E* can be used to refine the calculations. For the data shown, $A_T = 2.237 \pm 0.007$ meq kg$^{-1}$, $\Sigma CO_2 = 2.023 \pm 0.013$ m mol kg$^{-1}$, $pH_T = 7.731 \pm 0.019$, and $E^* = 407.6 \pm 1.1$ mv.

To make more accurate calculations it is necessary to make corrections for side reactions. The modified $F_2'$ is given by

$$F_2' = (V_0 + V)\{[H^+]_T + [HSO_4^-] + [HF] - [HCO_3^-]\} \quad (95)$$

$$F_1' = (V_2 - V)\{([H^+]^2 - K_1^*K_2^*)/(K_1^*[H^+] + 2K_1^*K_2^*)\} + (V_0 + V)\{[H^+] + [HSO_4^-] + [HF] - [B(OH)_4^-] - [OH^-] \times [H^+]^2 + K_1^*[H^+] + K_1^*K_2^*/N_{HCl}(K_1^*[H^+] + 2K_1^*K_2^*)\} \quad (96)$$

Computer programs make these calculations easier to do than one would expect from examining the complicated equation for $F_1'$.

Although it is possible to determine $\Sigma CO_2$ by titrations, more reliable values can be obtained by direct measurements. This is done by stripping the inorganic $CO_2$ from solution by adding acid. Nitrogen can be used as the stripping gas. The $CO_2$ can be collected in a liquid air trap and analyzed by gas chromatography, infrared spectroscopy, or by coulometry.

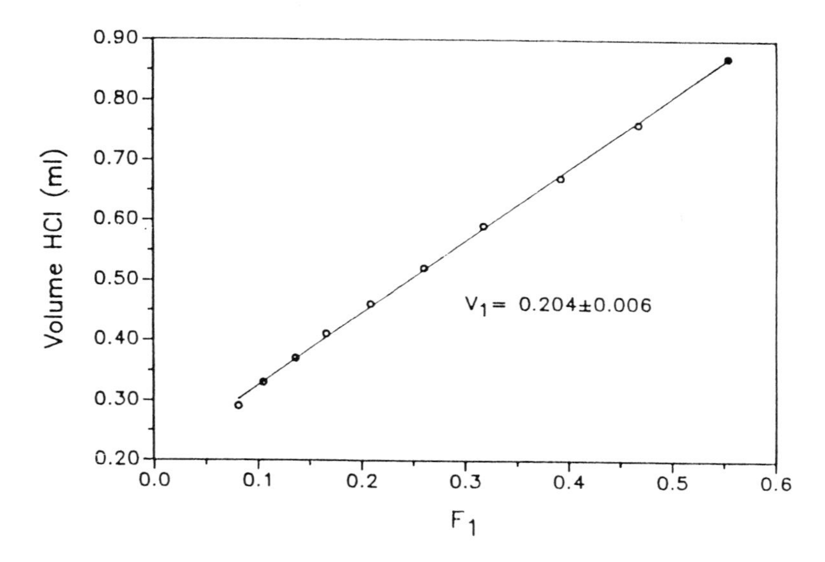


FIGURE 6.8. The Gran Function $F_1$ as a function of the volume of HCl added during the titration of seawater.

The partial pressure of $CO_2$ in seawater is determined by equilibrating the sample with air or nitrogen. The $CO_2$ in the equilibrated gas is measured by the same techniques used to measure $CO_2$. For more reliable measurements of $P_{CO_2}$ in seawater or the air above the surface of the oceans, standard gas volumetric methods developed by Keeling (University of California, San Diego) are used.

As mentioned earlier, any two combinations of the four observable parameters can be used to characterize the carbonate system. It is also possible to use three parameters. This gives a total of ten combinations that can be used. The investigator must make a selection based on his needs after considering both the desired analytical precision and area of interest. Park has given all the equations needed to determine the carbonate parameters. We will give the basic equations for the commonly determined parameters pH and $A_C$, and $\Sigma CO_2$ and $A_C$. For pH and $A_C$ we have

$$[HCO_3^-] = A_C/(1 + 2K_2^*/[H^+]) \tag{97}$$

$$[CO_3^{2-}] = A_C K_2^*/([H^+] + 2K_2^*) \tag{98}$$

$$[CO_2] = (A_C[H^+]/K_1^*)/(1 + 2K_2^*/[H^+]) \tag{99}$$

$$\Sigma CO_2 = [HCO_3^-] + [CO_3^{2-}] + [CO_2] \quad (100)$$

$$P_{CO_2} = [CO_2]/\alpha \quad (101)$$

The values of $K_1^*$, $K_2^*$, $\alpha$, and $[H^+]$ are determined for the *in situ* temperature, pressure, and salinity.

If the $A_C$ and $\Sigma CO_2$ are determined, the various components of the carbonate system can be determined from

$$[CO_2] = \Sigma CO_2 - A_C + (A_CK_R - \Sigma CO_2K_R - 4A_C + Z)/2(K_R - 4) \quad (102)$$

$$[HCO_3^-] = (\Sigma CO_2K_2^* - Z)/(K_2 - 4) \quad (103)$$

$$[CO_3^{2-}] = (A_CK_R - \Sigma CO_2K_R - 4A_C + Z)/2(K_R - 4) \quad (104)$$

where $K_R = K_1^*/K_2^*$ and Z is given by

$$Z = [(4A_C + \Sigma CO_2K_R - A_CK_R)^2 + 4(K_R - 4)A_C^2]^{1/2} \quad (105)$$

The value of $[H^+]$ needed to determine $A_C$ and $A_T$ can be measured or calculated by solving the cubic equation

$$[H]^3 + [H]^2\{[K_1^* (A - 1) + K_{HB}^* (A - B)]/A\} + [H^+]\{K_1^*K_{HB}^* (A - B - 1) + K_1^*K_2^* (A - 2)\}/A + K_1^*K_2^*K_{HB}^* (A - B - 2)/A = 0 \quad (106)$$

which is defined by combining the above equations. The values of A and B in this equation are given by

$$A = A_T/\Sigma CO_2 \approx 1.05 \quad (107)$$

$$B = [B]_T/\Sigma CO_2 \approx 0.18 \quad (108)$$

**TABLE 6.2**
**Changes in the $CO_2$ System Due to the Oxidation of Plant Material**

| | AOU (m*M*) | | | |
|---|---|---|---|---|
| | **Initial**[a] | **0.13** | **0.26** | **% Change** |
| $\Delta CO_2$ | 0 | 0.10 | 0.20 | — |
| $\Sigma CO_2$ | 2.200 | 2.300 | 2.400 | 9.1 ± 0.1 |
| $A_C$ | 2.487 | 2.487 | 2.487 | 0 |
| $P_{CO}$ | 350 | 610 | 1.160 | 231 ± 1.0 |
| pH | 8.200 | 8.001 | 7.753 | −5.5 ± 0.04 |
| $[CO_2]$ | 0.012 | 0.021 | 0.040 | 233 |
| $[HCO_3^-]$ | 1.889 | 2.072 | 2.234 | 18 |
| $[CO_3^{2-}]$ | 0.299 | 0.208 | 0.126 | −58 |

[a] All the concentrations are m*M*.

If the value of $\Sigma CO_2$ is not measured, it is estimated using a pH = 8.0 from the equation

$$\Sigma CO_2 = A_C(1 + [H^+]/K_1^* + K_2^*/[H^+])/(1 + 2K_2^*/[H^+]) \qquad (109)$$

The cubic equation for $[H^+]$ can be solved by using the solution to a cubic equation or using iterative techniques. The process is repeated until self-consistent values of $\Sigma CO_2$ and $[H^+]$ are determined.

To select the best parameters needed to study the carbonate system, we can examine how the system changes during the formation and breakdown of organic carbon and the dissolution and precipitation of $CaCO_3$. The largest changes in the $CO_2$ system in deep waters are due to the oxidation of organic carbon. This can be represented by the reaction

$$(CH_2O)_{106}\,(NH_3)_{16}\,H_3PO_4 + 138O_2 \rightarrow 106CO_2 + 122H_2O + 16HNO_3 + H_3PO_4 \qquad (110)$$

This oxidation can be followed by considering the changes in the apparent oxygen utilization (AOU). The effect of a change in AOU of 0.13 and 0.26 m*M* is shown in Table 6.2. The largest change occurs in $P_{CO_2}$ followed by

**TABLE 6.3**
**Changes in the $CO_2$ System Due to the Dissolution of $CaCO_3$**

| | $\Delta CaCO_3$ (m*M*) | | | |
|---|---|---|---|---|
| | **Initial**[a] | **0.05** | **0.10** | **% Change** |
| $\Delta CO_2$ | 0 | 0.05 | 0.10 | — |
| $\Sigma CO_2$ | 2.200 | 2.250 | 2.300 | 4.5 ± 0.1 |
| $A_C$ | 2.487 | 2.587 | 2.687 | 72 ± 0.05 |
| $P_{CO}$ | 350 | 310 | 290 | −17 ± 1.0 |
| pH | 8.200 | 8.264 | 8.321 | 1.5 ± 0.04 |
| $[CO_2]$ | 0.012 | 0.011 | 0.010 | −17 |
| $[HCO_3^-]$ | 1.889 | 1.892 | 1.844 | 0.3 |
| $[CO_3^{2-}]$ | 0.299 | 0.348 | 0.397 | 33 |

$\Sigma CO_2$ and pH. $A_C$, the carbonate alkalinity, does not change. If one considers the present capabilities of measuring $P_{CO_2}$ (±0.1%), $\Sigma CO_2$ (±0.17%), and pH (±0.04%), the best selection would be pH − $P_{CO_2}$ followed by $P_{CO_2}$ − $\Sigma CO_2$ and pH − $\Sigma CO_2$.

The changes in the $CO_2$ system due to the dissolution of $CaCO_3$ in deep waters are shown in Table 6.3. The greatest change occurs in $A_C$ with $P_{CO_2}$, with $\Sigma CO_2$ and pH following. The best combination is $A_C$-$\Sigma CO_2$, followed by pH-$A_C$ and $A_C$-$P_{CO_2}$. Thus, the classical pH-$A_C$ and $A_C$-$\Sigma CO_2$ obtained by acid titrations represent the best approach for studying changes in the carbonate system.

If one combines the AOU and $CaCO_3$ effects, one finds

$$(CH_2O)_{106}\,(NH_3)_{16}\,H_3PO_4 + 138O_2 + 124CO_3^{2-} \longrightarrow 16NO_3^- + HPO_4^{2-} + 230HCO_3^- + 16H_2O \quad (111)$$

The $CO_3^{2-}$ ions formed from the dissolution of $CaCO_3$ react with the protons formed from the oxidation of plant material. If x $\mu M$ of $CaCO_3$ and y $\mu M$ of organics are decomposed, the changes in $A_T$, $\Sigma CO_2$, and $NO_3^-$ are given by

$$\Delta A_T = 2x - 17y \quad (112)$$

$$\Delta\Sigma CO_2 = x + 106y \quad (113)$$

$$\Delta NO_3 = 16y \tag{114}$$

The changes in $Ca^{2+}$ are given by

$$\Delta Ca = 0.463\ \Delta A_T + 0.074\ \Delta\Sigma CO_2 = 0.5\Delta A_T + 0.53\ \Delta NO_3 \tag{115}$$

This equation has been shown by Chen to predict changes in $Ca^{2+}$ as a function of depth in the Pacific. The changes in inorganic carbon to organic carbon are given by

$$\text{Inorg C/Org C} = x/106y = 16\Delta Ca/106\Delta NO_3 =$$

$$(8A_T + 8.5\Delta\ NO_3)/106\Delta\ NO_3 \tag{116}$$

This equation has also been shown by Chen to predict reasonable ratios of Inorg C/Org C in ocean waters.

If one uses these equations to examine the changes in $\Sigma CO_2$ in old, deep waters and younger waters, it is possible to make an estimate of the changes due to the increase of $CO_2$ in the atmosphere due to the burning of fossil fuels. The estimated increase (~260 ppm for preindustrial values) are in reasonable agreement with other estimates (Figure 6.1).

## 4. DISTRIBUTION OF CARBONATE SPECIES

The distribution of the various components of the $CO_2$ system in the oceans has been studied by many workers. Skirrow has reviewed much of the earlier work. The extensive studies of the carbonate system during the GEOSECS program have yielded a wealth of information on the $CO_2$ system. The published Atlas gives detailed profiles. We have selected stations in the North Atlantic (#115, 28N, 26W) and the North Pacific (#204, 31N, 150E) to demonstrate depth dependence in the two major oceans. In this section we will discuss the distribution of the $CO_2$ parameters.

### 4.1. $P_{CO_2}$

The partial pressures of $CO_2$ in the atmosphere and the oceans have been studied by a number of workers. The classical measurements of $P_{CO_2}$ in the atmosphere were made by Keeling. His measurements at the Mauna Loa

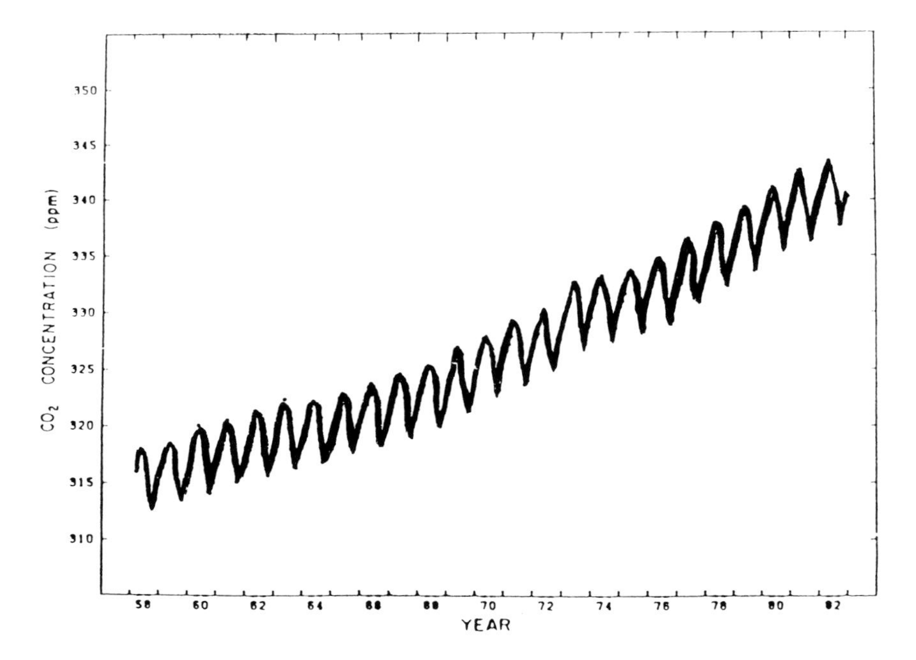


FIGURE 6.9. The annual cycling of carbon dioxide in the atmosphere over the last 24 years.

Observatory in Hawaii were started in 1958. These results are shown in Figure 6.9. The measurements clearly demonstrate the increase of $CO_2$ in the atmosphere due to the increased burning of fossil fuels. Although the rates of increase are the same as increases in the use of fossil fuels, the amounts are only one half of the expected values. This apparently is due to uptake by the oceans. In recent years, the values of $P_{CO_2}$ have been measured at other locations. These results are shown in Figure 6.10. The increases are the same as at Mauna Loa, but the yearly cyclic variations are larger in Alaska and smaller in Samoa and the South Pole. The average $CO_2$ growth is 1.5 ppm/year. This seasonal variability (shown in Figure 6.11) is caused by variations in the photosynthesis and respiration of land plants. The maximum occurs in April and May and the minimum occurs in September and October. Unfortunately, we have no historical data of sufficient accuracy to determine values of $P_{CO_2}$ in surface waters of the oceans over time.

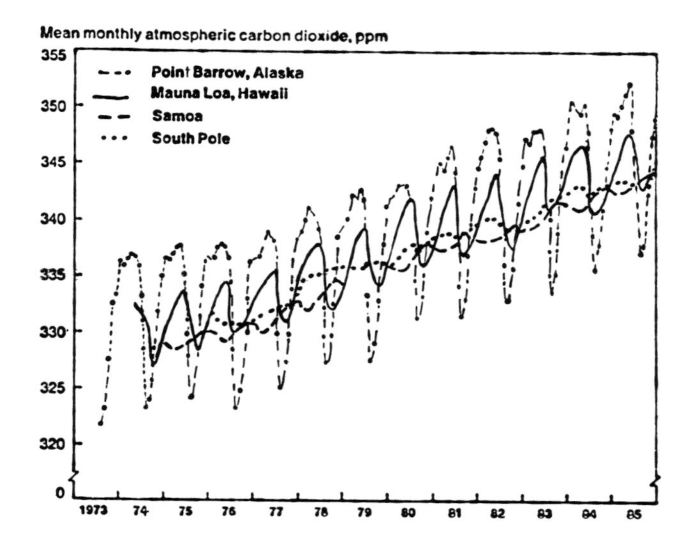


FIGURE 6.10. The difference in the annual cycling of carbon dioxide in the atmosphere over the last 24 years.

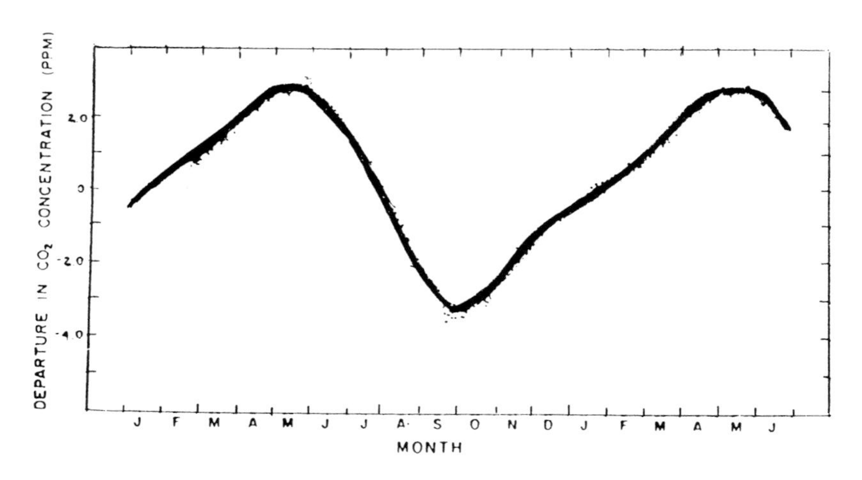


FIGURE 6.11. The mean annual cycling of carbon dioxide in the atmosphere.

Changes in $P_{CO_2}$ in surface waters can be caused by

1. Removal by photosynthesis
2. Removal by formation of $CaCO_3$
3. Removal by solar heating
4. Addition by oxidation of plant material
5. Addition by dissolution of $CaCO_3$
6. Addition by increases in $CO_2$ in the atmosphere from fossil fuel burning

Unravelling all these effects is even more difficult due to the sluggish response of the oceans to changes in the level of $CO_2$ in the atmosphere. As with other gases, the driving force of $CO_2$ across the air-sea interface is the differences between the concentrations in the atmosphere and oceans. If rapid exchange takes place, one would expect the $P_{CO_2}$ in the atmosphere to be equal to the values in the surface waters. If the exchange is sluggish, the $P_{CO_2}$ in surface waters will be higher in upwelling areas and lower in colder waters than the values in the atmosphere. Measurements in the Atlantic (Figure 6.12) show higher values of $P_{CO_2}$ near the equator apparently due to equatorial upwelling and lower values due to high productivity. The east-west section in the North Pacific (Figure 6.13) shows a large decrease in $P_{CO_2}$ due to downwelling. The north-south track in the Indian Ocean shows a gradual decrease (Figure 6.14) toward the colder waters in the south. The north-south profile in the western Pacific shown in Figure 6.15 indicates that the surface waters are in equilibrium with the atmospheric values except in the equatorial upwelling regions. The calculated values of $P_{CO_2}$ in surface waters using $A_T$ and $\Sigma CO_2$ are lower than the directly measured values and show more scatter; however, they also indicate that higher values exist near the equator.

The depth profile of $P_{CO_2}$ for the Atlantic and Pacific are shown in Figure 6.16. These values were calculated from $A_T$ and $\Sigma CO_2$ so they may not have the greatest accuracy. The general trends, however, are real and as expected. The surface values are similar to atmospheric values. The values increase to a maximum at 1 km due to the oxidation of plant material. The deep water values are higher in the Pacific due to the fact that the waters are older and due to the higher primary productivity in the surface waters.

### 4.2. pH

The pH of most surface waters in equilibrium with the atmosphere is 8.2 ± 0.1. Although early studies have been made on the pH on surface waters using NBS buffers, they show considerable scatter (Figure 6.17). In closed or small bodies of water, the pH can show diurnal variations and cycle between 8.2 to 8.9. The decrease occurs in the evening due to the respiration of organisms and the increase occurs in the afternoon due to photosynthesis (Figure 6.18). Changes in the pH with depth in the Atlantic and Pacific are shown in Figure 6.19. The pH goes through a maximum in surface waters due to photosynthesis. The loss of $CO_2$ increases the pH. The pH then decreases due to the oxidation of plant material and goes through a minimum at about 1000 m. This minimum coincides with the $O_2$ minimum and the maximum in $P_{CO_2}$. The pH increase in deep waters is due to the dissolution

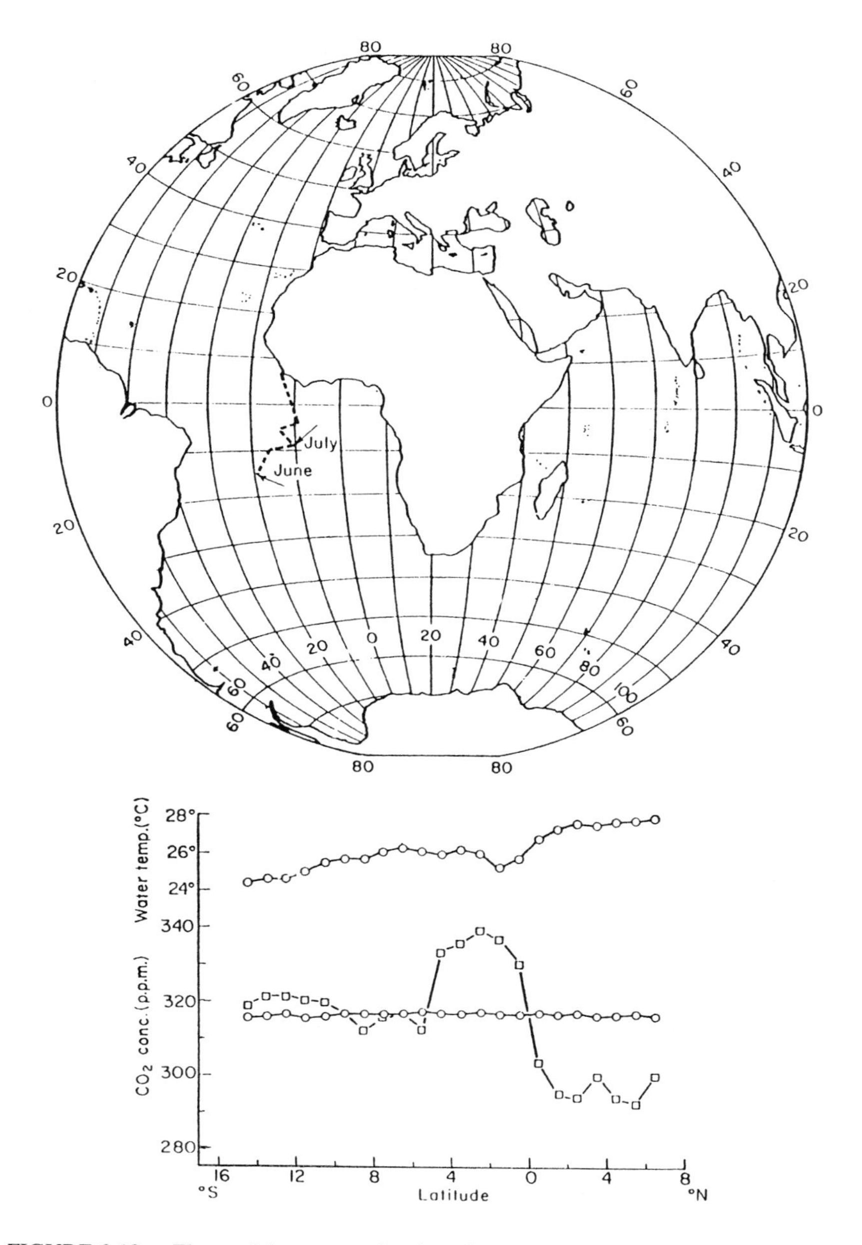


FIGURE 6.12. The partial pressure of carbon dioxide in the atmosphere (open circles) and surface waters (squares) of the Atlantic Ocean.

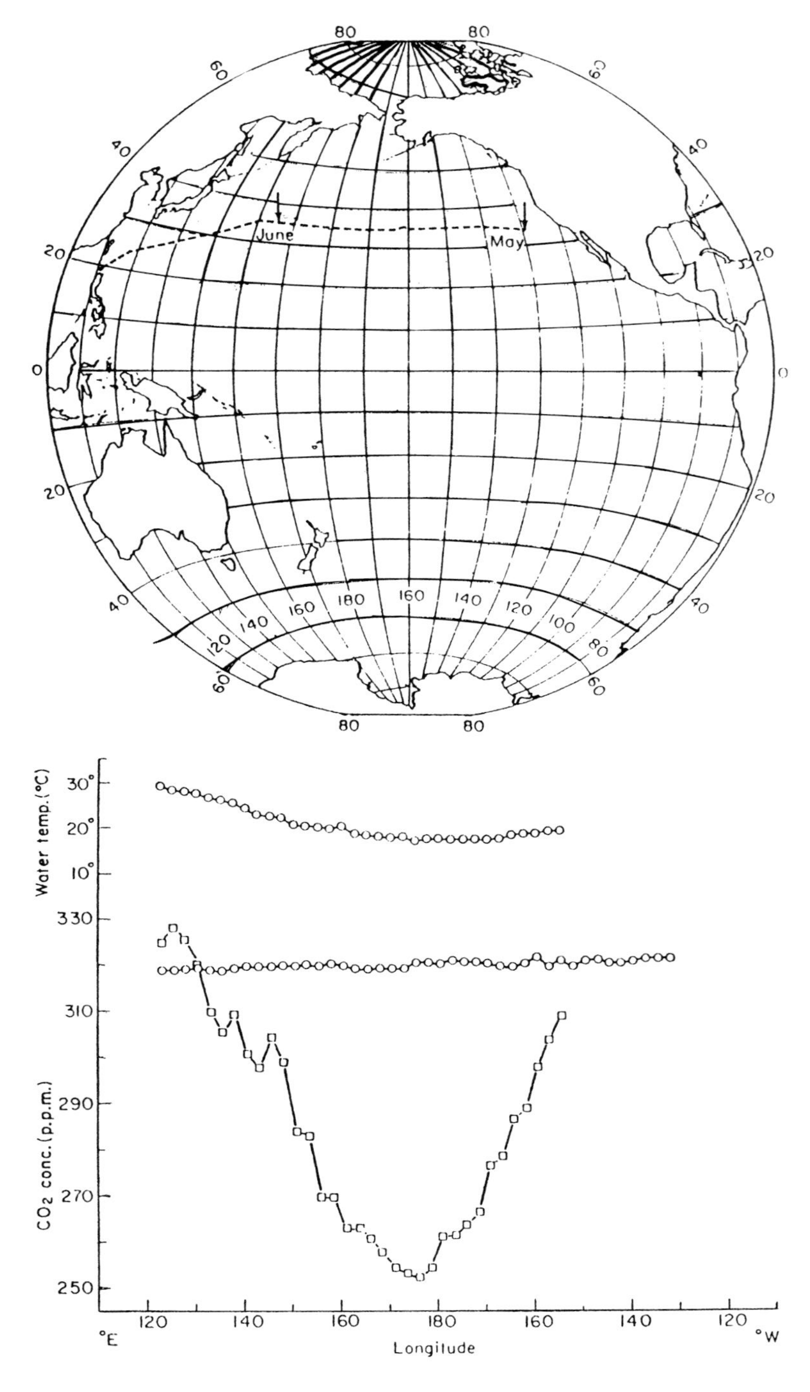


FIGURE 6.13. The partial pressure of carbon dioxide in the atmosphere (open circles) and surface waters (squares) of the Pacific Ocean.

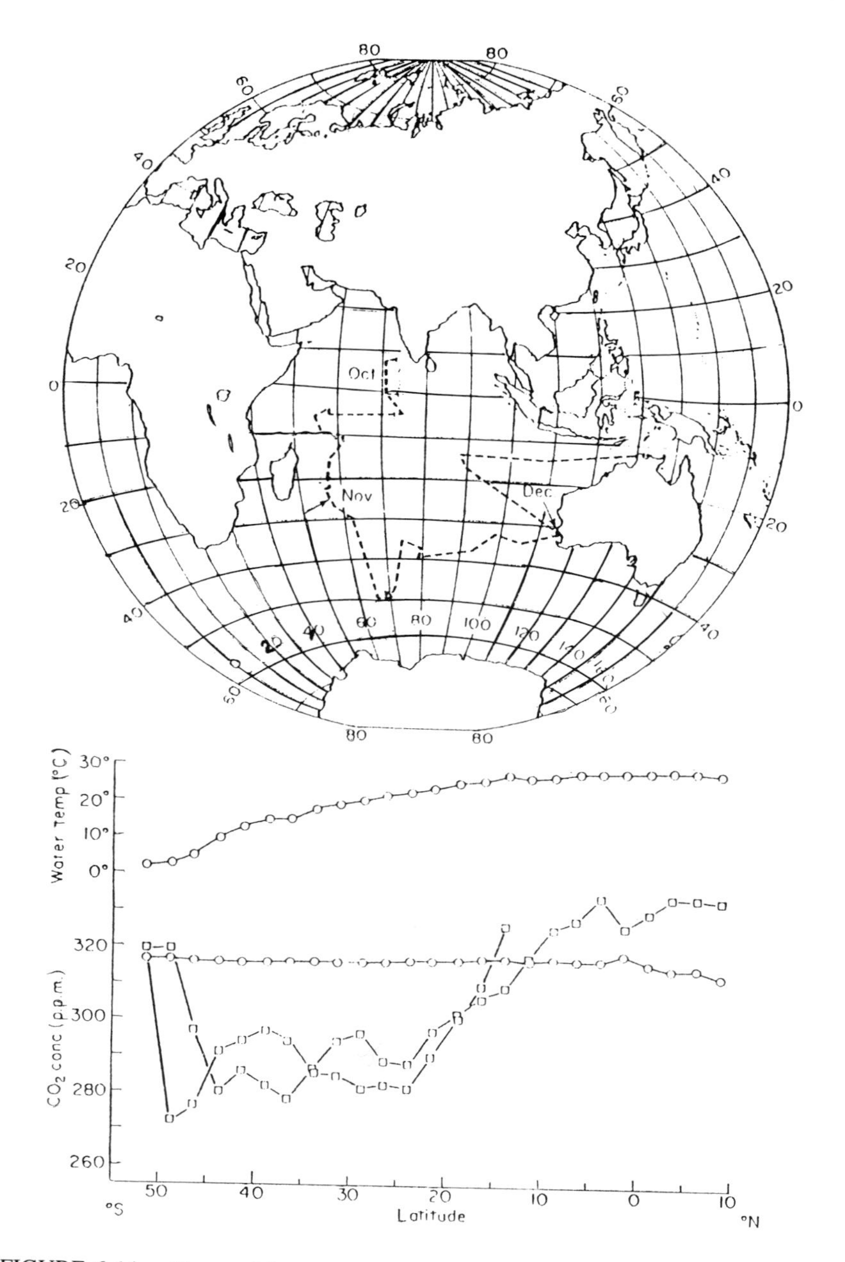


FIGURE 6.14. The partial pressure of carbon dioxide in the atmosphere (open circles) and surface waters (squares) of the Indian Ocean.

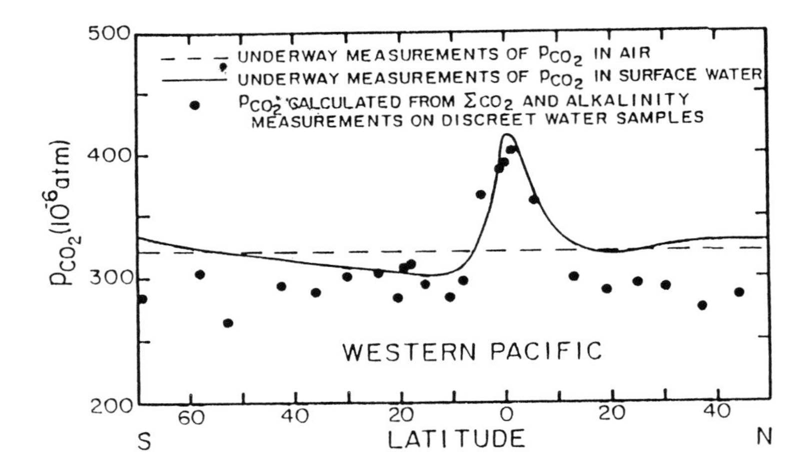


FIGURE 6.15. The partial pressure of carbon dioxide in the atmosphere and surface waters of the western Pacific.

of $CaCO_3$. The pH of deep waters can be as low as 7.5 near 1000 m. In very deep waters the pH can go through a maximum due to the effect of pressure on the ionization of carbonic acid. Park has used the Redfield model to calculate the pH as a function of depth. He attributed the changes to two factors

$$\Delta pH = \Delta pH(a) + \Delta pH(b)$$

where $\Delta pH(a) = -2.0$ AOU, the decrease due to the oxidation of plant material (AOU is the apparent oxygen utilization) and $\Delta pH(b) = 2.4\ \Delta Ca$ (where $\Delta Ca$ is the change in $Ca^{2+}$ due to the dissolution of $CaCO_3$).

## 4.3. $A_T$

The total alkalinity of surface and deep waters was measured extensively during the GEOSECS program. The surface values in the Atlantic are higher than in the Pacific (Figure 6.20). This is due to the higher salinities of these waters caused by evaporation. The results can be normalized by dividing by the chlorinity and calculating the specific alkalinity (S.A. $= A_T/Cl$). This is shown in Figure 6.21. The higher normalized value of alkalinity in deep waters is related to the dissolution of $CaCO_3$. The deep Pacific alkalinity values are higher than those in the Atlantic because they are older and have accumulated more $CO_3^{2-}$. Plots of the normalized alkalinity ($A_T \times S/35$ of Pacific waters are shown in Figure 6.22. The alkalinities of surface waters

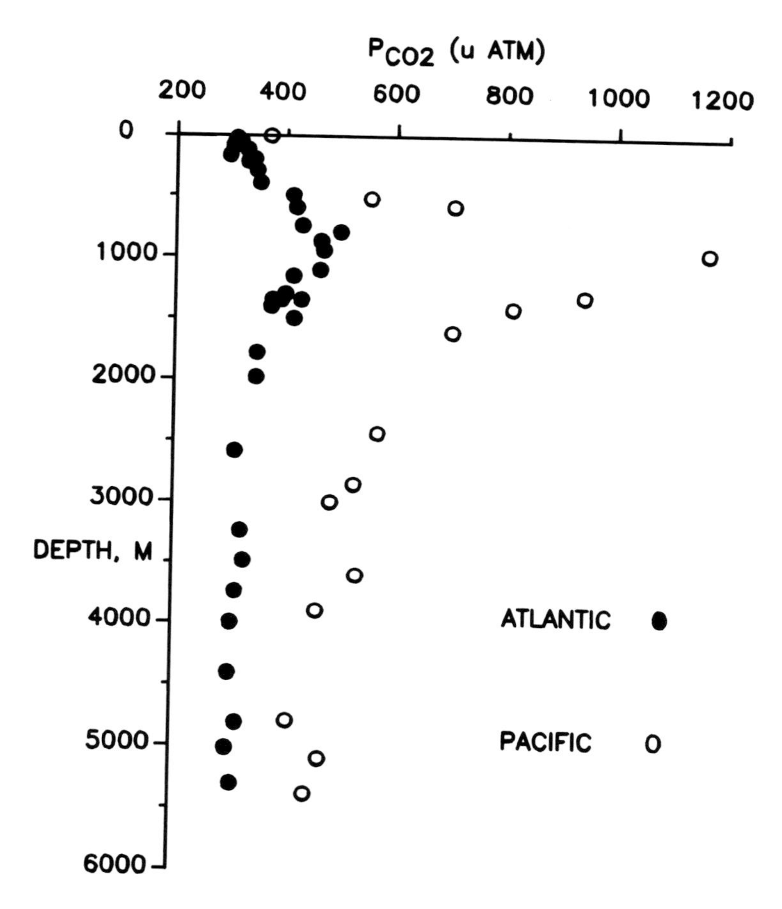


FIGURE 6.16. Depth profile for the partial pressure of carbon dioxide in the Atlantic and Pacific Oceans.

between 40S and 40N are quite constant ($A_T$ = 2.32 meq $kg^{-1}$ or S.A. = $A_T$/Cl = 0.120). This is related to the conservative behavior of $HCO_3^-$ in seawater. The values are higher in polar waters ($A_T$ = 2.38 meq $kg^{-1}$ or S.A. = 0.123). These differences could be related to the growth of organisms with shells of $CaCO_3$ (foraminifera and pterpods). These organisms prefer warm water, thus, the low S.A. could be related to the greater extent of formation of $CaCO_3$ shells. The section of $A_T$ for the Atlantic, shown in Plate 15*, follows the salinity of these waters.

* Plate 15 follows page 48.

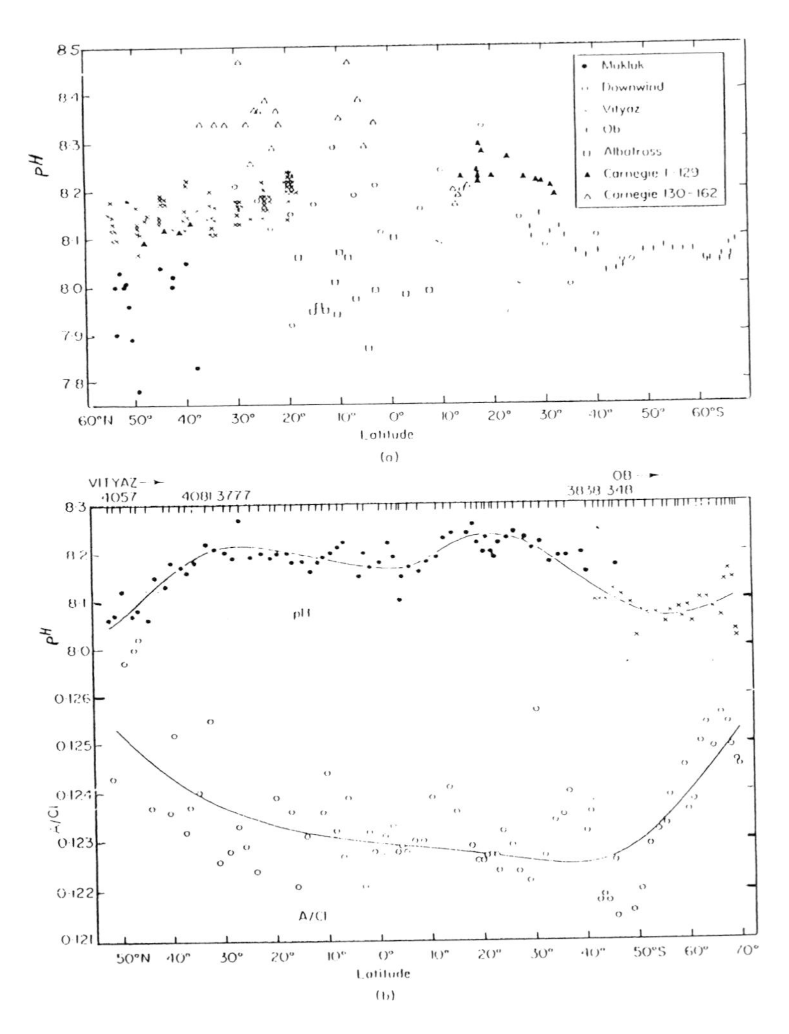


FIGURE 6.17. The pH and specific alkalinity (A/Cl) for surface waters.

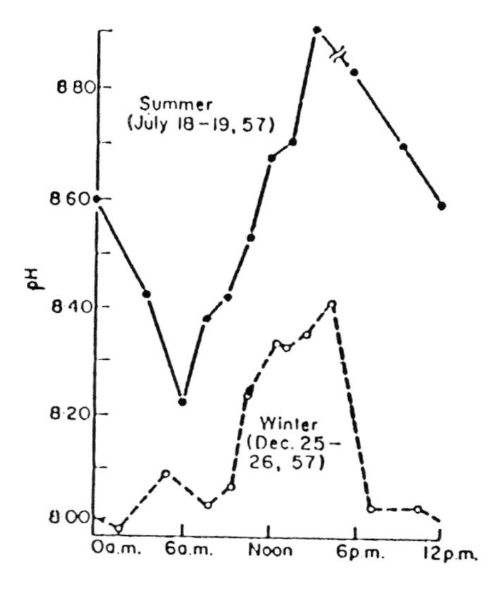


FIGURE 6.18. Diurnal variation of pH in surface waters of an enclosed bay.

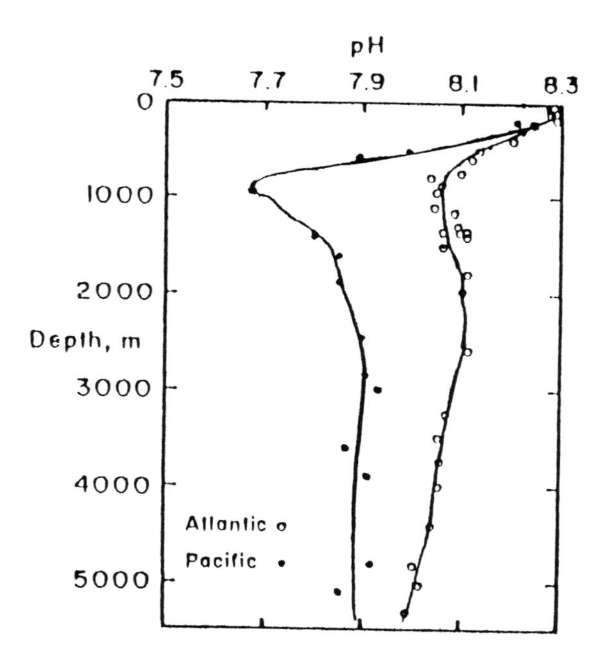


FIGURE 6.19. Depth profile of pH in the Atlantic and Pacific Oceans.

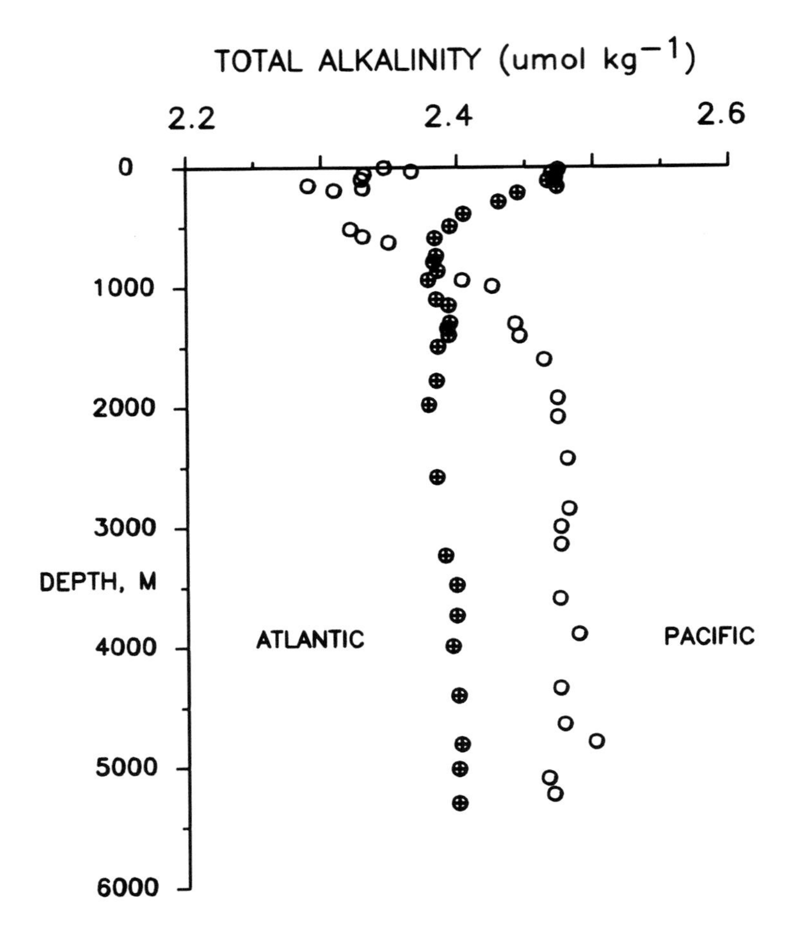


FIGURE 6.20. Depth profile of the total alkalinity ($A_T$) in the Atlantic and Pacific Oceans.

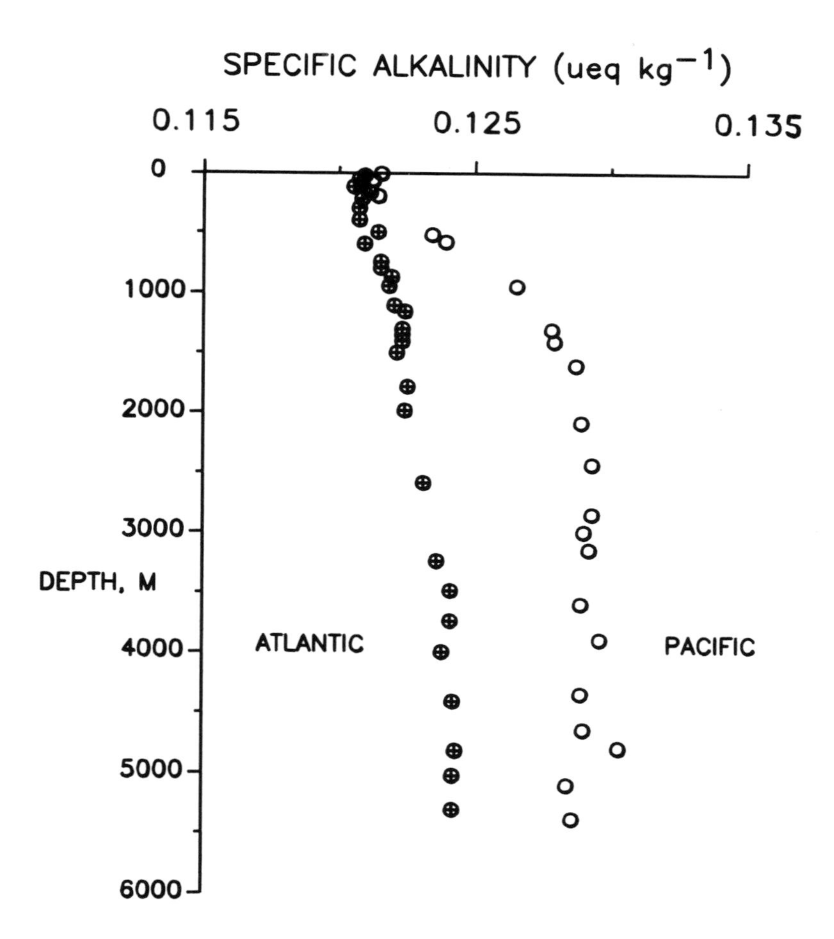


FIGURE 6.21. Depth profile of the specific alkalinity ($A_T/Cl$) in the Atlantic and Pacific Oceans.

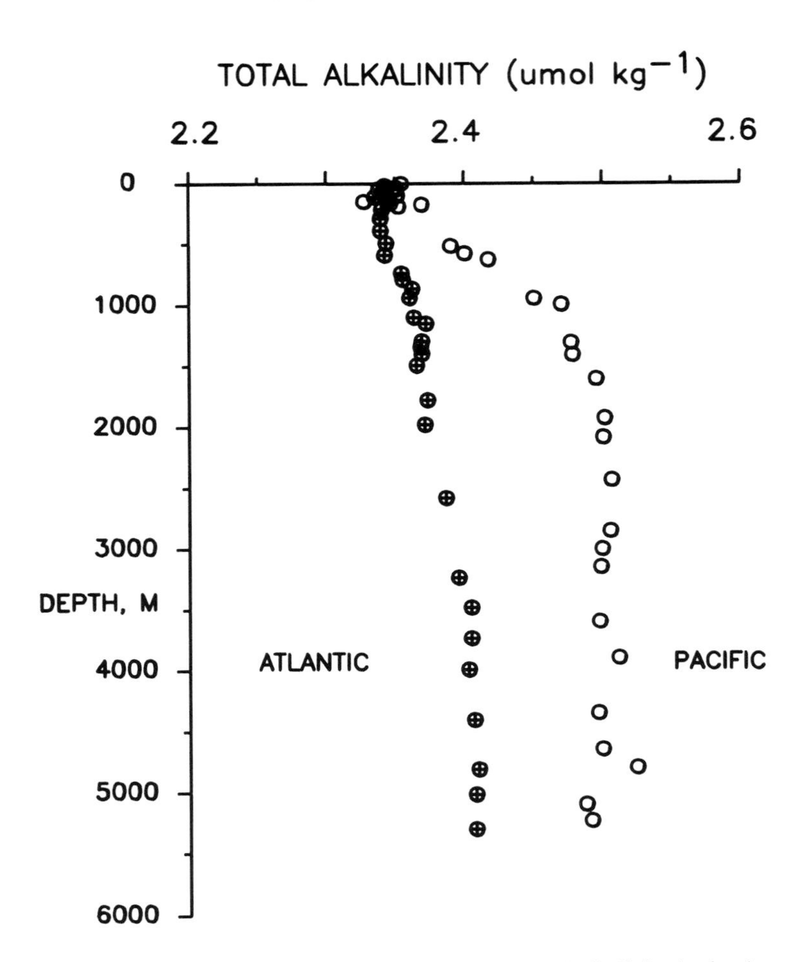


FIGURE 6.22. Depth profile of the normalized total alkalinity in the Atlantic and Pacific Oceans.

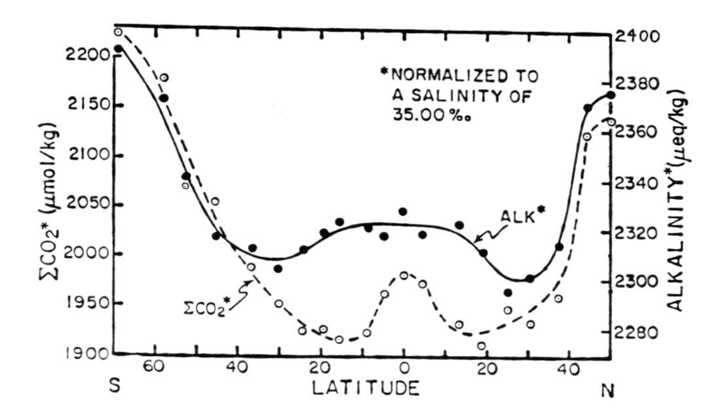


FIGURE 6.23. Surface values of the total carbon dioxide ($CO_2$) and normalized total alkalinity (Broecker and Peng, 1982).

## 4.4. $\Sigma CO_2$

The total dissolved inorganic carbon dioxide in surface oceans is shown in Figure 6.23. Unlike the alkalinity, the total $CO_2$ in the equatorial waters shows a large increase. This is due to equatorial upwelling. As discussed by Broecker and Peng, the levels of $\Sigma CO_2$ and $P_{CO_2}$ in surface waters are related to the exchange of $CO_2$ across the air-sea interface (Figure 6.24). Sluggish exchange causes $P_{CO_2}$ to be greater than the values in the atmosphere near the equator and lower in polar waters. The $\Sigma CO_2$ shows little latitudinal changes. For rapid exchange, the $P_{CO_2}$ in the water and air are similar and the $\Sigma CO_2$ is higher in polar regions.

Due to the buffering effect of seawater, only a small amount of $CO_2$ needs to be transferred to the oceans to restore the equlibrium between the atmosphere and surface. This buffering is called the Revelle factor(R) (Figure 6.25). It is the ratio of the fractional rise in the partial pressure of carbon dioxide in the atmosphere to the fractional increase of the total carbon dioxide in the ocean.

The increase in the total $CO_2$ near the equator is related to the Revelle factor (R)

$$R = (\Delta P_{CO_2}/P_{CO_2})/(\Delta \Sigma CO_2/\Sigma CO_2) \tag{117}$$

This value is about 14 for cold waters and 8 for warmer waters (the average is about 10). Thus, a 10% change in $P_{CO_2}$ only results in a 1% change in $\Sigma CO_2$. This factor is important when considering the effect that increases of $CO_2$ in the atmosphere have on the carbonate system.

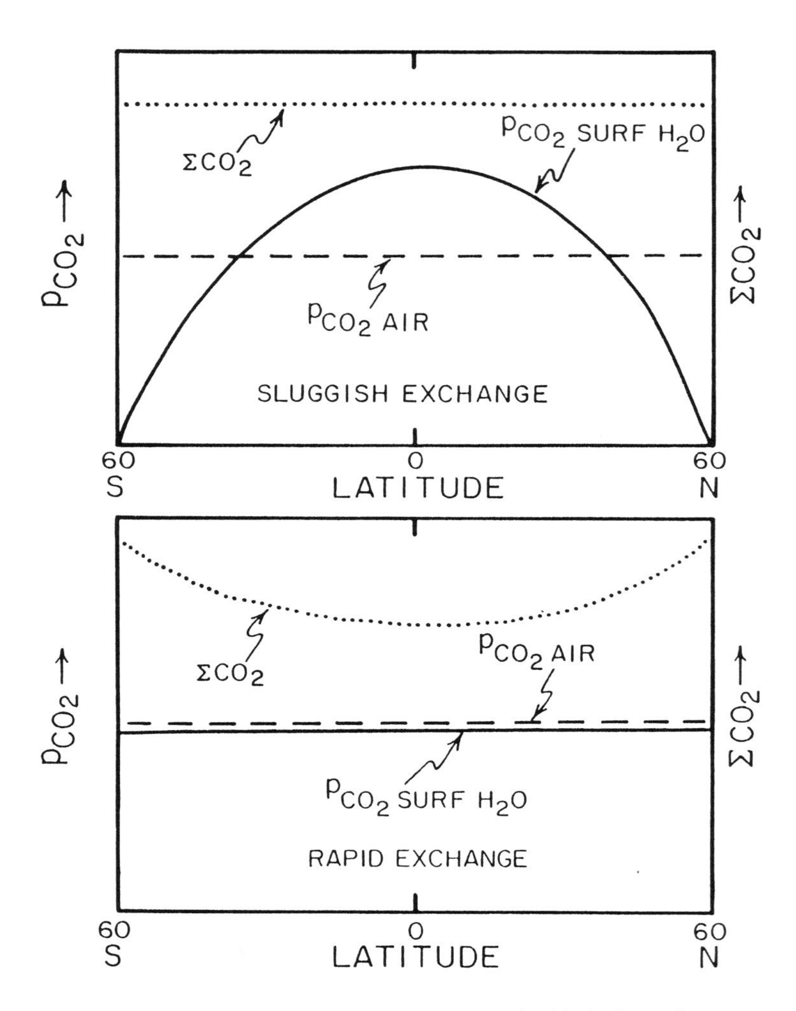


FIGURE 6.24. The partial pressure of carbon dioxide in the surface oceans with sluggish and rapid exchange with the atmosphere.

The flux (F) of $CO_2$ across the air-sea interface is related to the concentration difference ($\Delta C$) or partial pressure of $CO_2$ in the water and atmosphere

$$F = k\ \Delta C \tag{118}$$

The value k is called the transfer velocity. As discussed earlier, Liss has divided this transfer velocity into two terms

$$1/k = 1/\alpha k_w + 1/Hk_a \tag{119}$$

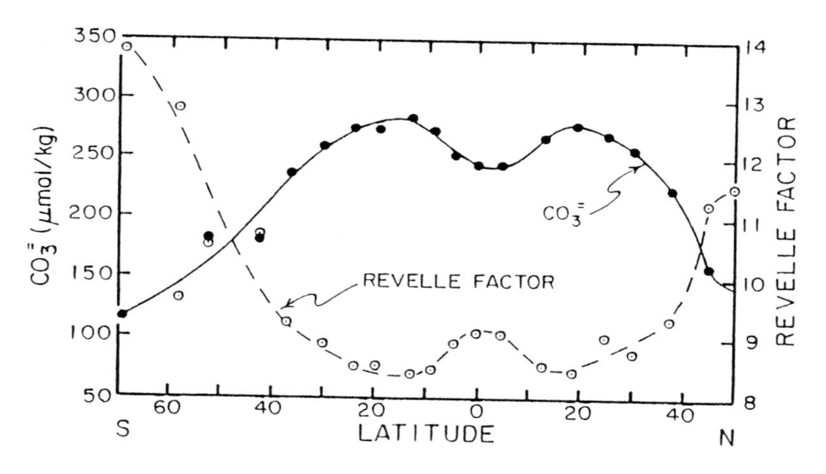


FIGURE 6.25. The concentration of carbonate ion and the Revelle Factor in the surface oceans.

where $k_w$ and $k_a$ are transfer velocities in the water and air, respectively; H is the Henry's law constant (a unitless value, the ratio of air to water concentrations at equilibrium), and α is a factor that accounts for any enhancement of the transfer on the water side due to chemical reactions between the gas and $H_2O$ ($CO_2 + H_2O \rightarrow H^+ + HCO_3^-$). The value of $\alpha = 1.02$ to $1.03$ for $CO_2$. From $^{14}C$ measurements, values of $k_w$ average about 20 cm $h^{-1}$ on a global basis. The transfer velocity increases with increasing wind speed (Figure 6.26). The depth profiles of $\Sigma CO_2$ in the Atlantic and Pacific are shown in Figure 6.27. The surface values (if normalized for salinity) are about 2.05 mmol $kg^{-1}$. The values decrease to a minimum in surface waters due to photosynthesis. In deeper waters the $\Sigma CO_2$ increases due to the oxidation of plant material. The $\Sigma CO_2$ values for deep Pacific waters are higher than those for the Atlantic because the waters are older and have had more time to accumulate $CO_2$ due to microbial oxidation. The values of $\Sigma CO_2$ and $A_T$ correlate very well with each other and can be used to characterize various water masses (Figure 6.28). A section of total carbon dioxide in the Atlantic Ocean is shown in Plate 16.*

* Plate 16 follows page 48.

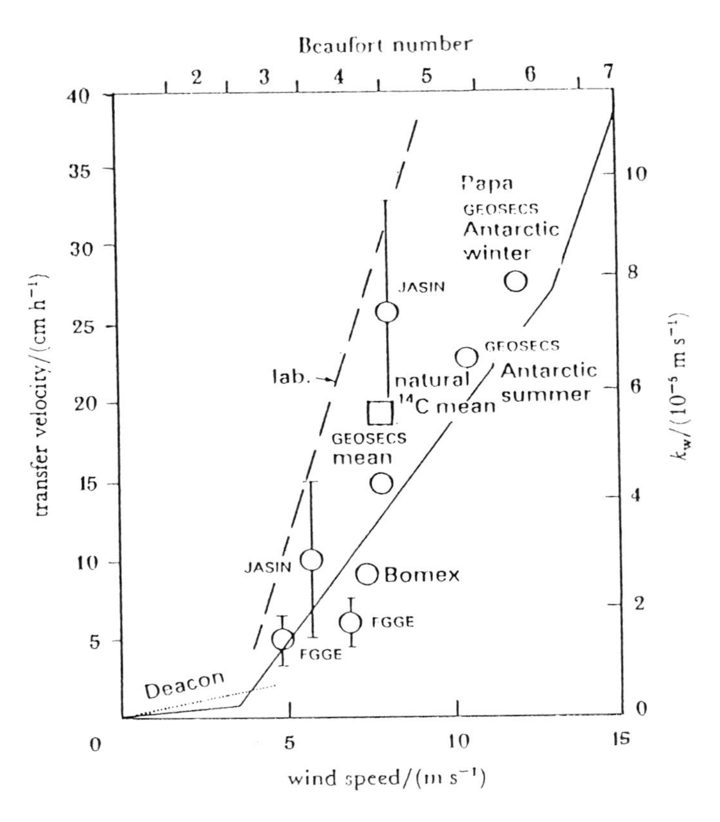


FIGURE 6.26. The transfer velocity as a function of the wind speed.

## 5. $CaCO_3$ DISSOLUTION IN SEAWATER

The precipitation or formation of solid $CaCO_3$ in surface waters and the dissolution of solid $CaCO_3$ in deep waters is very important in transferring $CO_2$ from surface waters to deep waters. $CaCO_3$(s) is also present in pelagic sediments of the world oceans (Figure 6.29). The saturation state of seawater with respect to $CaCO_3$ is determined from

$$\Omega = [Ca^{2+}][CO_3^{2-}]/K^*_{sp} \tag{120}$$

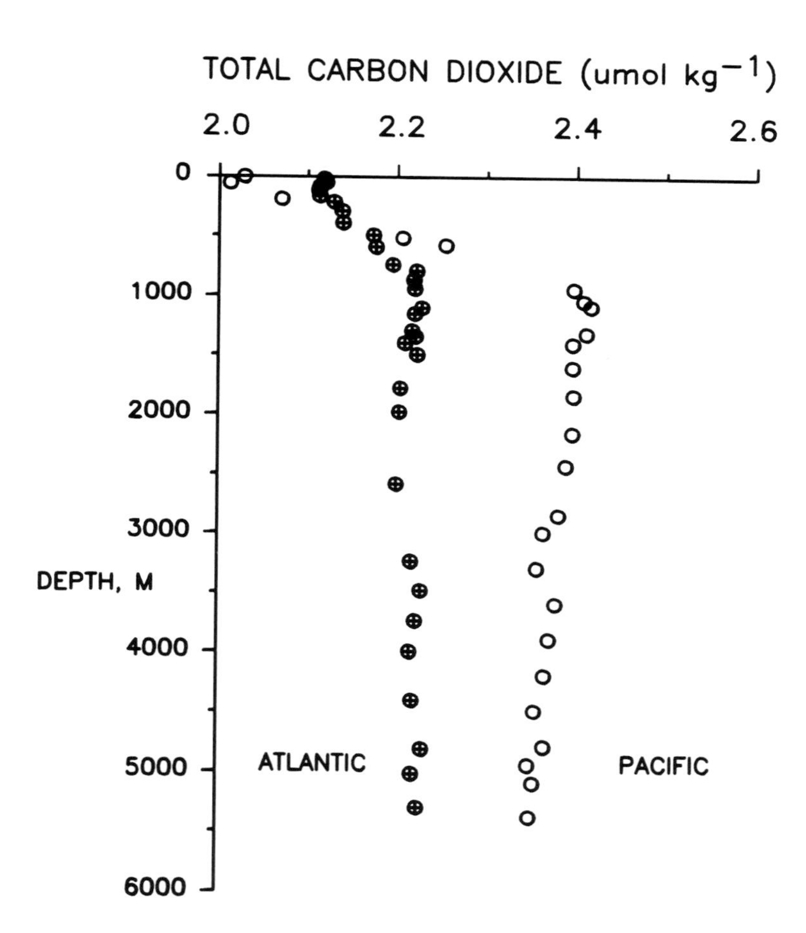


FIGURE 6.27. Depth profile of the total carbon dioxide in the Atlantic and Pacific Oceans.

where $[Ca^{2+}][CO_3^{2-}]$ is the ion product of the concentration of $Ca^{2+}$ and $CO_3^{2-}$, and $K^*_{sp}$ is the solubility product at the *in situ* conditions of S, t, and P. Since $Ca^{2+}$ is a major constituent of seawater (within 1%), its concentration (mol $kg^{-1}$) can be estimated from

$$[Ca^{2+}] = 2.934 \times 10^{-4}\ S \tag{121}$$

The solubility product for calcite formed by foraminifera and aragonite formed by pteropods can be determined from Equations 71 to 77. The values of $[CO_3^{2-}]$ can be determined from the measured carbonate parameters (pH and $A_T$ or $A_T$ and $\Sigma CO_2$).

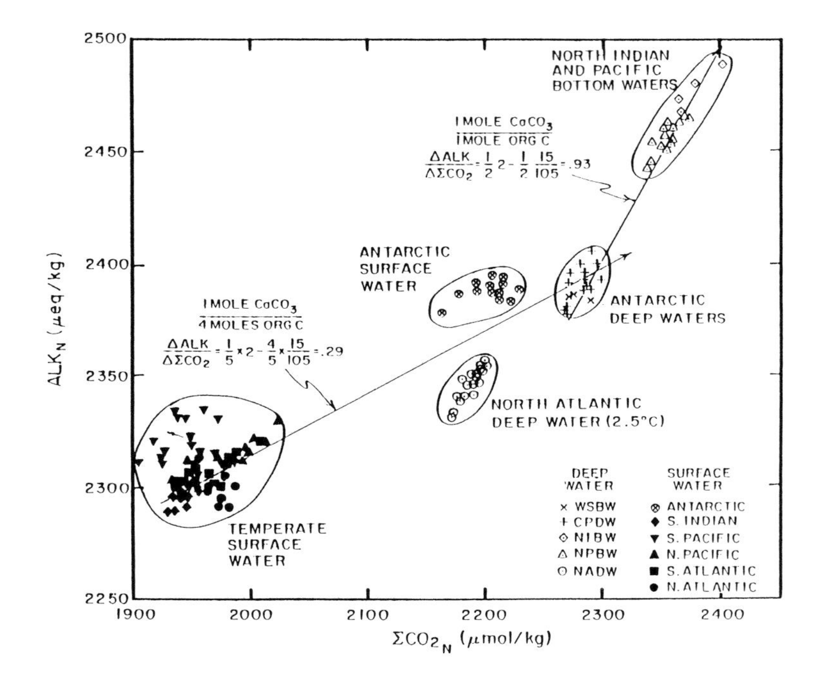


FIGURE 6.28. Values of the normalized total alkalinity and the total carbon dioxide in various water masses.

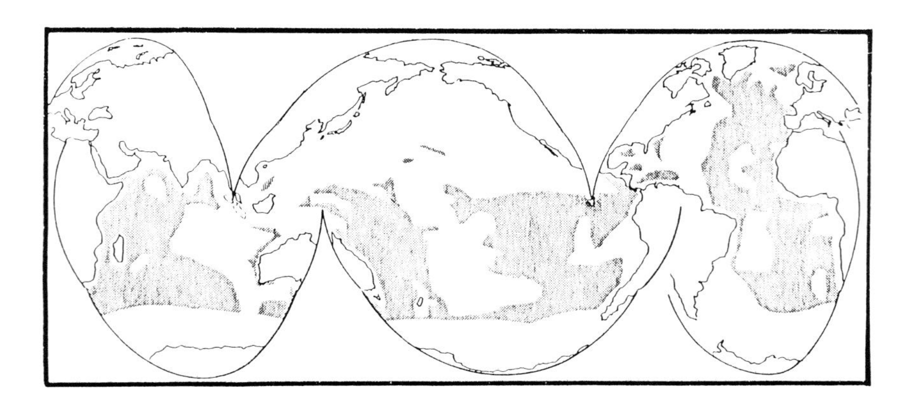

FIGURE 6.29. The areas of the oceans with sediments of calcium carbonate.

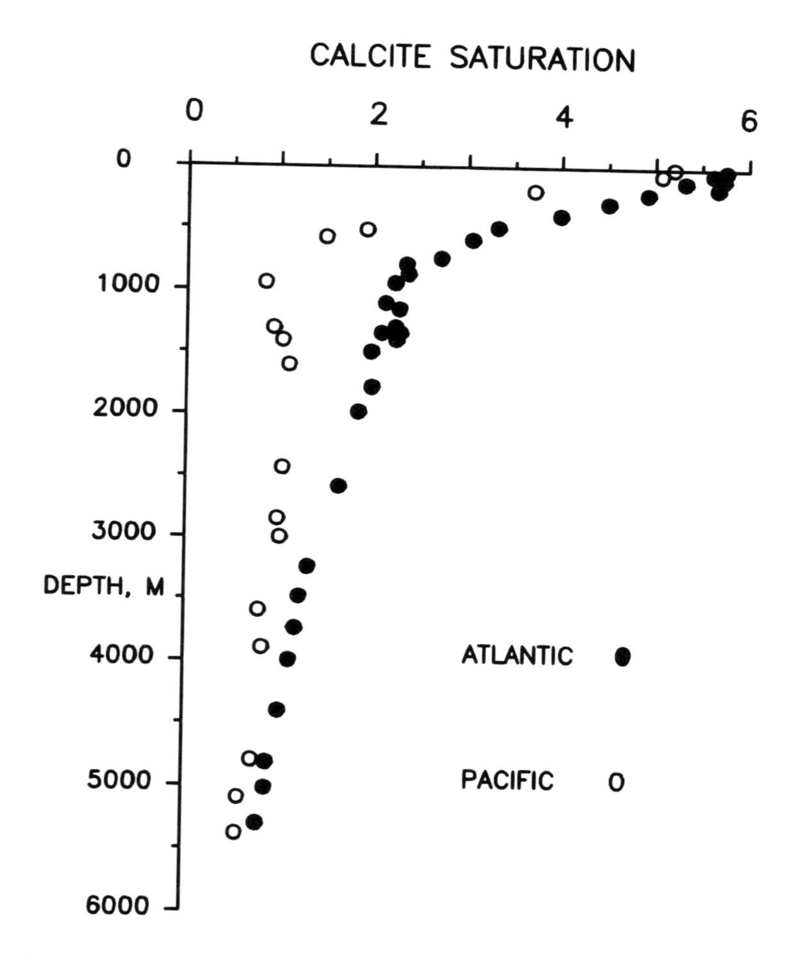


FIGURE 6.30. The depth profile of the calcite saturation state for the Atlantic and Pacific Oceans.

Values of $\Omega$ for calcite and aragonite for Atlantic and Pacific waters are shown in Figures 6.30 and 6.31. The surface values of $\Omega$ for calcite are near 5.0 and decrease below 1.0 in deep water. The surface water value of $\Omega$ is 3.0 for aragonite. Aragonite is 1.5 times more soluble than calcite at a given t, P, and salinity. The waters of the Pacific become undersaturated ($\Omega < 1.0$) at shallower depths than in the Atlantic. Approximate saturation levels in North Atlantic and North Pacific waters are given below.

| | **N. Atlantic** | **N. Pacific** |
|---|---|---|
| Calcite | 4300 m | 750 m |
| Aragonite | 1400 | 500 |

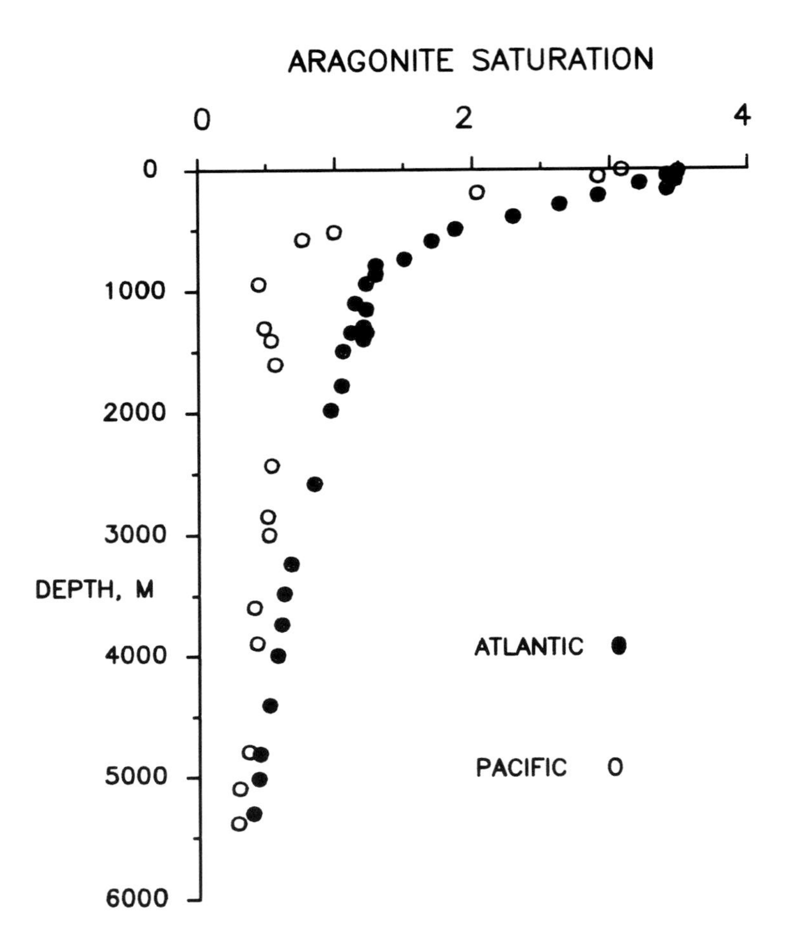


FIGURE 6.31. The depth profile of the aragonite saturation state for the Atlantic and Pacific Oceans.

The greater solubility of these minerals in deep waters is related to the effect of pressure on the solubility of $CaCO_3(s)$. Since two divalent ions are formed during the dissolution, the volume change is large and negative due to electrostriction. The Pacific deep waters become undersaturated at shallower depths due to the lower pH or higher $CO_2$ formed by the oxidation of plant material. This decreases the concentration of $CO_3^{2-}$ due to the shift in the equilibrium

$$CO_3^{2-} + H^+ \rightarrow HCO_3^- \tag{122}$$

The differences in $\Omega$ between the two oceans become smaller in the deep oceans due to the effect of pressure controlling solubility.

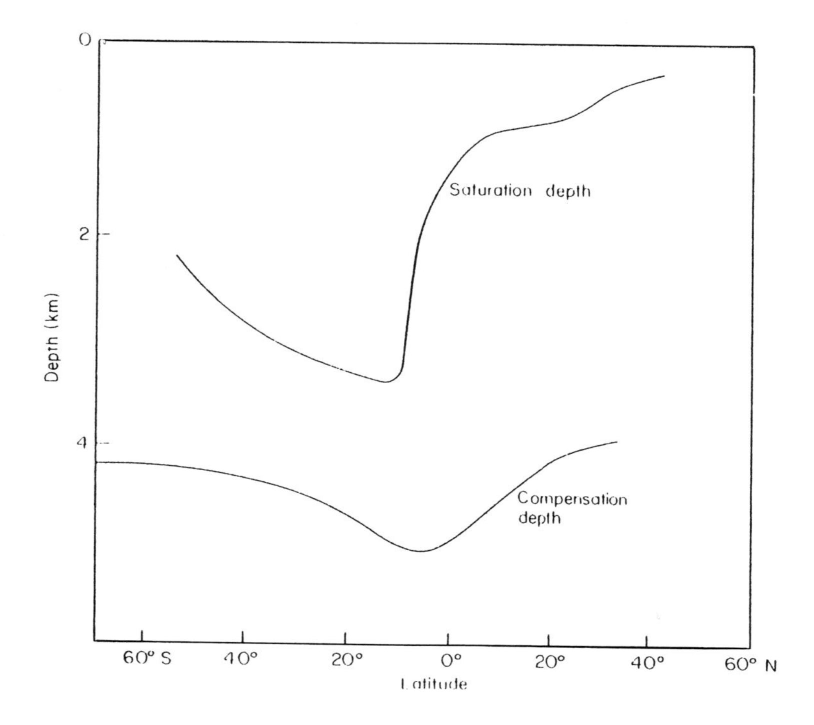


FIGURE 6.32. Comparisons of the saturation (thermodynamic) and compensation depths in the Atlantic Ocean.

Although much of the deep oceans are undersaturated with respect to $CaCO_3(s)$, large amounts of calcite are present in ocean sediments. Geologists call the layer where $CaCO_3(s)$ is above 5% of the sediments the compensation depth. As shown in Figure 6.32, the $CaCO_3$ compensation depth in the Atlantic is about 2 km below the saturation depth. These results indicate that the solubility of $CaCO_3(s)$ in seawater is not controlled by equilibrium but by kinetic constraints.

The first direct measurements made to examine the rates of dissolution of $CaCO_3(s)$ in the oceans were made by Peterson. He suspended calcite spheres in the Pacific on deep sea moorings for 250 days. His results are shown in Figure 6.33. At a depth of about 4000 m the solution rate dramatically increased. This depth of rapid increase in the rate of dissolution is called the lysocline. Honjo (WHOI) suspended a number of $CaCO_3$ solids for 79 days (coccoliths, foram shells, reagent calcite, and pteropod shells) in chambers through which seawater was drawn (Figure 6.33). The aragonite lysocline was found to be higher than the depth for calcite. The depth of the lysocline found by suspending $CaCO_3(s)$ agrees very well with the decrease

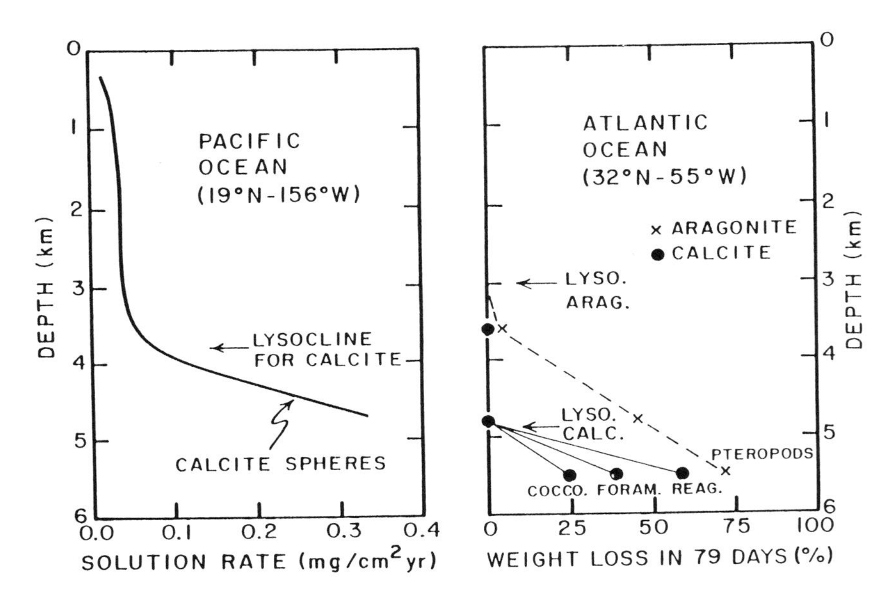


FIGURE 6.33. Depth profiles for the rates of solution of calcium carbonate in the Atlantic and Pacific Oceans.

in the mineral found in surface sediments at various depths in the same area (Figures 6.33 and 6.34). These results indicate that the lysocline and $CaCO_3$ compensation depth in sediments are frequently at the same depth. Thus, the causes of the compensation depth being deeper than the saturation depth are the variable rates of dissolution of various forms of $CaCO_3$. If the sedimentation rates are high, it is possible that $CaCO_3(s)$ could be preserved before it dissolves. This would cause the calcium carbonate compensation depth to be below the lysocline.

Values of the lysocline in the Atlantic and Pacific are shown in Figure 6.35. The lysocline is higher in the Pacific because of the greater undersaturation at lower depths. A comparison of the saturation horizon with the lysocline and calcium carbonate compensation depths (CCD) is shown in Figure 6.36. The values of the lysocline and the CCD are not affected by the saturation states. The CCD is close to the lysocline except in the equatorial region. This is due to the higher productivity of these waters. The higher the supply rate of $CaCO_3(s)$, the deeper the CCD will be (Figure 6.37).

Attempts to understand the causes of these increases in rates have been made by Morse (Texas A & M) using laboratory studies. He has shown that when the saturation of waters gets to a critical value, $CaCO_3(s)$ starts to dissolve. This critical value is about 30% undersaturation or at a $\Delta CO_3$ =

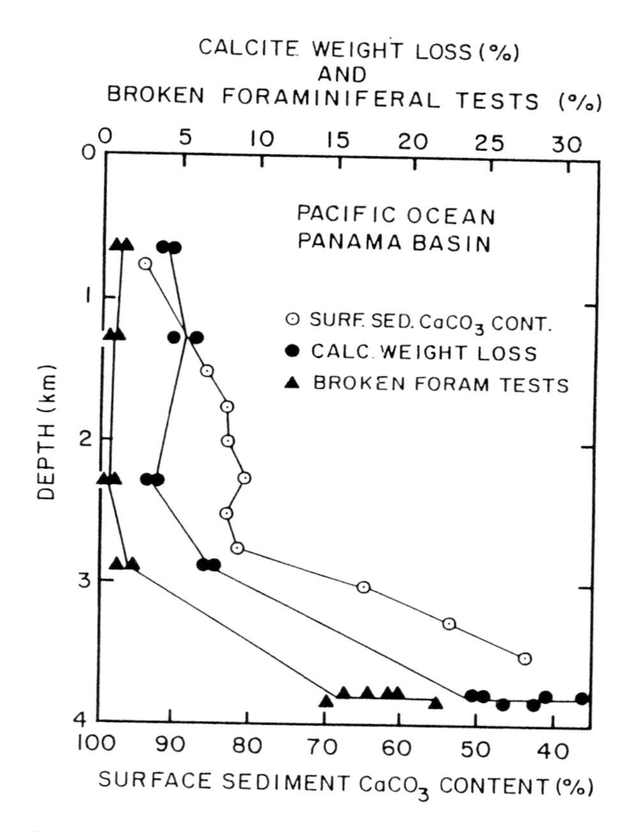


FIGURE 6.34. Depth profiles of the calcite loss and surface sediment concentration of $CaCO_3$ in the Pacific Ocean.

$CO_3^{2-}$ (sat) = $-10$ μmol kg$^{-1}$ (i.e., the solution could absorb another 10 μmol kg$^{-1}$ of $CaCO_3$). This critical $\Delta CaCO_3$ is shown by the percentage of $CaCO_3$(s) in sediment core tops by Broecker and Peng (Figure 6.38). It should be pointed out that this so-called critical value is strongly dependent upon the value selected for the solubility product of the $CaCO_3$(s). More recent studies by Byrne (University of South Florida) have concentrated on measuring solubility dissolution rates using minerals and waters collected in the oceans. He and his co-workers have found that the rates of dissolution of aragonitic $CaCO_3$(s) measured at sea could be described by

$$R = 130\,\{1 - [Ca^{2+}][CO_3^{2-}]/1.78\,K_{sp}\,(cal)\}^{3.1} \qquad (123)$$

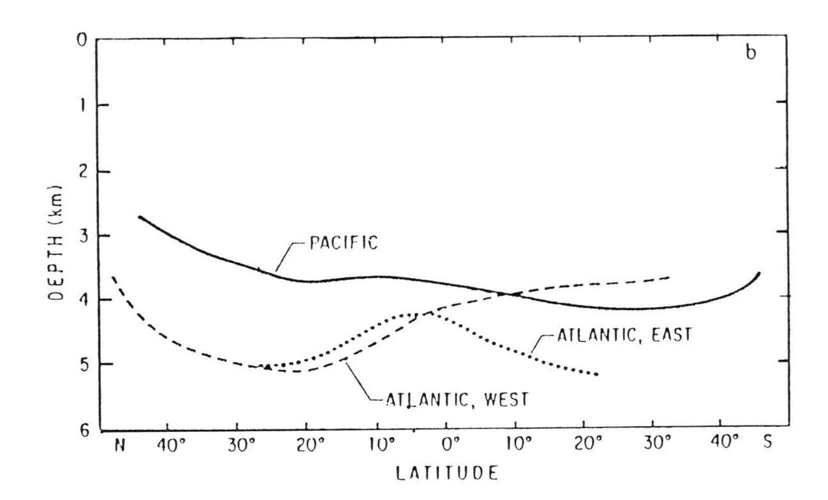


FIGURE 6.35. Depths of the lysocline in various oceans.

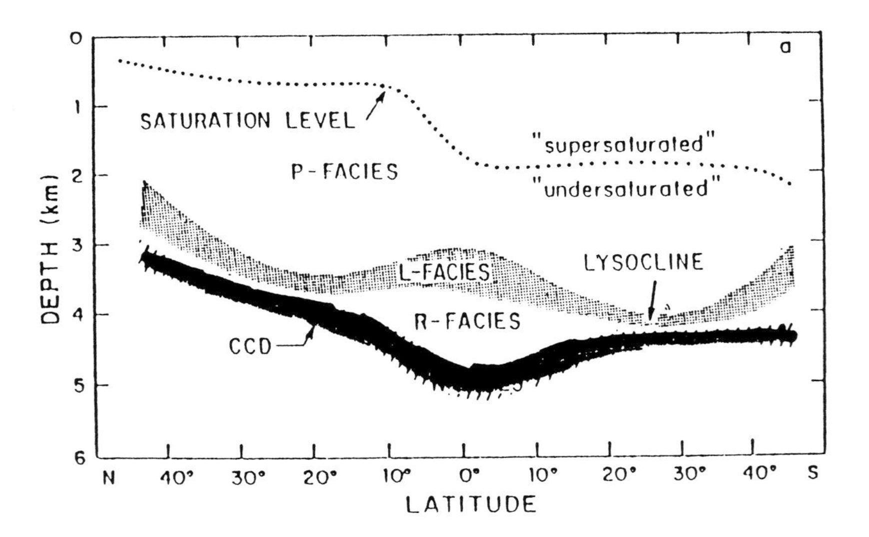


FIGURE 6.36. Comparisons of the saturation level, the carbonate compensation depth (CCD), and the lysocline at various latitudes.

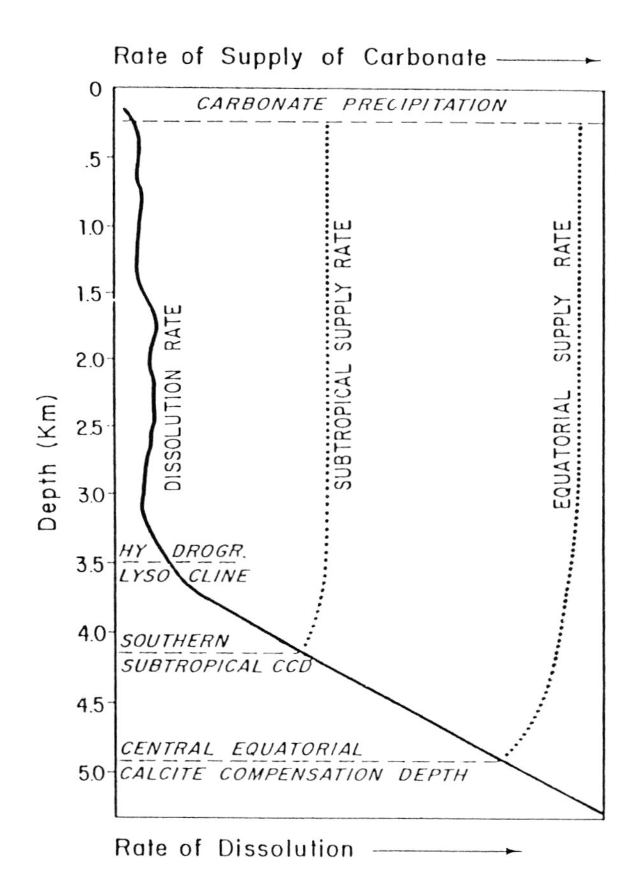


FIGURE 6.37. Depth profile of the dissolution rate and the rate of supply of carbonate to the oceans.

where $K_{sp}$ (aragonite) = 1.78 $K_{sp}$ (calcite). The factor of $\rho = 1.78$ is slightly higher than the theoretical value of 1.5. This is related to the changes in $K_{sp}$ (aragonite) and $\rho$ for aragonite solubility as a function of time (Figure 6.39).

The dissolution rates in the laboratory studies of Morse (Texas A & M) and co-workers resulted in two equations. For $\Omega_A \geq 0.44$

$$R\ (\%\ \text{per day}) = 110\ (1 - \Omega)^{2.93} \tag{124}$$

and for $\Omega_A \leq 0.44$

$$R\ (\%\ \text{per day}) = 1318\ (1 - \Omega)^{7.27} \tag{125}$$

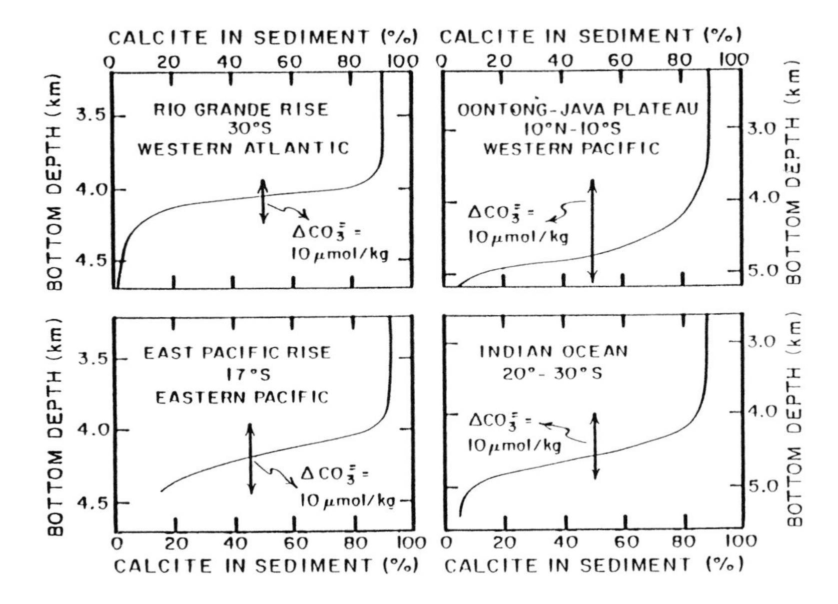


FIGURE 6.38. Calcium carbonate concentrations in the sediments of various oceans.

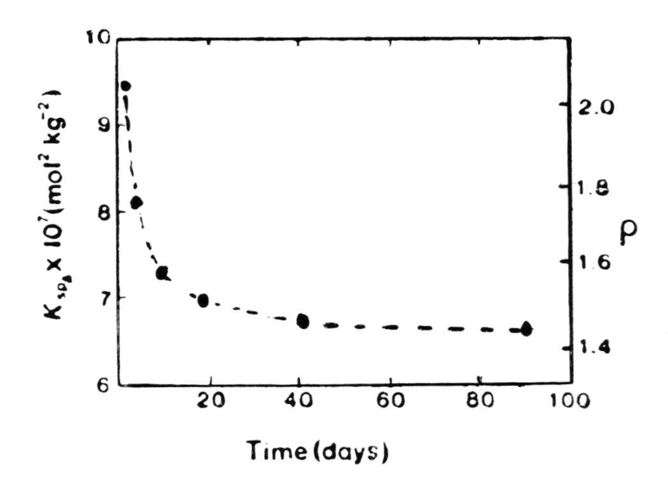


FIGURE 6.39. Values of the solubility product for aragonite in waters as a function of the equilibration times.

Although the equations are different, the calculated rates are in reasonable agreement. Since aragonite production appears to be high in the surface waters of the Pacific, and the deeper waters are undersaturated, the transport of the aragonite and dissolution could result in transporting carbon to deep waters. Betzer (University of South Florida) and co-workers have estimated that 90% of this aragonite flux would be dissolved in the upper 2.2 km of the water column.

## FURTHER READING

Anderson, N. and P. Malahoff, The Fate of Fossil Fuel $CO_2$ in the Oceans, Plenum Press, New York, (1977).

Berger, W.H., Biogeneous Deep Sea Sediments: Production, Preservation, and Interpretation, chap. 29, *Chemical Oceanography,* Vol. 5, 2nd ed., J.P. Riley and R. Chester, Eds., Academic Press, New York, 266-388 (1976).

Broecker, W.S. and T.H. Peng, *Tracers in the Sea,* Eldigio Press, New York, (1982).

Cloud, P.E., Carbonate Precipitation and Dissolution in the Marine Environment, chap. 17, *Chemical Oceanography,* Vol. 2, 1st ed., J.P. Riley and G. Skirrow, Eds., Academic Press, New York, 127-158 (1965).

Skirrow, G., The Dissolved Gases-Carbon Dioxide, chap. 9, *Chemical Oceanography,* Vol. 2, 2nd ed., J.P. Riley and G. Skirrow, Eds., Academic Press, New York, 1-192 (1975).

# 7 Table of Contents

**Micronutrients in the Oceans** .......... 323
1. *Phosphorus in Seawater* .......... 323
   1.1. *Determination of Phosphate* .......... 330
   1.2. *Distribution of Phosphate* .......... 331
2. *Nitrogen Compounds in Seawater* .......... 334
   2.1. *Determination of Nitrogen Compounds* .......... 335
      2.1.1. $NO_2^-$ .......... 335
      2.1.2. $NO_3^-$ .......... 335
      2.1.3. $NH_3$-$NH_4^+$ .......... 335
   2.2. *Distribution of Nitrogen Compounds* .......... 336
   2.3. *Nitrogen-Phosphorus Ratio* .......... 340
3. *Silicon in Seawater* .......... 345
   3.1. *Determination of Silicon* .......... 346
   3.2. *Distribution of Dissolved* $SiO_2$ .......... 348
4. *Use of Nutrients as Water Mass Tracer* .......... 349
*Further Reading* .......... 352

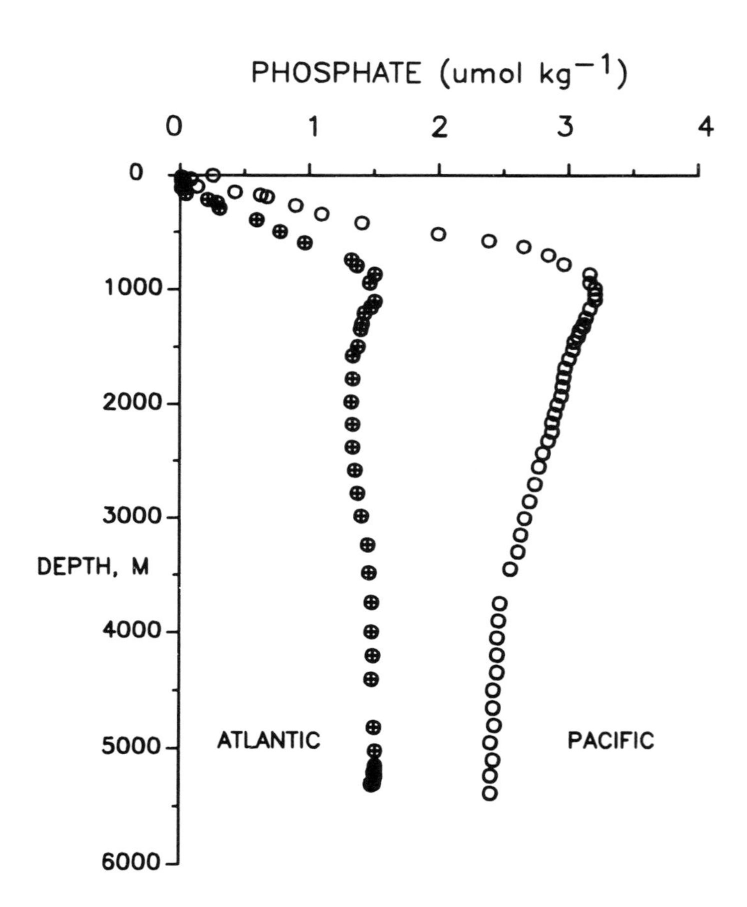


Profile of phosphate in the Atlantic and Pacific Oceans.

# 7 Micronutrients in the Oceans

Marine phytoplankton require certain trace elements for growth. These nutrients are used until they become limiting and further growth is inhibited. A crude way of assessing the availability and need of marine plants for various elements can be made by examining the composition of phytoplankton compared to the concentration in average seawater. Such a comparison is shown in Table 7.1. The most important micronutrients are nitrogen and phosphorus. Some organisms (diatoms) have siliceous frustules and require silica. Although other elements such as Fe, Mn, Cu, Zn, Co, and Mo are also essential for growth, it is generally thought that growth is not inhibited or limited by the concentration of these metals. An exception for certain waters may be Fe and Mn. In addition certain organic compounds, such as vitamins, are necessary for growth. In this chapter we will discuss the major micronutrients P, N, and Si.

## 1. PHOSPHORUS IN SEAWATER

Phosphorus occurs in seawater in dissolved and particulate forms. In surface waters dissolved and particulate organic phosphates occur due to the decomposition of plants. These dissolved organic phosphorus compounds

**TABLE 7.1**
**Distribution of Elements in Organism (N) and Seawater (A) — A Measure of Availability to Need**

| Element | N (g/100 g) | A (g/m$^3$) | A/N |
|---|---|---|---|
| H | 7 g | | |
| Na | 3 | 10.75 kg | 3600 |
| K | 1 | 390 g | 390 |
| Mg | 0.4 | 1.3 kg | 300 |
| Ca | 0.5 | 416 g | 830 |
| C | 30 | 28 g | 1 |
| Si[a] | 0.5 | 500 mg | 1 |
| Si[b] | 10 | 500 mg | 0.05 |
| N | 5 | 300 mg | 0.06 |
| P | 0.6 | 30 mg | 0.05 |
| O ($O_2$ + $CO_2$) | 47 | 90 g | 2 |
| S | 1 | 900 g | 900 |
| Cl | 4 | 19.3 kg | 4800 |
| Cu | 5 mg | 10 mg | 2 |
| Zn | 20 mg | 5 mg | 4 |
| B | 2 mg | 5 mg | 2500 |
| V | 3 mg | 0.3 mg | 0.1 |
| As | 0.1 mg | 15 mg | 150 |
| Mn | 2 mg | 5 mg | 2.5 |
| F | 1 | 1.4 g | 1400 |
| Br | 2.5 mg | 66 g | 26000 |
| Fe[a] | 1 mg | 50 mg | 0.05 |
| Fe[b] | 40 mg | 50 mg | 1.3 |
| Co | 0.05 mg | 0.1 mg | 2 |
| Al | 1 | 120 mg | 120 |
| Ti | 100 mg | — | — |

[a] Phytoplankton.
[b] Diatoms.

make up a significant, but variable portion of dissolved P in surface waters. Although they have not all been identified, they are undoubtedly related to the decomposition and excretion products of marine organisms.

Sugar phosphates, phospholipids, phosphonucleotides, and their hydrolyzed products occur. Phosphate esters (O-P bonds) and more stable aminophosphonic acids (C-P bonds) may also make up a considerable portion of the organic-phosphorus compounds. At present the details of the cycling of

**TABLE 7.2**
**Values of the $pK_i$ for $H_3PO_4$ in Water, NaCl, and Seawater**

| Media | $pK_1$ | $pK_2$ | $pK_3$ |
|---|---|---|---|
| $H_2O$ | 2.15 | 7.20 | 12.34 |
| NaCl(0.7) | 1.73 | 6.38 | 11.13 |
| Seawater (S = 35) | 1.57 | 5.86 | 8.69 |

these compounds in surface waters is largely unknown. Dissolved inorganic phosphorus exists entirely as the ionized products of $H_3PO_4$.

$$H_3PO_4 \rightarrow H^+ + H_2PO_4^- \quad (1)$$

$$H_2PO_4^- \rightarrow H^+ + HPO_4^{2-} \quad (2)$$

$$HPO_4^{2-} \rightarrow H^+ + PO_4^{3-} \quad (3)$$

The fraction of these forms are controlled by the pH and the composition of the waters. The ionization constants for the three step dissociation are defined by

$$K_1 = [H^+][H_2PO_4^-]/[H_3PO_4] \quad (4)$$

$$K_2 = [H^+][HPO_4^{2-}]/[H_2PO_4^-] \quad (5)$$

$$K_3 = [H^+][PO_4^{3-}]/[HPO_4^{2-}] \quad (6)$$

The pK values (= $-\log K$) for $H_3PO_4$ in water, NaCl (0.7 m), and seawater (S = 35) at 25°C and 1 atm are given in Table 7.2. The increase in the ionization as one goes from NaCl to seawater is related to the increase in ionic strength ($H_2O \rightarrow$ NaCl) and the formation of strong ion pairs (NaCl $\rightarrow$ seawater) of $Ca^{2+}$ and $Mg^{2+}$ with phosphate ions. For $M^{2+} = Mg^{2+}$ or $Ca^{2+}$, we have

$$M^{2+} + H_2PO_4^- \rightarrow MH_2PO_4^+ \quad (7)$$

$$M^{2+} + HPO_4^{2-} \rightarrow MHPO_4^0 \quad (8)$$

**TABLE 7.3**
**Values of $\log K_i^*$ for the Formation of $Mg^{2+}$ and $Ca^{2+}$ Phosphate Ion Pairs**

| Species (X) | $\log K^*_{MgX}$ | $\log K^*_{CaX}$ |
|---|---|---|
| $H_2PO_4^-$ | 0.14 | −0.15 |
| $HPO_4^{2-}$ | 1.23 | 0.97 |
| $PO_4^{3-}$ | 3.36 | 4.51 |

$$M^{2+} + PO_4^{3-} \rightarrow MPO_4^- \quad (9)$$

The association constants, $K_i^*$, for the formation of $Mg^{2+}$ and $Ca^{2+}$ phosphate ion pairs at I = 0.7 and 25°C are given in Table 7.3.

$$K_1^* = [MH_2PO_4^+]/[M^{2+}][2PO_4^-] \quad (10)$$

$$K_2^* = [MHPO_4^0]/[M^{2+}][HPO_4^{2-}] \quad (11)$$

$$K_3^* = [MPO_4^-]/[M^{2+}][PO_4^{3-}] \quad (12)$$

The fractions of the various forms of phosphate in a given media can be calculated from

$$\frac{[H_3PO_4]_T}{[P]_T} = \left[1 + \frac{K_1}{[H]} + \frac{K_1K_2}{[H]^2} + \frac{K_1K_2K_3}{[H]^3}\right]^{-1} \quad (13)$$

$$\frac{[H_2PO_4^-]_T}{[P]_T} = \left[1 + \frac{[H]}{K_1} + \frac{K_2}{[H]} + \frac{K_2K_3}{[H]^2}\right]^{-1} \quad (14)$$

$$\frac{[HPO_4^{2-}]_T}{[P]_T} = \left[1 + \frac{[H]}{K_2} + \frac{[H]^2}{K_1K_2} + \frac{K_3}{[H]}\right]^{-1} \quad (15)$$

$$\frac{[PO_4^{3-}]_T}{[P]_T} = \left[1 + \frac{[H]}{K_3} + \frac{[H]^2}{K_2K_3} + \frac{[H]^3}{K_1K_2K_3}\right]^{-1} \quad (16)$$

The calculated percent of the various forms of phosphate in $H_2O$, NaCl, and seawater as a function of pH are shown in Figure 7.1. At a pH = 8.1 the percent of each species is given in Table 7.4.

As one increases the pressure (or depth) the various forms of $H_3PO_4$ change due to the negative volume change that occurs for the ionization process. The ionized species are more highly charged and have a smaller molar volume. An increase in the pressure forces the equilibria to the smallest volume and the dissociation constants become larger. The effect of pressure on the ionization of $H_3PO_4$ is given in Table 7.5.

At a pH = 8.1 in seawater the various forms of $H_3PO_4$ change as given in Table 7.6. In deep waters the $PO_4^{3-}$ ion becomes a more important form (50% at P = 1000 bar, or 10,000 m).

In all the calculations shown above the fractions of $H_3PO_4$ refer to the total concentrations. The speciation of the various forms of $H_3PO_4$ can be calculated from the ion pairing constants given in Table 7.3. The results are given in Table 7.7. The $H_2PO_4^-$ is largely free (92%); the $HPO_4^{2-}$ is 49% free, 46% as $MgHPO_4$, and 5% as $CaHPO_4$; while $PO_4^{3-}$ is 27% as $MgPO_4^-$ and 73% as $CaPO_4^-$.

Little is known of the nature of particulate phosphorus in seawater. One might expect particulate forms of inorganic P to regulate the maximum concentration of $H_3PO_4$ in seawater. The solubility product of $Ca_3(PO_4)_2$ has been estimated to be $10^{-32}$. This value can be used to estimate the equilibrium concentration of $PO_4$ in seawater

$$K_{sp} = [Ca]_T^3[PO_4]_T^2\gamma_T^3(Ca)\ \gamma_T^2(PO_4) \tag{17}$$

using [Ca] = 0.0108, $\gamma_T(Ca) = 0.28$, and $\gamma_T(PO_4) = 3.7 \times 10^{-5}$ gives $[PO_4]_T = 0.02 \times 10^{-6}$ *M*.

One would also expect particulate organic phosphorus compounds due to the breakdown of plants in surface waters. Since $PO_4^{3-}$ can be absorbed on various surfaces, it may be associated with detrital material and clay minerals. In $CaCO_3$ environments, much of the phosphate is absorbed to carbonate minerals. Although one would expect concentrations of $PO_4^{3-}$ in pore waters of sediments receiving organic matter, the concentrations are nearly undetectable in carbonate sediments (for example, in the Bahama Islands sediments).

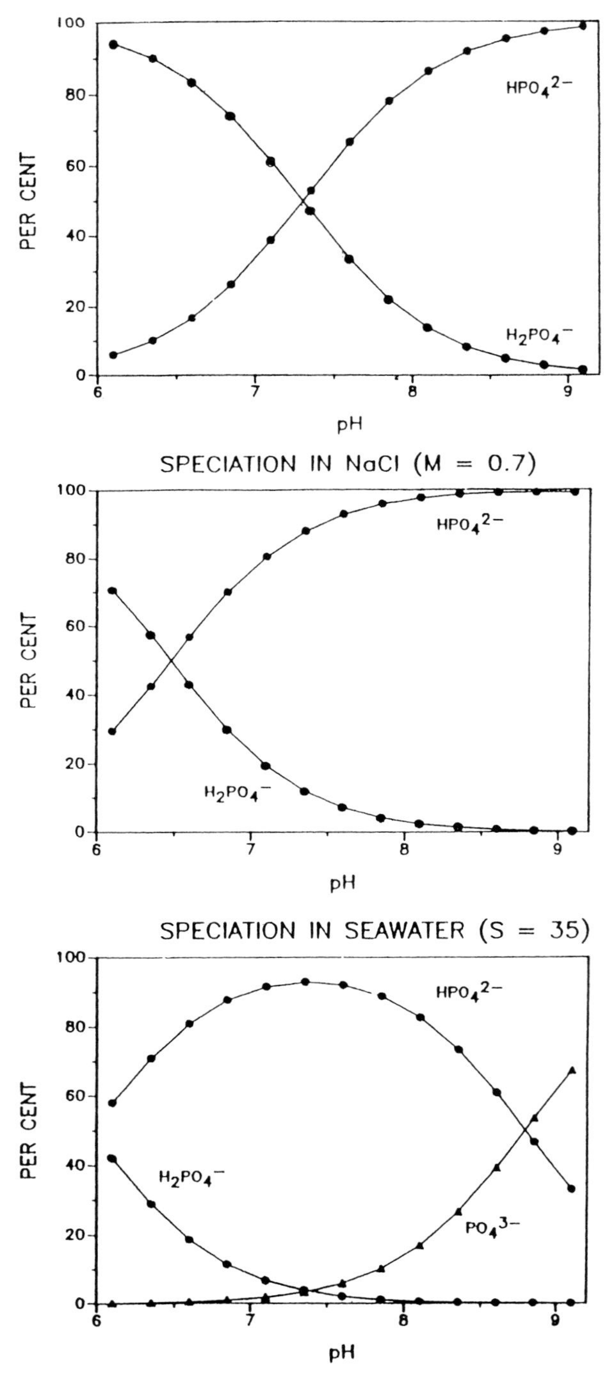


FIGURE 7.1. The various forms of phosphoric acid in water, NaCl (0.7 *M*), and seawater (S = 35).

**TABLE 7.4**
**The Percent of the Total Phosphate Species at 25°C and pH = 8.1**

| Species | $H_2O$ | NaCl | Seawater |
|---|---|---|---|
| $H_3PO_4$ | 0 | 0 | 0 |
| $H_2PO_4^-$ | 11.2 | 1.9 | 0.5 |
| $H_2PO_4^{2-}$ | 88.8 | 98.0 | 79.2 |
| $PO_4^{3-}$ | 0 | 0.1 | 20.4 |

**TABLE 7.5**
**The Effect of Pressure on the Ionization Constants of $H_3PO_4$ in Seawater (25°C)**

| P | $K_1^P/K_1^0$ | $K_2^P/K_2^0$ | $K_3^P/K_3^0$ |
|---|---|---|---|
| 0 bar | 1.00 | 1.00 | 1.00 |
| 500 | 1.36 | 1.65 | 1.98 |
| 1000 | 1.78 | 2.61 | 3.64 |

**TABLE 7.6**
**The Percent of the Various Forms of Phosphate in Seawater (S = 35 and 25°C) as a Function of Pressure (Bar)**

| Species | P = 0 | P = 500 | P = 1000 |
|---|---|---|---|
| $H_3PO_4$ | 0 | 0 | 0 |
| $H_2PO_4^-$ | 0.5 | 0.2 | 0.1 |
| $HPO_4^{2-}$ | 79.2 | 66.1 | 51.6 |
| $PO_4^{3-}$ | 20.4 | 33.7 | 48.3 |

**TABLE 7.7**
**The Percent of the Various Forms of $H_3PO_4$ in Seawater at 25°C**

| X | Free X | MgX | CaX |
|---|---|---|---|
| $H_2PO_4^-$ | 92.3 | 7.0 | 0.7 |
| $HPO_4^{2-}$ | 49.3 | 45.8 | 4.9 |
| $PO_4^{3-}$ | 0.2 | 26.6 | 73.2 |

## 1.1. Determination of Phosphate

The determination of phosphate is carried out by treating an aliquot of seawater with an acidic molybdate reagent containing ascorbic acid and a small amount of potassium antimonyl tartrate. The resulting phosphomolybdic acid is reduced to give a blue-purple complex. The absorbance is measured at 885 nm with a spectrophotometer. The reduced heteropoly acid has a ratio of 1:12:1 for P:Mo:Sb. Polyphosphates do not react, but can be determined after hydrolysis in acid media at 100°C. Before total P is determined, the organic compounds must be broken down by oxidation. This can be done by treating the sample with hydrogen peroxide and irradiating it for a few hours with high intensity UV radiation. Organic phosphorus concentrations are determined by difference. Particulate P can be determined by filtration through a 0.45 μm filter.

The chemical reactions for the analysis are given by two steps. First, a yellow ammonium molybdiphosphate complex is formed

$$H_2SO_4 + (NH_4)_2MoO_4 \cdot 4H_2O + PO_4^{3-} \rightarrow (NH_4)P(Mo_3O_{10})_4 \quad (18)$$

Second, upon treatment with a reducing agent, such as ascorbic acid, the complex yellow acid is reduced to molybdenum blue.

$$(NH_4)P(Mo_3O_{10})_4 + \begin{array}{l} O \\ \| \\ C\text{—} \\ | \\ HO\text{-}C \quad O \\ \| \\ HO\text{-}C \\ | \\ H\text{-}C\text{—} \\ | \\ HO\text{-}C\text{-}H \\ | \\ C\text{-}H_2OH \end{array} \longrightarrow \underset{\text{blue}}{H_3PO_4 \cdot (4MoO_3 \cdot MoO_2)_2} \cdot 4H_2O + \begin{array}{l} O \\ \| \\ C\text{—} \\ | \\ H\text{-}C \quad O \\ \| \\ O\text{-}C \\ | \\ H\text{-}C\text{—} \\ | \\ HO\text{-}C\text{-}H \\ | \\ C\text{-}H_2OH \end{array} \quad (19)$$

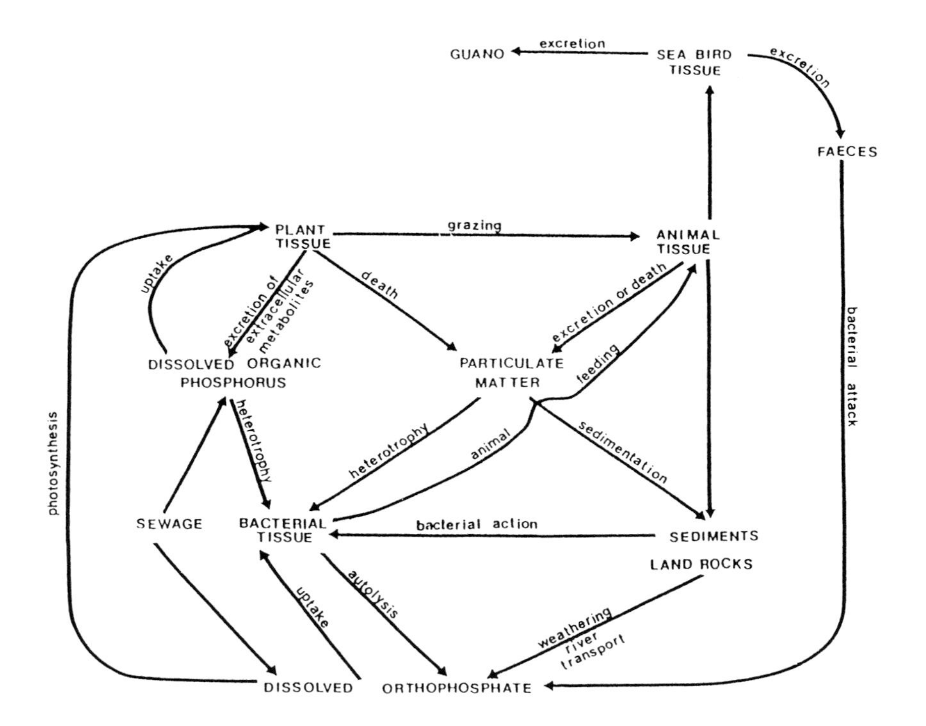


FIGURE 7.2. The phosphate cycle in ocean waters.

The amount of molybdenum blue formed is proportional to the concentration of phosphorus present in the seawater as orthophosphate. The intensity of the color can be measured with a spectrophotometer which relates concentration to light absorbance. Arsenate ions can interfere with the analysis. The measurement of $PO_4$ as well as other nutrients is normally measured with an autoanalyzer. This system allows one to make precise measurements in a short period of time. Recently, Johnson and Petty (U. Cal., Santa Barbara) have described a flow injection technique to measure $PO_4$. The precision of the technique is 1.5% at 3 $\mu M$ and the detection limit is 0.05 $\mu M$. The rates of analysis can be 90 samples per hour.

## 1.2. Distribution of Phosphate

The distribution of the various forms of phosphate in ocean waters is controlled by biological and physical processes. The phosphate cycle for the oceans is shown in Figure 7.2. In surface waters, $PO_4$ is taken up by phytoplankton during photosynthesis. Phosphorus compounds such as ATP and nucleotide coenzymes play key roles in photosynthesis and other processes in plants. Absorption and conversion to organic-P compounds proceed even

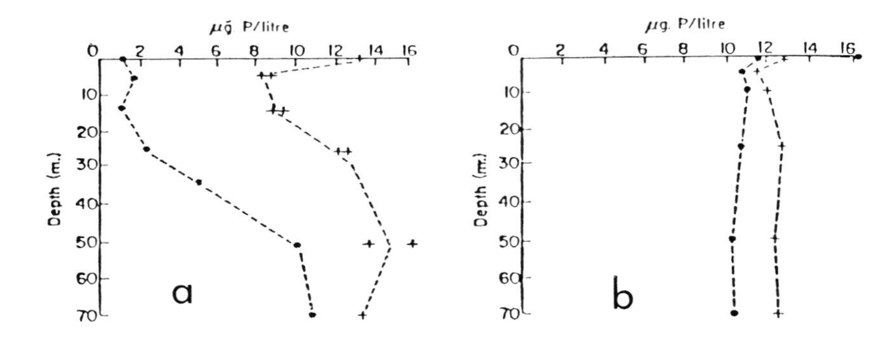


FIGURE 7.3. Typical profiles of total and inorganic phosphate in the English Channel in the summer (a) and winter (b).

in the dark. Above a concentration of 0.3 μ*M* the rate of growth of many species of phytoplankton is independent of the concentration of P. Below 0.3 μ*M* cell division becomes inhibited and P deficient cells are produced. This probably does not occur in the oceans since $NO_3$ is usually exhausted before $PO_4$ falls to a critical level.

When phytoplankton die, the organic-P is rapidly converted to $PO_4$. Much of the phytoplankton are consumed by zooplankton, which obtain $PO_4$ in the process. The unassimilated material is lost in fecal pellets which contain appreciable amounts of organic-P. Hydrolysis of organic-P occurs rapidly through the action of phosphorylases. The excretion of P by zooplankton is minimal when phytoplankton are abundant and maximal when phytoplankton are not abundant. The low rate of excretion when food is abundant arises because phospholipids are being stored or used for egg production. When food is scarce these trends are reversed and P is excreted.

Typical profiles of total and inorganic P in the English Channel are shown in Figure 7.3. In the summer the surface waters are nearly 50% as organic-P. In deeper waters most of the P is in an inorganic form. In the winter almost all the P is inorganic. The variations in coastal waters are due to upwelling and phytoplankton blooms. Surface mixing in the winter can cause linear profiles of P in nearshore waters. After spring and summer blooms the $PO_4$ is decreased considerably. In the Chesapeake Bay decreases from 0.4 to 0.1 μ*M* can occur at 3:00 pm. After sunset the values rapidly increase to a maximum at 2:00 am.

Typical profiles of $PO_4$ in the Atlantic and Pacific (GEOSECS) are shown in Figure 7.4. The surface values are nearly zero. As phytoplankton and other organisms die, the $PO_4$ is regenerated in the water column. A maximum occurs in both oceans near 1000 m, which is the same depth as the $O_2$ minimum layer. The maximum values in the Atlantic are about 1.5 μ*M* while in the

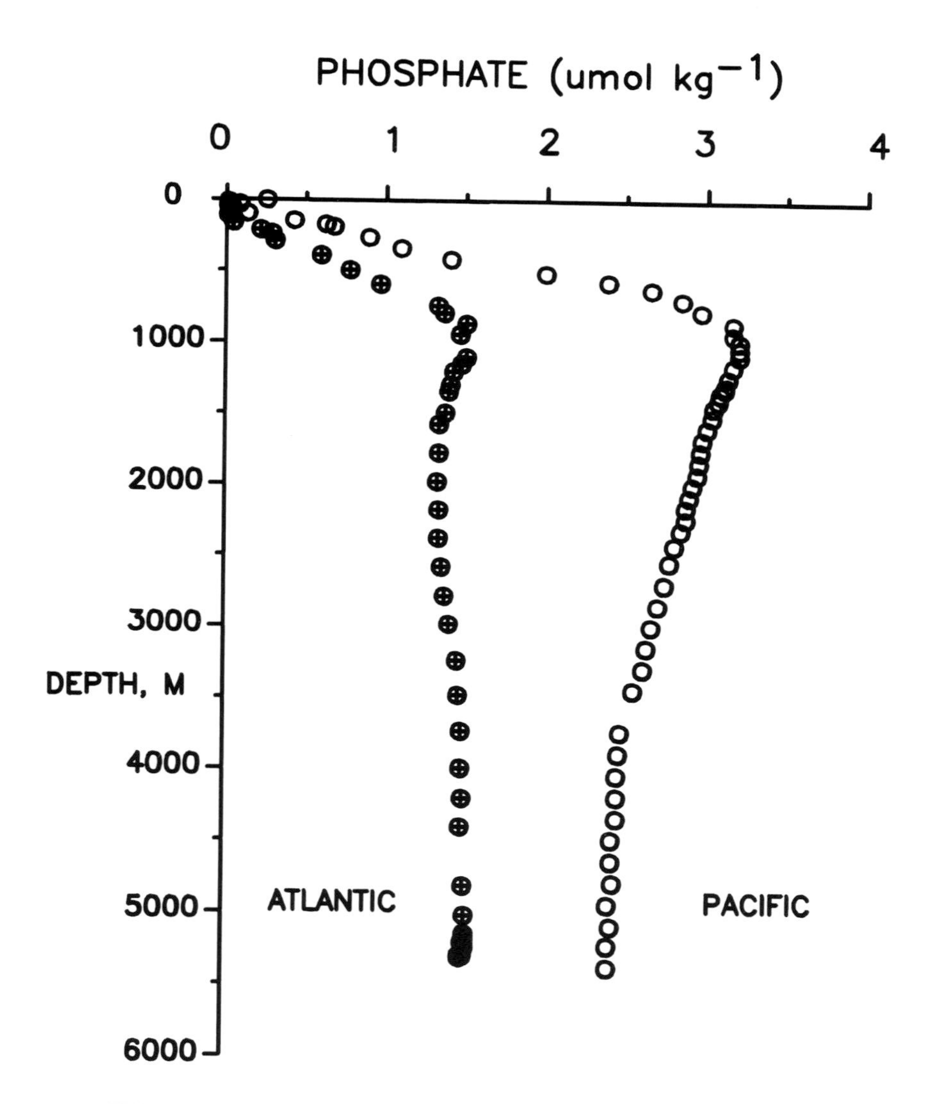


FIGURE 7.4. Profile of phosphate in the Atlantic and Pacific Oceans.

Pacific values of about 3.2 $\mu M$ occur. The higher values in the Pacific (and Indian) oceans compared to the Atlantic are due to the waters being older and, thus, accumulating more oxidized plant material.

Sections of $PO_4$ in the Atlantic and Pacific oceans are shown in Plates 17 and 18.* The values generally follow the movement of the major water masses.

* Plates 17 and 18 follow page 48.

**TABLE 7.8**
**The Various Oxidation States of Nitrogen**

| Oxidation State | Compound |
|---|---|
| +5 | $NO_3^-$, $N_2O_5$ |
| +4 | $NO_2$ |
| +3 | $HONO^a$, $NO_2^-$, $N_2O_3$ |
| +2 | $HONNOH^b$, $HO_2N_2^-$, $N_2O_2^{2-}$ |
| +1 | $N_2O$ |
| 0 | $N_2$ |
| −1 | $H_2NOH$, $HN_3$, $N_3^-$, $NH_2OH$ |
| −2 | $H_2NNH_2$ |
| −3 | $RNH_4$, $NH_3^c$, $NH_4^{+\,c}$ |

[a] pK = 3.35.
[b] $pK_1 = 7.05$, $pK_2 = 11.0$.
[c] $pK_B = 4.75$, $pK_A = 9.48$.

## 2. NITROGEN COMPOUNDS IN SEAWATER

In addition to $N_2$, the sea contains small amounts of inorganic and organic nitrogen compounds (about 1/10 the concentration of $N_2$). The forms can be dissolved or particulate and subdivided into organic and inorganic. The principal inorganic forms are $NO_3^-$ (1 to 500 $\mu M$), $NO_2^-$ (0.1 to 50 $\mu M$), and ($NH_3$ + $NH_4^+$) (1 to 50 $\mu M$). Small amounts of nitrous oxide, hydroxlyamine, and hyponitrite ion also occur. The difficulty with nitrogen compounds is that they can exist in nine oxidation states (Table 7.8).

The ammonia ion can exist in two forms depending on the pH. The dissociation of $NH_4^+$ is given by

$$NH_4^+ \rightarrow H^+ + NH_3 \quad (20)$$

and has a pK = 9.5 at 25°C in seawater. At a pH = 8.1, 95% of the total ammonia is $NH_4^+$ and 5% is $NH_3$.

## 2.1. Determination of Nitrogen Compounds

Before looking at nitrate cycles in the oceans, it is of value to briefly mention the methods used to determine $NO_3^-$, $NO_2^-$, and $NH_4^+$. The inorganic forms of N are normally determined by colorimetric methods using an auto analyzer.

### 2.1.1. $NO_2^-$

$NO_2^-$ is determined by treating the water sample with a solution of sulphanilamide. The resultant diazonium ion is coupled with *N*-(1 naphthy)lethylenediamine to give a pink azo dye. The absorbance is measured at 543 nm with a spectrophotometer. The reactions are given below.

$$\underset{\text{sulphanilamide}}{NH_2\text{–}C_6H_4\text{–}SO_2NH_2} + NO_2^- + 2H^+ \rightarrow$$

$$\underset{\text{diazonium ion}}{^+N{=}N\text{–}C_6H_4\text{–}SO_2NH_2} + H_2O \qquad (21)$$

$$\underset{\text{naphthyl-ethylenediamine}}{NH_2CH_2CH_2NH\text{–}C_{10}H_7} + \underset{\text{diazonium ion}}{^+N{=}N\text{–}C_6H_4\text{–}SO_2NH_2} \rightarrow$$

$$\underset{\text{pink azo dye}}{NH_2CH_2CH_2NH\text{–}C_{10}H_6N{=}N\text{–}C_6H_4\text{–}SO_2NH_2} + H^+ \qquad (22)$$

### 2.1.2. $NO_3^-$

$NO_3^-$ is determined by reducing it to $NO_2^-$, which is analyzed as given above. The reduction is carried out by treating the sample with $NH_4Cl$ or EDTA and passing it through a glass column packed with amalgamated or copper-coated cadmium filings.

### 2.1.3. $NH_3$-$NH_4^+$

No completely satisfactory method exists for the routine analysis of $NH_3$-$NH_4^+$, which are at very low concentrations in oxygenated seawater. At present two methods are being used (certain organic N compounds can cause interference).

1. $NH_3$ is oxidized to $NO_3^-$ by alkaline hypochlorite. The excess hypochlorite is then reduced with arsenite and the $NO_2$ determined as above.
2. $NH_3$ is oxidized in an alkaline citrate medium with Na hypochlorite and phenol in the presence of catalytic amounts of Na nitroprusside. A blue indophenol dye is produced and is measured with a spectrophotometer.

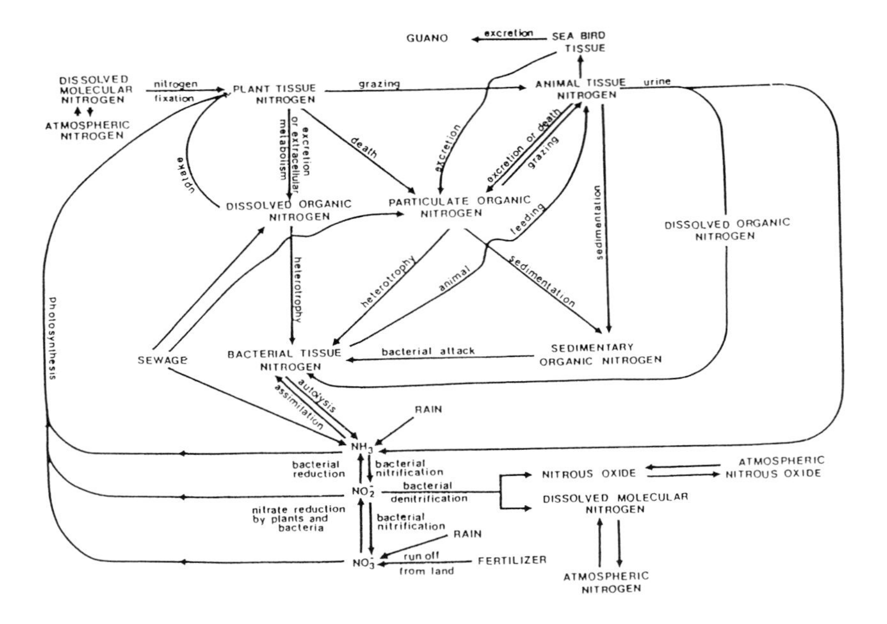


FIGURE 7.5. The nitrogen cycle in ocean waters.

## 2.2. Distribution of Nitrogen Compounds

The nitrogen cycle in the oceans is shown in Figure 7.5.

There are three major inputs of nitrogen to ocean waters: (1) volcanic activity ($NH_3$), (2) atmospheric ($NO_2$ from nitrogen fixation), and (3) rivers (fertilizers). Nitrates are taken from surface waters during primary productivity. When the plants die and decompose, nitrogen compounds are regenerated to the water column. Marine birds can also cause a loss of nitrogen as $NaNO_3$ in guano. The large deposits of $NaNO_3$ in Chilean deserts could also have been formed by bacteria fixation or volcanism. Nitrogen can be lost to the atmosphere as $N_2O$. As discussed later, this gas can react with ozone. In summary, the nitrogen cycle is controlled by biological organisms.

Nitrogen fixation ($N_2 \rightarrow NO_3$) is carried out by nodular bacteria of land plants. Fresh water bacteria, molds, and yeasts can also fix nitrogen. Some counterparts have been found in seawater (blue-green algae). Nitrogen fixation, however, does not have significant effect in ocean waters. The assimilation of fixed nitrogen ($NH_3$, $NO_2^-$, and $NO_3$) by phytoplankton takes place in the euphotic zone during photosynthesis. The $NH_3$ or $NH_4^+$ form is usually preferred. The uptake shows a direct hyperbolic behavior which increases to a maximum rate at a given nutrient concentration. When nitrate is below 0.7

$\mu M$, nitrogen deficient cells are produced before cell division stops. These cells can take up $NH_3$ and $NO_3^-$, but not $NO_2^-$ in the dark. Some phytoplankton can utilize amino acids, as can some diatoms with the help of bacteria. In polluted waters a significant amount of $NO_3^-$ can be obtained from organic nitrogen. Urea is also used in restricted coastal and estuarine areas.

The conversion of $NO_3^-$ to amino acids requires the formation of $NH_3$

$$NO_3^- + 2H^+ + 2e \rightarrow NO_2^- + H_2O \tag{23}$$

$$2NO_2^- + 4H^+ + 4e \rightarrow N_2O_2^{2-} + 2H_2O \tag{24}$$

$$N_2O_2^{2-} + 6H^+ + 4e \rightarrow 2NH_2OH \tag{25}$$

$$NH_2OH + 2H^+ + 2e \rightarrow NH_3 + H_2O \tag{25a}$$

where $N_2O_2^{2-}$ is hyponitrite and $NH_2OH$ is hydroxylamine. The first step is catalyzed by coenzyme II. The $NH_3$ is converted to glutamic acid by

$$\begin{aligned} &HOOC\text{–}CO\text{–}(CH_2) + NH_3 + 2NADPH \rightarrow \\ &\qquad HOOC\text{–}CH(NH)2CH_2CH_2COOH + 2NADP + H_2O \end{aligned} \tag{26}$$

The acid reacting with $NH_3$ is called ketoglutaric acid. Twenty other amino acids are formed from glutamic acid by transamination. The formation of alanine, for example, is formed from pyruvic acid

$$\begin{aligned} &CH_3COCOOH + HOOC\text{–}CH(NH_2)CH_2CH_2COOH \rightarrow \\ &\qquad CH_3CH(NH_2)COOH + HOOC\text{–}CO(CH_2)_2COOH \end{aligned} \tag{27}$$

Proteins are formed by the linking of the various amino acids by reactions involving RNA and DNA and using energy from ATP.

The regeneration of $NO_3$ is done mainly by bacterial oxidation of organic nitrogen. When cells die they undergo rapid autolysis, releasing $NH_3$ and $PO_4^{3-}$. The decomposition of organic nitrogen compounds to $NO_3^-$ takes place in a number of steps. This is shown in Figure 7.6 for the bacterial oxidation of dead organisms. A small percentage of the particulate organic nitrogen is resistant to bacterial attack and can be accumulated in the sediments. Nitrification is the oxidation of $NH_3$ to $NO_3^-$. This process yields $NO_2^-$ as an

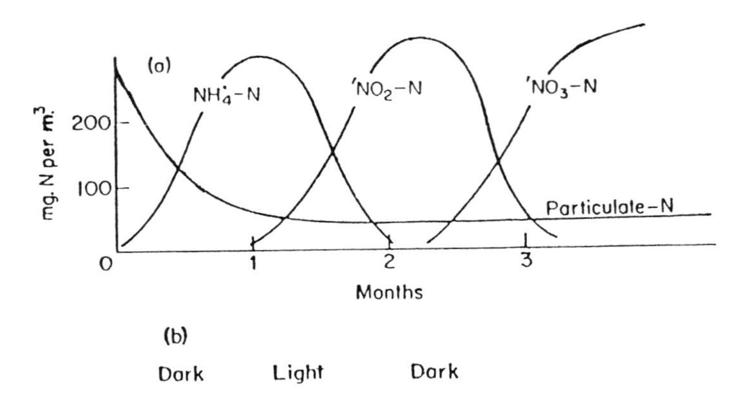


FIGURE 7.6. The decomposition of organic nitrogen by bacterial oxidation.

intermediate. This process is carried out by bacteria in the water column and in the sediments. It may lead to $NO_2^-$ in upwelled waters off Peru.

Bacteria are also capable of reducing $NO_3^-$ to $NO_2^-$. This reduction was thought to occur in waters with high organic levels, but it is now thought to occur in low oxygen waters.

Most denitrification ($NO_3^- \rightarrow N_2$ or $N_2O$) occurs by bacterial growth in anoxic waters. The $NO_3^-$ is used as an electron acceptor instead of $O_2$. It also appears to occur in waters with low levels of $O_2$. $N_2O$ is also produced by this process. Typical profiles of $NO_3^-$ in the Atlantic and Pacific are shown in Figure 7.7.

Surface waters have very low levels while the deep waters have significant concentrations. The higher concentrations in the deep Pacific are due to the waters being older and accumulating more $NO_3^-$. Sections of $NO_3^-$ in the Atlantic and Pacific shown in Plates 19 and 20* follow the major water masses. In coastal areas there is frequently a seasonal variation of inorganic nitrogen (Figure 7.8). In the spring rapid removal occurs due to phytoplankton growth. The phytoplankton are consumed by zooplankton and fish returning the $NH_4^+$ and $NO_3$ to the water. In the summer the thermocline prevents vertical mixing or replenishment of surface waters. This can deplete the nutrients enough to prevent further primary productivity. In upwelling areas $NO_3^-$ is not limiting. The oxidation of plant material causes the $NO_3^-$ to maximize just below the oxygen minimum.

The distribution of $NO_2^-$ in ocean waters shows two maxima (Figure 7.9), one just below the photosynthetic compensation depth. This is probably due

* Plates 19 and 20 follow page 48.

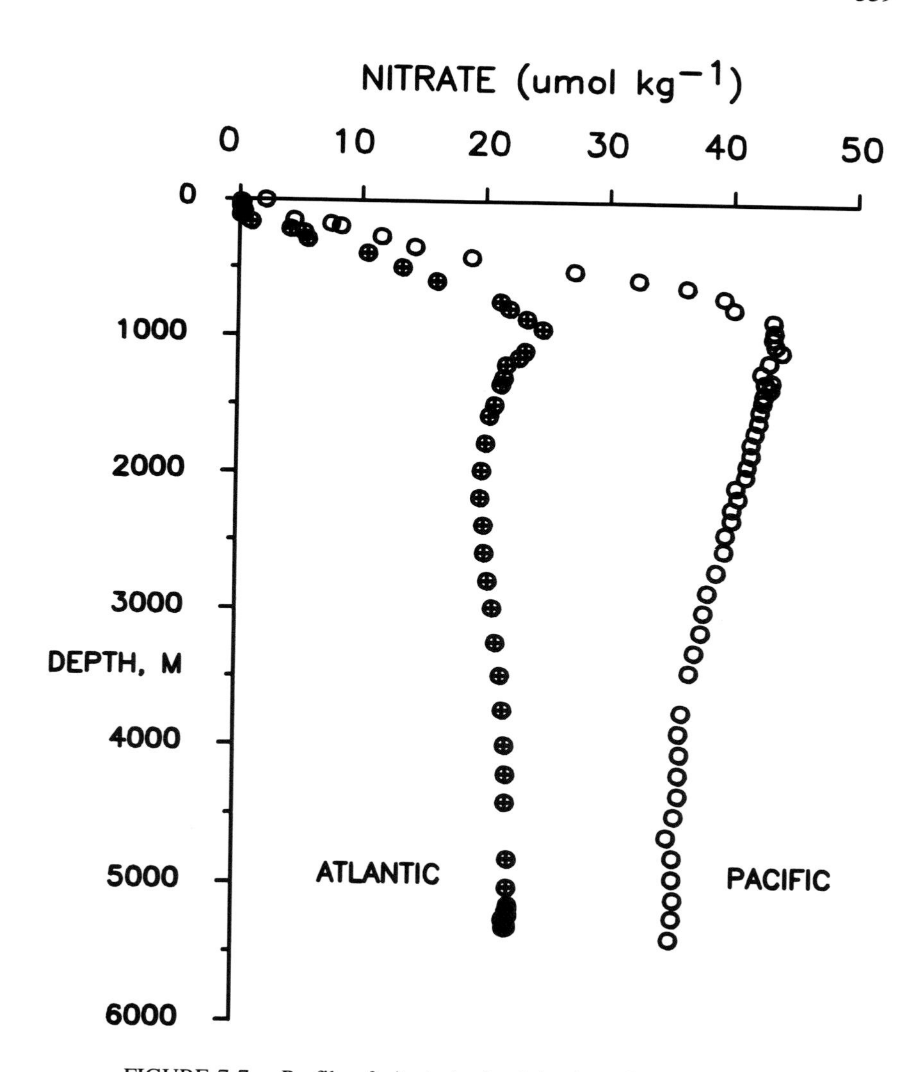


FIGURE 7.7. Profile of nitrate in the Atlantic and Pacific Oceans.

to the oxidation of $NH_3$ ($NH_3 \rightarrow NO_2^- \rightarrow NO_3^-$). In certain areas of the Pacific, a second maximum occurs at the oxygen minimum layer. This is thought to be caused by the bacterial reduction of $NO_3^-$.

As discussed earlier, $N_2O$ is formed in the oxygen minimum layer. This is shown in Figures 7.10 and 7.11. The excess $N_2O$ produced is proportional to the AOU (oxygen utilized). One molecule of $N_2O$ is produced for every 10,000 molecules utilized. When the waters have low concentrations of oxygen, the relationship between $N_2O$ and $O_2$ is broken (Figure 7.12). This is caused by the utilization of $N_2O$ as a source of oxygen when $O_2$ is limiting.

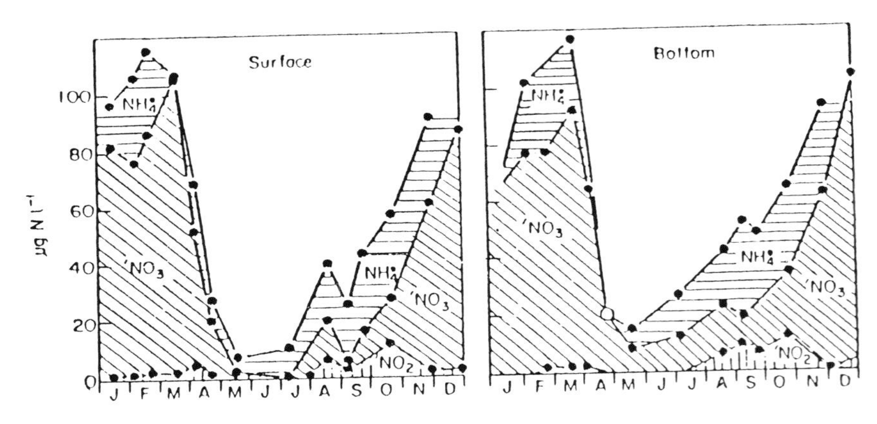


FIGURE 7.8. Seasonal distribution of nitrogen species in the English Channel.

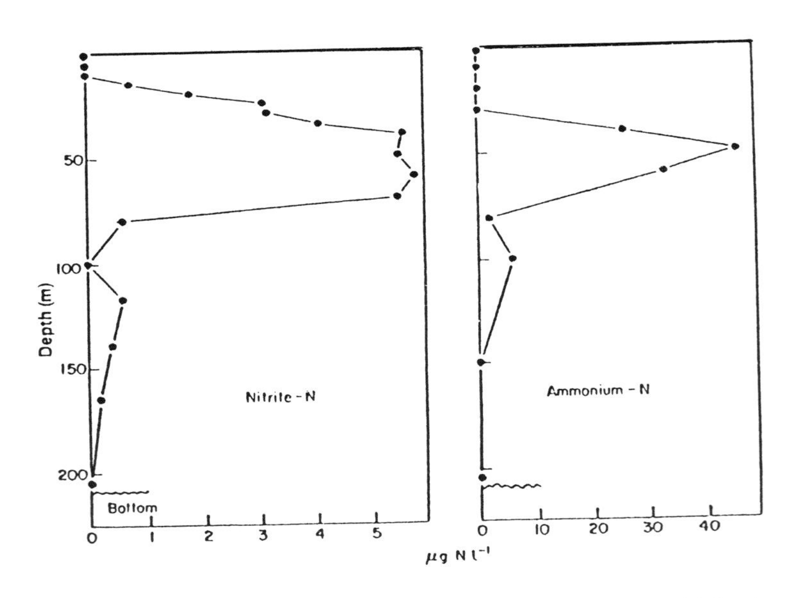


FIGURE 7.9. Profiles of nitrite and ammonium in the Gulf of Maine in the summer.

## 2.3. Nitrogen-Phosphorus Ratio

The N:P ratio assimilated from seawater is about 15:1 by phytoplankton as they grow. Lower values are found in coastal waters. For example, in the English Channel, winter values are 10.5:1 while in the summer they are 19:1. The P:N:C ratio in phytoplankton is 1:16:106. From Liebig's law of minimum (the concentration present in the smallest quantity relative to the requirement

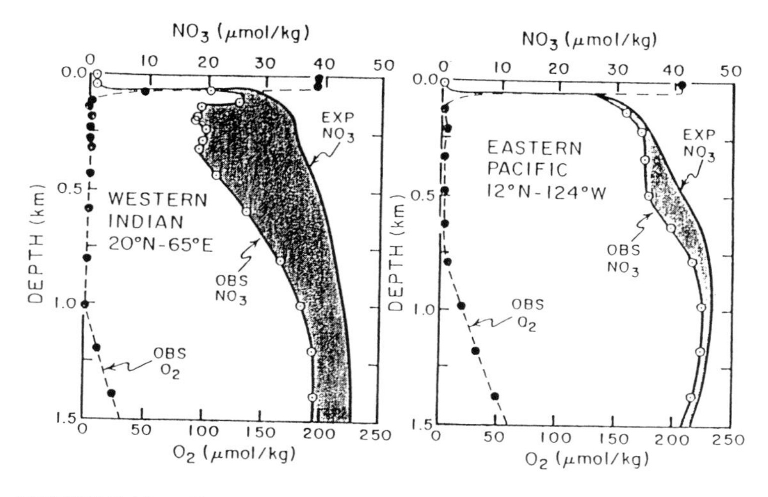


FIGURE 7.10. Comparison of the observed and estimated nitrate in low oxygen waters in the Indian and Pacific Oceans.

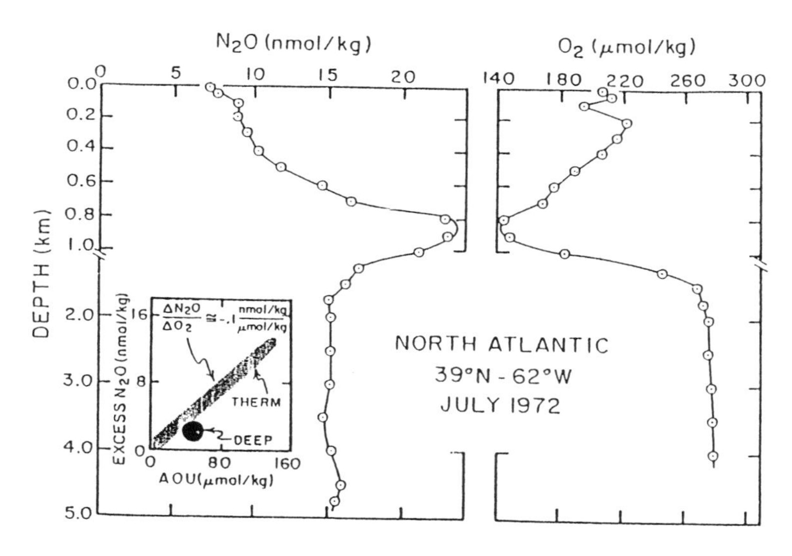


FIGURE 7.11. Profiles of nitrous oxide and oxygen in the Atlantic Ocean.

for growth of an organism will be the limiting factor) it is clear that the P:N is limiting compared to carbon (the actual P:C ratio is 1000). Typical values of N:P in the Atlantic and Pacific are shown in Figure 7.13.

The waters in the Atlantic vary from 16:1 at the surface to 15:1 in deep waters. The deep waters in the Pacific are also close to 15:1, however, the waters in the $O_2$ minimum layer are as low as 14:1. This is due to the

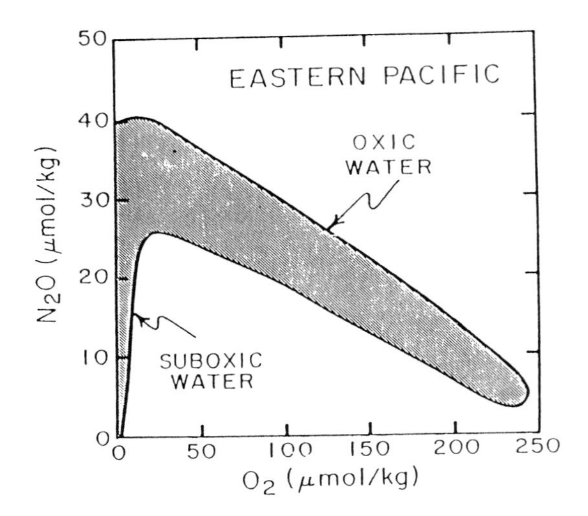


FIGURE 7.12. The relationship of nitrous oxide and oxygen in Pacific Ocean waters.

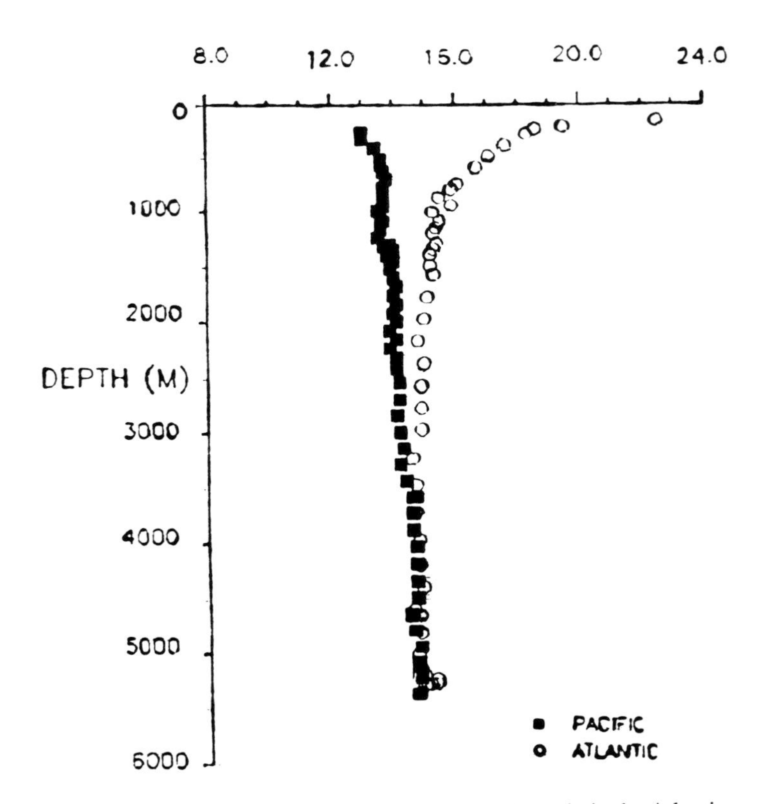


FIGURE 7.13. The nitrogen to phosphate molar ratio in the Atlantic and Pacific Oceans.

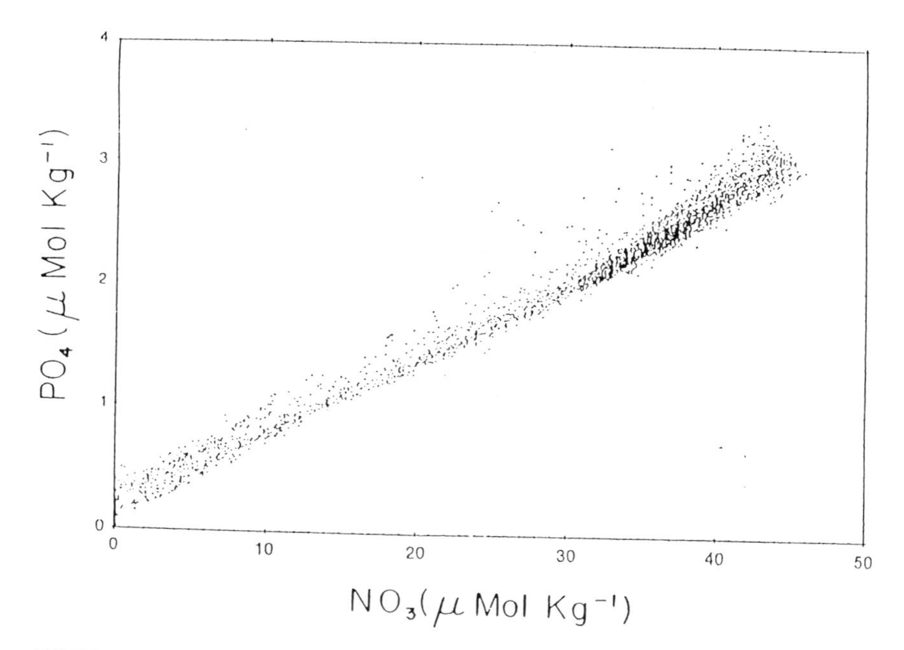


FIGURE 7.14. Correlation of the concentration of nitrogen to phosphate in the Atlantic Ocean.

conversion of $NO_3^-$ to $N_2O$ in these low oxygen waters. All of the GEOSECS $PO_4^{3-}$ and $NO_3^-$ data is plotted in Figure 7.14. The slope through all the data yields a N/P ratio close to 16.

In 1934, Redfield examined the relationship between the concentrations of $O_2$, $CO_2$, $NO_3^-$, and $PO_4^{3-}$ in seawater based on the average chemical composition of plankton. This relationship predicts the ratio of oxygen consumption to nutrient production caused by biological oxidation. In the North Atlantic he found a N:P ratio of 20:1. He also found an apparent $O_2$ utilization to $NO_3^-$ of 6:1. He found variations that he attributed to preformed concentrations in the deep waters which originated before the waters sank. He estimated a N:C ratio of 1:7. He proposed an oxidation ratio of P:N:C of 1:20:140 accompanied by 120 units of $O_2$. These ratios supported the plankton sample ratios of 1:18:137. In 1963, Redfield, Ketchum, and Richards proposed a revised ratio of 1:16:106:138 for P:N:C:$O_2$. These results lead to the following equation for the oxidation of plant material

$$(CH_2O)_{106}(NH_3)_{16}(H_3PO_4) + 138O_2 \rightarrow 106CO_2 + 122H_2O + 16HNO_3 + H_3PO_4 \quad (28)$$

**TABLE 7.9**
**Molecular Ratios of P, N, C, $O_2$, and $CaCO_3$ Changes in the Atlantic and Indian Oceans**

| Location | $\sigma_1$ Surface | P | N | $CO_2$ | ($O_2$-2N) | $O_2$ | $CaCO_3$ |
|---|---|---|---|---|---|---|---|
| N. Atlantic | 27 | 1 | 17.6 ± 0.6 | 97 ± 9 | 130 ± 6 | 165 ± 7 | 15 ± 4 |
| | 27.2 | 1 | 16.8 ± 0.5 | 88 ± 6 | 139 ± 6 | 173 ± 6 | 8 ± 3 |
| S. Atlantic | 27 | 1 | 16.7 ± 0.7 | 102 ± 7 | 131 ± 6 | 165 ± 6 | 8 ± 2 |
| | 27.2 | 1 | 16.7 ± 1.2 | 95 ± 10 | 150 ± 2 | 182 ± 9 | 8 ± 4 |
| Mean Atlantic | | 1 | 17 ± 0.4 | 96 ± 6 | 138 ± 9 | 171 ± 8 | 10 ± 4 |
| S. Indian | 27 | 1 | 15.2 ± 0.6 | 112 ± 6 | 138 ± 7 | 169 ± 8 | 15 ± 4 |
| | 27.2 | 1 | 14.5 ± 0.5 | 125 ± 7 | 145 ± 5 | 174 ± 6 | 19 ± 6 |
| Mean Indian | | 1 | 14.9 ± 0.4 | 119 ± 5 | 142 ± 5 | 172 ± 5 | 17 ± 4 |
| Overall Mean | | 1 | 16.3 ± 1.1 | 103 ± 14 | 140 ± 8 | 172 ± 7 | 12 ± 5 |

where the first compound is a hypothetical organic molecule containing the C:N:P ratio of average plankton. This equation assumed a C:$O_2$ ratio of 1:1, N:$O_2$ ratio of 1:2, and a computed P:$O_2$ ratio of 1:138. The $O_2$ ratio is not based on observations made in the oceans.

To test the Redfield model one normally examines a plot of apparent oxygen utilization (AOU) vs. the concentrations of nutrients arising from the oxidation. A number of workers have tested this model by examining values of $O_2$:$PO_4$ and $O_2$:$NO_3$ in the Atlantic and Pacific. These studies showed that the ratios were consistent with the Redfield model. The variations found with depth, latitude, and time were attributed to mixing between different water types with different preformed concentrations of $O_2$, $PO_4^{3-}$, and $NO_3^-$. They were unaware at the time of the conversion of $NO_3^-$ to $N_2O$ in $O_2$ minimum zones.

More recently, Takahashi et al. (Lamont) have used GEOSECS data to examine the Redfield ratio along isopycnal (constant density) surfaces in the Atlantic and Indian oceans. Their values corrected for mixing are given in Table 7.9. The authors attempted to correct the $CO_2$ values for the dissolution of $CaCO_3$. The mean values of P:N:C:$O_2$ were found to be 1:17:100:165 at $\sigma_T$ = 27.0 and 1:17:92:178 at $\sigma_T$ = 27.2. The small differences between horizons cannot be explained. They suggest values of 1:16:103:172. If the C value is assumed to be represented by the oxygen utilization minus the $O_2$ used for oxidation of $NH_3$ with 2 mol of $O_2$, a ratio of 1:16:140:172 is obtained. The P:$CO_2$ ratio lies between 1:103 to 1:140. This difference may be due to the increase of $CO_2$ due to the anthropogenic $CO_2$ or an excess demand for oxidation of organic molecules without N. The decomposition of fatty acids would require more $O_2$

$$C_2H_4 + 3O_2 \rightarrow 2CO_2 + 2H_2O \qquad (29)$$

than assumed by Redfield

$$C + O_2 \rightarrow CO_2 \qquad (30)$$

The true P:C ratio cannot be obtained at present. The ratio of P:$CaCO_3$ has been estimated to be 1:12. This indicates that the $CO_2$ produced by the oxidation of organic carbon to that derived from $CaCO_3$ is about 10:1. Since the density depth horizons lie above the lysoclines of calcite and aragonite, they propose that the dissolution takes place in the gut of the zooplankton. This, however, could be due to the dissociation of aragonite (Byrne, University of South Florida). The new results lead to the following equation

$$(CH_2O)_{103}(NH_3)_{16}(H_3PO_4) + 135O_2 \rightarrow 103CO_2 + 119H_2O + 16HNO_3 + H_3PO_4 \qquad (31)$$

## 3. SILICON IN SEAWATER

Silicon in seawater can be both dissolved and particulate. The solubility of solid $SiO_2$ gives

$$SiO_2(s) + 2H_2O \rightarrow Si(OH)_4(aq) \qquad (32)$$

Since $Si(OH)_4$ is a weak acid it can ionize in aqueous solutions.

$$Si(OH)_4 \rightarrow H^+ + Si(OH)_3O^- \qquad (33)$$

$$Si(OH)_3O^- \rightarrow H^+ + Si(OH)_2O_2^{2-} \qquad (33a)$$

The $pK_1^* = 9.47$ and $pK_2^* = 12.60$ for the ionization of $Si(OH)_4$ in 0.6 *M* NaCl at 25°C. At a pH = 8.1 in seawater, these pK values give

$$[Si(OH)_4]/[Si]_T = \{1 + K_{HA}/[H^+]\}^{-1} = 95.9\% \quad (34)$$

$$[Si(OH)_3O^-]/[Si]_T = \{1 + [H^+]/K_{HA}\}^{-1} = 4.1\% \quad (35)$$

Polymerized forms of $Si(OH)_4$ and $Si(OH)_2O_2{}^{2-}$ are not important for seawater solutions. This is due to the low concentrations of $SiO_2$ in natural waters. If $Mg^{2+}$ or $Ca^{2+}$ ions form strong complexes with $Si(OH)_3O^-$ and $Si(OH)_2O_2{}^{2-}$, the charged forms could be in higher concentrations.

Seawater contains a wide variety of finely divided siliceous material. Much of this material is produced by the weathering of rocks and is transported to the oceans by rivers and by the wind. The materials include quartz, feldspar, and clay minerals. As these minerals sink through the water column to the sediments, they can react with the components of seawater to form secondary minerals. Recent studies have shown that hydrothermal vents can also contribute a considerable amount of $SiO_2$ to the oceans.

In surface waters are found diatoms and radiolarian, which have skeletons composed of opal — a noncrystalline form of hydrated $SiO_2$. When these organisms die, they sink to the sediments forming diatom oozes. These diatom oozes are quite prevalent in Antarctic waters. The concentration of suspended material is quite variable. On the average 50% is inorganic and Si can make up 15 to 60% of the inorganic material (the remainder being mostly $CaCO_3$). Concentrations as high as 100 μg/l of biogenic $SiO_2$ are present in Antarctic surface waters during diatom blooms. Since ocean waters are undersaturated with respect to $SiO_2$, the sinking particulate silica will dissolve in deep waters. The breakdown of the diatom in deep waters contributes to this increase of $SiO_2$ to the water column. This leads to depth profiles given in Figure 7.15. Since the release of $SiO_2$ is a slow process, the profiles of dissolved $SiO_2$ do not show the maximum at 1000 km seen in the $NO_3^-$ and $PO_4^{3-}$ profiles. The values of $SiO_2$ are higher in the deep Pacific than the Atlantic because the waters are older and have had a longer time to accumulate $SiO_2$.

### 3.1. Determination of Silicon

The determination of silicon dissolved in seawater is made by forming a yellow silicomolybdic complex. Other molybdate complexes are formed with phosphate and arsinate. These interfering complexes are decomposed by the

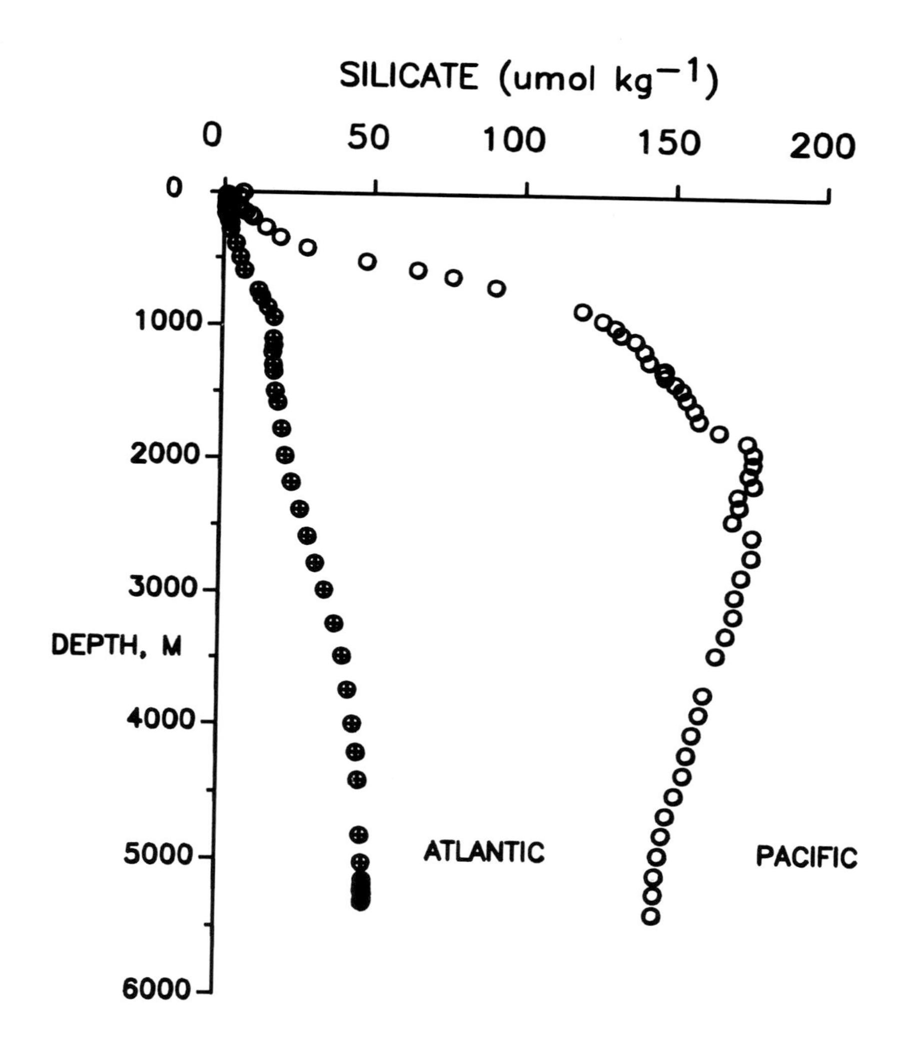


FIGURE 7.15. Profile of silicate in the Atlantic and Pacific Oceans.

addition of oxalic acid. The silicomolybdate complex is reduced by the addition of a solution containing metal (*p*-methyaminophenol sulfate). This forms a blue compound that is determined spectrophotometrically at 812 nm. It is generally reduced to a stable and more absorbant molybdenum blue complex. The reduction can be carried out with metal (*p*-methylaminophenol sulfate) and Na sulfite. Phosphate produces a similar blue complex, but its formation is prevented by incorporating oxalic or tartaric acid in the reducing reagent.

The values of $SiO_2$ vary from 0 to 200 $\mu M$ in seawater. It is an essential part of the solid structure of diatoms, radiolarian, and sponges. Up to 60% of the inorganic material in diatoms is $SiO_2$. These plants can completely exhaust the dissolved $SiO_2$ in surface waters. This process is the principal stripping process of $SiO_2$ from seawater. The great majority is deposited as diatomaceous oozes in Antarctica. The $SiO_2$ coming into the oceans from rivers can be removed in the estuary before the waters reach the oceans. This is thought to be due to diatom production, but interactions with other minerals could be important. Much of the particulate $SiO_2$ coming in from rivers is deposited at the river mouth. The finely suspended minerals, however, can remain in the water column for years. These suspended clay minerals can affect the concentration of trace organic and inorganic species due to absorption and ion-exchange processes. As much as 70 to 99% of the particles have diameters less than 10 $\mu$m.

Not a lot is known about the mechanisms of how diatoms take up $SiO_2$ and deposit it as hydrated silica. Proteins are involved in the absorption of Si on the cytoplanic membrane. The process is fast and it spreads from particular centers. As much as 50% of the dry weight of a diatom can be $SiO_2$ depending upon the species. If diatoms are grown in depleted media, the cells become Si deficient. Such cells are viable for several weeks. They will take up added Si even in the dark. If deficient cells are illuminated, they photosynthesize for a limited period but they soon die. The silica in diatoms is insoluble when living, but dissolves rapidly when they die. An organic or inorganic (Al or Fe) skin may protect them when they are alive. The treatment of dead cells with EDTA has been shown to accelerate the dissolution.

### 3.2. Distribution of Dissolved $SiO_2$

The distribution of $SiO_2$ in coastal waters is generally higher than in the open oceans due to river runoff. In regions where diatom blooms occur, the seasonal variations are similar to $PO_4$. The concentrations decrease in the spring and increase in the summer when growth slackens and then increases to a maximum in early winter. The concentrations in surface waters is low except in upwelling areas. A typical depth profile was shown earlier (Figure 7.15). The lateral distribution follows the general water mass circulation as shown in Plate 21.* The high values in the Antarctic region are from the large diatom production in the surface waters. The $SiO_2$ is also added to the deeper waters from the flux from the sediments. The glacial weathering may also lead to higher concentrations of $SiO_2$ in these regions. The high con-

* Plates 21 and 22 appear following page 48.

centration of $SiO_2$ in the Antarctic bottom waters can be used to trace this water mass. A vertical section in the Pacific is shown in Plate 22. The deep water values increase toward the north and reach concentrations of 220 $\mu M$ in the Bering Sea.

## 4. USE OF NUTRIENTS AS WATER MASS TRACER

Traditionally temperature and salinity have been used as a tracer of water masses. More recently, isotopes such as tritium have been used. From time to time a number of workers have used nutrients to follow various water masses. To use nutrients it is necessary to make a correction to the amount of nutrients added due to the oxidation of plant material. The concentration of a nutrient present in the water when it was at the surface is called the preformed value. This preformed value should be conservative if other processes are not involved in the addition or subtraction from the water mass. Although both $PO_4^{3-}$ and $NO_3^-$ have been used to trace water masses, only $NO_3^-$ will be discussed in this section. As shown by Broecker (Lamont), the combination of $O_2$ and $NO_2^-$ can lead to a conservative water mass tracer. The oxidation that occurs is estimated from the apparent oxygen utilized (AOU). The oxidation of plant material is given by

$$CH_2O + 3O_2 \rightarrow CO_2 + H_2O \quad (36)$$

$$NH_3 + OH^- + 2O_2 \rightarrow NO_3^- + 2H_2O \quad (37)$$

The change in the moles of oxygen used is given by

$$\Delta O_2 = -(2\,\Delta NO_3 + \Delta CO_2) \quad (38)$$

The change in the $NO_3^-$ to $CO_2$ can be estimated from the Redfield ratio

$$\Delta NO_3/\Delta CO_2 = 16/106 \quad (39)$$

The substitution into Equation 38 gives

$$-\Delta O_2 = (2 + 106/16)\,\Delta NO_3 = 9\Delta NO_3 \quad (40)$$

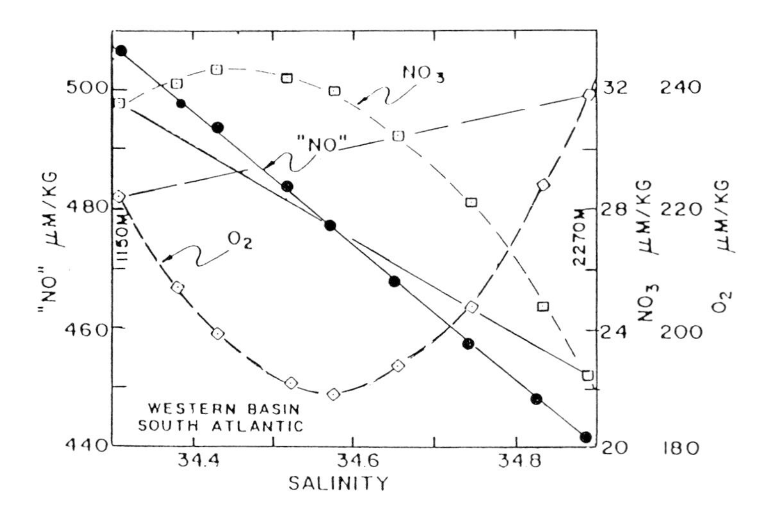


FIGURE 7.16. Values of "NO", $NO_3$, and $O_2$ in the Atlantic Ocean.

The preformed $NO_3^-$ concentration is thus related to the measured value by

$$[NO_3^-]_p = [NO_3^-]\text{meas} - 1/9 \text{ AOU} \tag{41}$$

Substituting the AOU = $[O_2^-]$calc − $[O_2]$meas into Equation 42 gives upon rearrangement

$$NO = 9[NO_3^-]\text{meas} + [O_2]\text{meas} = 9[NO_3]_p + [O_2]\text{calc} \tag{42}$$

where "NO" is a conservative tracer. By a similar argument, it is possible to describe a conservative trace using phosphate

$$PO = 135[PO_4^{3-}]\text{meas} + [O_2]\text{meas} \tag{43}$$

called "PO". Since the concentration of $NO_3^-$ can be measured more accurately than the $PO_4^{3-}$, "NO" is a better conservative tracer than "PO".

The use of NO as a tracer for the mixing of North Atlantic Deep Water (NADW) and Antarctic Intermediate Water (AAIW) is shown in Figure 7.16. The concentration of $O_2$ and $NO_3^-$ show a nonlinear variation as a function of salinity. The quantity "NO", however, is a linear function of salinity or a conservative tracer of the mixing of the two water masses. The values of "NO" vary from 440 $\mu M$ for NADW to 510 $\mu M$ for AAIW. A plot of "NO" for various water masses is shown in Figure 7.17.

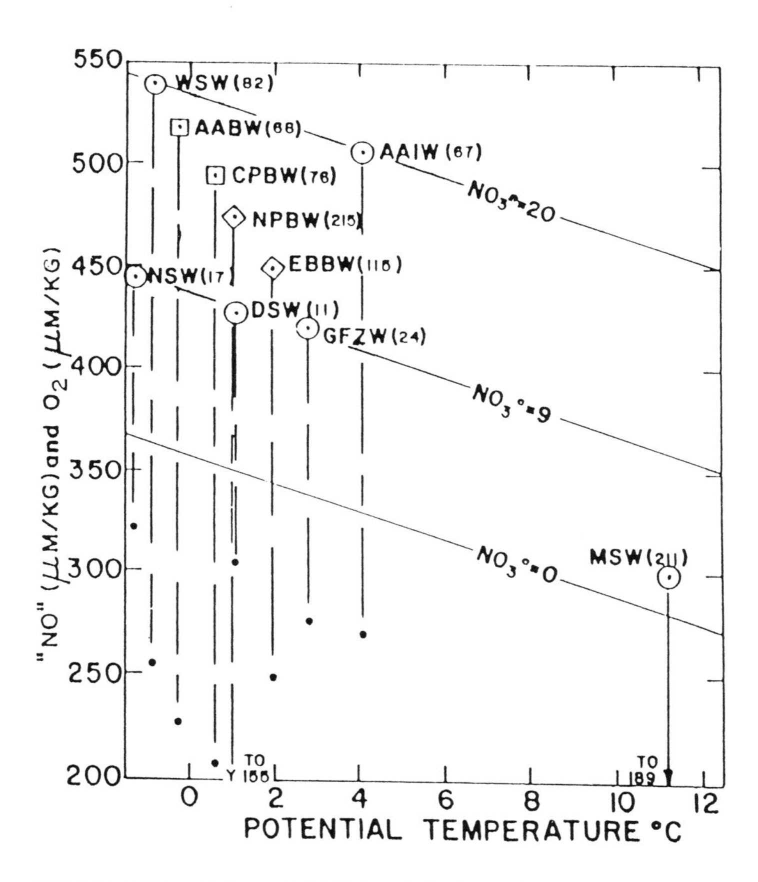


FIGURE 7.17. Values of "NO" and $O_2$ for various water masses as a function of potential temperature.

WSW — Weddell Sea Water
AAIW — Antarctic Intermediate Water
AABW — Antarctic Bottom Water
CPBW — Circumpolar Bottom Water
NPBW — North Pacific Bottom Water
EBBW — Eastern Basin Bottom Water (Atlantic)
NSW — Norwegian Sea Surface Water
DSW — Denmark Straits Overflow Water
GFZW — Gibbs Fracture Zone Water (mainly overflow from Norwegian Sea)
MSW — Mediterranean Overflow Water

## FURTHER READING

Spencer, C.P., The Micronutrient Elements, chap. 11, *Chemical Oceanography,* Vol. 2, 2nd ed., J.P. Riley and G. Skirrow, Eds., Academic Press, New York, 245-300 (1975).

### Phosphorus

Armstrong, F.A.J., Phosphorus, chap. 8, *Chemical Oceanography,* Vol. 1, 1st ed., J.P. Riley and G. Skirrow, Eds., Academic Press, New York, 323-364 (1965).

### Nitrogen

Vaccaro, R.F., Inorganic Nitrogen in Sea Water, chap. 9, *Chemical Oceanography,* Vol. 1, 1st ed., J.P. Riley and G. Skirrow, Eds., Academic Press, New York, 365-408 (1965).

### Silicon

Armstrong, F.A.J., Silicon, chap. 10, *Chemical Oceanography,* Vol. 1, 1st ed., J.P. Riley and G. Skirrow, Academic Press, Eds., New York, 409-432 (1965).

# 8 Table of Contents

**Organic Compounds** ... 355
1. *Dissolved and Particulate Organic Matter* ... 355
   1.1. *Dissolved Organic Matter* ... 360
   1.2. *Particulate Organic Matter* ... 360
2. *Organic Matter Sources* ... 362
   2.1. *Terrestrial Input by Rivers* ... 363
   2.2. *Terrestrial Input from the Atmosphere* ... 364
   2.3. *Additional Sources of Organic Matter* ... 366
3. *Carbohydrates* ... 366
4. *Amino Acids and Proteins* ... 371
5. *Hydrocarbons* ... 376
   5.1. *Volatile Hydrocarbons* ($<C_{14}$) ... 377
   5.2. *Higher Molecular Weight* ($>C_{14}$) *Hydrocarbons* ... 379
   5.3. *Saturated Hydrocarbons* ... 379
      5.3.1. *Normal Alkanes* ... 382
      5.3.2. *Branched and Cyclic Alkanes* ... 384
   5.4. *Unsaturated Hydrocarbons* ... 385
      5.4.1. *N-Alkenes* ... 385
      5.4.2. *Branched and Cyclic Alkenes* ... 386
   5.5. *Aromatic Hydrocarbons* ... 387
6. *Carboxylic Acids* ... 388
   6.1. *Fatty Acids* ... 388
   6.2. *Hydroxy, Dicarboxylic, and Aromatic Acids* ... 390
      6.2.1. *Hydroxy Acids* ... 390
      6.2.2. *Dicarboxylic Acids* ... 391
      6.2.3. *Aromatic Acids* ... 391

7. *Humic Substances and Kerogen* .............................. 391
7.1. *Humic Substances* .............................. 392
7.2. *Kerogen* .............................. 398
8. *Trace Compounds* .............................. 399
8.1. *Terpenoids* .............................. 399
8.2. *Steroids* .............................. 404
8.3. *Aldehydes and Ketones* .............................. 406
8.4. *Wax Esters* .............................. 408
8.5. *Alcohols* .............................. 408
8.6. *Porphyrins* .............................. 409
8.7. *Purines, Pyrimidines, and Nucleic Acids* .............................. 411
8.8. *Organosulfur Compounds* .............................. 412
*Further Reading* .............................. 414

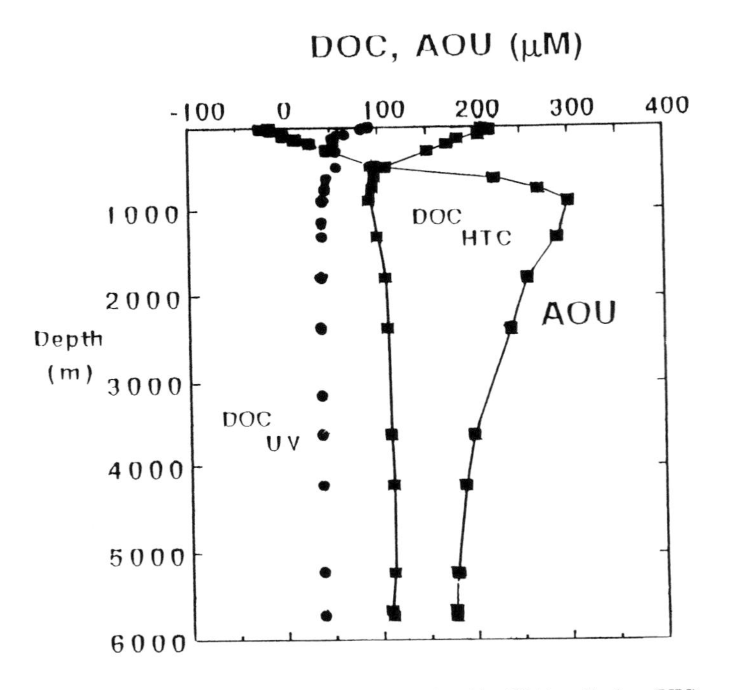


Profile of dissolved organic carbon determined by UV irradiation (UV) and high temperature combustion (HTC). A profile of apparent oxygen utilization (AOU) is given for comparison.

# 8 Organic Compounds

## 1. DISSOLVED AND PARTICULATE ORGANIC MATTER

In one sense, the ocean is akin to an enormous reaction vessel containing a very dilute solution/suspension of an enormous number of different organic substances present at very low levels in a slightly basic buffer solution. Inorganic matter is approximately 30,000 times more concentrated than organic matter in seawater. The ultimate source of organic matter is of course photosynthetically derived, i.e., from primary production. In coastal environments a significant proportion of the total organic matter may be terrestrially derived, while in open ocean environments most of the organic matter is at least initially produced by the fixation of inorganic carbon dioxide by marine phytoplankton in the euphotic zone.

The organic matter formed during photosynthetic processes is partially metabolized in order to satisfy the energy requirements of the primary producers. Other fractions of that organic matter will be assimilated by the higher trophic levels and eventually released into the water column by the autolysis of dead cells or actively excreted by such diverse organisms as benthic algae, copepods, sea urchins, as well as planktonic species.

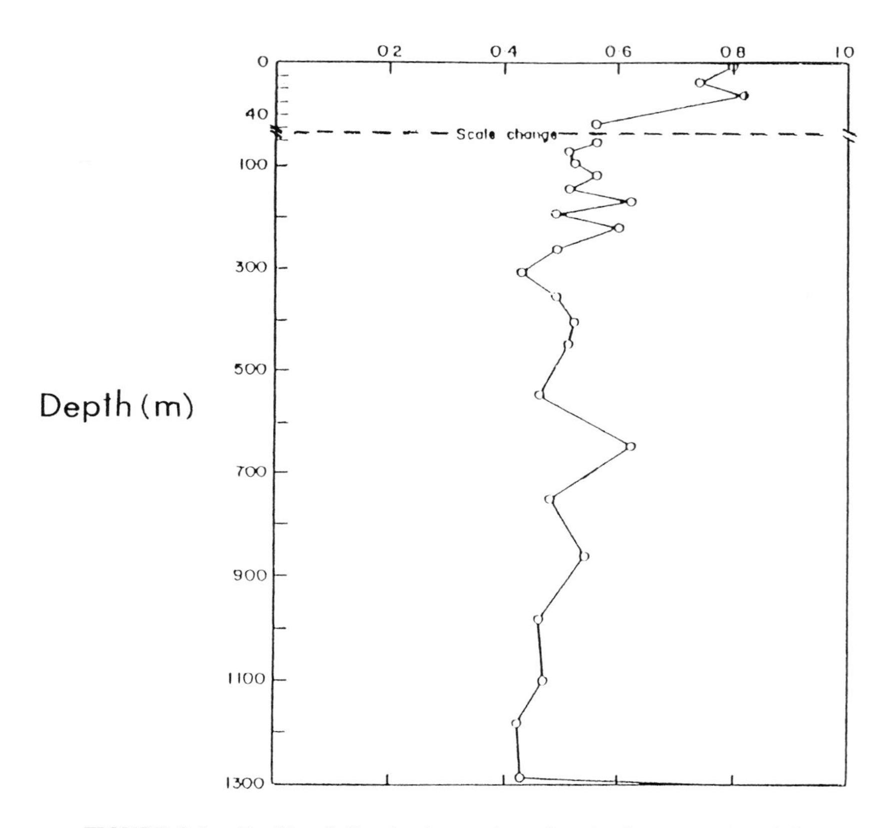


FIGURE 8.1. Profile of dissolved organic carbon in the oceans (mg $l^{-}$).

In aquatic systems it is traditional to divide organic matter into the two major categories of dissolved and particulate (DOM and POM). POM includes those species which will be retained by a glass fiber filter of 0.45 μm. When seawater is passed through a 0.45 μm filter, some colloids are retained as well as suspended matter. Various workers have used filters with pore sizes ranging from 0.40 to 1.0 μm to achieve this separation.

Dissolved organic carbon (DOC) is about $^1/_2$ the value of DOM and, likewise, particulate organic carbon (POC) is about $^1/_2$ the value of POM. The concentration of dissolved organic carbon (DOC) in seawater is typically significantly greater than that of POC. While the typical range of DOC concentrations in shallow (near surface, i.e., less than 300 m) seawater is 0.3 to 2.0 mg/l, the average concentration of POC in surface seawater is only 0.1 mg/l. Typical profiles of dissolved and particulate carbon in the oceans are shown in Figures 8.1 and 8.2. Seasonal variations in DOC and POC occur primarily in these shallow waters and are similar to those for primary pro-

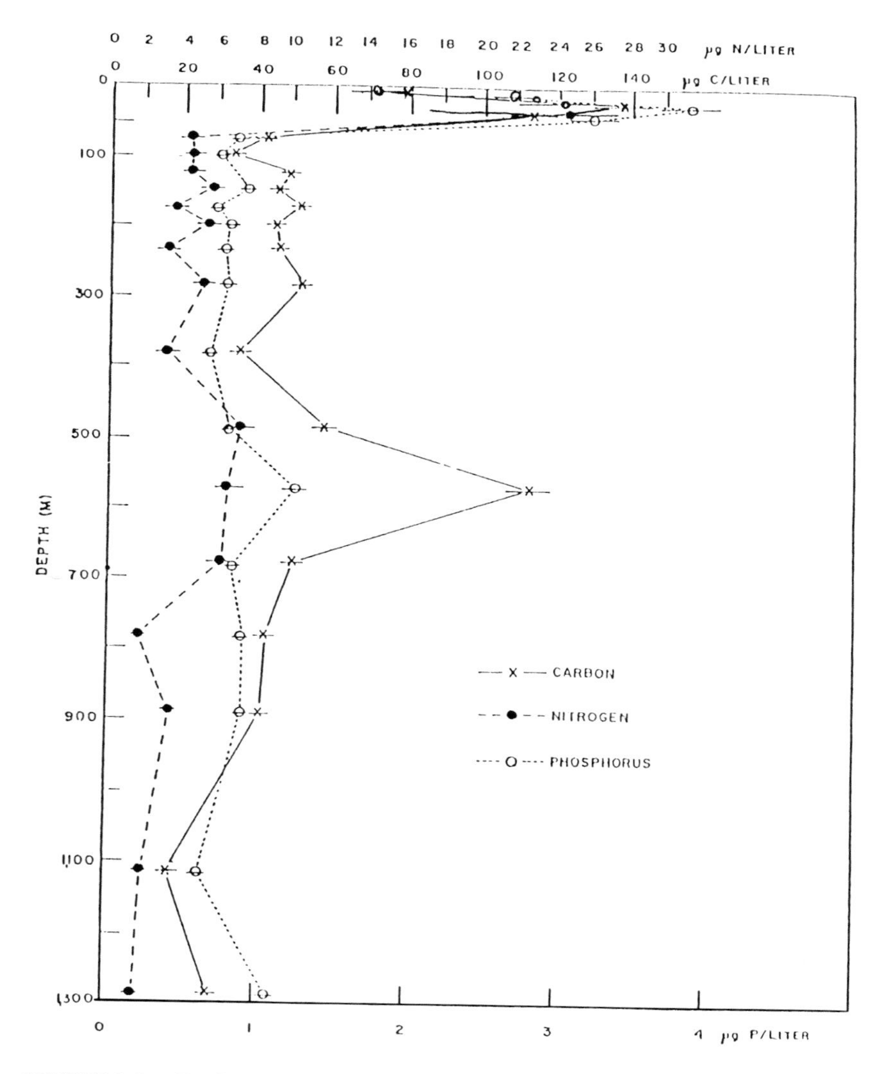


FIGURE 8.2. Profile of particulate organic carbon, nitrogen, and phosphorus in the oceans.

ductivity. At depths below a few hundred meters, the average DOC is 0.5 mg/l, while the average POC is 0.01 mg/l. Although the contents of inanimate organic matter in the oceans is less than 0.01% of the total amount of salts, these substances are important modifiers of many of the biological and chemical reactions that take place in the sea. They provide a nutritional base for micro- and macroorganisms, have a major impact on the speciation of many inorganic constituents by such processes as complexation and adsorption, and

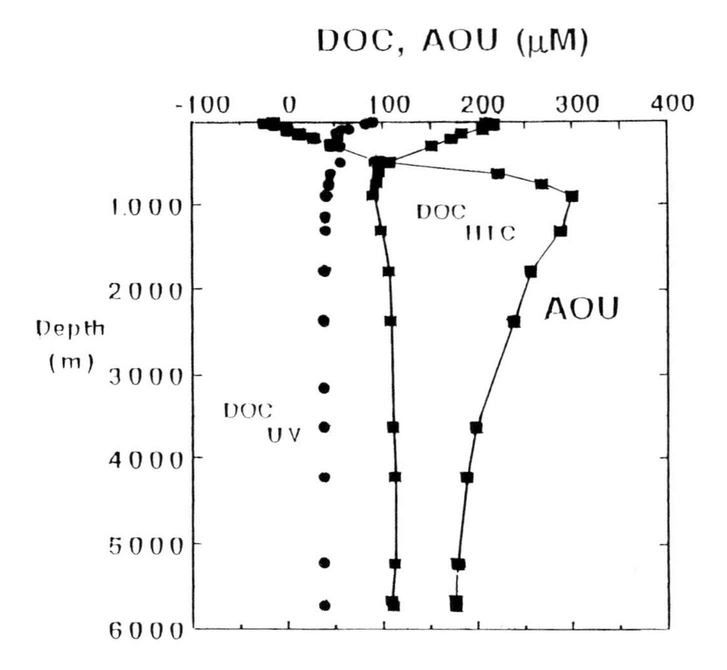


FIGURE 8.3. Profile of dissolved organic carbon determined by UV irradiation (UV) and high temperature combustion (HTC). A profile of apparent oxygen utilization (AOU) is given for comparison.

are precursors of fossil fuels such as petroleum and oil shale. Thus, in recent years much research activity has been directed towards identifying the specific components of organic matter in the seas and untangling the complex web of biogeochemical relationships between various compound classes associated with microbial processes. These classes include the amino acids, carbohydrates, fatty acids, etc. that are found in varying concentrations in both dissolved and particulate phases. They are discussed individually in subsequent sections of this chapter. Recent measurements of dissolved organic carbon by Sugimura and Suzuki have revealed that the previously measured values may be lower by 50 to 400% (Figure 8.3). The new measurements have been made using a high temperature oxidation. The new values are not only larger than the earlier results, but show an inverse correlation with the apparent oxygen utilization (AOU). This is shown in Figures 8.4a and 8.4b. These new measurements accepted by some and considered as heresay by others may change our understanding of the organic carbon system in the oceans.

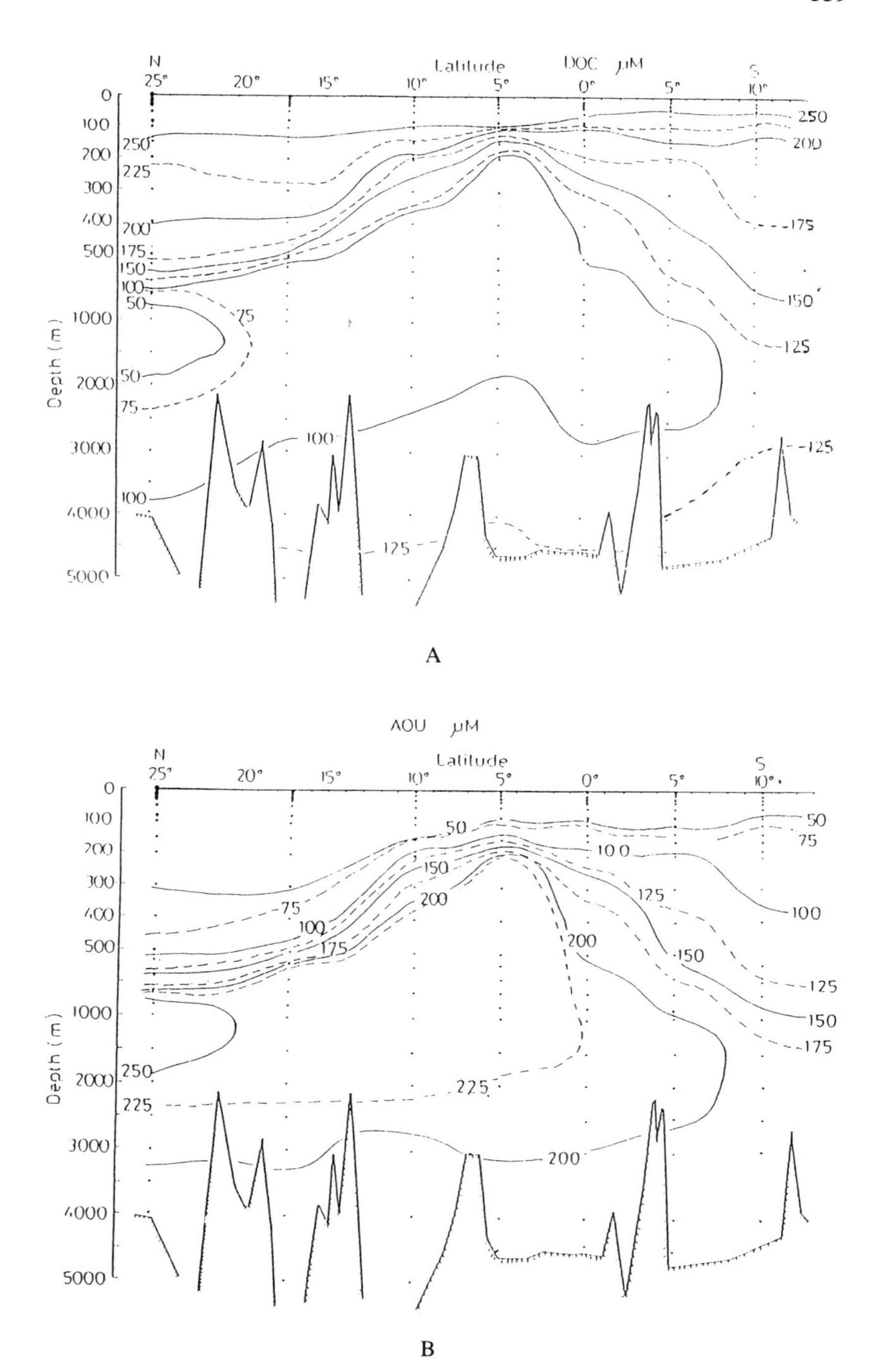


FIGURE 8.4. Section of dissolved organic carbon and apparent oxygen utilization in the Pacific Ocean.

### 1.1. Dissolved Organic Matter

Dissolved marine organic matter is such an extremely complex and dilute mixture of compounds that only 10 to 20% can be fully characterized. Much of the data on individual organic compounds in seawater has been obtained by gas chromatography/mass spectrometry (GC/MS) or by capillary GC. In all current chromatographic methods, much of the total DOM appears as an undifferentiated hump on the chromatogram that is often referred to as an "unresolved complex mixture (UCM)". Thin layer chromatography coupled to flame ionization detection (FID) is a promising new technique which can be used to separate the dissolved organic matter from seawater into compound classes such as lipids, base mobile (amines, alcohols, ketones), and acid mobile fractions. The large number of unidentified organic compounds as seen on a chromatogram is indicative of how much we still have to learn before we can hope to begin to form realistic models of the organic chemistry of the oceans.

The dissolved organic matter in seawater consists largely of humic substances and more labile compounds from the major biochemically important compound classes such as carbohydrates, steroids, alcohols, amino acids, hydrocarbons, and fatty acids (as well as fatty acid esters and waxes). These compound classes are treated individually in later sections of this chapter.

### 1.2. Particulate Organic Matter

Of the total organic carbon (TOC) in the oceans, it has been estimated that particulate organic carbon constitutes only 1 to 10%. Frequently, POM is evaluated from measurements of POC made by carbon analyses of organic-free glass fiber filters (0.45 μm pore size) used to separate POM and DOM. POM is obtained by multiplying POC by a factor of two, based on the approximation that carbon constitutes roughly 50% by weight of marine organic matter.

Particulate organic matter consists of a mixture of living and dead phytoplankton and zooplankton, bacteria, their degradation and exudation products, and macroscopic aggregates often termed "marine snow". The distribution and nature of particulate organic matter has been found to be quite variable both geographically, vertically, diurnally, and seasonally and is influenced by a complex set of equilibria between sources, sinks, and circulation patterns. In the euphotic zone, the major portion of the POM is due to phytoplankton, the chemical composition of which will vary with species as well as with environmental conditions, hence, leading to significant variations in the nature of the POM. For example, the marine algae *Phaeodactylum* exhibits an increased rate of protein synthesis under conditions of decreasing temperature at the expense of carbohydrate synthesis. It is well known that the major and minor metabolic products of phytoplankton will vary with changes in temperature, light intensity, and nutrient availability. Much less is known about the chemical content and biogeochemistry of particulate organic matter than is known about dissolved organic matter. However, assays indicate that in the euphotic zone the majority (~90%) of POC is due to living matter, while at depths of 2400 m and more, less than 1% of the POC is due to living organisms.

In recent years, much research activity has focused on the vertical flux of particulate organic matter from the surface waters of the ocean to the seafloor. The chemical and biochemical changes that occur as particles sink through the water column significantly affect the composition of both the sediments and the overlying water. Thus, the physical and biogeochemical processes that affect the composition of sinking particulate matter also affect biological productivity in the water column as well as modify the organic matter which reaches the ocean floor, which serves in turn as a major source of nutrition for benthic organisms and as a source of precursors to marine derived fossil fuels.

Sediment trap experiments conducted at various depths have recently yielded valuable information on the chemical content and fluxes of marine particulate organic matter. Sediment traps, as employed by Gagosian et al., consist of PVC cones with a 1.5 $m^2$ opening fitted with a Lucite sample cup and numerous baffles and employ sodium azide as a bacterial decomposition retardant. Several studies, including those by Honjo, indicate that the downward flux of POC seems to be dominated by rapidly sinking particles, i.e., particles with diameters of $>62$ $\mu$m. The organic components of these particles have been found to consist largely of zooplankton fecal pellets, zooplankton and phytoplankton remains, and humic substances. The chemical content of these particles has been shown to be affected by a very complex set of reactions including degradation and transformation processes which result in the increased removal of some labile organic compounds with depth, while, other labile organic compounds (often unsaturated) as reported by Wakeham and others (Skidway) were found at depths as great as 5068 m. The feeding activity of meso- and bathypelagic zooplankton on these rapidly sinking particles was invoked by Wakeham et al. to explain the selective removal of some labile organic compounds from sinking POM and the selective incorporation of other labile organic matter into sinking POM as fecal material excretion products.

Much of the work on the biogeochemistry related to sinking marine particles has focused on lipids (fats, oils, and fat-soluble compounds). Lipids constitute a minor but important fraction of the DOC and POC due to their involvement in energy storage and use, reproduction, regulation of metabolic processes, membrane structure, etc. A recent study by Kennicutt and Jeffrey indicated that the chemical composition of marine particulate lipids was much more complex than that of marine dissolved lipids. Although the major components of dissolved and particulate lipids were similar (*n*-alkanes, pristane, phytane, and fatty acid esters), particulate lipids included a substantial contribution from unresolved compounds as well as olefins, alkylated benzenes, quinones, and minor contributors not found in the dissolved fractions.

## 2. ORGANIC MATTER SOURCES

The levels of DOC and TOC in seawater are governed primarily by the rates of input processes and by the rates of decomposition. The major source or input of organic matter to the oceans is of course from the photosynthetic fixation of inorganic carbon dioxide by phytoplankton. Primary production by marine phytoplankton has been estimated to be responsible for the production of approximately $2 \times 10^{16}$ g C per year. Most of this organic matter is consumed by organisms, but a significant amount is decomposed at the surface, while 1 to 12% of the annual production ($\sim 6 \times 10^{16}$ g C per year remains as a refractory fraction quite resistant to biological and chemical transformations. The residence time of this residual dissolved organic carbon in the sea has been estimated by Menzel Skidway as 3300 years, which accentuates its inertness to chemical and biological degradation. More recent studies by Williams (Scripps) and Druffel (WHOI) give radiocarbon dates of 1310 years for surface and 6240 years for deep dissolved organic matter. Most of this residual organic carbon consists of humic substances which are discussed in detail in Section 7 of this chapter.

The other sources of input of organic matter to seawater, in addition to primary productivity by marine plankton, include terrestrial input by rivers and by the atmosphere, excretion by marine organisms, and the resuspension of organic matter from marine sediments. The direct addition of organic matter due to oil spills is an additional source. These sources are considered in greater detail in subsequent paragraphs.

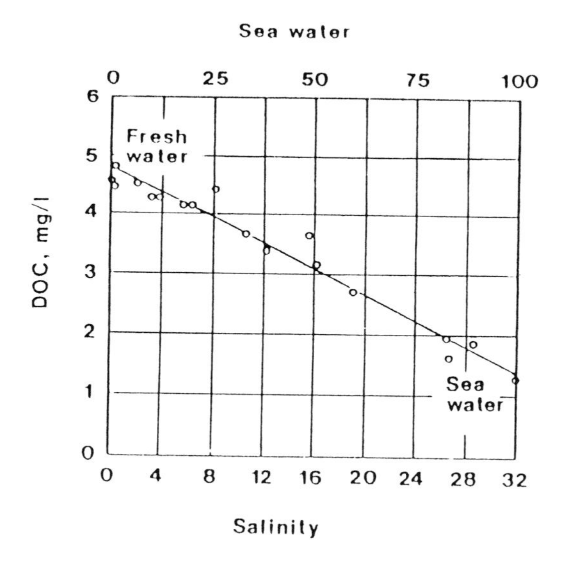


FIGURE 8.5. Dissolved organic carbon as a function of salinity in a typical estuary.

## 2.1. Terrestrial Input by Rivers

The typical concentration of dissolved organic carbon in an estuary in Figure 8.5 show that rivers can be a source of dissolved organic carbon to the oceans. Based on levels of POC and DOC of major river systems, it has been estimated by Meybeck (University of Marseilles) and others that the amount of total organic carbon (TOC) transported to the seas by river systems amounts to about $33 \times 10^{12}$ mol per year. Approximately half of this carbon is due to POC and the other half to DOC.

Terrestrially derived carbon is often referred to as allochthonous carbon, which means that it is derived from some system outside of the system being considered, which in this case is the oceans. In fresh water systems, the two major sources of organic carbon are plant and soil derived carbon. Rainwater that penetrates soils can leach organic soil and plant debris, as well as inorganic constituents, into dissolved and suspended phases. Numerous studies on the leaching of plant organic matter into streams have shown that within 24 h as much as 40% of the organic matter of fresh plant litter can be solubilized by distilled water. At least half of the DOC of the leachates from plant debris consists of carbohydrates. The remaining DOC consists of a mixture of substances which are similar to and either contributors and/or precursors of fulvic acids. The major portion of organic matter leached from soils consists of humic substances which may account for as much as 70% of the organic matter of soil. Other terrestrial plant and animal decomposition products are also transported from the soil into streams and rivers and eventually into the marine environment. Most studies on the organic matter of the deep ocean basins indicate that virtually all of it is derived from autochthonous (within the system) sources. Most of this data is based on carbon isotope ratios which are significantly different for terrestrial vs. marine vegetation. The $^{13}C$ value of dissolved marine organic matter was found to range from −21 to 24.0 ‰ by Williams and Gordon, while that of the Amazon River was evaluated at −28 to −5‰. Various studies have indicated the flocculation and subsequent precipitation of organic matter in fresh waters upon mixing with higher salinity water in estuaries. This apparent "salting out" of organic matter results in the restriction of allochthonous river-transported organic matter largely to coastal regions.

In addition to naturally derived terrigenous sources of organic matter, streams and rivers also deliver an increasing amount of organic substances derived from sewage and industrial effluents. Although these forms of pollution have not yet caused any open-ocean disasters, their introduction into fresh water and estuarine environments have caused significant changes in population diversities of microorganisms, invertebrates, and vertebrates and have resulted in negative changes in water quality.

### 2.2. Terrestrial Input from the Atmosphere

The introduction of organic matter to seawater from the atmosphere takes place at the air-seawater interface which has been the subject of much recent research. It has been well known for many years that the chemistry of the sea surface microlayer is very different from that of bulk seawater. For example, the DOC of this interfacial layer, which is less than a micrometer in thickness, can be as much as ten times the DOC concentration in surface seawater below the microlayer. The chemical composition of the organic matter in the seawater microlayer is not yet well known, but appreciable quantities of macromolecular species similar to humic substances (see Section 7 of this chapter) constitute a major fraction of the DOC and POC of this interface. Mono and polysaccharides as well as fatty acids have also been isolated from surface microlayers by various workers.

The transfer of organic material across the air-seawater interface can be accomplished by either wet (liquid) or dry (gas or solid) deposition, although these categories are sometimes difficult to effectively separate from each other.

For example, when liquid deposition takes place (rainwater), the dissolved gases and particulate material in rainwater are simultaneously being deposited. Williams (Scripps) has estimated that $2.2 \times 10^{14}$ g C per year is a reasonable estimate for the amount of organic carbon contributed to the oceans by rainfall. This figure is similar in value to the amount of organic carbon input by rivers as discussed in the preceding section ($3.3 \times 10^{13}$ mol C/year $\times$ 12 g C/1 mol C $= 4.0 \times 10^{14}$ g C/year). However, as noted by Hunter and Liss (East Anglia), it is important to consider that this flux of organic carbon into the oceans may not represent a net input due to the input of marine organic matter into the atmosphere, primarily by the bursting of bubbles in whitecaps and breaking waves. Blanchard has determined that bubble bursting ejects a significant amount of organic matter from the seawater microlayer into the atmosphere and that rising bubbles also harvest organic matter from subsurface waters. Thus it is possible that a significant portion of the organic matter transferred to the sea from the atmosphere by rain is organic matter recycled from seawater. Work by Blanchard (WHOI), Garrett (NRL), and others on the chloroform soluble fraction of organic matter isolated from surface seawater and overlying air samples indicate similarities in the lipid compositions, lending support to a marine source for marine aerosols.

Other recent studies on marine aerosols, such as those by Simoneit (U. Oregon), found that the organic matter in the aerosols was dominated by waxes of terrestrial origin. Gagosian and Peltzer (WHOI) report that four lipid compound classes (aliphatic hydrocarbons, fatty alcohols, fatty acid esters, and fatty acid salts) all indicated a terrestrial vascular plant source for organic matter isolated from atmospheric samples collected from the tropical North Pacific. Their results indicated that these aerosol lipids originate from wind erosion of soils and direct emission from vegetation, and that the major fluxes to seawater result from rain rather than from dry deposition.

On the other hand, definite net fluxes of organic matter from the atmosphere to seawater have been established for anthropogenic substances such as DDT and polychlorinated biphenyls (PCB). DDT is a chlorinated hydrocarbon pesticide which was used extensively following the Second World War. It is no longer used in the U.S. because it is extremely persistent and tends to accumulate in food chains. PCBs are also very persistent in the environment and until the manufacture of these compounds was discontinued in 1977, PCBs were used industrially as plasticizers, in paints, as coolants and transformers. Hunter and Liss estimate that 10 and 0.5% of the total annual production of PCBs and DDT, respectively, are transported into the world oceans through atmospheric-seawater fluxes. Both PCBs and DDT are extremely widespread in the marine environment and have been found in the tissues of many marine organisms. Further discussions on anthropogenic organic contaminants in the marine environment, including petroleum derived pollutants, are presented in a later section of this chapter.

### 2.3. Additional Sources of Organic Matter

In addition to the sources of organic matter described above, marine organisms are also often considered an ''internal source'' of organic matter by the excretion of metabolites and the release of substances upon the death and decay of cells. However, marine organisms, except for the case of primary producers, are not really sources of organic matter, but essentially transformers which modify the organic substances that enter the marine environment by the processes already discussed.

Another ''internal source'' of marine organic matter which must be considered is marine sediments. Although marine sediments act as a sink for organic matter, in local situations they may act as an important source due to resuspension.

## 3. CARBOHYDRATES

A carbohydrate can be defined as a polyhydroxy carbonyl compound, which produces such material upon hydrolysis, or a closely related derivative. The general formula for a carbohydrate is $C_n(H_2O)_m$, although there are exceptions to this formula.

Carbohydrates can be classified by the number of separate carbon chains per molecule, the simplest being a monosaccharide with a single continuous chain of carbon atoms. Disaccharides are composed of two monosaccharide units per molecule, while the more general term oligosaccharide indicates a carbohydrate composed of two to six monosaccharide units. Polysaccharides are carbohydrates which are composed of many mono or disaccharide units per molecule. The term sugar (or saccharide) refers to monosaccharides, disaccharides, or lower oligosaccharides.

Monosaccharides are themselves classified by the number of carbon atoms in the chain and also by the type of carbonyl group present. The prefixes tri, tetra, etc. indicate the number of carbon atoms per chain while the suffix ''-ose'' indicates that it is a carbohydrate. Thus the term hexose indicates a monosaccharide composed of a chain of six carbon atoms. The functionality

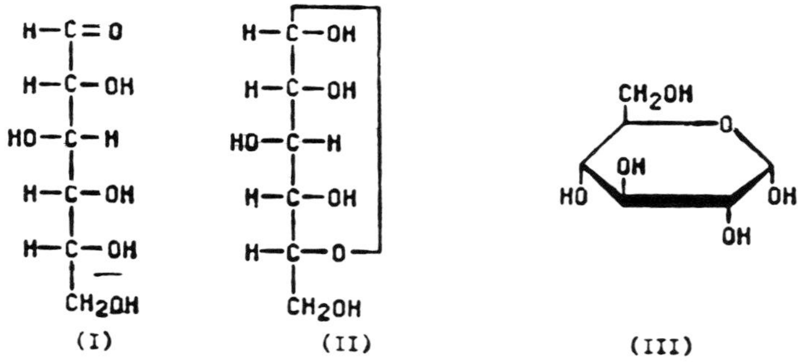

FIGURE 8.6. Structure of sugar (D-glucose).

of the carbonyl group present is also important in classifying a carbohydrate. If the carbonyl group is terminal (R-CO-H) and thus protonated (an aldehyde), the monosaccharide may be referred to as an aldose. If the carbonyl functionality is that of a ketone (R-CO-R, i.e., nonterminal) the monosaccharide is a ketose. Thus an aldohexose is a monosaccharide composed of a backbone of six continuous carbon atoms including an aldehyde group.

Virtually all monosaccharides are optically active, i.e., rotate the plane of polarized light. The direction in which light is rotated depends upon the relative configuration of the molecule on a 3-dimensional basis. The letters D and L are used to indicate molecules of the same structural formulas which rotate plane-polarized light in equal but opposite directions.

The most common simple sugar is D-glucose. This monosaccharide can be represented in several different ways (Figure 8.6). Structure (I) is the straight chain structure with an aldehyde group. However, most of the molecules of glucose at any given time actually exist in the form of Structure (II), which is known as the hemiacetal form, and are often drawn as in Structure (III).

Carbohydrates comprise one of the major classes of naturally occurring organic substances and are actually the most abundant class produced in the biosphere. This important class of compounds is the central source of energy for mechanical work and chemical reactions in all living cells, while carbohydrate derivatives such as ATP (adenosine triphosphate) and nucleic acids control energy transformations and the transfer of genetic material, respectively. On a dry weight basis, plants are composed of 50 to 80% carbohydrate. The main constituent of plant cell walls is the polysaccharide cellulose which composes 40 to 50% of wood by weight. Cellulose consists of a linear sequence of glucose units linked by bonds formed involving the acetal carbonyls (Figure 8.7). In plants, cellulose is the main structural component in asso-

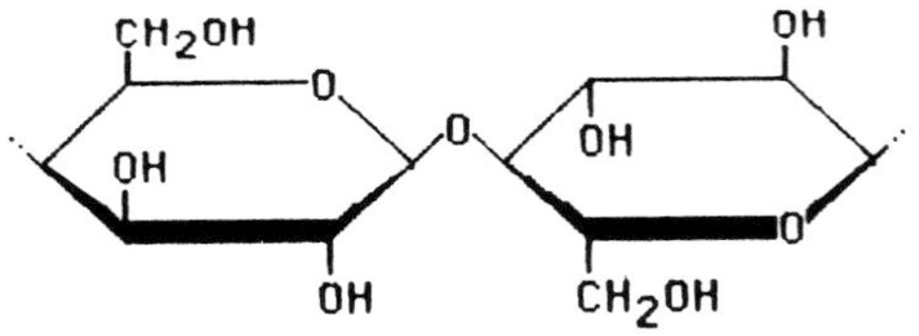

FIGURE 8.7. Structure of glucose units in cellulose.

ciation with hemicelluloses (various polysaccharides of the cell wall exclusive of cellulose) and lignins (methoxy-phenolpropanoid polymers). Starches serve as major energy reservoirs in plants and consist of a water soluble fraction (amylose) and a residual amylopectin. Amylose consists of a straight chain polymer of D-glucose with 60 to 300 glucose units bonded together. Plants also contain many noncarbohydrate organic compounds that are conjugated with sugars to form substances known as glycosides (a carbohydrate derivative in which the carbonyl group has formed an acetal or ketal linkage to a noncarbohydrate moiety).

In higher animals, proteins are the principal structural material but are frequently found in association with carbohydrates. The amorphous substance between animal cells is composed largely of the polysaccharide hyaluronic acid (uronic acid is an aldose in which the terminal $CH_2OH$ group has been oxidized to a carboxylic acid, i.e., -COOH group), and glycogen (a polymer with branched chains of glucose) is a primary energy reservoir for animals.

Carbohydrates are ubiquitous in the marine environment and occur as monosaccharides, oligosaccharides, and polysaccharides. They have been found to vary geographically, seasonally, diurnally, and with depth and are primarily derived from marine phytoplankton. Monosaccharides and oligosaccharides occur principally within the cell sap of phytoplankton, while polysaccharides serve as food storage reservoirs and as structural components of the cell wall. Studies on carbohydrate compositions within the biota, DOC, POC, and sediments have yielded data which in many cases are contradictory. Much of the confusion is due to inconsistent analytical methodologies and there is an enormous need to standardize the extraction, identification, and quantitation of marine carbohydrates.

Many studies have focused on the interrelationships between total dissolved carbohydrate (TCHO), polysaccharides (PCHO), monosaccharides (MCHO), and the DOC and POC of seawater. Studies on a particular class of compounds, such as carbohydrates, may provide novel insights into the chemistry and cycling of the total DOC and POC in marine environments.

However, the study of carbohydrates has been hindered by the use of inconsistent methods of extraction, derivatization, and quantification, making it almost impossible to compare the results of the various studies.

Most of the early studies in which TCHO was measured employed a colorimetric technique such as the *N*-ethylcarbazol method of Erdman and Little or a modification of this procedure. This method was later shown to suffer from interferences from noncarbohydrates. Alternate methods employing androne or phenol/sulfuric acid largely replaced the *N*-ethylcarbazole method in the 1960s, but produced dramatically different absorbance readings with different carbohydrates. Various workers have also employed different extractants (distilled water, ethanol) and hydrolysis techniques (HCl, $H_2SO_4$), thus making it very difficult to compare data from different laboratories. In recent years, significant advances have been made by Mopper and co-workers (Washington State) in identifying monosaccharides by gas or liquid chromatography after desalting by electrodialysis.

Glucose and fructose were found to be the dominant monosaccharides in a study involving over 150 seawater and sediment pore water samples from a wide variety of open ocean (deep and surface) and nearshore samples by Mopper et al. The dominating presence of glucose in seawater is not surprising since glucan (a glucose polymer) is a major storage polysaccharide of marine phytoplankton. The water extractable carbohydrates from different diatom samples yield mainly glucose upon acid hydrolysis, indicating the presence of glucan. The presence of fructose at comparable levels to those of glucose in the water samples described by Mopper et al. was tentatively attributed mainly to a spontaneous abiotic conversion of glucose to fructose which will occur in glucose spiked distilled water buffered at pH 8.1, although fructose may be involved in plankton food storage as well. Under these conditions 30% of the added glucose had been converted to fructose after 24 h at 80°C. Total MCHO levels varied from 0.015 *M* (deep Sargasso Sea) to as great as 2.0 *M* at the sediment-water interface of a near-shore Baltic Sea sample. Other identified monosaccharides included fucose, ribose, xylose, arabinose, mannose, rhamnose, and galactose.

Phytoplankton were found to be the major source of glucose in a *Spartina alterniflora* salt marsh estuary, although additional inputs from *Spartina,* benthic algae, and possible adjacent marsh soils are evident. The major glucose utilizers were found to be the microheterotrophs, namely bacteria, yeasts, and possibly some algae. Glucose concentrations were found to be highly variable with respect to tidal cycle, season, and geographic location. Microheterotrophs were found to utilize glucose rapidly even at low concentrations and, thus, it was deduced that glucose exchange between water masses is probably negligible and that rapid recycling occurs within each water mass.

Although glucose is a dominant monosaccharide in seawater, it is typically present at higher levels in vascular plants. Plankton, on the other hand, are relatively enriched in ribose as compared to vascular plants. Thus, several studies, including those by Mopper and Degens, have employed glucose/ribose ratios as a source indicator of terrestrial vs. marine contributions to organic matter. Hedges and co-workers (University of Washington) found that plankton, bacteria, and vascular plants yielded consistent compositional carbohydrate patterns which could be used to distinguish between marine and terrestrial input. Their marine study of Dabob Bay (a deep arm of Puget Sound, Washington) indicated a predominantly marine origin for the carbohydrates in sediments and POC with increased relative proportions of terrestrially derived carbohydrates during periods of low phytoplankton productivity.

Most studies on marine carbohydrate chemistry have dealt with total dissolved carbohydrates or polysaccharides. Several studies have used the 3-methyl-2-benzothiazolinone hydrazone hydrochloride (MBTH) method to determine the concentration of total dissolved monosaccharide (MCHO) and then coupled this technique with a hydrolysis step to determine total dissolved carbohydrate (TCHO). The value of total dissolved polysaccharide (PCHO) can then be estimated from the difference between TCHO and MCHO.

In an early study on the vertical distribution of dissolved carbohydrates in the Sargasso Sea off Bermuda, the TCHO increased from a concentration of 0.27 mg $l^{-1}$ at the surface to 0.75 mg $l^{-1}$ at a depth of 75 m. TCHO then decreased rapidly until a depth of 100 m, suggesting that much of the TCHO was due to the overproduction by phytoplankton, since the greatest concentration of chlorophyll-a was previously found at a depth of 100 m while the rate of uptake at that depth was only 35% of the uptake at 75 m.

In a later study in which total dissolved sugars (TDS) were determined during a plankton bloom in the North Sea, TDS varied from 40 to 400 $\mu g$ $l^{-1}$. Glucose and mannose exhibited maxima during the bloom, while galactose, rhamnose, xylose, arabinose, and fucose maximized after the bloom, indicating that variations in TDS were directly related to phytoplankton activity. High glucose concentrations observed towards the end of the bloom were attributed to the release of glucan, while variations in the levels of other sugars were probably due to bacterial destruction of detrital material and bacterial release into the water column.

Significant correlations were found between TCHO and PCHO concentrations and microbial plankton population which suggested that the $<20\ \mu m$ microbial plankton population actively regulated PCHO and TCHO in the western Sargasso Sea. Furthermore, those results suggested that carbohydrate levels were regulated by the combined activities of phototrophic nanoplankton, heterotrophic nanoplankton, and planktonic bacteria. Further studies on carbohydrate concentrations during diel cycles in a salt marsh highlighted the role played by heterotrophic plankton (zooplankton) as an important source of dissolved carbohydrate. It was also determined that planktonic bacteria were able to respond rapidly to carbohydrate fluctuations and appeared to be able to control natural concentrations on a time scale of only a few hours. Thus, dissolved carbohydrates which are largely released by phytoplankton, with smaller contributions from zooplankton and bacterial excretion, as well as other minor sources, are important bacterial substrates and their levels are controlled largely by biological processes.

Carbohydrates are also an important constituent of marine particulate matter. An early study revealed the presence of glucose, galactose, mannose, arabinose, and xylose in particulate matter isolated from the northeastern Pacific Ocean. In a later study, rhamnose, fucose, ribose, arabinose, xylose, mannose, galactose, and glucose were identified as constituents of particulate matter from the northwest Pacific Ocean after acid hydrolysis. Analyses of phytoplankton carbohydrate compositions inidicated that these organisms and the detritus largely determine the carbohydrate composition of marine particulate matter. Water extractable carbohydrate, which yielded mainly glucose upon hydrolysis, was attributed to food storage materials such as glucan of diatoms and was found to decrease with depth in these waters. This decrease in glucose content with depth is believed to be due to its utilization by the respiration of diatoms as they sink out of the euphotic zone. This utilization of food storage carbohydrates results in the relative enrichment of POM in more resistant water-insoluble polysaccharides that are largely structural carbohydrates.

## 4. AMINO ACIDS AND PROTEINS

Proteins are present in all major regions of living cells and serve a wide variety of functions. In addition to serving as a major contributor to cell

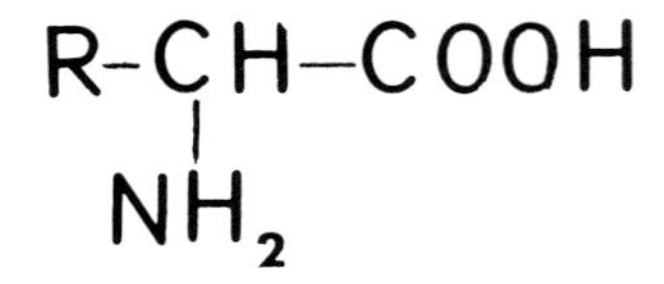


FIGURE 8.8. Structure of amino acid.

structure, they also function as metabolic regulators and are instrumental in motion and defense against foreign substances, i.e., as antibodies. Proteins are high molecular weight materials and like the polysaccharides discussed in Section 3 of this chapter, proteins are composed of recurring units of molecular structure. The fundamental unit in the case of proteins is the amino acid. Proteins which yield only amino acids upon hydrolysis are termed simple proteins. Those which yield other molecular types besides amino acids are termed conjugated proteins. Whereas proteins may range in molecular weight from about 6000 (insulin) to 20 million (keratin), peptides consist of several amino acids linked together to form lower molecular weight units.

Amino acids are simply organic acids with amino groups (Figure 8.8). Virtually all amino acids are alpha ($\alpha$) aminocarboxylic acids in which the amino group ($-NH_2$) is attached to the carbon which is next to (or alpha to) the carboxylic acid group (-COOH). Three different types of $\alpha$-amino acids are found to occur naturally: (1) monoamino-monocarboxylic acids or neutral amino acids, (2) diamino-monocarboxylic acids or basic amino acids, and (3) monoaminodicarboxylic acids or acidic amino acids. Table 8.1 lists some of the common amino acids discussed later in this chapter. Although a wide variety of proteins exist in living systems, they are composed of a rather small number of amino acids.

All $\alpha$-amino acids, except glycine, contain at least one asymmetric carbon atom. Thus, except for glycine, amino acids isolated from the acid hydrolysis of proteins are optically active, indicating their biological origin. Amino acids which are synthesized in the laboratory from inorganic starting materials are racemic mixtures (contain equal amounts of D and L isomers) and, therefore, show no optical activity. Amino acids recovered from natural proteins by acid hydrolysis are virtually always L-amino acids, although D-amino acids have been identified as metabolic products of some lower organisms.

Amino acids are linked together to form proteins by peptide (amide) bonds (Figure 8.9). Although different proteins differ in their amino acid components, the nitrogen content is nearly constant at 16% by weight and, thus, nitrogen content is often used as the basis to determine the protein content

**TABLE 8.1**
**Common Amino Acids**

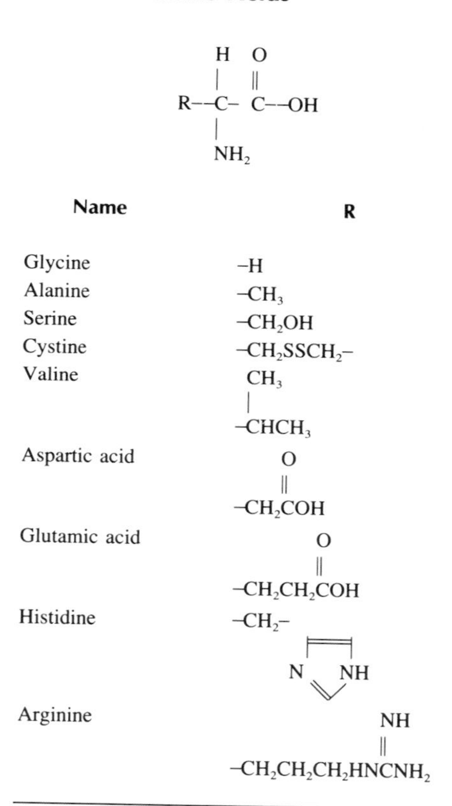

$$R{-}\overset{\displaystyle H}{\underset{\displaystyle NH_2}{\overset{|}{\underset{|}{C}}}}{-}\overset{\displaystyle O}{\overset{\|}{C}}{-}OH$$

| Name | R |
|---|---|
| Glycine | $-H$ |
| Alanine | $-CH_3$ |
| Serine | $-CH_2OH$ |
| Cystine | $-CH_2SSCH_2-$ |
| Valine | $-\underset{}{CH}(CH_3)CH_3$ (CH$_3$ above, –CHCH$_3$) |
| Aspartic acid | $-CH_2\overset{O}{\overset{\|}{C}}OH$ |
| Glutamic acid | $-CH_2CH_2\overset{O}{\overset{\|}{C}}OH$ |
| Histidine | $-CH_2-$ (imidazole ring: N, NH) |
| Arginine | $-CH_2CH_2CH_2HN\overset{NH}{\overset{\|}{C}}NH_2$ |

of proteinaceous materials. Most of the nitrogen compounds in living systems consist of protein and, in fact, protein can often account for over 50% of the bulk weight of organisms. Free amino acids, on the other hand, account for only a small fraction of nitrogenous compounds present in living organisms.

In view of the fact that amino acids and proteins make up such a substantial portion of living matter, it is not surprising that they are found ubiquitously in the marine environment. Since the initial report by Erdman on the presence of amino acids in recent marine sediments, many investigators have studied the distribution of amino acids in marine organisms, sediments, particulate matter, and DOM. In the marine environment, as elsewhere in the biosphere, combined amino acids constitute the majority of organic nitrogen compounds. Combined amino acids are found in dissolved, colloidal, and precipitated humic substances; in peptides, proteins, and enzymes; and in living cells as

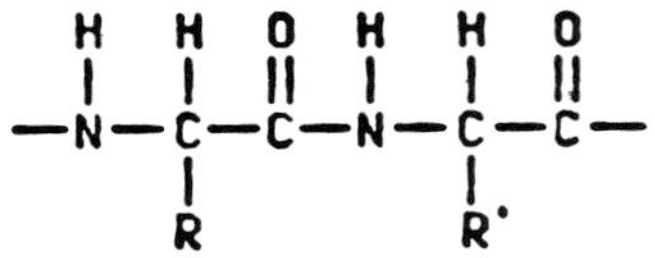

```
   H   H   O   H   H   O
   |   |   ||  |   |   ||
 —N—C—C—N—C—C—
       |           |
       R           R'
```

FIGURE 8.9. Peptide bonds in proteins.

well as extracellular material. Free amino acids, although present as such in extracellular and intracellular materials, are also produced by the hydrolysis of peptides and proteins *in situ* in seawater by bacterial and other microbial enzymatic activity.

The median concentration of total dissolved amino acids in seawater (free and combined) is approximately 50 $\mu g\ l^{-1}$ (range of 20 to 250 $\mu g\ l^{-1}$) which accounts for 2 to 3% of the DOC of seawater. The concentration of free amino acids in seawater is normally 4 to 10 times less than that of the combined forms. In marine sediment samples, typically 90% or more of the amino acids present are released by 6 *N* HCl hydrolysis which would hydrolyze peptides and proteins as well as extract amino acids combined in the humic fraction.

Due to the low concentration of dissolved free amino acids (DFAA) in seawater and other marine phases, sample contamination is a constant problem and knowledge of the amino acid content of extractants and other reagents is crucial. Several studies of the concentration of DFAA with depth have revealed very little variation. Thus changes in the concentration of total hydrolyzable amino acids (THAA) with depth are mainly due to changes in the concentration of combined amino acids. The concentration of THAA tends to decrease with depth until about 200 m and often mirrors the trend of DOC with depth. Thus, as is true for DOC, it is believed that most of the THAA in the marine environment is autochthonously derived from marine phytoplankton with smaller additional inputs from heterotrophic activity and some allochthonous sources. The limited variation in concentration of DFAA is most likely due to microbial utilization. In other words, DFAA are maintained at a low constant level because they are utilized by microbes very efficiently.

The most commonly reported components of the DFAA fraction of marine sediments and seawater include glutamic acid, aspartic acid, glycine, serine, histidine, alanine, and ornithine, with glycine, alanine, and serine typically most abundant.

Studies on the distribution of particulate amino acids (PAA) have revealed that the total PAA tends to be highest in nearshore euphotic zones with a maximum at the surface, except at shallow water locations where suspended bottom sediment is influential. This study also revealed that for open-ocean areas, PAA concentrations are highest in surface waters at the equator where greater phytoplankton growth is observed and, likewise, lower in colder, nutrient-depleted water. As was observed earlier with respect to dissolved THAA, PAA tends to decrease with depth down to about 200 m, while the overall amino acid composition is roughly similar to that of the dissolved fraction, although relative amounts of glycine and serine reportedly increased at greater depths.

In a recent study, hydrolyzable amino acids have been found to be enriched in the sea surface microlayer for both dissolved and particulate phases as compared to subsurface water samples from the same locations. Consistent with other studies of sedimentary amino acids, it was found that DFAA formed a small component of the THAA and, thus, the THAA is largely composed of peptide, protein, and humic amino acid.

In marine sediments, the amount of amino acids decreases quite rapidly with depth, and again, very little of the THAA is due to FAA. As reported by Whelan with respect to a sediment core of the Atlantic abyssal plain, most of the amino acids are released with 6 *N* HCl hydrolysis. Whelan also reports very little resemblance between water and sediment amino acid compounds, with high levels of nonprotein amino acids in sediments confirmed by GC-MS analysis. This suggests rather extensive reworking of organic nitrogen throughout the water column and at the water-sediment interface.

The influence of environmental conditions of deposition on the distributions of amino acids in sediments was revealed in a study by Gonzalez et al. The authors found glutamic acid, hydroxyproline, glycine, and threonine as the major amino acids in reducing areas, while ornithine, lysine, phenylalanine, serine, and valine were dominant in sediment samples from more oxidizing offshore locations. The predominance of basic amino acids over acidic amino acids in offshore samples, in spite of the predominance of acidic over basic amino acids in plankton, also indicates an intense reworking of these substances in the water column and surface sediments. The analysis of core samples revealed that basic amino acids seem to be the most stable and within this group, ornithine (a nonprotein amino acid) is produced at the expense of arginine. The least stable groups were found to be the sulfur containing and neutral branched-chain amino acids.

The distribution of DFAA in interstitial waters from coastal Atlantic cores was investigated by Henrichs and Farrington. Concentrations were found to be two orders of magnitude greater than those found in nearshore open-ocean seawater and distributions were significantly different from those of seawater.

$$CH_3{-}CH_2CH_2{-}CH_3$$

FIGURE 8.10. Structure of *n*-butane.

In particular, high levels of glutamic acid and β-aminoglutaric acid (a glutamic acid isomer) were apparent. Distributions suggested the β-aminoglutaric acid had formed from glutamic acid. An amino acid maximum within the upper 6 cm of sediment interstitial water and a gradual decrease with depth was observed. Burdige and Martens also report the presence of β-aminoglutaric and glutamic acids, along with alanine and glycine, as dominent DFAA in pore waters squeezed from anoxic sediments of Cape Lookout, Bight, North Carolina. While the highest levels of DFAA were found at the sediment-water interface, a secondary maximum was found at depth, suggesting that DFAA is an important intermediate in organic matter remineralization and that reactions involving amino acid production and consumption are likely coupled to bacterial sulfate reduction and methanogenesis.

## 5. HYDROCARBONS

Only two elements, namely hydrogen and carbon, are needed to generate the thousands of compounds known collectively as hydrocarbons. The simplest hydrocarbon is methane ($CH_4$), but, by bonding carbon atoms to each other in various patterns, it is possible to generate a limitless number of larger hydrocarbon molecules, many of which, such as natural rubber, contain over a thousand carbon atoms in a chain.

Many natural hydrocarbons are "straight chain" structures in which no carbon atom is attached to more than two other carbon atoms. An example of such a compound is *n*-butane ($C_4H_{10}$), in which "*n*-" indicates a straight-chain structure (Figure 8.10).

A structural isomer of *n*-butane which has a branched chain structure but the same formula ($C_4H_{10}$) is isobutane (Figure 8.11). These two formulas represent two distinctly different compounds with different melting points, boiling points, and densities. Many hydrocarbons have one or more asymmetric carbon atoms and, thus, as discussed earlier with respect to carbohydrates and amino acids, have the potential to be optically active if of the biological origin and not present as a racemic mixture.

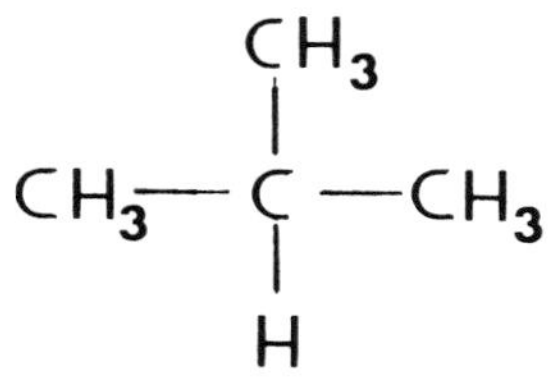

$$\begin{array}{c} CH_3 \\ | \\ CH_3-C-CH_3 \\ | \\ H \end{array}$$

FIGURE 8.11. Structure of isobutane.

There are several different ways in which the hydrocarbons can be classified. When dealing with an aqueous system, such as seawater, hydrocarbons are often initially separated into those of low molecular weight and high volatility ($<C_{14}$) vs. higher molecular weight and lower volatility ($>C_{14}$, i.e., hydrocarbons with more than 14 carbon atoms per molecule). The distinction between these two groups is based on the different biological and chemical roles of these two classes in the marine ecosystem as well as on the different methods used to study them. This secton on hydrocarbons will focus primarily on the higher molecular weight ($>C_{14}$), "low volatility" hydrocarbons.

Although the hydrocarbons as a group typically account for less than 1% of the dissolved organic matter of seawater (hydrocarbon concentrations in seawater typically vary from 1 to 50 $\mu g\ l^{-1}$, with an average of about 10 $\mu g\ l^{-1}$, they are considered an important constituent of the marine ecosystem. One reason for this is based on the enormous specificity of biologically produced and transformed molecules with respect to particular organisms and environmental conditions. Due to this biological metabolic specificity and the enormous stability of many hydrocarbon molecules, this class of molecules contains many excellent biological and geochemical markers. Hydrocarbon research has also intensified in recent years because of the need to establish background concentrations of hydrocarbon pollutants (petroleum spills) and halogenated hydrocarbons (PCBS, DDT, chloroform, etc.) and because of the need to study the concentration and effects of these compounds on living organisms. In addition, hydrocarbons have been extensively used in petroleum exploration as source rock indicators.

### 5.1. Volatile Hydrocarbons ($<C_{14}$)

Methane ($CH_4$) is the structurally simplest hydrocarbon and is an ubiquitous gas found at trace levels in all marine and freshwaters. The reported range for methane concentrations in seawater is 10 to 100 $\mu g\ C\ l^{-1}$. In open waters, the concentration of volatile compounds is normally limited by their volatility rather than solubility.

Wakeham et al. performed mesocosm experiments to study the persistence of various volatile organic compounds in coastal seawater and found that several classes of low molecular weight hydrocarbons and their derivatives are primarily removed from seawater by volatilization, although biodegradation and adsorption played important roles for some hydrocarbon classes. In marine sediments and interstitial waters, especially in nearshore and anoxic environments, the concentration of methane is usually significantly higher than in open ocean waters due to higher production rates. The decomposition of complex organic substances by marine microbial populations results in the production of biogenic methane. Methanogenesis involves the decomposition of organic substances under anaerobic conditions and the production of methane via carbon dioxide reduction or acetate dissimilation. Thus, within the water column, methane concentrations are typically highest at the sediment-water interface and the majority of the water column methane is probably produced from the bubbling of methane from the sediment. In a suite of biologically produced low molecular weight hydrocarbons, methane is overwhelmingly predominant and methane to ethane plus propane ratios — $CH_4/(C_2H_6 + C_3H_8)$ — have values over 100. Natural petrogenic sources generally yield much lower ratios. In coastal waters, anthropogenic inputs may also be significant and result mainly from petroleum operations such as production, transportation, and refining.

The biological production of methane is currently of interest, not only because of its accumulation in quantities which are commercially significant, but also because of possible climatic effects. Polar ice core studies indicate an approximate doubling of concentrations of atmospheric methane during the past few hundred years, and because methane strongly absorbs infrared radiation, atmospheric methane may be a significant potential contributor to global warming. A recent study on volatile organic substances found in sediments from the Peru upwelling region revealed that methane, ethane, propane, $C_4$-$C_7$ alkanes, and alkenes (hydrocarbons with at least one carbon-carbon double bond) were among the hydrocarbons detected in ng/g dry weight quantities. In cores from oxygenated bottom waters, $C_4$-$C_6$ alkenes along with branched and cycloalkanes showed maxima at the sediment-water interface and are believed to represent metabolic products of living organisms.

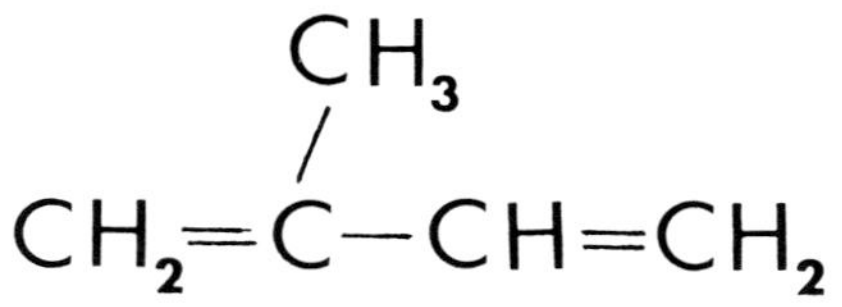

$$CH_2=C(CH_3)-CH=CH_2$$

FIGURE 8.12. Structure of isoprene.

### 5.2. Higher Molecular Weight ($>C_{14}$) Hydrocarbons

Hydrocarbons can also be classified as either saturated, unsaturated, or aromatic. Saturated hydrocarbons are also referred to as paraffins or alkanes and although they are trace components of biological lipids, they are geochemically quite stable and are a major constituent of the organic matter of terrestrial sedimentary rocks. Unsaturated hydrocarbons contain one or more multiple carbon-carbon bonds. The most common unsaturated hydrocarbons found in the marine environment are alkenes (also called olefins) and they contain at least one double carbon-carbon bond. A molecule containing two carbon-carbon double bonds is called a diene, three a triene, and so forth. Isoprene is a diene which can be isolated from many plant and animal sources and is often referred to as "one of nature's favorite building blocks" (Figure 8.12). Many complex natural products can be broken down into isoprene units by what is called "isoprene rule". Terpenes, for example, are isoprenoids that are found in the essential oils of many plants and have carbon skeletons made up of isoprene units jointed in a regular "head-to-tail" way. Terpenes are discussed in greater detail later on in this chapter. Aromatic hydrocarbons contain at least one benzene ring (6 carbon atoms arranged in a ring with alternating single and double bonds, ($C_6H_6$)) and, except for very minor inputs from organisms, they are considered to be derived mainly from abiotic processes such as the pyrolysis of organic matter at high temperatures (forest fires and fossil fuel combustion among others).

### 5.3. Saturated Hydrocarbons

The alkanes which have been isolated from organisms and sediments are mainly isoprenoids, normal paraffins, and steranes, many of which are structurally similar to biological acids and alcohols. Steranes are alkanes which are classified under the general category of steroids, the general structure of which is shown in Figure 8.13.

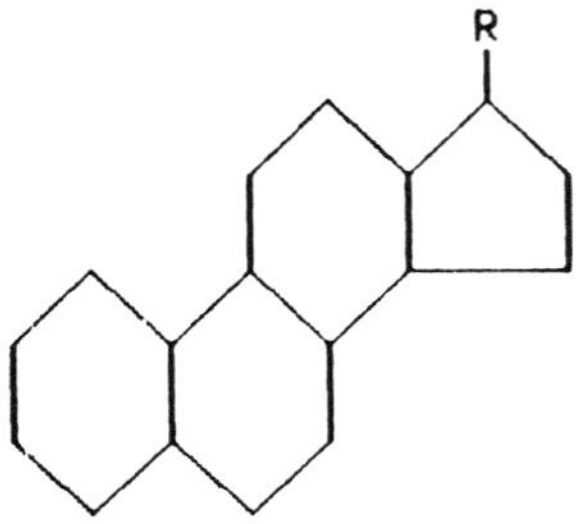

R

FIGURE 8.13. Structure of steroid.

Steranes and other steroids are discussed in greater detail in a later section of this chapter.

Hydrocarbons are major constituents of petroleum and, therefore, interest in petroleum generation, diagenesis, and utilization have led to increased interest in hydrocarbon geochemistry. Many hydrocarbons in petroleum, sediments, and in various fossil fuels besides petroleum resemble compounds produced by biological systems. Since the isolation and identification of metalloporphyrins in bitumens by Treibs in the 1930s, and their correlation with porphyrins of chlorophylls, implying a biological origin for petroleum, a great deal of research has been done in the field of biomarkers. Porphyrins are molecules based on structures of four pyrrole rings connected by methine (CH) groups. Although porphyrins are not hydrocarbons, since they contain nitrogen and commonly oxygen in addition to carbon and hydrogen, many useful biomarkers which will be discussed later are hydrocarbons. One classic example is the hydrocarbon pristane (Figure 8.14).

Pristane (2,6,10,14-tetramethylpentadecane), a common isoprenoid, is found in trace levels in several marine organisms including phytoplankton, benthic algae, and zooplankton, as well as in seawater and marine sediments. Pristane may also be derived from diagenesis of the phytol side chain of chlorophyll-a under reducing conditions (Figure 8.15).

An interesting review of pristane in the marine environment by Blumer et al. in 1964 discusses the enrichment of pristane in the zooplankton *Calanus* as compared to other genera. The authors report that under starvation conditions the concentration of pristane in *Calanus* increases relative to other lipids used as metabolic substrates. The retention of pristane relative to other metabolic lipids may assist in the organism in maintaining its position in the water column because of the low specific gravity of pristane (0.78), hence conserving energy which would otherwise be used in active swimming. However, because pristane is associated with many different organisms, it is not

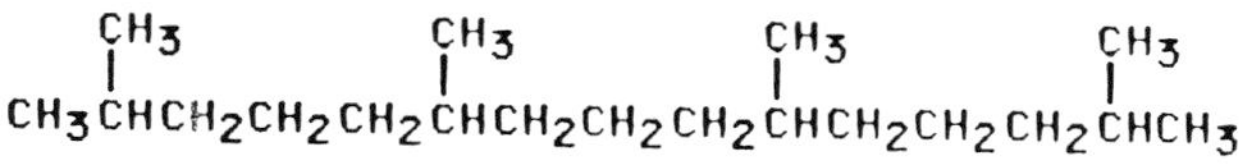

$$CH_3\overset{\overset{CH_3}{|}}{C}HCH_2CH_2CH_2\overset{\overset{CH_3}{|}}{C}HCH_2CH_2CH_2\overset{\overset{CH_3}{|}}{C}HCH_2CH_2CH_2\overset{\overset{CH_3}{|}}{C}HCH_3$$

FIGURE 8.14. Structure of pristane.

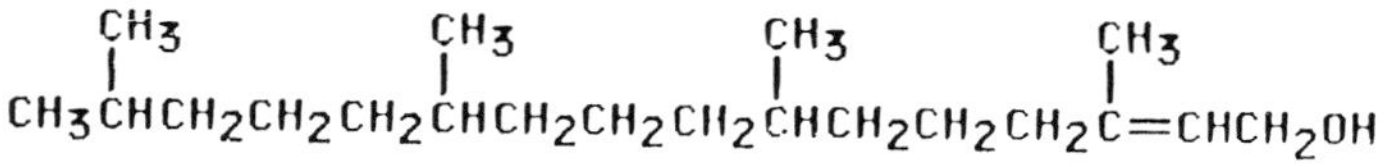

$$CH_3\overset{\overset{CH_3}{|}}{C}HCH_2CH_2CH_2\overset{\overset{CH_3}{|}}{C}HCH_2CH_2CH_2\overset{\overset{CH_3}{|}}{C}HCH_2CH_2CH_2\overset{\overset{CH_3}{|}}{C}{=}CHCH_2OH$$

FIGURE 8.15. Structure of phytol.

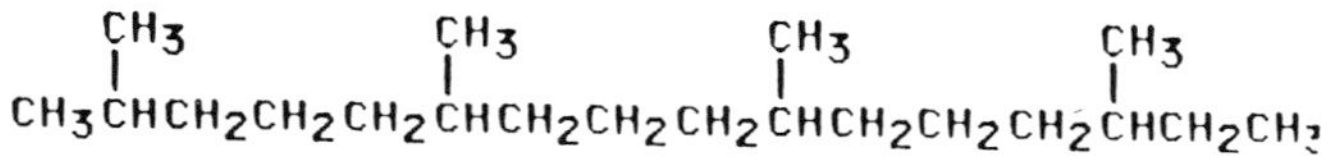

$$CH_3\overset{\overset{CH_3}{|}}{C}HCH_2CH_2CH_2\overset{\overset{CH_3}{|}}{C}HCH_2CH_2CH_2\overset{\overset{CH_3}{|}}{C}HCH_2CH_2CH_2\overset{\overset{CH_3}{|}}{C}HCH_2CH_3$$

FIGURE 8.16. Structure of phytane.

a useful specific biomarker. It is more commonly used in association with the related structure phytane (Figure 8.16) as an indicator in redox conditions: as mentioned above, pristane may form from phytol under reducing conditions, while oxidizing conditions lead to the production of phytane from phytol. Thus, pristane to phytane ratios are useful indicators of redox conditions in sediments. Phytane, unlike pristane, has not been isolated in organisms, but is found in older sediments and rocks.

In order to be a useful biomarker, a compound must retain enough of its original structure to be identified as a modified version of the original biological parent structure. The retention of the carbon backbone of the alcohol phytol from chlorophyll-a in the alkanes pristane and phytane is an example of the phenomenon. Other examples will be cited in the ensuing sections of this chapter.

Another commonly used source indicator of hydrocarbons is the carbon number distribution. For example, *n*-alkane distributions with carbon number maxima in the 17-21 range ($C_{17}$-$C_{21}$) are typical of aquatic algae, while terrestrial plant sources typically produce alkanes with $C_{23}$-$C_{33}$ maxima. In addition, these higher molecular weight *n*-alkanes derived from terrestrial plant sources tend to be found with a predominance of an odd number of carbons relative to even number carbon backbones. This odd over even predominance (OEP) is due to the predominance of odd carbon numbers in the composition of higher plant cuticular waxes. On the other hand, plants of lower organizational levels, such as aquatic algae, produce almost equivalent amounts of odd- and even-numbered homologues and thus show an OEP of approximately one in the $n$-$C_{15}$ to $n$-$C_{23}$ range. However, even when a distinct odd to even predominance exists, it has been found that there is a gradual lowering of the OEP index with the maturity of a deposit, apparently due to the preferential degradation of odd carbon number alkanes. The isotopic composition of hydrocarbons and other classes of compounds can also be used as a general guide or source input indicator. Isotopic fractionation of carbon and hydrogen ($^{13}C$, $^{12}C$ and $^{1}H$, $^{2}H$) resulting from different metabolic pathways involved in the anabolism (metabolic synthesis) of biological compounds, when preserved in a diagenetic product, is frequently used to distinguish between terrestrial and aquatic sources. This topic is discussed further with respect to humic substances.

The majority of alkanes found in organisms and sediments are isoprenoids, *n*-paraffins, and steranes, which are structurally similar to biological acids and alcohols and, in many instances, may be derived from these biological precursors. On the other hand, the relatively high concentration of alkanes in sedimentary deposits, as compared to those found in organisms, is at least partially due to their low chemical and catabolic activity which results in a preferential preservation and consequent relative enrichment of these compounds in older deposits.

### 5.3.1. Normal Alkanes

*n*-Alkanes are the dominant natural hydrocarbons in the marine environment and, of the various types of alkanes, *n*-alkanes are the most abundant in organisms. This compound class has been isolated from bacteria, phytoplankton, zooplankton, benthic and pelagic algae, and from higher level organisms of terrestrial and aquatic origin. Marine phytoplankton typically show a predominance of either *n*-pentadecane ($n$-$C_{15}$) or *n*-heptadecane ($n$-$C_{17}$) over the *n*-paraffin distribution. Benthic and pelagic algae can be grouped into two categories on the basis of *n*-alkane distributions. Red and green algae show a predominance of $n$-$C_{17}$, while $n$-$C_{15}$ is found as a natural maximum in brown algae. Although *n*-alkanes have also been isolated from zooplankton samples, whether these alkanes are produced by the zooplankton themselves or are due to the consumption of phytoplankton is questionable. *n*-Alkanes have also been isolated from both photosynthetic and nonphotosynthetic bacteria. The alkanes extracted range from $C_{13}$ to $C_{21}$ with maxima in the $C_{17}$ to $C_{20}$ range.

Normal alkanes ranging from about $C_{14}$ to $C_{36}$, without any major odd or even carbon number predominance, have been isolated from surface and deep ocean waters. Most studies indicate that the *n*-alkane distribution of open ocean water is fairly similar to that of phytoplankton and algae with differences due to evaporative losses of lower molecular weight (more volatile) components and some anthropogenic input, including petroleum.

Studies of *n*-alkanes in recent marine sediments indicate a predominance of odd-carbon number compounds in the *n*-$C_{15}$ to *n*-$C_{36}$ range. In nearshore sediments, a significant terrestrial influence on the *n*-alkane profile is often indicated by high proportions of long-chain alkanes (>*n*-$C_{23}$) with a great predominance of odd over even carbon numbers. Conversely, lower molecular weight *n*-alkanes ($C_{12}$-$C_{22}$) with an even carbon number preference have recently been reported in a relatively wide variety of depositional environments. Nishimura and Baker found this even-to-odd predominance in *n*-alkanes ranging from $C_{16}$ to $C_{24}$ with a maximum at $C_{18}$ in marine surface sediments from three different locations. Various possible sources were considered including the possible reduction of the accompanying *n*-fatty acids to *n*-alkanes in the sediments. Pristane to phytane ratios in the three samples were all suggestive of anoxic conditions which would be conducive to the reduction of fatty acids to alkanes. However, the distribution patterns of the fatty acids were very dissimilar to those of the alkanes and, thus, this possibility was ruled out. An alternative hypothesis was considered in which diagenetic processes resulted in the preferential removal of odd-carbon number *n*-alkanes, resulting in a relative enrichment of even-number alkanes. Although the odd predominance is known to diminish with maturity, as discussed above, there is no evidence which suggests that early diagnesis can cause the dramatic even carbon number predominance which was found in these samples. The possibilities of major anthropogenic and oil seepage inputs were eliminated on the basis of a lack of any other evidence for these phenomena as well as the unlikelihood of the sources containing such an outstanding even carbon number preference. The possibility of a terrigenous source was also eliminated due to the lack of any significant concentration of higher molecular weight *n*-alkanes ($C_{25}$-$C_{33}$) with an odd-carbon number predominance, which suggests a marine source for these compounds. The concept of a significant marine contribution was reinforced by the $\delta^{13}C$ values which ranged from $-20$ to 22 ‰, suggestive of a major marine plankton source. The authors concluded that the most likely source for this suite of *n*-alkanes was marine bacteria based on the previously reported occurrence of even-carbon number predominant *n*-alkanes in some bacterial species (especially *n*-$C_{18}$). Grimalt and Albaiges found similar distributions in recent and Miocene sediments and

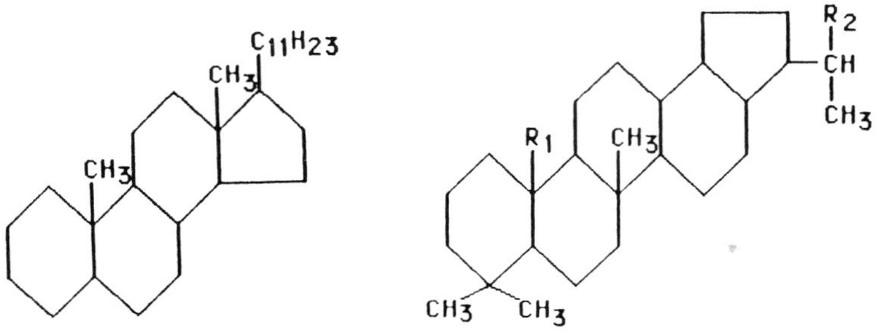

$C_{30}$-alkane type

hopane

FIGURE 8.17. Structure of $C_{30}$-steranes.

suggest that fungi and yeast species, in addition to bacterial sources, may contribute to this unusual suite of sedimentary *n*-alkanes.

### 5.3.2. Branched and Cyclic Alkanes

Many of the compounds that are branched and/or cyclic alkanes are isoprenoids, i.e., can be broken down into $C_5$ isoprene type units by applying the isoprene rule which was discussed earlier in this section. Pristane is an isoprenoid or ''regular branched'' alkane which is widespread in the marine food web and may be derived from the oxidative diagenesis of the phytol chain of chlorophyll. The usefulness of the pristane to phytane ratio as an indicator of redox depositional environment was described. Other branched alkanes which have been isolated from marine organisms and sediments include 7- and 8-methylheptadecane, which are found in algae and seawater, and are the major hydrocarbon components of some blue-green algae.

Cyclic alkanes, also referred to as chcloalkanes or alicyclic hydrocarbons, are proving to be excellent biomarkers. Included in this class of compounds are a group of polycyclics known as hopanes, the general structure of which is shown below (Figure 8.17). The correlations found to exist between the presence of hopanes and bacterial activity have established these compounds as biological markers. Hopanes have also proved to be useful geochemical indicators with respect to studies on the diagenesis and biodegradation of crude oils. It has been found that, in heavily degraded crudes, regular fossil fuel hopanes were transformed by biodegradation to demethylated hopanes and that this effect, together with an *n*-alkane depletion, resulted in an increase in optical activity of the oils. Others report the usefulness of another group of cyclic alkanes, $C_{30}$-steranes of the structural type shown in Figure 8.17, as a generalized biomarker. The study of about 40 oils from both marine and lacustrine sources revealed that the $C_{30}$-steranes were not detected in any lacustrine oils, but were detected in all marine derived oils with the exception of those dating from the Cambrian or Precambrian. The lack of these compounds in older oils is believed to be due to molecular evolutionary trends. The most likely biological precursors of these $C_{30}$-steranes are $C_{30}$ marine sterols (steroid alcohols) and, thus, these $C_{30}$-steranes should prove useful as a general indicator for marine biota input.

## 5.4. Unsaturated Hydrocarbons

The unsaturated hydrocarbons, or olefins, are more abundant in biological lipids than the saturated hydrocarbons, but are typically found in relatively lower concentrations in sedimentary deposits. The reason for this is most likely due to the greater chemical reactivity of the olefins which are associated with the presence of one or more double carbon-carbon bonds.

### 5.4.1. *n*-Alkenes

Normal alkenes have been isolated from marine phytoplankton, zooplankton, benthic algae, and bacteria as well as from terrestrial sources. The distribution of normal alkenes in marine phytoplankton is typically dominated by odd carbon number olefins, often $n$-$C_{21}$:6, where the 6 indicates that six carbon-carbon double bonds exist within the $C_{21}$ backbone. Other predominant *n*-alkenes of marine phytoplankton include $n$-$C_{16}$:1 and $n$-$C_{19}$:1. In contrast, the *n*-alkenes of the nonphotosynthetic diatom *Nitzschia alba* are found to be dominated by even carbon number species such as $n$-$C_{16}$:2, $n$-$C_{18}$:2, and $n$-$C_{28}$:2. Both odd and even carbon number alkenes have been found to be abundant in zooplankton, while marine bacteria appear to be characterized by an abundance of odd number carbon alkenes such as $n$-$C_{17}$:1 and $n$-$C_{17}$:2. For example, $n$-$C_{17}$:1 accounted for 50% of the total hydrocarbons in the sulfur bacterium *Chlorobium*.

Normal alkenes are also characteristic of terrestrial sources. The simplest alkene, ethylene ($C_2H_4$), is released by ripening fruit. Olive oils contain polyolefinic linear hydrocarbons in the $C_{13}$ to $C_{28}$ carbon number range, and *n*-alkenes of plant waxes are typically characterized by small amounts of alkenes ranging from $C_{23}$ to $C_{31}$.

Alkenes, because of their more chemically reactive nature as compared to alkanes, for example, are not found as abundantly in seawater and sedimentary deposits as in organisms. Heneicosahexane ($n$-$C_{21}$:6) has been isolated from seawater and is considered to be indicative of phytoplankton con-

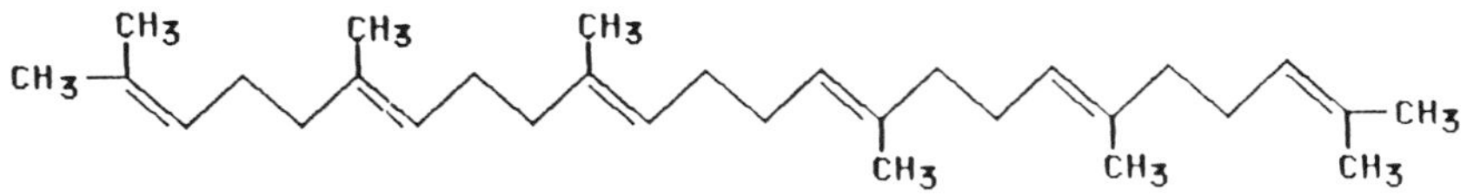

FIGURE 8.18. Structure of squalene.

tribution. Monounsaturated *n*-alkenes have been isolated from Cariaco Trench and Norwegian Sea sediments and are believed to originate from algae, terrestrial plants, and the early diagenesis of other biological lipids.

### 5.4.2. Branched and Cyclic Alkenes

Many of the branched and cyclic alkenes that have been isolated from organisms and from water and sediment samples are isoprenoids. Terpenes (also called essential oils) are isoprenoids and are the most abundant biological hydrocarbon. Terpenes may be either open chain or cyclic structures, the simplest of which are the monoterpenes ($C_{10}$), and increase in molecular weight through the more common structures such as the $C_{15}$ (sesqui-), $C_{20}$ (di-), $C_{30}$ (tri-), and $C_{40}$ (tetraterpenes). Interestingly, although the terpenes consist of an integral number of isoprene ($C_5$) units, compounds composed of five and seven isoprene units ($C_{25}$ and $C_{35}$) are not found in organisms.

Monoterpenes are found in resins and oils of plants, while diterpenes are the most common isoprene derivative in both living and fossil organisms. Phytol (the alcohol which esterifies the carboxylic acid group in chlorophyll-a) is a diterpenoid precursor of phytane and pristane.

Squalene ($C_{30}H_{50}$, Figure 8.18) is the most commonly produced triterpene in animals and occurs in most marine organisms. It is a precursor of triterpenoids, such as sterols, and is the major constituent of the branched and cyclic hydrocarbon fraction of several species of photosynthetic bacteria. The counterparts of the triterpenoid steroids found in organisms are the steranes, and tetra- and pentacyclic triterpanes found in sediments and crude oils. Much of the optical activity of crude oil is due to triterpanes derived from squalene.

The dominant triterpanes found in nature are the carotenes. Carotenes are precursors to vitamin A which is itself a diterpenoid alcohol. Carotenoids are plant pigments composed of mono- and dicyclic tetraterpenes and occur in all green plant tissues and in some nonphotosynthetic bacteria and fungi. Both carotenes and carotane have been isolated from sediments.

An unusual set of structurally related alkenes containing 20, 25, and 30 carbon atoms and varying degrees of unsaturation have been identified as virtually ubiquitous components of estuarine and coastal sediments and often comprise the dominant resolvable hydrocarbons on gas chromatograms. Although their molecular structures have not been determined, the homologous series differing by five carbon units is suggestive of terpenoids and thus suite of compounds is often referred to as "biogenic alkenes". A series of structurally related $C_{25}$ and $C_{30}$ biogenic alkenes have been described in sediments of the Narragansett Bay estuary. Four $C_{25}$ mono- and dienes and one $C_{30}$ diene were found to comprise 73 to 91% of the total alkenes in the surface (upper 2.5 to 5 cm) sediments of the estuary. A comparison of alkene distributions with $\delta^{13}C$ values of the sedimentary organic matter indicated that the geographic variations of two of the $C_{25}$ alkenes suggest a marine source (probably planktonic) for these compounds. In contrast, the concentrations of the $C_{30}$ diene were found to be relatively constant and exhibited no dependence on the origin of the sedimentary organic matter, implying an *in situ* production of the diene throughout the estuary that may be of bacterial origin. Vertical sedimentary profiles were characterized by a uniform concentration decrease to a depth of about 25 cm, suggestive of the degradation of these alkenes in the sediments.

### 5.5. Aromatic Hydrocarbons

Aromatic hydrocarbons are present in, at most, trace concentrations in the lipid fractions of living organisms and yet are far more abundant than alkenes, though less abundant than alkanes, in sedimentary deposits. Thus, most of the aromatic components found in marine sediments are not derived directly from biological sources.

Minor amounts of aromatic hydrocarbons have been attributed to algae, bacteria, and higher plants. For example, workers have demonstrated the conversion of a $^{14}C$ labeled acetate medium by algae into various polycyclic (polynuclear) aromatic hydrocarbons (PAH, aromatic compounds in which rings are fused). With the exception of minor inputs from these sources, aromatic hydrocarbons in seawater and marine sediments are due to *in situ* production from possible precursors such as isoprenoids and sterols and also are associated with high temperature pyrolysis of various fuels and with petroleum contamination.

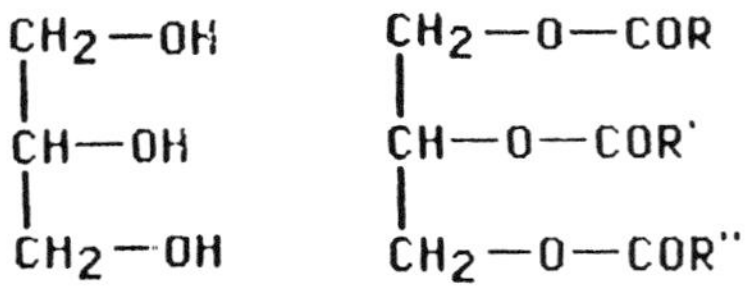

```
CH2—OH          CH2—O—COR
|               |
CH—OH           CH—O—COR'
|               |
CH2—OH          CH2—O—COR"

glycerol        fat or oil

                "triglyceride"
```

FIGURE 8.19. Structure of glycerol and fatty acids.

## 6. CARBOXYLIC ACIDS

Carboxylic acids contain at least one carboxyl group (-COOH, carbonyl + hydroxyl). Humic and fulvic acids also contain carboxyl groups but are covered separately in Section 7 because they behave quite differently from the discrete molecular species discussed in this section. Amino acids (Section 4) contain at least one amino group, in addition to at least one carboxyl group, and are also treated separately in this chapter because their biogeochemistry is distinctly different from that of the carboxylic acids covered here. This section will discuss the marine biogeochemistry of fatty acids, hydroxy acids, dicarboxylic and aromatic acids.

### 6.1. Fatty Acids

Fatty acids consist of a carboxyl group attached to a long hydrocarbon chain. Examples of common fatty acids include stearic acid ($CH_3$-$(CH_2)_{16}$-COOH) and palmitic acid ($CH_3$-$(CH_2)_{14}$-COOH). Although fatty acids are major components of most organisms, they have been found to be useful as biomarkers because of the variety of different fatty acids in different organisms and because of their chemical stability which allows them to persist over geological time periods.

Fatty acids, in living organisms, are found mainly as fats in animals and waxes in plants and have 4 to 36 carbon atoms. Fats and oils are esters (-COOR) of glycerol and fatty acids, as shown in Figure 8.19, where R represents a fatty acid.

The term ''fat'' is normally applied to solid glycerol esters, while ''oils'' are liquid esters. The term ''lipid'' refers to fats, oils, and fat soluble compounds. Most of the fatty acids arise from the hydrolysis of triglycerides (triaclglycerols) during the degradation of derital organic matter. Triglycerides comprise a major fraction of the lipid component of plant and animal storage tissue, and the fatty acids account for approximately 90% of the weight of triglycerides. Thus, fatty acids are a major component of lipids derived from the degradation of organic matter.

The most common fatty acids found in living organisms are saturated, straight-chain acids of even carbon number. Palmitic acid (16.0) is often the first or second most abundant acid in many organisms. However, small amounts of saturated, straight-chain fatty acids of odd carbon number are also found in most organisms. Straight-chain even carbon number acids with one or more double bonds are another major group of fatty acids found in organisms and are especially abundant in aquatic organisms. Unsaturated fatty acids with 16 and 18 carbons are the most common olefinic fatty acids found in aquatic organisms. Palmitoleic acid ($CH_3$-$(CH_2)_5$-CH=CH-$(CH_2)_7$-$CO_2H$) is a typical fatty acid isolated from natural waters.

Low molecular weight (volatile) fatty acids ($<C_5$) are also present in natural waters and sediments and are produced by microbial degradation of organic matter. Sansone and Martens measured volatile fatty acid concentrations in organic rich marine sediments off the coast of North Carolina and found that acetic ($CH_3COOH$) and butyric ($CH_3$-$(CH_2)_2$-COOH) were the most abundant and were found at 2.5 to 20 mg $l^{-1}$ levels individually. This study and others have indicated that volatile fatty acids appear to be largely associated with bacterial activity in anaerobic environments. The concentration of low molecular weight ($<C_5$) fatty acids in seawater ranges from 5 to 50 $\mu g\ l^{-1}$ and tends to vary with the DOC.

Nonvolatile fatty acids ($>C_5$) are found in seawater at concentrations of 10 to 200 $\mu g\ l^{-1}$ and typically account for about 1% of the DOC. Although straight-chain fatty acids are the most commonly found in the biosphere, fatty acids which are branched are often of special interest to organic geochemists. Because these structures are rather rare in natural systems, they can sometimes be assigned to specific organisms and hence have the potential to be excellent biomarkers. For example, Goossens et al. conducted a systematic study of 13 bacterial species and found that methanogenic and extremely halophilic archaebacteria are characterized by an absence of fatty acids, while gram-positive eubacteria primarily contain iso- and anteiso-fatty acids. Acids which contain a single branched methyl group furthest from the carboxyl group are referred to as iso-acids, while those containing a methyl branch closest to the carboxyl group are described as anteiso. Kawamura and Ishiwatari found a maximum of branched fatty acids at a depth of 3 to 15 m in a sediment from an ancient lake (Lake Biwa, Japan) and interpreted this as a geologic record of increased bacterial activity in the water column and surface sediments sometime in the past. Fatty acids are also believed by many to be important precursors of sedimentary alkanes.

The common fatty acids found in seawater range from $C_{10}$ to $C_{22}$ with $C_{16}$ and $C_{18}$ commonly the most abundant and originating mainly from planktonic sources. Fatty acids extracted from the surface microlayer typically show characteristics of planktonic lipids, i.e., a carbon number range of $C_{10}$ to $C_{22}$, centered around $C_{16}$, and a high even-to-odd carbon number predominance. Unsaturated fatty acids have been isolated from seawater, recent and ancient sediments, and petroleum. Most commonly found are $C_{16:1}$ and $C_{18:1}$ fatty acids. Studies by Gillan et al. on monounsaturated fatty acid distributions of marine bacteria and sediments support the use of specific monounsaturated acids as bacterial markers and found bacterial contributions to the acids to be as high as 10 to 80%.

## 6.2. Hydroxy, Dicarboxylic, and Aromatic Acids

Much less is known about the biogeochemistry of hydroxy, dicarboxylic, and aromatic acids in aquatic systems than is known about fatty acids. This is mainly due to the fact that these substances occur at even lower levels than those of fatty acids.

### 6.2.1. Hydroxy Acids

Hydroxy acids account for about 1% of the DOC of natural waters. Although they are important intermediates of metabolic pathways such as the Kreb's cycle, they are rapidly degraded and thus have a large flux which results in a low concentration level in natural waters and sediments. Peltzer and Bada discuss the production of lactic acid ($CH_3CHOHCOOH$), an α-hydroxy acid, by anaerobic fermentation and its assimilation by sulfate reducing bacteria. Most sulfate reducing bacteria are incapable of metabolizing glucose and may use lactic acid as their sole source of organic carbon. They then convert lactic acid to acetate and carbon dioxide which are themselves available substrates for methane producing bacteria. Thus, hydroxy carboxylic acids are part of a "recyclable carbon source" of bacterial populations and their concentrations in aqueous systems typically remain low because of rapid utilization by microbes. Peltzer and Bada report that the subsurface maxima and enantiomeric ratios of lactic acid also suggest a microbial origin and control of the distribution of this compound.

Hydroxy acids were also isolated from pure cultures of Chlorophyta, Rhodophyta, and Cyanophyta by Matsumoto and Nagashima. Normal and branched (iso and anteiso) hydroxy acids ranging from $C_8$ to $C_{26}$ were found in all samples studied at concentrations up to 2.3 mg $g^{-1}$ of dry sample. There was a predominance of even-carbon numbered, normal hydroxy acids with carbon chain lengths shorter than $C_{20}$. Thus, microalgae and cyanobacteria may be significant hydroxy acid sources in aquatic systems.

### 6.2.2. Dicarboxylic Acids

Dicarboxylic acids are fatty acids which contain two carboxylic acid groups per molecule. Dicarboxylic acids, like hydroxy acids, are produced by a variety of organisms and are common end products of anaerobic respiration and fermentation. Peltzer and Bada (Scripps) discuss the distribution of dicarboxylic acids in reducing sediments from the Santa Barbara Basin and Cariaco Trench. Oxalic acid (HOOC-COOH) and succinic acid ($COOH(CH_2)_2COOH$) were identified by their retention times as methyl esters on a gas chromatogram and confirmed by gas chromatography-mass spectrometry. Concentrations were found to be less than 100 μmol $g^{-1}$ and generally decreased with depth. The gradual concentration decrease with depth was explained by biological utilization of these extremely labile compounds.

### 6.2.3. Aromatic Acids

Aromatic carboxylic acids contain at least one benzene ($C_6H_6$) ring in addition to the carboxylic acid group. Most studies of aromatic acids in natural aquatic systems report the presence of these compounds in association with humic substances, which are discussed in detail in the following section of this chapter.

## 7. HUMIC SUBSTANCES AND KEROGEN

Naturally occurring organic matter found in soils, sediments, and in the aqueous phase consists of an extremely diverse combination of compounds of various origins. One way of classifying this organic matter would be to consider two very broad categories such as (1) nonhumic substances and (2) humic substances and kerogen.

Included in the nonhumic substances category would be the types of compounds discussed earlier in this chapter (carbohydrates, amino acids, etc.) as well as the natural and pollutant trace organics described later in this chapter (Section 8). These nonhumic substances retain the characteristics of their compound classes and chemically react as individual compounds. Humic substances, on the other hand, in many respects act as a polymeric unit, although they are composed of mixtures of individual compound classes and tend to be quite resistant to extensive microbial degradation as compared to nonhumic compounds. Humic substances are precursors of the geopolymeric material known as kerogen (dispersed organic matter found in sedimentary rocks and insoluble in organic solvents, mineral acids, and bases). Petroleum is primarily produced from the geochemical maturation of kerogen derived primarily from aquatic organisms. There is presently a significant gap in our understanding of the processes involved in the diagenesis of plant and animal remains to form humic substances which are precursors of petroleum-forming aquatic kerogen. Kerogen is distinguished from bitumen on the basis of the solubility of bitumen in organic solvents (typically benzene or carbon tetrachloride). Since humic substances consist of a mixture of plant and animal debris at various stages of decomposition, the chemical composition will vary with source of input and environment of deposition.

### 7.1. Humic Substances

Humic substances have been studied by soil and water chemists for many years because of the important roles they play with respect to water quality, soil fertility, soil porosity, and their association with both inorganic and organic nutrients and pollutants through complexation and adsorption processes. Aquatic humic substances have been shown to be important precursors of chloroform and other chlorinated organics in drinking water. Geochemically, humic substances are also of interest because they are important contributors to the precursors of fossil fuels, as mentioned above, and they compose the bulk of organic matter in natural systems. In seawater, the concentration of humic substances ranges from 60 to 600 μg C/l, which accounts for 10 to 30% of the dissolved organic carbon of seawater.

Early studies on marine humic substances describe the isolation of "gelbstoff", a yellowish-brown acid material which could be extracted from seawater with organic solvents. On the basis of solubility behavior, humic substances are commonly classified into three categories: (a) humic acids are base soluble but not acid insoluble; (b) fulvic acids are acid and base soluble; and (c) humin is both acid and base insoluble. Humic and fulvic acids are typically extracted together into a basic solution (−0.5 *M* NaOH or $NA_4P_2O_7$), and the humic acids are precipitated out by acidification of the solution (addition of HCl until a pH of 2.0, or the use of an ion-exchange resin in the $H^+$ form) leaving the acid-soluble fulvic acid fraction in solution. Humin is chemically very similar to humic acid, but is more tightly bound to the inorganic soil/sediment constituents.

In terms of elemental composition, humic substances consist of approximately 40 to 60% carbon, 30 to 45% oxygen, 1 to 5% nitrogen, and 2% or less of sulfur. Humic acids tend to be richer in carbon and poorer in oxygen as compared to fulvic acids. Historically, the chemical compositions of humic substances from various environments were studied by wet chemical and instrumental methods of analysis. The major structural components include aliphatic chains with varying degrees of branching (aromatic rings, phenolic, hydroxyl, methoxyl, and carboxylic acid groups). As mentioned earlier, humic substances consist of a complex polymeric mixture of compounds and, thus, their functional group content varies with sources of input and geochemically important factors such as depositional environment.

Many of the classical methods of analysis proved to be only partially successful with respect to qualitative and quantitative analysis of humic substances. For instance, infrared analysis of humic acids (Figure 8.20) yields spectra with broad overlapping bands reflecting a complex mixture of functional groups and making specific group assignments rather difficult. The chemical modification of groups, such as methylation and acetylation of carboxylic and phenolic groups, has proved useful in the assignment of specific peaks to absorbing groups. In the ultraviolet and visible regions of the electromagnetic spectrum, humic substances yield featureless spectra exhibiting decreasing absorbance with increasing wavelength and typically show no characteristic absorption maxima (Figure 8.21). The fractionation of humic substances into fractions based on molecular weight ranges has yielded spectra of greater definition although still very complex. Molecular weight fractionation by gel filtration chromatography using Sephadex gels (an artificially crosslinked polysaccharide derived from dextran, which is a natural glucose polymer) yields molecular weights ranging from the hundreds to hundreds of thousands, although molecular weights obtained by this method are only approximate due to the calibration of gels by the use of molecules of specified geometries.

One of the most successful and promising instrumental techniques that has provided conclusive information of the structure of humic substances is $^{13}C$ nuclear magnetic resonance (NMR) spectroscopy. Early studies employed

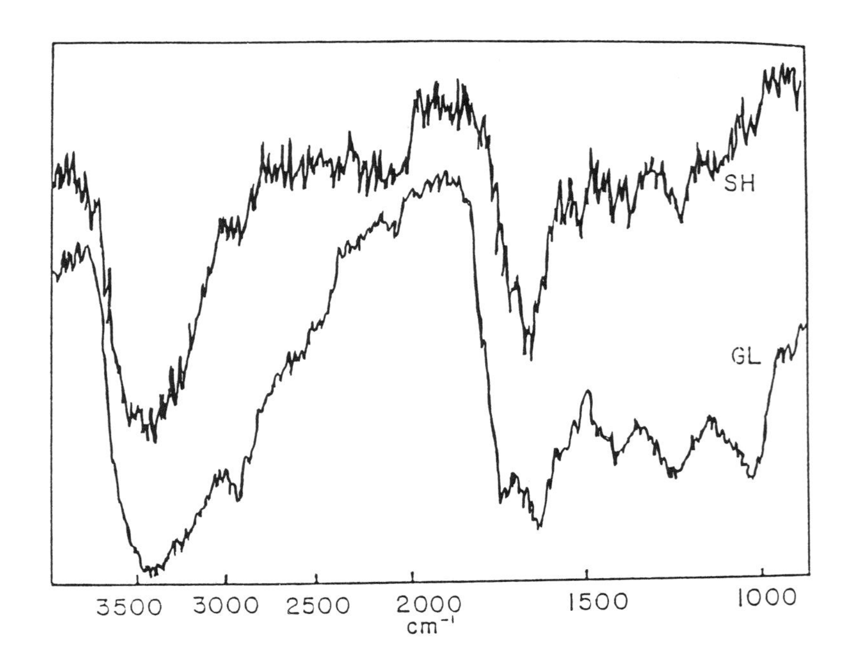


FIGURE 8.20. Infrared spectra of Stone Harbor (SH) and Grass Lake (GL) humic acids.

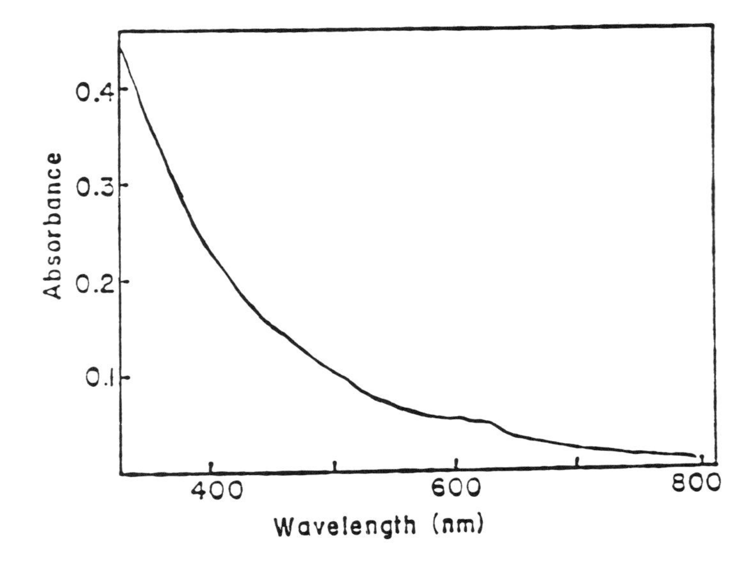


FIGURE 8.21. Visible spectrum of Grass Lake humic acid.

solution NMR spectroscopy, but the development of techniques such as cross polarization (CP) and magic angle spinning (MAS) has allowed for the structural elucidation of solid samples. The analysis of solid samples allows for direct structural comparisons between humic fractions of very different solubilities, including insoluble humin. Cross polarization is a sensitivity enhancing technique allowing for a transfer of polarization from the abundant $^{1}H$ population to the less abundant $^{13}C$ population. Magic angle spinning is a method of minimizing line broadening and thus allows one to achieve sharper signal peaks by spinning the solid sample very rapidly (several thousand rps) to compensate for the rapid tumbling motion of molecules in the liquid state that does not occur in the solid state. The additional technique of dipolar dephasing allows one to distinguish between protonated and nonprotonated carbons and has been extremely useful in structural comparisons between humic substances of terrestrial and aquatic origin. Dipolar dephased spectra will contain signals from nonprotonated carbons and methyl carbons.

Marine and estuarine humic substances appear to be quite different with respect to their functional group composition as compared to their terrestrial counterparts, although in coastal environments a significant terrestrial contribution is often apparent. Terrestrial humic substances typically contain a recognizable contribution from lignin. Lignins are polymeric methoxyphenolpropanoids which comprise 17 to 33% of wood and assist in increasing the mechanical strength of vascular plants. Lignins are unique to vascular plants and their presence is thus indicative of terrestrial plant input. Thus $^{13}C$-NMR analysis of terrestrial humic acids, for example, will typically show the presence of aromatic phenolic and methoxyl carbon. In the CP-MAS $^{13}C$-NMR spectrum (Figure 8.22) of a soil humic acid, there are strong resonances from aromatic carbon (110 to 160 ppm), methoxyl carbon (54 ppm), and phenolic carbon (151 ppm). Additional resonances from carboxyl carbon (171 ppm) and aliphatic carbon (0 to 50 ppm) are also apparent.

The estuarine humic acid spectrum shown in Figure 8.23 provides a striking visual contrast which reflects the lack of a lignin contribution. Marine and estuarine humic acids are highly aliphatic (low aromaticity) in nature and the alkyl chains show a high degree of branching. The branching of the alkyl groups is best detected by use of dipolar dephasing which, as noted earlier, allows one to differentiate between protonated and nonprotonated carbons. In the dipolar dephased spectrum of the marine humic acid shown in Figure 8.24, the presence of methyl groups (19 ppm) and substituted aliphatic carbon (30 ppm) suggests rather extensive branching as compared to terrestrial coun-

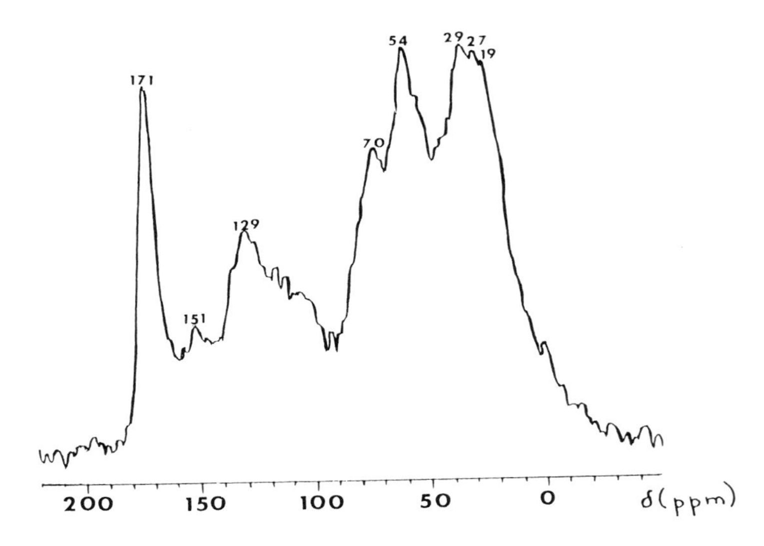


FIGURE 8.22. CP/MAS $^{13}C$ nuclear magnetic resonance spectrum of a soil humic acid.

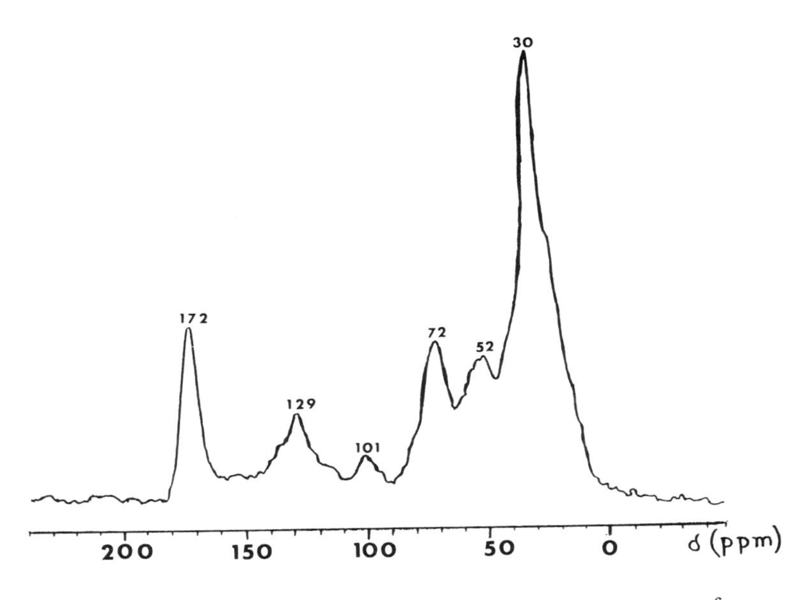


FIGURE 8.23. CP/MAS $^{13}C$ nuclear magnetic resonance spectrum of an estuarine humic acid.

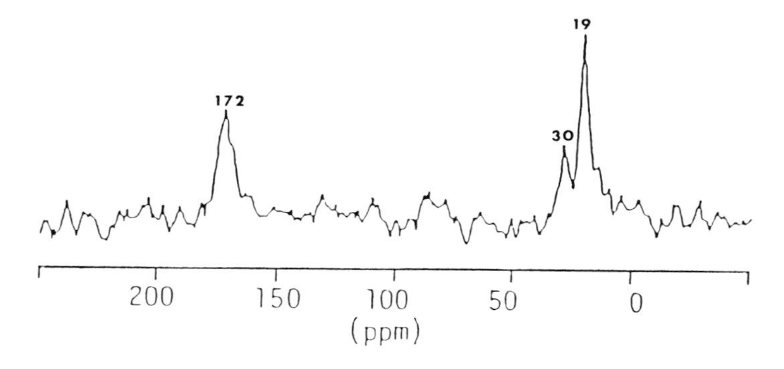


FIGURE 8.24. Dipolar dephased CP/MAS $^{13}C$ nuclear magnetic resonance spectrum of a marine humic acid.

terparts. The other prominent peak in this spectrum at 172 ppm is due to carboxylic acid carbon. The aliphatic structures are believed to be derived from marine microorganisms. NMR and gas chromatography-mass spectrometry (GC-MS) studies indicate strong similarities between the composition of marine diatom extracellular materials and dissolved marine humic substances, indicating that algal exudates and their biodegradation products are probable precursors of marine humic substances. As mentioned earlier, petroleum is formed from an aquatically derived biomass which tends to be fairly lignin free, while coal, which is more highly aromatic in nature than petroleum, contains a more substantial contribution from lignin-rich biomass of terrestrial origin.

Since humic acids from aquatic sources are more aliphatic than their terrestrial counterparts, the H/C ratio of a sample may be indicative of its major source of origin. Humic acids derived from organic matter of vascular plant input have low H/C ratios because of a greater aromatic character as compared to aquatic humic acids. Stuermer et al. studied 12 different humic acids and "proto-kerogens" from various sample locations and found that the H/C ratio, combined with the $\delta^{13}C$ value, provided a reliable source indicator which allows one to differentiate between marine derived and terrestrially derived organic matter. The definition of $\delta$ value is:

$$\delta = [R(\text{sample})/R(\text{standard}) - 1] \times 1000$$

where $\delta$ is $^{13}C/^{12}C$. The $\delta^{13}C$ value of soil humic acid is usually from $-25$ to $-26$‰ and reflects values found for terrestrially derived vascular plants, while marine humic acids have $\delta^{13}C$ values of $-20$ to $-22$‰. These $\delta^{13}C$

values are also further evidence that marine humic acids are formed *in situ* and are not terrestrially derived. In addition, since marine humates tend to have higher nitrogen contents than terrestrial counterparts, they also tend to have higher N/C ratios.

The diagenetic interrelationships among marine humic fractions (humin, humic acid, and fulvic acid) are an area of current research interest. Much of the original research on humic substances was done on soil samples and it was generally accepted that fulvic acids (the most soluble humic fraction) closely resembled microbiological degradation products of plants (carbohydrates, amino acids, etc.), and these substances reacted (or condensed) in a complex series of reactions, such as the Maillard reaction, to form humic acids which then became associated with mineral matter and eventually evolved into humin and kerogen. This condensation model may not be an accurate model for marine systems. More recently, the ''selective preservation'' model, as developed by Hatcher and others, has been gaining support as a realistic picture of the relationships between humic substances. In this model, the humification pathway is viewed as one of degradation with preservation of nonlabile components rather than of condensation.

Marine humin is believed to consist largely of a refractory residue of aquatic algae which is selectively preserved because of the low biodegradability of branched paraffinic substances. Humin is slowly degraded to form the more soluble humic acid while fulvic acids are products of more intense degradation and also include the degradation products of more labile sedimentary organic detritus such as carbohydrates.

## 7.2. KEROGEN

Originally the term ''kerogen'' was proposed in 1912 as a term describing organic matter found in oil shales. Presently, this term is used to designate that part of the organic matter of sedimentary rocks that is nonextractable with organic solvents, mineral acids, or bases. It is believed to be the largest reservoir of organic carbon in existence on the earth and it has been estimated that approximately $3 \times 10^{15}$ tons of kerogen exist on this planet as compared to $5 \times 10^{12}$ tons of coal in the earth's crust. In spite of its great abundance, a great deal still remains to be learned about its structure. As is true with humic substances, the structure of any particular kerogen sample will vary according to the types of organic matter input during formation and will also be dependent on the geochemical environment during and after its formation.

Aquatic kerogen, including marine derived kerogen, undoubtedly contains a major contribution from humic substances and therefore contains many of the same types of chemical structures found in humic acids and humin, which are, however, more tightly bound to the mineral matter of sedimentary rocks and hence more difficult to extract. In particular, aquatic kerogens are often referred to as algal kerogens and are highly aliphatic in nature, as are algal humic substances, and contain polyethylene chains. These aliphatic structures are believed to represent selectively preserved structures from algae and, very probably, bacteria. Kerogen-like material can be isolated from recently deposited algal mats and the oxidative degradation of this material yields products that are very similar to products of the degradation of algal kerogens. The presence of these polymethylene chains in algal kerogens is also reflected in the findings that the release of long-chain carboxylic acids occurs from an algal kerogen upon pyrolysis (200 to 400°C).

## 8. TRACE COMPOUNDS

The compound classes discussed in preceding sections of this chapter collectively compose the bulk of authigenic marine organic matter. This section presents a brief survey of some of the minor contributors to marine organic matter, although occasionally, they may compose a more significant fraction. These compound classes are typically found at ng $l^{-1}$ to $\mu g\ l^{-1}$ concentrations, although as mentioned, occasionally they are more abundant under specific, atypical conditions. Although minor contributors to the TOC of seawater or sediments, the biogeochemistries of these trace substances, when studied, have often provided valuable insights into diagenetic processes. Also, many of these trace components have proven to be or have the protential to be used as biomarkers and/or geochemical indicators. Such is the case with steroids, the first class of compounds to be discussed in this section.

### 8.1. Terpenoids

Terpenoids are natural products built up of five-carbon atom isoprene ($CH_2CHCH_3CHCH_2$) units. While the rather rare hemiterpenes consist of a single isoprene unit, the more common monoterpenes are composed of ten carbon atoms (two isoprene units). The related compounds known as sesquiterpenes, diterpenes, triterpenes, etc. have fifteen, twenty, thirty, or more multiples of the terpene framework. Terpenoids may be acyclic, mono-, bi-, or polycyclic.

Most terpenoids are biosynthesized by plants, although related structures such as steroids are associated with the animal kingdom. Terpenoids are characterized by a wide variety of oxygen containing functional groups and may either be found as oxygen deficient hydrocarbons or as oxygen containing esters, aldehydes, ketones, alcohols, carboxylic acids, etc. These polyunsaturated compounds tend to be quite reactive and diagenetically undergo oxidation, reduction, or polymerization reactions depending upon geochemical conditions. When these processes result in aromatic or saturated derivatives, these diagenetic products are typically extremely stable chemically and may persist as biological and geochemical marker compounds for many millions of years in geological formations.

Terpenes have been isolated from various marine organisms including algae, sponges, coelenterates, echinoderms, and mollusks. However, there is some evidence that in the case of coelenterates the true terpenoid source may be algal symbionts, while terpenoids isolated from mollusks may be derived from ingested algae.

Monoterpenes are rather highly volatile compared to higher molecular weight terpenes and account for the characteristic odors of many plant species. Limonene is a monocyclic monoterpene which is largely responsible for the odor of lemons. Pinene, a bicyclic monoterpene and the major constituent of turpentine (the oil secreted by conifers), is the dominant volatile component of pine resins and is also a molecule of much historical significance. Lavoisier was the first to analyze pinene and the correct formula of pinene was included in the early results of Dumas' molecular weight determination method. In addition, the first observation of optical activity by Biot in 1815 was done on terpentine.

More often than sesquiterpenoids themselves are found reports of products which are believed to be derivatives of these compounds. In fact, until the work of Ruzicka in the early 1920s, the correct formula of only one sesquiterpene (farnesol) was reported in the literature. However, by 1971, at least one thousand different sesquiterpenes were identified in the literature, and most of these were isolated from terrestrial plants. These rapid advances in terpenoid chemistry are largely due to advances in analytical instrumentation and analogy. Although research into the natural products of marine algae have resulted in the identification of many new sesquiterpenoids, this group of terpenoids is not widely distributed among marine phyla. In addition to marine algae, these compounds have thus far been found in sponges, mollusks, and coelenterates, although the sesquiterpenoids found in coelenterates may actually be produced by symbiotically associated unicellular algae.

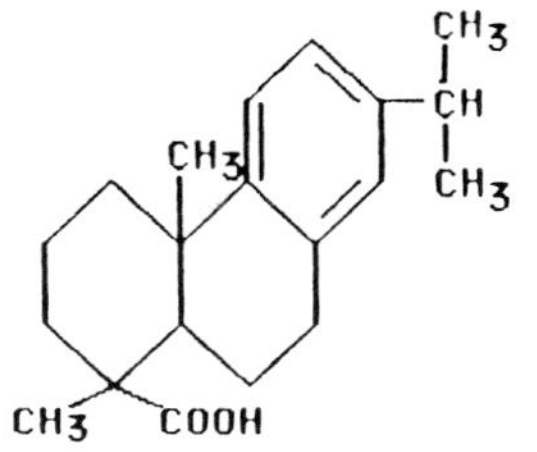

FIGURE 8.25. Structure of dehydroabietic acid.

Diterpenoids are the major constituents of tree resins, especially conifers. Resins are often found in a fossilized state which is known as amber. Diterpenoids are also important components of the supportive tissues of higher plants. One of the diterpenes associated with fossilized tree amber remains is fichtelite ($C_{19}H_{34}$), which forms the main component of a white or yellowish paraffin-like deposit associated with the remains of fossilized conifers. Retene ($C_{18}H_{18}$), along with the aforementioned fichtelite, is a dehydrogenation and decarboxylation product of diterpenic acids and has been isolated from marine sediments. Diterpenic acids, also referred to as "resin acids", are a major component of plant resins. Dehydroabietic acid ($C_{20}H_{28}O_2$) is a tricyclic diterpenic acid which is the predominant diterpenoid isolated from marine sediments (Figure 8.25). Simoneit has estimated the concentration range of dehydroabietic acid in marine sediments as 0.2 to 20 $\mu g\ g^{-1}$, while lower concentrations of retene and dehydroabietin (the decarboxylation product of dehydroabietic acid) are also present. Because cyclic diterpenoids are not endogenous to marine biota, but are associated with higher plants, they may be used as molecular markers for the contribution of terrigenous organic matter to marine sediments.

Triterpenes are the most widespread terpenes in plants and are almost always found to exist as pentacyclic structures with few sidechain functional groups. Examples of pentacyclic terpenoids include compounds having hopane type skeletons. These and similar compounds are found to occur ubiquitously in the geosphere and exist predominantly as saturated and unsaturated hydrocarbons as well as carboxylic acids. Their alcohol and ketone derivatives are also found, but less commonly and in lower abundances. Other examples of commonly reported pentacyclic triterpenoids include diplotene and moretane (Figure 8.26). Diplotene has been isolated from bacteria and blue-green algae as well as from recent and older (Cretaceous) marine sediments.

The triterpenes are biosynthetically derived from the cyclization of the folded squalene chain. Squalene is an aliphatic isoprenoid which was discussed

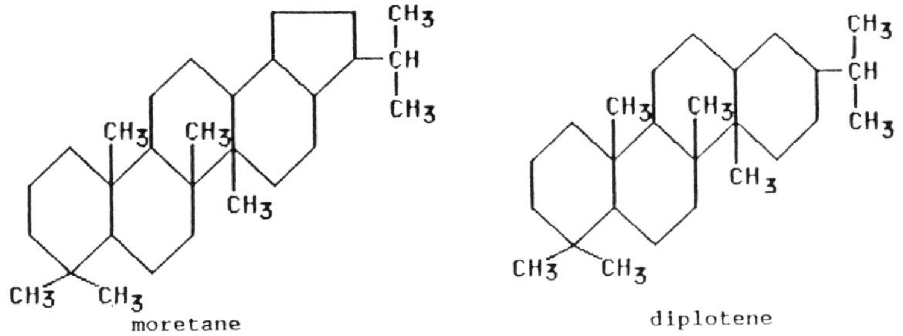

FIGURE 8.26. Structures of moretane and diplotene.

earlier. Work by Ruzicka and his colleagues has demonstrated that all known triterpenes can be derived *in vivo* from various spatial coilings of the acyclic squalene chain prior to cyclization followed by proton and methyl shifts. In fact, even the stereochemical configurations of the triterpenes are determined by the manner in which the squalene chain is folded.

Included in the group of compounds known as tetraterpenoids are the fat-soluble pigments known as carotenoids. Carotenoids are an important class of natural pigments (yellow to red) composed of isoprene units (typically eight). The bright colors of these pigments are associated with a series of conjugated double bonds which constitutes the chromophore system (the chemical functionality with which color is associated). Carotenoids may be subdivided into two groups: the carotenes and the xanthophylls. Carotenes are hydrocarbon pigments while the xanthophylls are closely related structures which contain oxygen in addition to carbon and hydrogen. This division of the carotenoids is based historically on solubility differences of the two groups.

The colors of many plants, flowers, and fruits such as tomatoes and carrots are due to the presence of carotenoids. β-carotene ($C_{40}H_{56}$), shown in Figure 8.27, is one of the principal carotenoids associated with the chloroplasts of higher plants and is also the dominant carotene of many freshwater and marine algae. In addition to green plants, carotenoids have also been isolated from nonphotosynthetic bacteria, fungi, and many other organisms. Carotenoids have also been isolated from about twenty species of sponges. The presence of β-carotene in sponges was first noted by Krukenberg in 1880 and echinenone (Figure 8.28), one of the simplest xanthophylls, was extracted from the sea urchin *Paracentrotus lividus*. Astaxanthin (Figure 8.29) is one of the major xanthophylls found in the marine environment. It is the characteristic xanthophyll of copepods which are a major component of marine zooplankton. Yamaguchi discovered a novel class of sponge carotenoids containing aro-

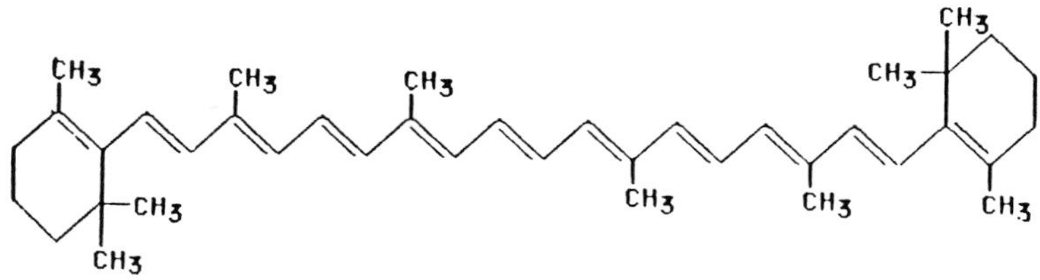

FIGURE 8.27. Structure of β-carotene.

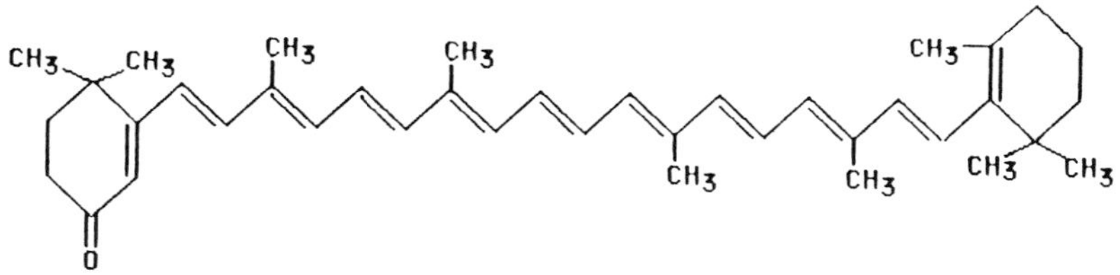

FIGURE 8.28. Structure of echinenone.

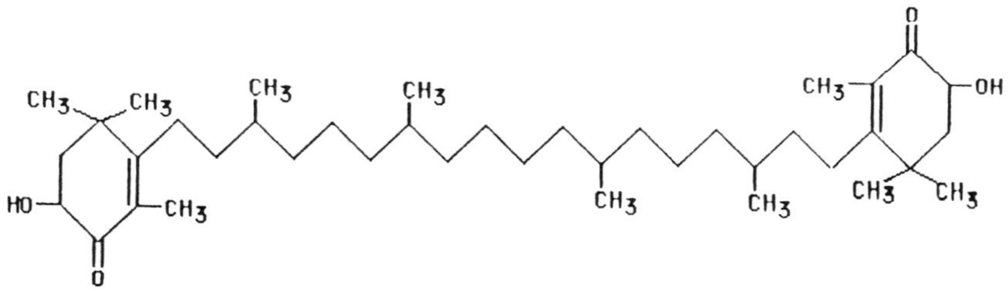

FIGURE 8.29. Structure of astaxanthin.

matic end groups. Mammals require carotenoid consumption for vitamin A production. These pigments are converted to vitamin A in the human liver.

In natural waters, carotenoids are found at sub-μg $l^{-1}$ concentrations. Extractions of recent marine sediments have yielded carotenoids that are both fully and partially reduced. Watts and Maxwell demonstrated the progressive reduction of carotenoids in Black Sea and Cariaco Trench sediments and found that reductive rates differed for various carotenoids. Carotenoids generally disappear fairly rapidly in sediments with time. The oldest sediments from which carotenoids have been extracted are approximately 100,000 years old.

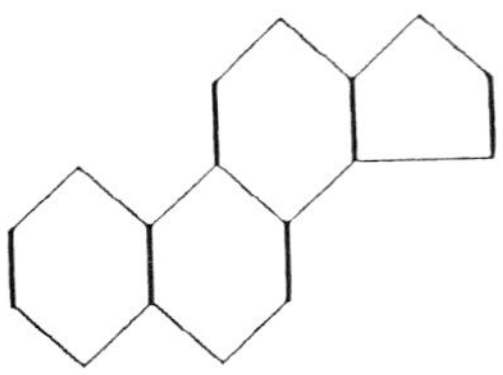

FIGURE 8.30. Structure of steroid ring.

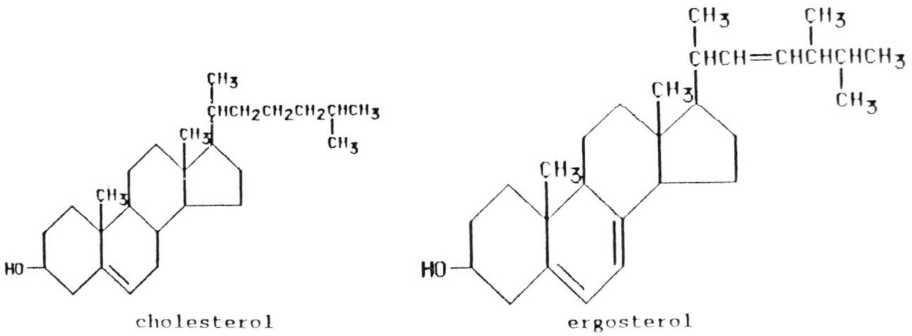

cholesterol

ergosterol

FIGURE 8.31. Structure of cholesterol and ergosterol.

## 8.2. STEROIDS

Steroids are closely related structurally and biochemically to terpenes and are sometimes classified as a type of terpenoid. The triterpene squalene is a biochemical intermediate in the synthesis of the steroids lanosterol and cholesterol.

Steroids (Figure 8.30) can be classified with respect to their principal functional groups present. For example, cholesterol (mentioned above), is classified as a sterol since it contains an alcohol (ROH) group. All steroids have the same basic carbon skeleton in common, with methyl groups typically at carbon positions 10 and 13 and a chain structure at carbon 17. Steroids typically contain 27 to 29 carbon atoms and are only very slightly soluble in water.

Cholesterol (Figure 8.31) is the most abundant animal sterol while ergosterol (Figure 8.31) is the principal plant sterol and is also an important source of vitamin D for animals. Cholesterol is also a major sterol produced by phytoplankton and zooplankton in surface waters.

Steroids are the most extensively studied organic chemicals of animal origin. Many hormones (metabolic regulators produced in endocrine glands) are steroids. The biomedically important steroid cortisone is a hormone which balances against insulin in hexose utilization and has been used medically to reduce inflammation. In addition to acting as regulators of growth and reproduction, steroids are also structural components of cell membranes.

Sterols can further be subdivided into groups such as stanols (saturated steroid), stenols (monunsaturated sterols), and di-stenols (diunsaturated sterols). Steranes (saturated aliphatic steroids) and stanones (ketone-containing steroids) are other classes of steroids which are geochemically interesting. Data on marine sediments indicate that steranes, stanols, and sterenes (unsaturated aliphatic steroids) are the major steroid classes present. Due to the rather stable steroid carbon skeleton and the numerous variations of the basic structure which are possible (variations in side chain alkylation, positions of unsaturation, and stereochemistry), many studies have been done on the use of steroids as biological and geochemical markers. However, many steroids are rather widely distributed in terms of their biological production, which makes it difficult to assign their source to a specific type of organism. For example, a large proportion of $C_{29}$ steranes in sediments is often used as an indicator of terrestrial organic matter input because of the abundant presence of precursor $C_{29}$ sterols in vascular plants. Volkman's work has indicated that in oligotrophic waters off the east Australian coast, these same $C_{29}$ sterols were biosynthesized by marine phytoplankton. Thus, in sediments and in crude oils, a large proportion of $C_{29}$ steranes is not necessarily indicative of substantial terrestrial input. In addition to $C_{29}$ sterols, algae also produce significant amounts of $C_{27}$ and $C_{28}$ sterols while higher animals usually produce cholesterol ($C_{27}$) as a major sterol, and marine invertebrates contribute a complex mixture of $C_{27}$, $C_{28}$, and $C_{29}$ sterols.

The marine biogeochemistry of steroids has been extensively studied by researchers from Woods Hole Oceanographic Institution since the mid-1970s. Numerous research papers by Gagosian, Lee, Farrington, and co-workers have unraveled some of the very complex chemistry of this important class of compounds. In order to utilize sterols and other steroids as indicators of organic matter source, changes in steroid structure due to geological, physiochemical, and biological factors must be considered. For instance, the reduction of sterols to their saturated analogs, the stanols, is one of the more commonly studied geochemical steroid reactions. Studies on seawater from the Black Sea and from anoxic waters of Walvis Bay revealed no evidence of sterol to stanol conversion at either the $O_2/H_2S$ interface or in deeper anoxic waters despite highly favorable reducing conditions. However, a rapid metabolism of stenols and stanols was observed in the water column at depths above the $O_2/H_2S$ interface and, thus, it appears that these metabolic decomposition reactions occur much more quickly than do the stenol to stanol

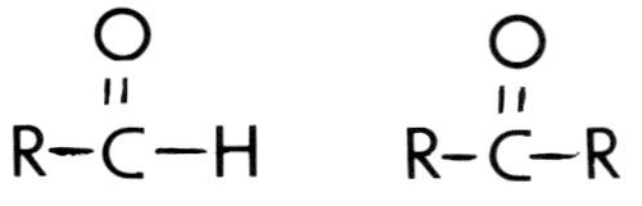

FIGURE 8.32. Structures of aldehydes and ketones.

reduction reactions in the water column. In contrast, stanols and stanones (believed to be microbially produced intermediates in the stenol to stanol reduction scheme) were isolated from the sediment-water interface of samples taken from the southwest African shelf and slope in an area of upwelling off Walvis Bay. Thus, the conditions found in these slope and shelf sediments were conducive to the rapid diagenesis of stenols, while the conditions in the water columns sampled were not. Several studies have also provided substantial evidence for a terrestrial contribution to sterols isolated from deep sea sediments. Further studies by Gagosian and Nigrelli have revealed correlations between total sterol concentrations and POC and PON, indicating that phytoplankton and zooplankton may be an important sterol source in seawater. Because of the low concentration of total and individual steroids in the water column and sediments, they are typically determined by gas chromatography-mass spectrometry after forming their trimethyl silyl ether derivatives from a hexane extraction.

## 8.3. Aldehydes and Ketones

Aldehydes and ketones both contain the carbonyl group (C=O) and are often referred to collectively as carbonyl compounds because it is the presence of this group that determines their chemistries. Aldehydes have the general formula RCHO, while ketones have the general formula RR′CO. The R and R′ groups may be aliphatic or aromatic (Figure 8.32).

Aldehydes and ketones have many similar chemical properties due to the carbonyl presence, however, the aldehydes have a hydrogen atom attached to the carbonyl carbon, whereas ketones have an organic carbon group instead. This structural difference results in a greater reactivity of aldehydes relative to ketones towards certain types of reactions including oxidations. Hence, aldehydes are quite easily oxidized whereas ketones are not.

Both aliphatic and aromatic aldehydes have been isolated from seawater and marine sediments. The aliphatic aldehydes are at least partially of biogenic origin, but some fraction appears to be generated from the photochemical oxidation of other biogenic compounds including humic substances. The aliphatic aldehydes most commonly found in seawater are $C_6$ to $C_{10}$ aldehydes at 10 to 100 ng $l^{-1}$, although trace amounts of formaldehyde ($CH_2O$) and acetaldehyde ($C_2H_4O$) have recently been reported. A recent geochemical study of long-chain *n*-aldehydes by Prahl and Pinto revealed the presence of a series of $C_{20}$ to $C_{32}$ aldehydes with a strong even-to-odd carbon preference in a mixture of solvent extractable lipids isolated from various Washington coastal sediments. The authors also noted a similar series of *n*-aldehydes intrinsic to the surface waxes of foliage and cuticular waxes of pollen from terrestrial vegetation indigenous to the Pacific Northwest. Thus, these long-chain *n*-aldehydes are introduced to the coastal sediments by erosion of terrestrial sources. In addition to the low molecular weight aldehydes known to be biosynthesized by marine organisms and the long-chain aldehydes contributed from terrestrially derived lipids, phenolic aldehydes have also been isolated from the oxidation products of marine particulate matter and are believed to be associated with terrestrially derived lignin.

Marine organisms are also known to biosynthesize ketones in trace amounts, but relatively little research has been directed towards unraveling the marine biogeochemistry of these compounds. Methyl ketones ($RCOCH_3$) are the most prevalent ketones reported to be present in marine sediments. Ikan and Simoneit have reported the presence of branched isoprenoid methyl ketones in Recent sediments and in deposits dating back to the Lower Cretaceous. These methyl ketones are most likely microbial oxidation products of phytol. Normal (unbranched) methyl ketones have been isolated from older sedimentary sequences, such as Black Sea Pleistocene sediments (by Simoneit), and are believed to be microbial oxidation products of either *n*-fatty acids or *n*-alkanes. Long-chain unsaturated ketones ($n$-$C_{37}$ to $n$-$C_{39}$) have been isolated from many marine sediments varying in age from Recent to Eocene and from coccolithophorid and microalgal sources. These long straight-chain ketones were separated into methyl and ethyl ketones and within each group only diunsaturated and triunsaturated ketones were found. Identical compounds were isolated from the marine coccolithophore *Emiliana huxleyi* with similar compound distributions to those of the sediments. Feeding these coccolithophores to copepods resulted in the assimilation of only a small proportion of the ketones, with the major fraction of these compounds being excreted in fecal pellets due to the inability of the copepod to digest these ketones.

The presence of long-chain ($n$-$C_{37}$ to $n$-$C_{39}$) unsaturated ketones has also been reported in several species of microalgae belonging to the class Prymnesiophyceae, which do not bear coccoliths. All were found to be methyl and

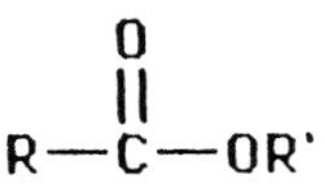

$$R{-}\overset{\displaystyle O}{\overset{\|}{C}}{-}OR'$$

FIGURE 8.33. Structure of wax ester.

ethyl ketones with two, three, or four positions of unsaturation. Thus, these long-chain unsaturated ketones may be fairly widely distributed in algae and may be useful as indicators of algal input.

## 8.4. Wax Esters

Organic esters are compounds of the formula shown in Figure 8.33 and are formed typically by the reaction of a carboxylic acid with an alcohol. Wax esters are simple esters of long-chain fatty alcohols (commonly $C_{14}$ to $C_{22}$, even-carbon number homologues) with long-chain fatty acids (usually even-carbon number homologues of $C_{14}$-$C_{24}$). They are important cell wall components of many higher plants and comprise a significant fraction of soil lipids, while in marine organisms they often function as cellular food reserves and are particularly abundant in zooplankton such as the calanoid copepods. In fact, wax esters have been reported to be major components of surface sea slicks reported in the North Sea and in the northwest Pacific Ocean and are associated with a high mortality of marine copepods.

Wax esters have been isolated from the uppermost 2 to 3 m of Recent marine sediments and in certain areas were found to comprise approximately 50% of the total sedimentary lipids. Studies of wax esters isolated from particulate matter from the Peru upwelling region indicate a quite variable diel flux. The wax esters isolated from daytime traps were significantly different from those isolated from nighttime traps. The night traps showed a 60 fold difference in wax ester flux, which may be due to differences in zooplankton composition. The most abundant chain lengths found were $C_{30}$, $C_{32}$, and $C_{34}$, composed of fatty acids and fatty alcohols of mainly $C_{14}$, $C_{16}$, and $C_{18}$.

## 8.5. Alcohols

Alcohols are compounds with a hydroxyl group (-OH) on a nonaromatic carbon atom. When the hydroxyl group is attached to an aromatic ring, strikingly different chemical properties are observed and the compound is referred to as a phenol.

Low molecular weight alcohols like methanol ($CH_3OH$) and ethanol ($CH_3CH_2OH$) are present in natural waters because of their extensive water

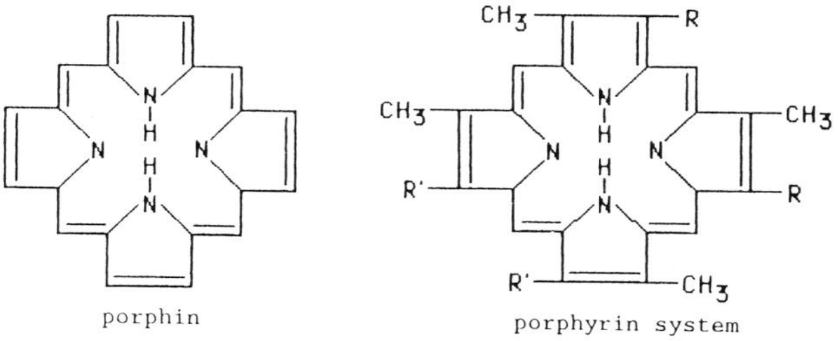

porphin

porphyrin system

FIGURE 8.34. Structure of porphyrins.

solubilities. More complex alcohols are also commonly found in natural waters and sediments. In an earlier section of this chapter dealing with steroids, the biogeochemistry of marine sterols (alcohol derivatives of steroids) is discussed. Cholesterol is an example of a sterol which is virtually ubiquitous in the marine environment and is associated with phytoplankton, zooplankton, and more complex marine animals. Isoprenoid alcohols, such as phytol, the alcoholic side-chain of chlorophyll-a, and diagenetic products of these compounds, are widespread in marine waters, particulate matter, living organisms, and sediments.

## 8.6. PORPHYRINS

The compound porphyrin, from which porphyrins derive their name, consists of four nitrogen-containing pyrrole rings (tetrapyrrole) linked by methine (-CH-) groups to form a resonating ring system (Figure 8.34).

The porphin ring system forms the nucleus of a group of physiologically important substances called porphyrins. Metal derivatives of porphyrins include the cytochromes hemoglobin and chlorophyll. In each of these metalloporphyrins, a metal ion takes the place of the two central hydrogen atoms in the porphyrin system shown in Figure 8.34 and is coordinated by the nitrogens which act as ligands. Cytochrome-450 is a chromoprotein which contains ferriporphyrin groups and functions as an enzyme which catalyzes metabolic oxidation reactions. Ferric cytochromes are widely distributed in plants and animals. Hemoglobin is also a ferriporphyrin-containing chromoprotein while chlorophylls are magnesioporphyrins.

In addition to the widespread occurrence of porphyrins in living matter and in recently deposited organic matter, porphyrins and other diagenetically related tetrapyrroles are consistently isolated from ancient deposits such as bitumen and petroleum. Instead of magnesium or iron, petroporphyrins typically contain vanadyl or nickel as the chelated cation. It was the presence

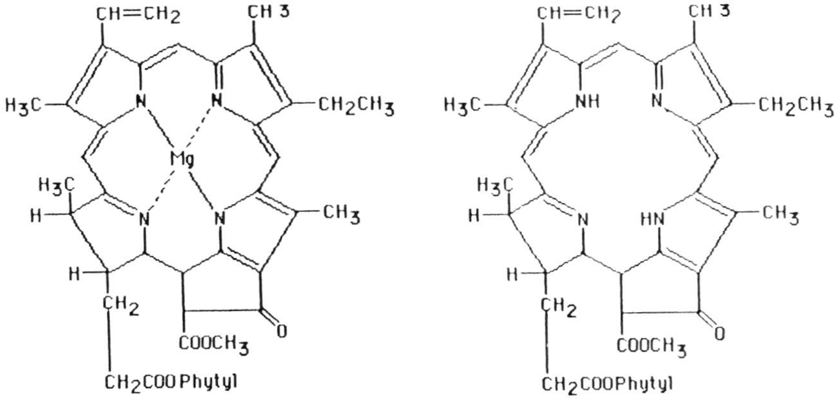

FIGURE 8.35. Structures of chlorophyll-a and pheophytin-a.

of porphyrins in a wide variety of bitumens and petroleums which led Treibs in the 1930s to conclude that these ancient remains of living systems, or petroporphyrins, were the diagenetic products of chlorophyll and established the concept of a biological origin for these fossil fuels. Indeed, porphyrins are proving to be useful geochemical markers because of their inherently stable aromatic nucleus which ensures their geologic persistence coupled to a variety of peripheral functional groups which respond quite readily to changes in the geochemical environment.

The diagenetic pathway from chlorophyll (Figure 8.35) to petroporphyrins has been studied in detail and is quite complex involving a number of steps. The initial step is believed to be the loss of magnesium, resulting in the formation of pheophytin, and occurs very rapidly either at the time of burial or within the water column itself. The loss of the chlorophyll phytol chain by hydrolysis is also a rapid process resulting in the product chlorophyllide (Figure 8.36), or pheophorbide if both magnesium and phytol are lost. Pheophytin-a and pheophorbide-a (Figure 8.36) are the primary chlorophyll-a derivatives deposited in marine and sedimentary environments. Various subsequent steps in the diagenetic process involve hydrogenations, dehydrogenations, reductions, decarboxylations, and chelation.

Chlorophyll and many of its tetrapyrrole derivatives have been studied in recent and in ancient marine environments. Chlorin (a dihydroporphyrin chlorophyll derivative) is widely distributed in recent marine sediments and is a known precursor of petroporphyrin. Chelation of nickel or vanadyl ion by petroporphyrins, which apparently stabilizes the porphyrin structure, re-

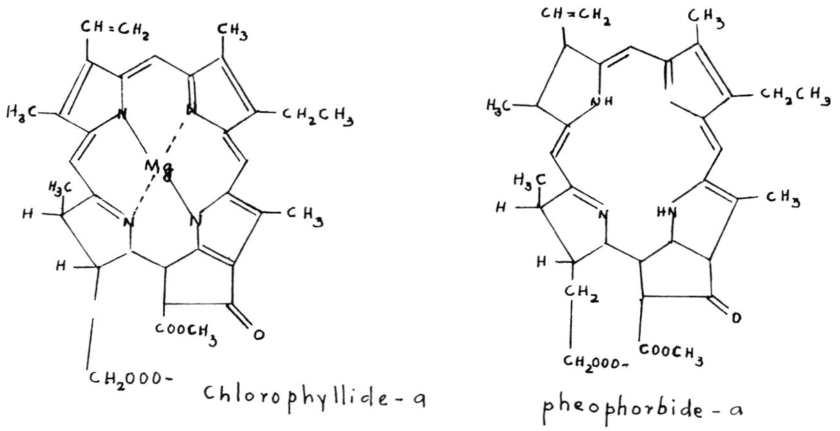

Chlorophyllide-a

pheophorbide-a

FIGURE 8.36. Structures of chlorophyllide-a and pheophorbide-a.

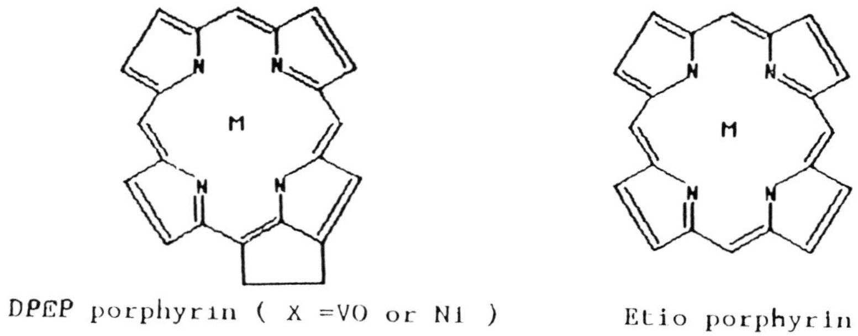

DPEP porphyrin ( X =VO or Ni )

Etio porphyrin

FIGURE 8.37. Structures of DPEP porphyrin and Etio porphyrin.

sults in what is commonly called the DPEP porphyrin series which can then be thermally rearranged to yield the Etio porphyrin series (Figure 8.37). As one might expect, tetrapyrrole systems are ubiquitous in marine sediments. However, although virtually ubiquitous in marine life, these compounds account for less than 1% of the DOC in the water column because of their extremely low solubilities.

## 8.7. Purines, Pyrimidines, and Nucleic Acids

Nucleic acids are high molecular weight components of nucleoproteins, so named because of their occurrence in cellular nuclei. They are composed of pentoses, phosphate groups, and purine and pyrimidine bases and are

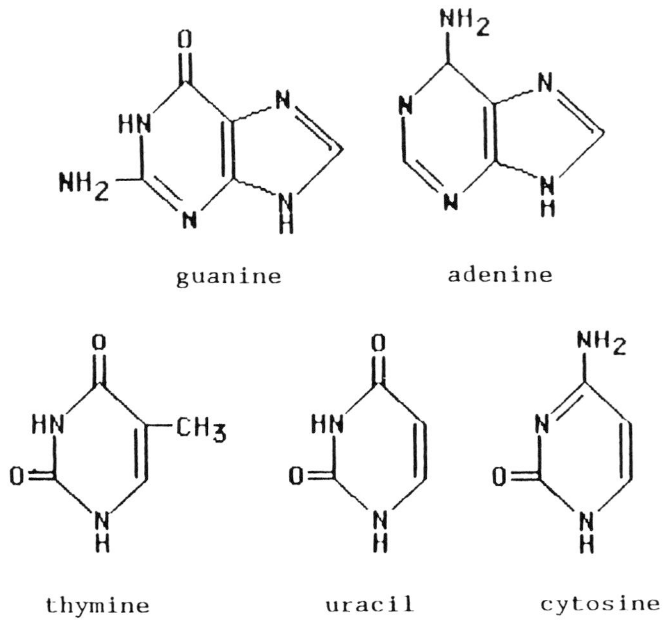

FIGURE 8.38. Structures of the major bases of nucleic acid.

considered to be the biological building blocks of plants and animals. Deoxyribonucleic acid (DNA) and ribonucleic acid (RNA) are responsible for the coding of each organism's characteristics. Genes consist almost exclusively of nucleoproteins. The structures of the major bases (purines) of nucleic acids are shown in Figure 8.38.

Although these bases have not been detected in marine waters or sediments, these remains of genetic material have been reported in lake sediments and are probably present at $\mu g\ l^{-1}$ concentration levels and should be adsorbed into particulate matter. DNA concentrations have been measured in marine waters and may be useful as a biomass indicator.

## 8.8. Organosulfur Compounds

A wide variety of organosulfur compounds are present in fresh water, estuarine, and marine waters and sediments which include sulfur oxidation states which range from $+6$ to $-2$. Although many of these compounds are present at only ng $l^{-1}$ levels, they are briefly discussed here because they are typically involved in a number of interesting biogeochemical processes. For instance, the speciation of many transition metals is affected by the presence of sulfur-containing ligands in seawater.

The sulfur content of living organisms is quite variable with a mean value of 0.5% for plants and 1.3% for animals. Cysteine, cystine, and methionine are common sulfur-containing amino acids and the production of other forms of organosulfur compounds by living organisms is well documented in the scientific literature. For example, methyl mercaptan ($CH_3SH$) is known to be produced by various bacteria and fungi from methionine.

Dimethylsulfide ($CH_3$-S-$CH_3$), or DMS for short, is a major contributor to the organosulfur population and to the odor of waters which contain decaying algae. The release of DMS from phytoplankton populations has been tied to metabolic secretion and algal cell senescence. Recent studies by J. Dacey indicate that the grazing of zooplankton on phytoplankton populations and subsequent cell rupture results in the release of approximately ten times as much DMS as is associated with normal algal metabolic release. Although DMS concentrations can be significantly greater, especially in anoxic waters, the average seawater concentration of DMS in open ocean waters has been estimated to be about 100 ng $l^{-1}$.

In addition to very volatile sulfur-containing organics, such as DMS, recent research has focused on less volatile compounds such as the thiols (R-SH). Thiols, like many other organosulfur species, arise in sediments due to biosynthesis (assimilatory sulfate reduction), microbial degradation of organic matter, and from nonmicrobial diagenetic processes due to the interaction of organic matter with $H_2S$ and with elemental sulfur. Thiols have been shown to be important ligands affecting metal speciation in natural waters and also function biochemically as coenzymes and as important components of macromolecular structures. In a recent study by Mopper and Taylor (U. Miami) on thiols in the intertidal sediments of Biscayne Bay, Florida, it was found that thiols were present at concentrations up to 100 $\mu M$ and over 30 different thiols were identified. This study also indicated that bound thiols which were released by the addition of specific cleaving reagents to sediments were present in at least 20 times greater concentrations as compared to free dissolved thiols or disulfides, implying that the binding of thiols to sediment particles may be an important geochemical process in the incorporation of sulfur into organic geopolymers such as humic substances, and ultimately fossil fuels.

## FURTHER READING

Degens, E. T. and K. Mopper, Factors Controlling the Distribution and Early Diagenesis of Organic Material in Marine Sediments, chap. 31, *Chemical Oceanography,* Vol. 6, 2nd ed., J. P. Riley and R. Chester, Eds., Academic Press, New York, 60-113 (1976).

Duursma, E. K., The Dissolved Organic Constituents of Sea Water, chap. 11, *Chemical Oceanography,* Vol. 1, 1st Ed., J. P. Riley and G. Skirrow, Eds., Academic Press, New York, 433-475 (1965).

Duursma, E. K. and R. Dawson, *Marine Organic Chemistry,* Elsevier, New York, (1981).

Lee, C. and S. G. Wakeham, Organic Matter in Sea-Water: Biogeochemical Processed, chap. 49, *Chemical Oceanography,* Vol. 9, 2nd ed., J. P. Riley, Ed., Academic Press, New York, 1-52 (1989).

Parsons, T. R., Particulate Organic Carbon in the Sea, chap. 13, *Chemical Oceanography,* Vol. 2, 2nd Ed., J. P. Riley and G. Skirrow, Eds., Academic Press, New York, 365-383 (1965).

Sackett, W. M., Suspended Matter in Sea-Water, chap. 37, *Chemical Oceanography,* Vol. 7, 2nd ed., J. P. Riley and R. Chester, Eds., Academic Press, New York, 127-172 (1978).

# 9 Table of Contents

**Processes in the Oceans** .......... 417
1. *Atmospheric Chemistry* .......... 417
    1.1. *Greenhouse Gases* .......... 430
    1.2. *Loss of Ozone* .......... 433
2. *Primary Productivity* .......... 442
    2.1. *Phytoplankton Production* .......... 445
    2.2. *Standing Crop or Biomass* .......... 445
    2.3. *$O_2$ Liberation Method of Measurement* .......... 446
    2.4. *Uptake of $CO_2$ Method of Measurement* .......... 446
    2.5. *Growth and Distribution of Phytoplankton in the Sea* .......... 449
3. *Hydrothermal Vent Chemistry* .......... 453
4. *Photochemical Processes in Seawater* .......... 468
    4.1. *Principles* .......... 469
5. *Anoxic Waters* .......... 486
    5.1. *The Black Sea* .......... 489
    5.2. *Framvaren Fjord* .......... 491
    5.3. *Cariaco Trench* .......... 495
    5.4. *Saanich Inlet* .......... 506
*Further Reading* .......... 508

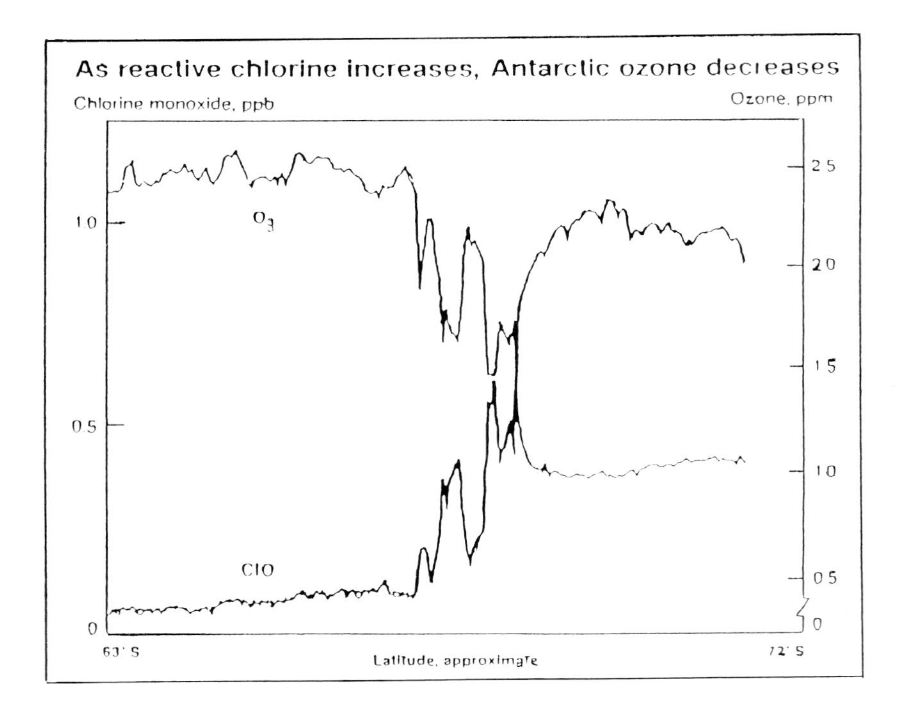


The relationship between $O_3$ and ClO in the Antarctic.

# 9 Processes in the Oceans

## 1. ATMOSPHERIC CHEMISTRY

In recent years there has been an increasing interest in atmospheric chemistry. Since the oceans are in intimate contact with the atmosphere and may act as a source or sink for gases, it is appropriate to briefly examine this area of science. The early interest in the atmosphere was related to the formation of photochemical smog in various cities. Smog is formed by complicated interactions of unburned hydrocarbons from automobiles and power plants and nitrogen oxides. The sun provides the necessary energy to furnish reactive species. More recently, interest in atmospheric chemistry has focused on the formation of acid rain ($HNO_3$ and $H_2SO_4$) from the oxidation of $NO_x$ and $SO_2$ gases that result from the oxidation of fossil fuels. The decrease in the ozone layer due to the use of chloro-fluoro hydrocarbons (CFC) has also provided new interest in atmospheric chemistry.

Interest has also been focused on the increasing concentration of gases that can absorb infrared (IR) energy. These gases ($CO_2$, $CH_4$, etc.) contribute to the warming of the atmosphere or the so-called "greenhouse" effect. Since many of the chemical reactions of interest occur near the surface of the earth, it is important to briefly examine the various layers of the atmosphere. The

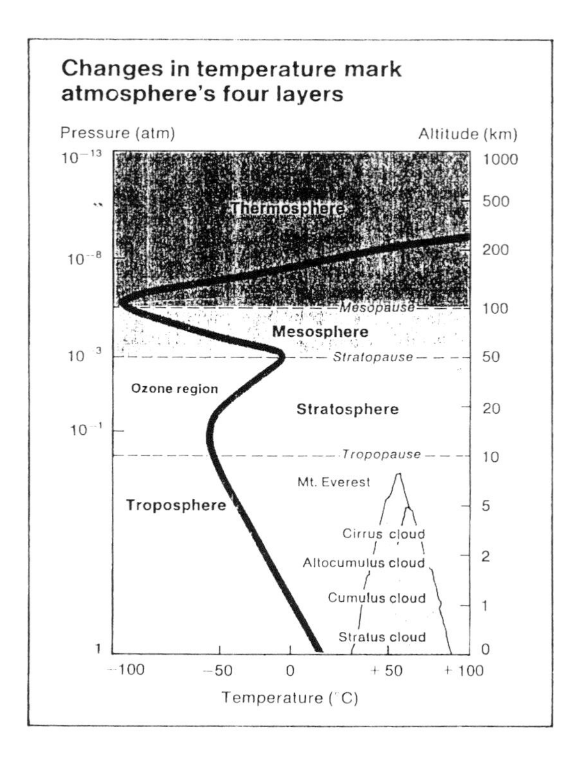


FIGURE 9.1. Changes in the temperature in the layers of the atmosphere.

atmosphere can be divided into four layers (Figure 9.1):* the troposphere (0 to 10 km), the stratosphere (10 to 50 km), the mesosphere (50 to 100 km), and the thermosphere (100 to 1000 km). These layers are marked by changes in the temperature of the atmosphere. From the surface to 10 km, the temperature decreases to a minimum at the tropopause. In the stratosphere, the temperature increases to a maximum at the stratopause. In the mesosphere, the temperature decreases to a minimum at the mesopause. In the thermosphere, the temperature increases again as the atmosphere diminishes.

Chemical reactions between the major constituents of the atmosphere occur at slow rates. The formation of active species occurs due to the influence of light and causes a number of rapid reaction chains. The chemistry of the lower atmosphere is often referred to as tropospheric photochemistry since it is driven by the absorption of photons.

* Figures 1.1 to 1.5, 1.7, and 1.8 from Chameides and Davis, 1982.

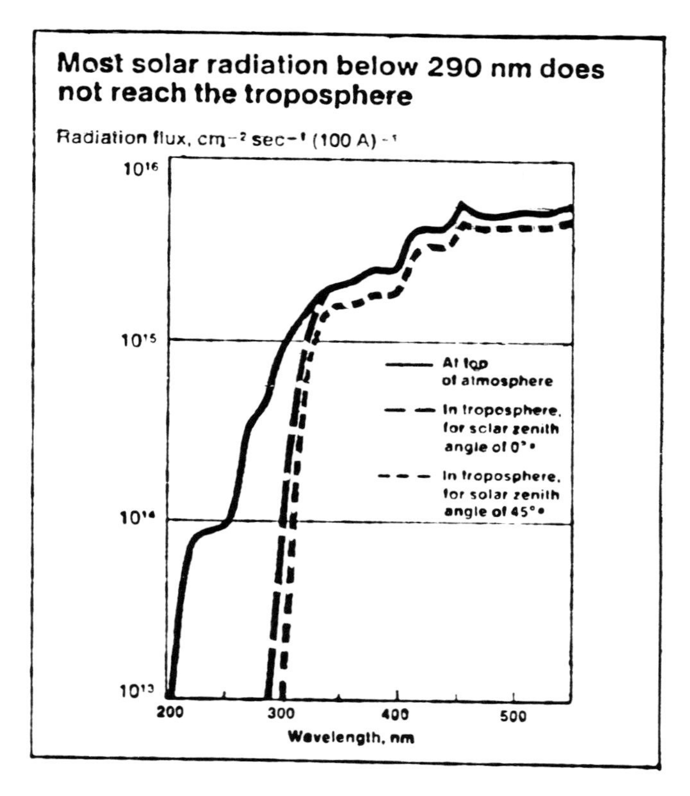


FIGURE 9.2. The solar radiation that reaches various layers in the troposphere.

The flux of solar photons as a function of wavelength is shown in Figure 9.2. Most of the radiation below 290 nm does not reach the troposphere. Photons with wavelengths shorter than 240 nm are absorbed by $O_2$ and $N_2$ molecules in the thermosphere. This results in the formation of ozone, $O_3$

$$O_2 + h\nu \rightarrow 2O \tag{1}$$

$$O_2 + O + M \rightarrow O_3 + M \tag{2}$$

where M represents another molecule of oxygen or nitrogen that is unchanged in the reaction. The concentration of $O_3$ formed in this manner peaks in the lower stratosphere (Figure 9.3). The ultraviolet radiation can also stimulate dissociation of $O_3$

$$O_3 + h\nu \rightarrow O_2 + O(^1D) \tag{3}$$

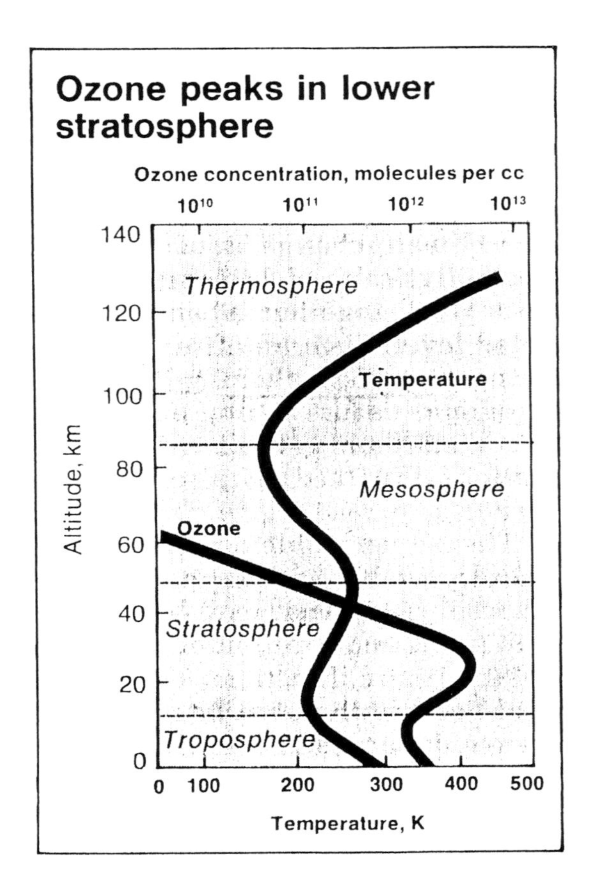


FIGURE 9.3. The concentration of ozone in various layers of the atmosphere.

This reaction is responsible for the absorption of light between 240 to 300 nm. The electronically excited oxygen $O(^1D)$, formed in the dissociation of $O_3$, is responsible for the formation of odd nitrogen radicals and odd hydrogen radicals

$$N_2O + O(^1D) \rightarrow 2NO \quad (4)$$

$$H_2O + O(^1D) \rightarrow 2OH \quad (5)$$

More will be said about these reactions later. The $O_3$ that is broken apart by photons will form oxygen atoms that will quickly reform $O_3$ by reactions with

$O_2$. This leads to a steady state buildup of $O_3$. The $O_3$ dies off when it collides with an O atom forming two oxygen molecules

$$O_3 + O \rightarrow 2O_2 \tag{6}$$

Other reactions that destroy $O_3$ in the stratosphere (such as reactions with atomic Cl) will be discussed later.

Since the light reaching the troposphere has wavelengths above 300 nm, there is not enough energy available to break O-O bonds which have a bond strength of 120 kcal mol$^{-1}$. This means that $O_2$ cannot oxidize reduced gases in the troposphere. At one time it was thought that $O_3$ and $H_2O_2$ were oxidizers in the troposphere. Now the OH radical is thought to be the oxidizer.

The production of OH radicals is initiated by the photolysis of $O_3$. Ozone is present in the troposphere at concentrations from 10 to 100 ppb and has a bond energy of 26 kcal mol$^{-1}$. Solar photons with wavelengths between 315 to 1200 nm can dissociate $O_3$ and produce an oxygen atom in its ground electronic state

$$O_3 + h\nu\ (1200 > \lambda > 315\ \text{nm}) \rightarrow O_2 + O(^3P) \tag{7}$$

The $O(^3P)$ atom rapidly transforms ozone by reaction with $O_2$ in a three body reaction

$$O(^3P) + O_2 + M \rightarrow O_3 + M \tag{8}$$

where M is $N_2$ or $O_2$. This sequence results in no net chemical effect. When ozone reacts with wavelengths shorter than 315 nm, an electronically excited oxygen atom is produced

$$O_3 + h\nu\ (\lambda < 315\ \text{nm}) \rightarrow O(^1D) + O_2 \tag{9}$$

The $O(^1D)$ to $O(^3P)$ transition is forbidden and results in a relatively long lifetime for $O(^1D)$ of 100 s. The $O(^1D)$ most often collides with $N_2$ or $O_2$ (M)

$$O(^1D) + M \rightarrow O(^3P) + M \tag{10}$$

This ultimately reacts with $O_2$ to form $O_3$ which also results in no net chemical

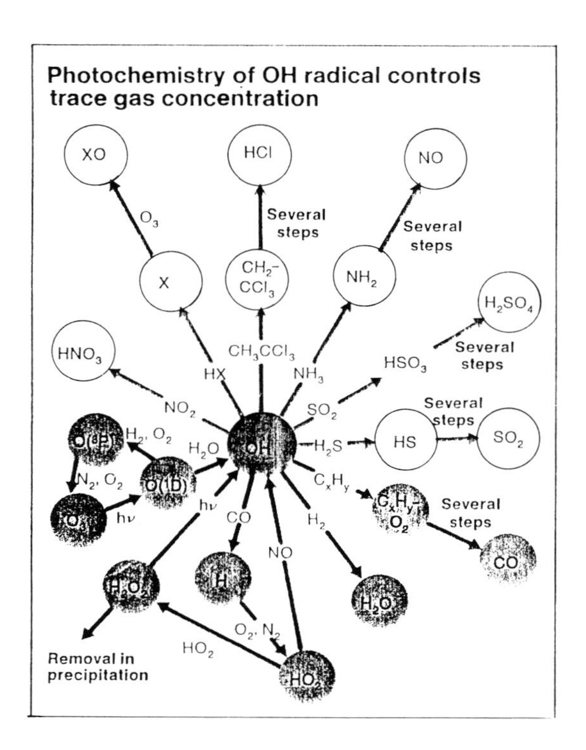


FIGURE 9.4. The photochemical control of OH radicals on trace gases.

change. Occasionally $O(^1D)$ collides with water to generate two hydroxyl radicals

$$O(^1D) + H_2O \rightarrow 2OH \tag{11}$$

This reaction sequence is the primary source of hydroxyl radicals in the troposphere. The OH radical formed in this manner is thought to control the concentration of many trace gases (Figure 9.4).

The removal of OH from the atmosphere results from the reactions

$$CO + OH \rightarrow CO_2 + H \tag{12}$$

$$CH_4 + OH \rightarrow CH_3 + H_2O \tag{13}$$

Both the H and the $CH_3$ radicals combine rapidly with $O_2$ to form hydroperoxyl

($HO_2$) and methyperoxyl ($CH_3O_2$) radicals. The hydroperoxyl radical, however, can regenerate OH radicals.

$$HO_2 + NO \rightarrow NO_2 + OH \quad (14)$$

$$HO_2 + O_3 \rightarrow 2O_2 + OH \quad (15)$$

It can also lead to the chain termination

$$HO_2 + OH \rightarrow H_2O + O_2 \quad (16)$$

$$HO_2 + HO_2 \rightarrow H_2O_2 + O_2 \quad (17)$$

The hydrogen peroxide ($H_2O_2$) is removed in rain. The chemistry of the methyperoxyl radical ($CH_3O_2$) is quite complicated and not all of its reactions are known at present. Mathematical models have been used to simulate these reactions and the average OH concentration is about 2 to 20 $\times$ $10^5$ radicals per $cm^3$ with the highest levels in the tropics. Model calculations predict that about 20% more OH radicals should be found in the Southern Hemisphere. This is caused by the higher CO concentrations in the Northern Hemisphere. Direct measurements of OH radicals have not been made at present.

The distribution of gases in the atmosphere is related to their lifetimes. The atmospheric lifetimes vary from seconds to hundreds of years (Figure 9.5). These lifetimes can be compared to interhemispheric mixing times of years to intrahemispheric mixing times of months. Gases such as methane and carbon monoxide that have continental sources will have different distributions between hemispheres due to their different lifetimes (Figure 9.6). The long lifetime of methane (7 years) results in a nearly uniform distribution between hemispheres, while the more reactive carbon monoxide (65 d) is more concentrated near its sources in the north. The slow movement of gases between hemispheres is caused by the intertropical convergence zone (ITCZ). The ITCZ is caused by air rising near the equator. This prevents mixing across the two hemispheres and results in interhemispheric mixing times of 1 to 2 years. The fast intrahemispheric mixing was recently demonstrated by following the movement of dust and particles from the El Chichon volcanic eruption in 1982 (Figure 9.7). The eruption occurred April 4th and made the movement around the northern hemisphere by April 25th.

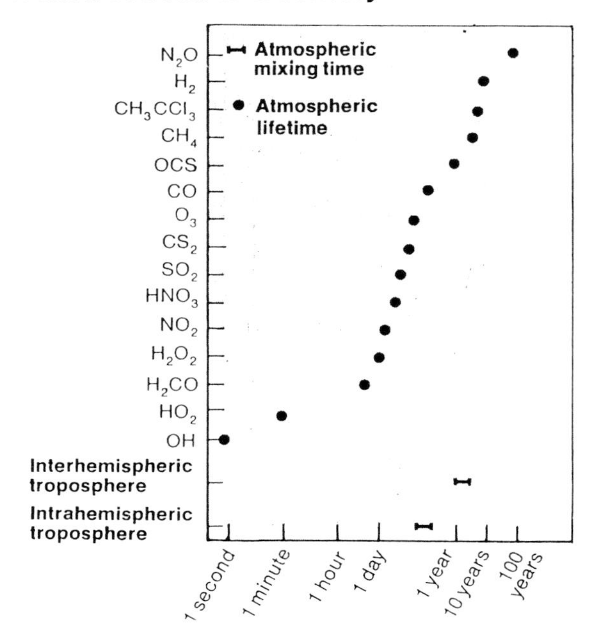


FIGURE 9.5. The atmospheric lifetimes of trace gases.

The concentrations of trace gases in the atmosphere are controlled by a complex combination of processes. Most trace gases are at higher concentrations than would be expected based on thermodynamic calculations. The principal sources of these gases are:

1. Biogenic ($CH_4$, $NH_3$, $N_2O$, $H_2$, $CS_2$, OCS)
2. Photochemical (CO, $O_3$, $NO_2$, $HNO_3$, $H_2$, OH, $HO_2$, $H_2O_2$, $H_2CO$)
3. Lightning (NO, $NO_2$)
4. Volcanic ($SO_2$)

The sinks of these gases are largely photochemical, with the exception of $H_2O_2$, $HNO_3$ and $H_2SO_4$, which are lost due to rainout.

The reactive nitrogen species in the lower atmosphere are NO, $NO_2$, and $HNO_3$. These species are coupled by a series of reactions that cycle the various species. The various reactions of these species are shown in Figure 9.8. The reaction of nitric oxide with hydroperoxyl radicals is of special interest

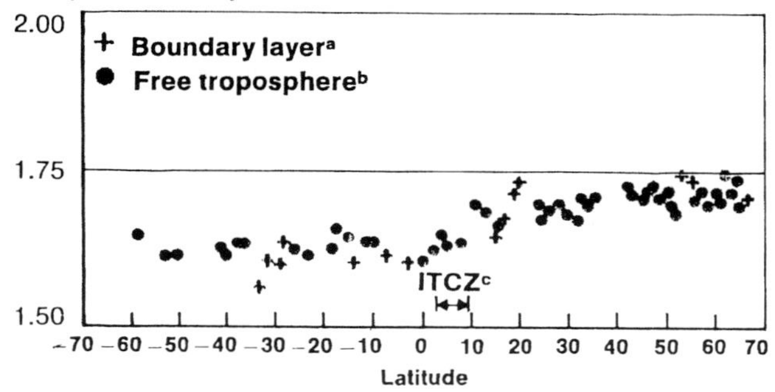


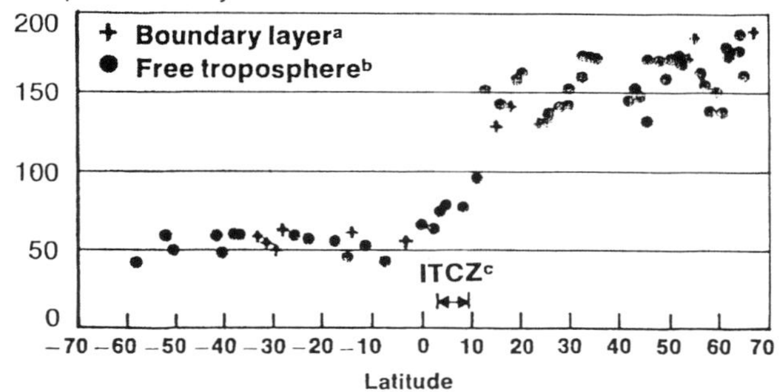


FIGURE 9.6. The distribution of methane (top) and carbon monoxide (bottom) between the hemispheres.

$$NO + HO_2 \rightarrow NO_2 + OH \qquad (18)$$

This reaction regenerates OH from $HO_2$. It also leads to the generation of ozone by the following

$$NO_2 + h\nu \rightarrow NO + O \qquad (19)$$

$$O + O_2 + M \rightarrow O_3 + M \qquad (20)$$

where $M = N_2$ or $O_2$. As with CO, large amounts of NO are produced from the burning of fossil fuels (e.g., in automobiles). Although CO causes the OH levels to decrease, NO can enhance the OH levels especially in remote areas. The nitrogen oxides are rapidly removed from the atmosphere as $HNO_3$,

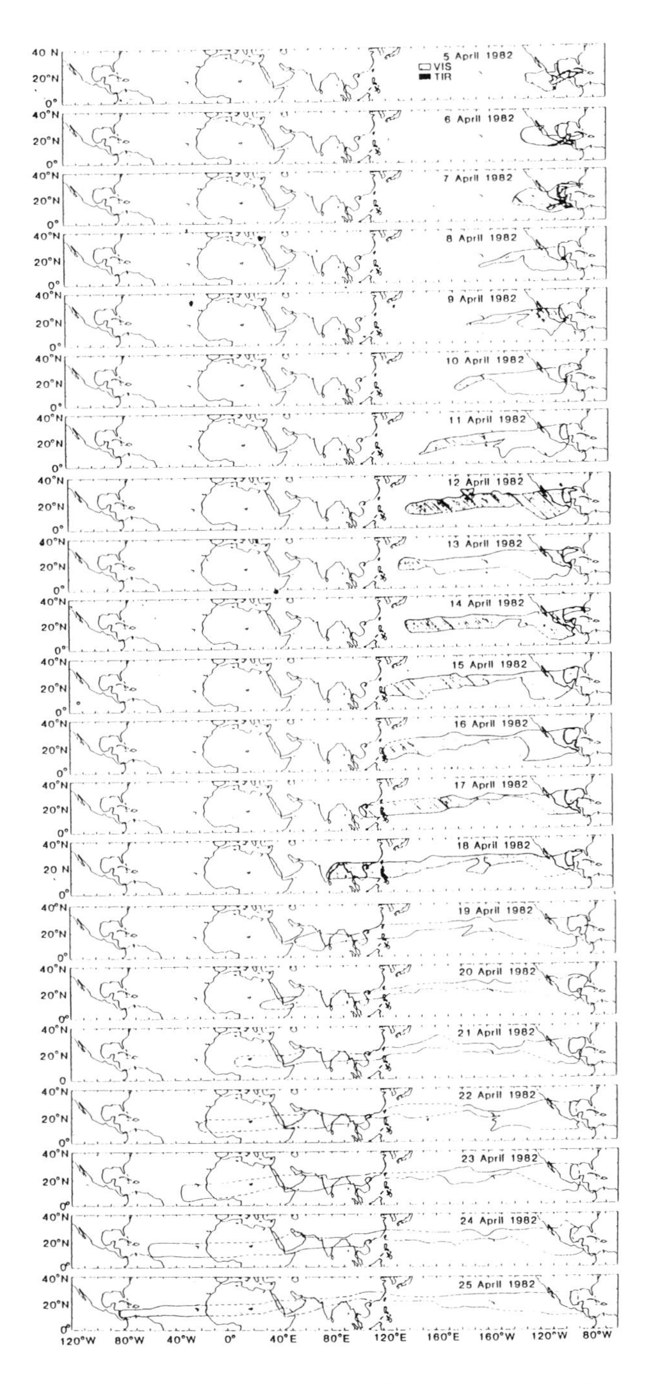


FIGURE 9.7. The movement of dust and particles in the northern hemisphere after the El Chichon volcano eruption.

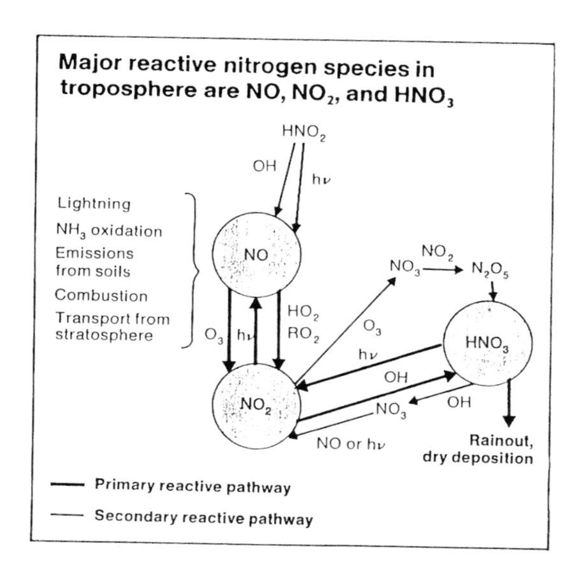


FIGURE 9.8. The major reactive nitrogen species in the troposphere.

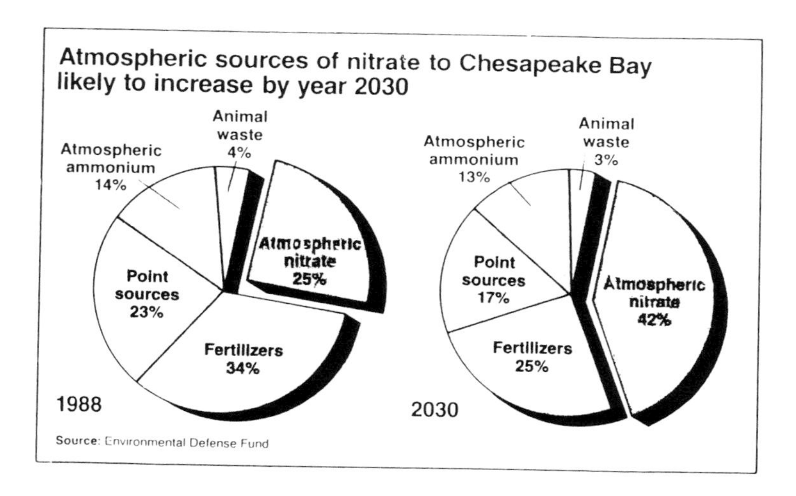


FIGURE 9.9. The atmospheric sources of nitrate to the Chesapeake Bay.

which is soluble in rainwater. It also can be attached to aerosols and particles and removed as dry deposition. The $HNO_3$ in rainwater is one of the components of acid rain. Although the importance of $HNO_3$ as a component of acid rain has been known to affect the pH of rain, recent work (Figure 9.9) indicates that atmospheric sources of nitrate may also be contributing to the eutrophication of the Chesapeake Bay.

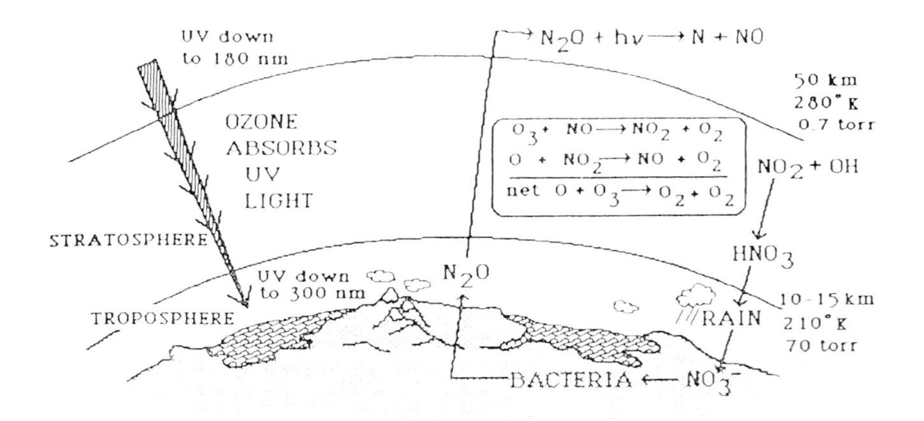


FIGURE 9.10. The effect of nitrogen oxides on the concentration of stratospheric ozone.

The $N_2O$ in the atmosphere can also reduce the stratospheric ozone (Figure 9.10).

Bacterially produced $N_2O$ can break down at high altitudes by the absorption of light

$$N_2O + h\nu \rightarrow N + NO \qquad (21)$$

The NO formed can react with $O_3$ in a chain reaction

$$O_3 + NO \rightarrow NO_2 + O \qquad (22)$$

$$O + NO_2 \rightarrow NO + O_2 \qquad (23)$$

the net reaction being

$$O + O_3 \rightarrow 2O_2 \qquad (24)$$

The concern that sulfur species are causing acid rain has led to an interest in the sulfur cycle. Biogenic processes emit a number of sulfur species ($H_2S$, $CH_3SCH_3$, OCS). These reduced species are oxidized to $SO_2$ by OH radicals. $SO_2$ can also be directly injected into the atmosphere as a by-product of the oxidation of fossil fuels. The lifetime of $SO_2$ is between a few days to a month. Removal occurs in rainout, and wet and dry deposition (Figure 9.11)

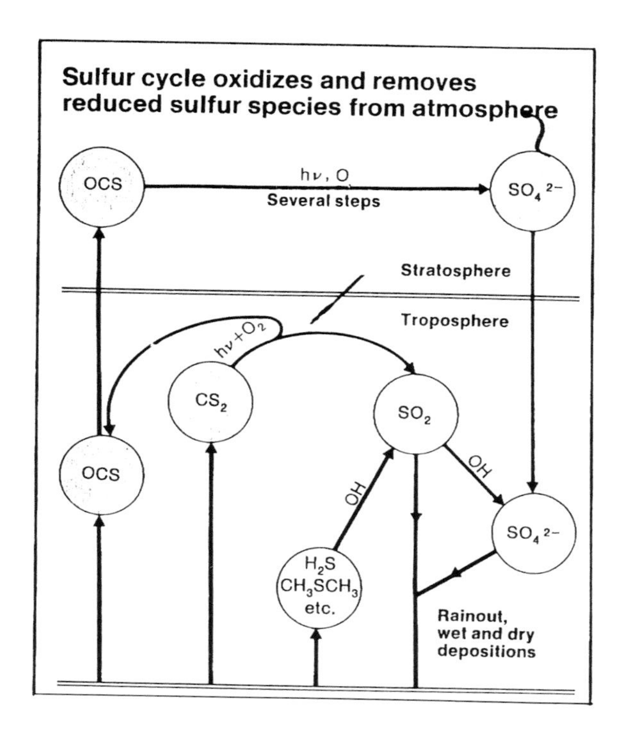


FIGURE 9.11. The sulfur cycle in the atmosphere.

after the oxidation of $SO_2$ to $H_2SO_4$. The $H_2SO_4$ is incorporated into cloud droplets and aerosols. The oxidation can occur in the gas phase, the solution phase, or on particles (Figure 9.12).

The term acid rain was first used by Angus Smith to describe the effect industrial emissions had on the precipitation in Great Britain. The pH of water in equilibrium with atmospheric $CO_2$ has a value of 5.6. In noncontaminated areas the pH is closer to 5.0 because of natural levels of acids. In most urban areas the pH is normally lower than 5.0. In Europe and North America, 90% of the sulfur comes from the burning of fossil fuels. The lower pH of the rain (pH = 4.6 to 4.7) is normally attributed to the concentrations of $HNO_3$ and $H_2SO_4$ formed by the oxidation of $NO_x$ and $SO_2$. Organic acids may also be important components of acidic rain especially in remote areas. A number of areas are sensitive to acid rain. These include many northern lakes that have low alkalinity ($<50\ \mu M$) and vast forest areas. The effects of acid rain have led to lower alkalinity to $Mg^{2+}$ and $Ca^{2+}$ ratios. Fish can tolerate values of pH as low as 5.5. Juvenile fish and many organisms can be affected at pH above 5.5. This can cause starvation of predatory fish. Acidic streams

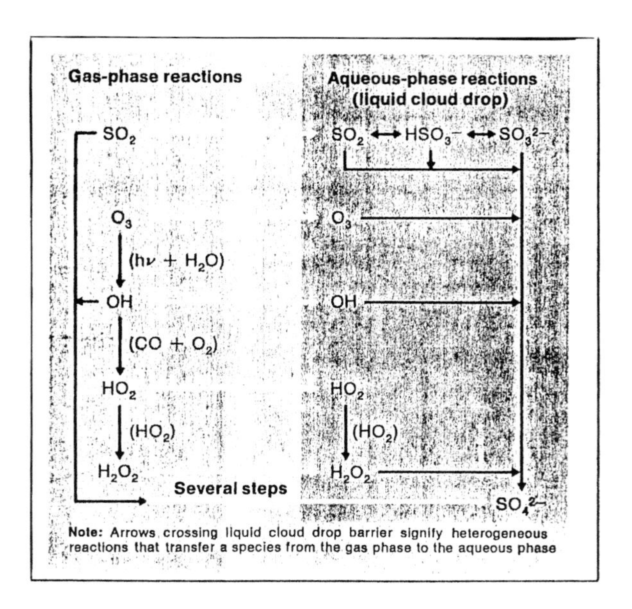


FIGURE 9.12. The gas and liquid phase reactions of sulfur in the atmosphere.

have frequently had fish kills due to pulses of acid waters during the spring snow melt. The lower pH of lakes and streams can also cause aluminum to be released from the sediments. Ionic aluminum, $Al^{3+}$, appears to be toxic while the hydrolyzed forms are nontoxic ($Al(OH)^{2+}$, $Al(OH)_2^+$, etc.). Recent studies have shown that exchange reactions of $H^+$ with sediments may lessen the effect of acid rain on a lake. The recovery of a lake may be fast if the basic components of the soil are not lost. It is unlikely, however, that the original pH will be reached for years and widespread stocking will be necessary to bring back game fish and the resultant food chains.

## 1.1. Greenhouse Gases

The likelihood that $CO_2$ increases will cause global climate changes is now well known. Less is known, however, of the effects of other trace gases on the climate. Gases such as $CH_4$, $O_3$, $N_2O$, and chlorofluorocarbons (CFCs) such as $CCl_3F$ (CFC-11) and $CCl_2F_2$ (CFC-12) can contribute to the increasing temperature of the troposphere. Solar radiation is absorbed by the atmosphere providing energy for many processes. The amount of energy that affects the earth is shown in Figure 9.13. For the global climate to be balanced, the absorbed radiation must be equal to the outgoing radiation. The trapping of

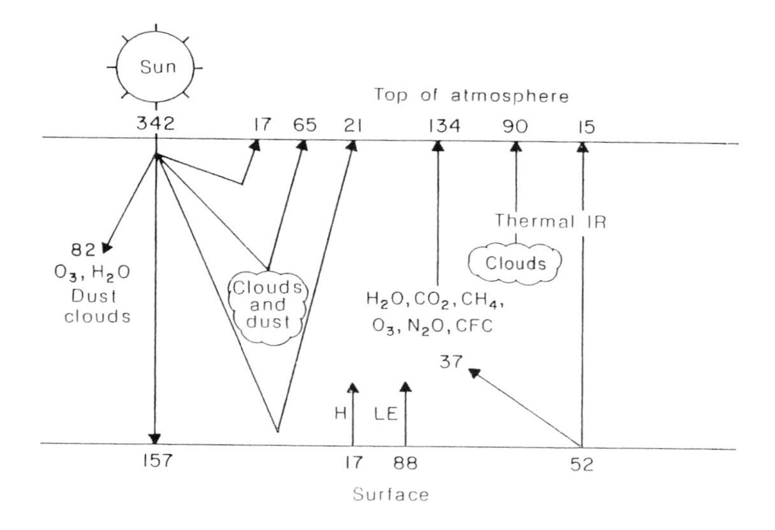


FIGURE 9.13. The amount of energy that affects the earth.

some of the thermal radiation by particles and molecules increases the surface temperature by 10 to 15°C compared to what it would be without these molecules. This process of absorption of (IR) energy radiated from the earth is called the ''greenhouse'' effect. Clouds and water vapor are dominant contributors to this process. Other atmospheric gases, although present in small amounts, can also contribute to this trapping of thermal radiation. Without these active constituents the earth would lose thermal radiation as a black body (~387 W $m^{-2}$). The actual radiative flux at the top of the atmosphere is 239 W $m^{-2}$, so ~148 W $m^{-2}$ (38%) is trapped. Changes in the trapping of 1 W $m^{-2}$ (0.3%) can change the balance sufficiently to change the climate. The current trapping by trace constituents of the atmosphere is given in Table 9.1. The preindustrial estimates of the trace gas concentrations and the change in thermal trapping is given in Table 9.2.

One would expect from comparing the two tables that there has been approximately a 200% increase in the $\Delta Q$ (change in thermal radiation) due to industrial processes. It is difficult, however, to translate the increases in $\Delta Q$ to an increase in the global temperature. Although some measurements (Figure 9.14) indicate that the temperature has increased, it is hard to attach a reliable estimate of changes in $\Delta Q$ with changes in $\Delta T$.

We discussed earlier the documented increase of $CO_2$ over the last 40 years. Recent work has also demonstrated that $CH_4$ has increased dramatically as well (Figure 9.15). Recently various workers have attempted to estimate

**TABLE 9.1**
**Trapping of IR Radiation by Trace Gases ($\Delta Q$, W $m^{-2}$)**

| Gas | Present level | Present $\Delta Q$ |
|---|---|---|
| $CO_2$ | 345 ppm | 2.0 |
| $CH_4$ | 1.7 ppm | 1.7 |
| $O_3$ | 10—100 ppb | 1.3 |
| $N_2O$ | 340 ppb | 1.3 |
| CFC-11[a] | 0.22 ppb | 0.06 |
| CFC-12[b] | 0.38 ppb | 0.12 |
| | | 6.5 |

[a] $CCl_3F$
[b] $CCl_2F_2$

**TABLE 9.2**
**Preindustrial Trapping of IR Radiation of Trace Gases**

| Gas | Past level | Past $\Delta Q$ |
|---|---|---|
| $CO_2$ | 275 ppm | 1.3 |
| $CH_4$ | 0.7 ppm | 0.6 |
| $O_3$ | 0—25% less | 0—0.2 |
| $N_2O$ | 285 ppb | 0.05 |
| CFC-11 | 0 | 0 |
| CFC-12 | 0 | 0 |
| | | 2.2 |

the long term effects of the addition of greenhouse gases to the atmosphere. The estimated increases of the gases in the year 2050 are given in Table 9.3. A number of workers have attempted to estimate the effect the projected increase of trace gases will have on the temperature in the next century. The models used to make these estimates are quite complicated and may not be reliable; however, at present this is the best one can do. The increases in temperature estimated by assuming various increases in trace gases are given in Table 9.4. Although $CO_2$ makes up to 75% of the increase, other gases are important and will become more important in the future (Plate 23)*.

* Plate 23 follows page 48.

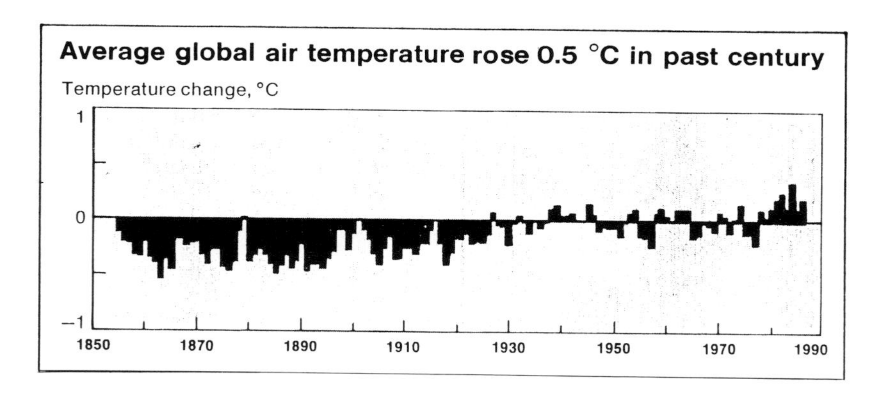


FIGURE 9.14. The average global air temperature over the last century.

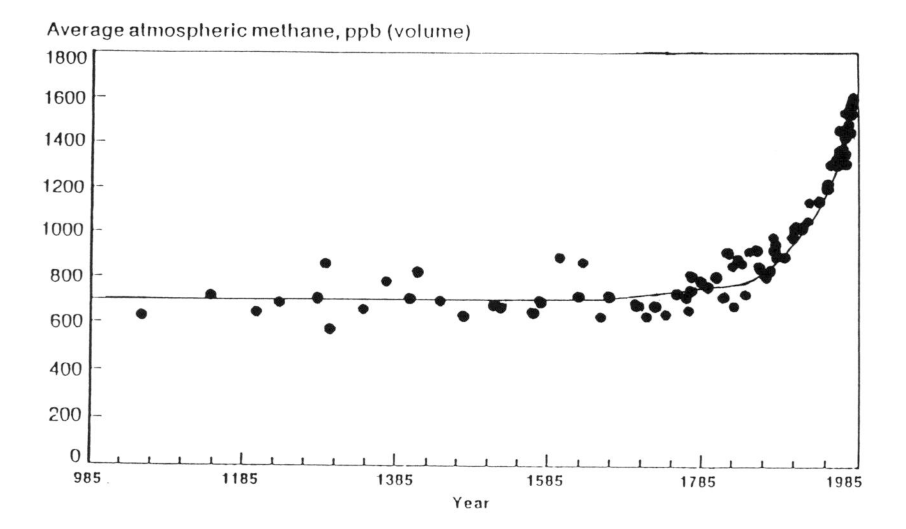


FIGURE 9.15. The increase in the atmospheric methane concentrations.

## 1.2. Loss of Ozone

It has been known for some time that gases produced by man can contribute to the loss of ozone in the stratosphere. The effect of $N_2O$ was discussed earlier (Figure 9.10). Chlorine can also reduce $O_3$ by the following reactions (Figure 9.16)

$$O_3 + Cl \rightarrow ClO + O_2 \tag{25}$$

$$O + ClO \rightarrow Cl + O_2 \tag{26}$$

**TABLE 9.3**
**Trapping of IR Radiation by Trace Gases in 2050**

| Gas | 2050 Levels | ΔQ |
|---|---|---|
| $O_2$ | 440—600 ppm | 0.9—3.2 |
| $CH_4$ | 2.1—4.0 ppm | 0.2—0.9 |
| $O_3$ | 15—50% more | 0.2—0.6 |
| $N_2O$ | 350—450 ppb | 0.1—0.3 |
| CFC-11 | 0.7—3.0 ppb | 0.23—0.7 |
| CFC-12 | 2.0—4.8 ppb | 0.6—1.4 |
| | | 2.2—7.2 |

**TABLE 9.4**
**Estimated Increase in Temperature due to Increases in the Concentration of Trace Gases**

| Gas | Assumed increase | Temp. change (°C) |
|---|---|---|
| $CO_2$ | 2 | 2.6 |
| $H_2O$ | 2 | 0.65 |
| $O_3$ | 0.75 | −0.4 |
| $N_2O$ | 2 | 0.65 |
| $NH_3$ | 2 | 0.12 |
| $HNO_3$ | 2 | 0.08 |
| $CH_4$ | 2 | 0.26 |
| $SO_2$ | 2 | 0.02 |
| CFCs | 20 | 0.65 |
| | | 4.63 |

The net reaction is

$$O + O_3 \rightarrow 2O_2 \tag{27}$$

Free atomic chlorine is produced by the photodissociation of chlorofluorocarbons (CFC-11 and CFC-12)

$$CCl_2F_2 + h\nu \rightarrow CClF_2 + Cl \tag{28}$$

$$CCl_3F + h\nu \rightarrow CCl_2F + Cl \tag{29}$$

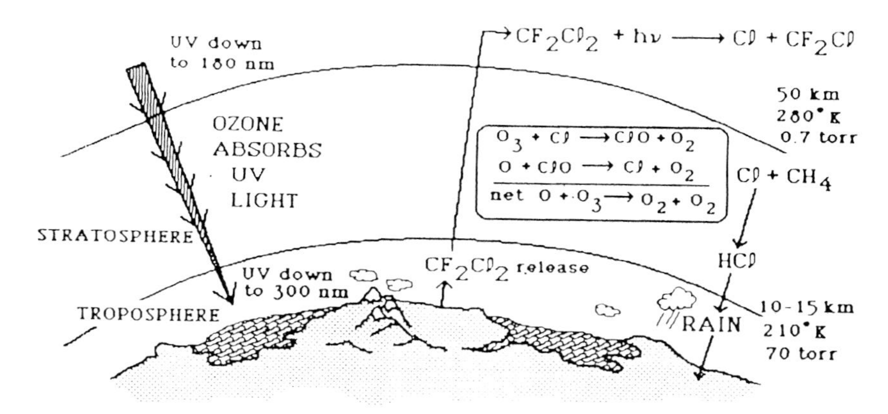


CHLORINE ALSO CAN REDUCE STRATOSPHERIC OZONE

FIGURE 9.16. The effect of chlorine on the concentration of ozone in the stratosphere.

These chlorofluorocarbons reach the high altitudes where light of sufficient energy can cause dissociation due to their long lifetimes (CFC-11), $\tau = 75$ years and CFC-12, $\tau = 110$ years). As mentioned earlier, NO and OH radicals can also reduce ozone

$$NO + O_2 \rightarrow NO_2 + O_2 \tag{30}$$

$$NO_2 + O \rightarrow NO + O_2 \tag{31}$$

$$OH + O_3 \rightarrow HO_2 + O_2 \tag{32}$$

$$HO_2 + O \rightarrow OH + O_2 \tag{33}$$

The net effect of both reaction paths is the destruction of two ozone molecules (the primary reaction with $O_3$ and the destruction of O that would have formed $O_3$).

The reactive species are removed by reactions that produce stable species or water soluble species that are rained out

$$Cl + CH_4 \rightarrow HCl + CH_3 \tag{34}$$

$$Cl + HO_2 \rightarrow HCl + O_2 \tag{35}$$

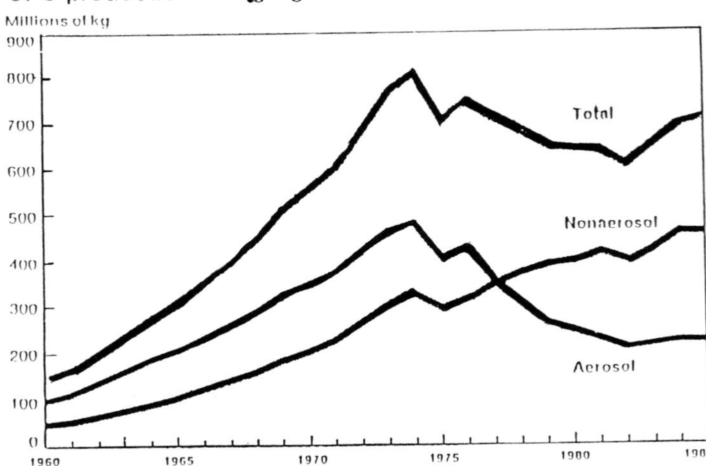


FIGURE 9.17. The production of cloro-fluoro-carbons (CFC) over the last 25 years.

$$NO_2 + OH \rightarrow HNO_3 \tag{36}$$

$$OH + HO_2 \rightarrow H_2O + O_2 \tag{37}$$

Photodissociation or other free radical reactions can reverse these reactions

$$OH + HCl \rightarrow H_2O + Cl \tag{38}$$

$$HNO_3 + h\nu \rightarrow OH + NO_2 \tag{39}$$

$$HO_2 + NO \rightarrow OH + NO_2 \tag{40}$$

$$OH + CO \rightarrow CO_2 + H \tag{41}$$

With increasing CFC production (Figure 9.17), one would expect that the formation of Cl by photodissociation would be more important in removing $O_3$ in future years. Even though the aerosol usage of CFCs has decreased, the nonaerosol usage has increased. The increase of CFCs, methylchloroform, carbon tetrachloride and nitrous oxides since 1978 has been documented by direct measurements (Figure 9.18).

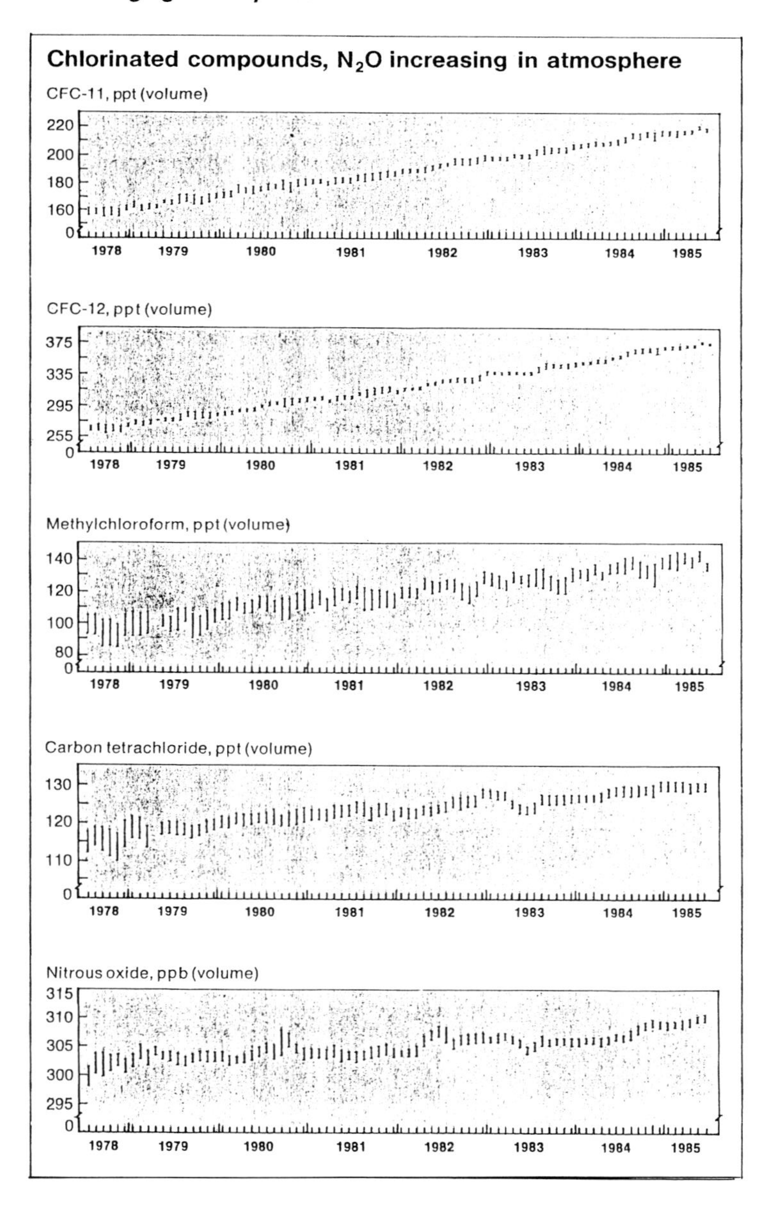


FIGURE 9.18. The increase in the concentration of chlorinated compounds and nitrogen oxide in the atmosphere.

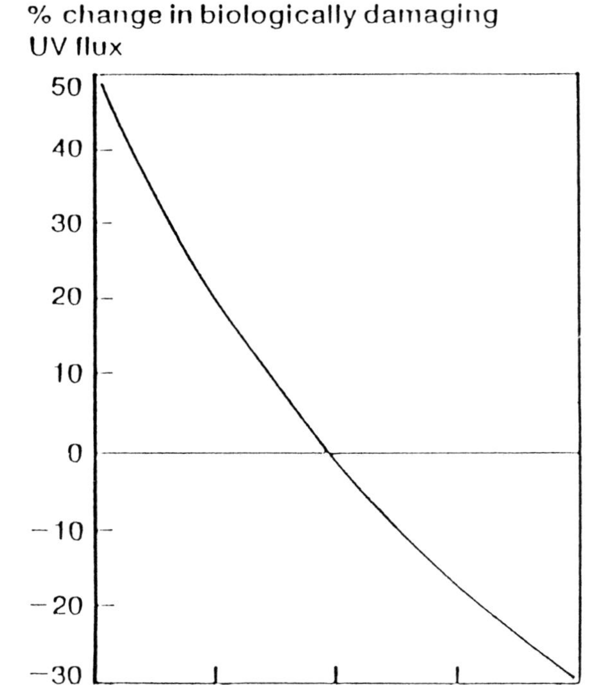


FIGURE 9.19. The changes in the biologically damaging ultraviolet (UV) flux as a function of changes in the ozone levels.

The decrease in $O_3$ levels will increase the amount of UV radiation that reaches the earth. This is shown in Figure 9.19. This plot includes all the wavelengths weighted according to how damaging they are to biological systems (300 to 345 nm). A 100% decrease of $O_3$ will increase the damaging UV flux by about 18%.

Recent attention to the destruction of $O_3$ has concentrated in the Antarctic region over the last 30 years (Figure 9.20) in the springtime. This has caused the so-called ozone hole over the South Pole. This springtime loss of ozone extends well north of Antarctica (Figure 9.21). It should be pointed out that one Dobson unit equals a concentration of $O_3$ of one molecule in every $10^9$ total molecules. Since the global concentrations of $O_3$ show large variations (Figure 9.22), it is impossible to distinguish the trends over time.

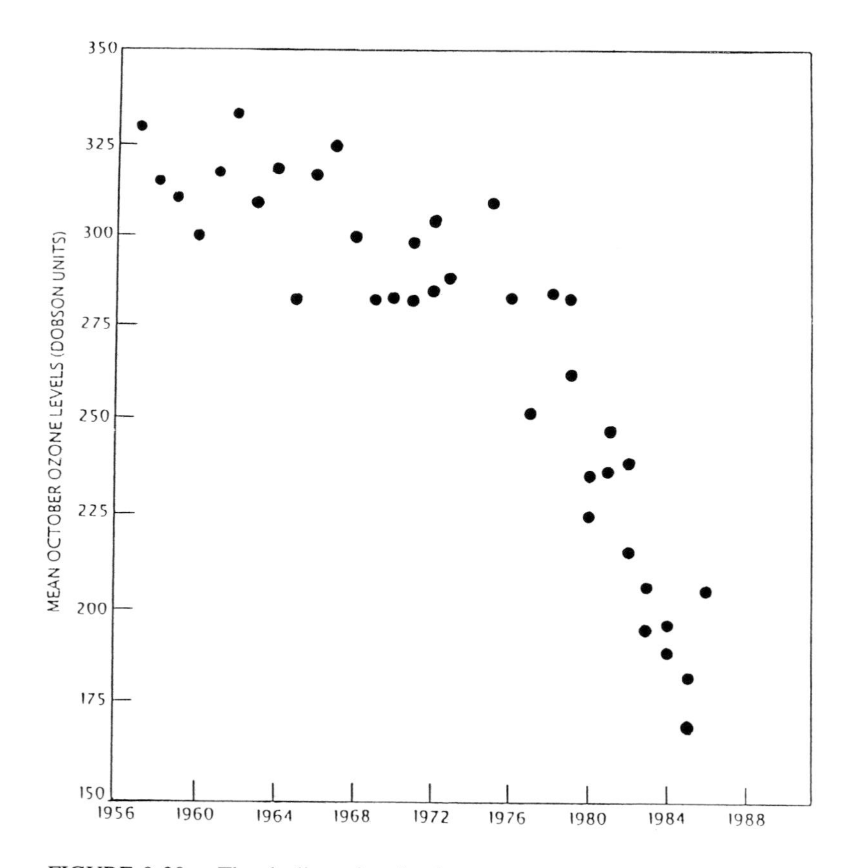


FIGURE 9.20. The decline of springtime ozone over the Antarctic over the last 33 years.

The loss of ozone in Antarctica in the springtime has been attributed to the formation of ClO (Figure 9.23). The large effects in the springtime have been attributed to a dynamic uplifting of the lower stratosphere as well as catalytic cycles involving Cl. One potential cycle is

$$ClO + ClO \rightarrow Cl_2O_2 \tag{42}$$

$$Cl_2O_2 + M \rightarrow Cl_2 + O_2 + M \tag{43}$$

$$Cl_2 + h\nu \rightarrow 2Cl \tag{44}$$

$$2Cl + 2O_3 \rightarrow ClO + O_2 \tag{45}$$

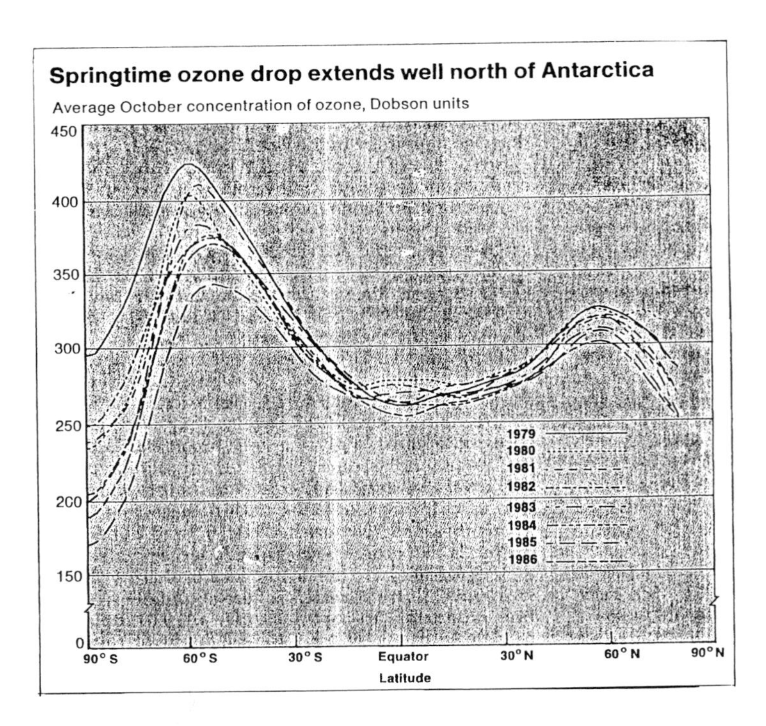


FIGURE 9.21. The levels of ozone as a function of latitude.

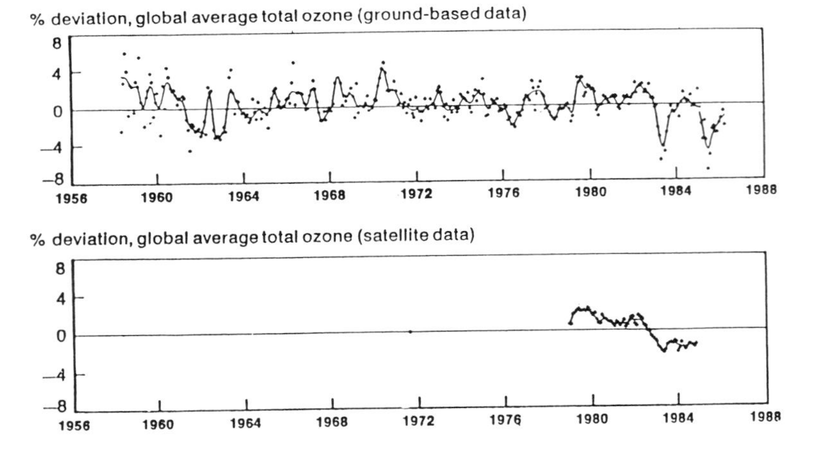


FIGURE 9.22. The fluctuations of ozone as a function of time.

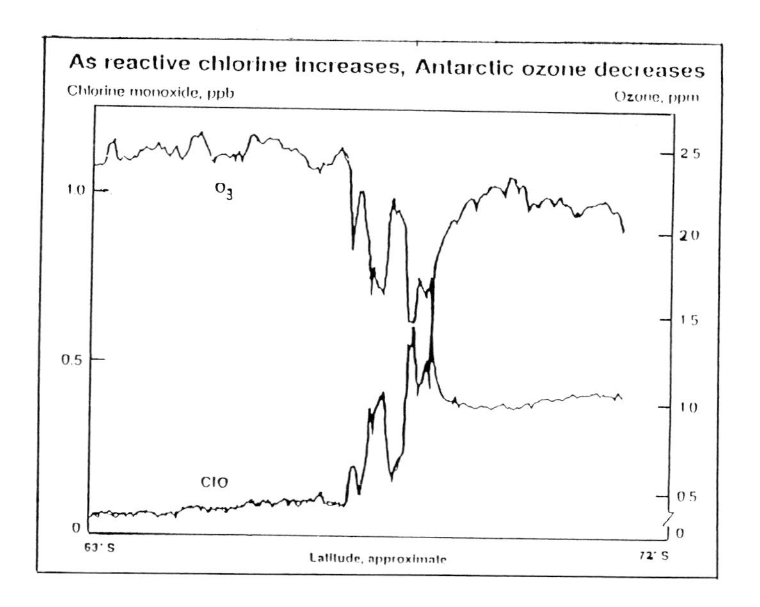


FIGURE 9.23. The relationship between $O_3$ and ClO in the Antarctic.

The net reaction is

$$2O_3 \rightarrow 3O_2 \quad (46)$$

The reaction of Cl or ClO with $CH_4$ and $NO_2$ can produce inert compounds

$$Cl + CH_4 \rightarrow HCl + CH_3 \quad (47)$$

$$ClO + NO_2 \rightarrow ClONO_2 \quad (48)$$

Low levels of $NO_2$ are needed to prevent the formation of $ClONO_2$. Reactions on ice have been proposed to explain how the inert compound becomes active

$$ClONO_2 + HCl \rightarrow Cl_2 + HNO_3 \quad (49)$$

$$ClONO_2 + H_2O \rightarrow HOCl + HNO_3 \quad (50)$$

The unique meteorology of Antarctica results in a circulation of air in the stratosphere that is different from the Northern Hemisphere. A stream of air called the polar vortex tends to circle the South Pole in the winter. Air trapped in the vortex gets very cold (−90°C) and forms clouds even in very dry conditions. These clouds provide ice crystals that have been shown by laboratory studies to cause molecules of HCl and $ClONO_2$ to react. The $Cl_2$ gas formed is released to the gas phase, whereas $HNO_3$ remains in the ice. The $Cl_2$ reacts with light to produce 2 Cl that further react with $O_3$ to produce more ClO. The total reaction sequence is

$$ClONO_2 + HCl \rightarrow Cl_2 + HNO_3 \tag{51}$$

$$Cl_2 + h\nu \rightarrow 2Cl \tag{52}$$

$$2Cl + 2O_3 \rightarrow 2ClO + 2O_2 \tag{53}$$

$$ClO + NO_2 \rightarrow ClONO_2 \tag{54}$$

The net reaction is

$$HCl + 2O_3 + NO_2 \rightarrow HNO_3 + 2O_2 + ClO \tag{55}$$

The HCl gets rid of $NO_2$ and forms ClO that further removes $O_3$. Although these reactions explain the formation of the ozone hole formed in Antarctica, further work is needed to predict the loss of $O_3$ in the global atmosphere. As man continues to add gases to the atmosphere that upset the natural balance, there will continue to be changes in the climate that affect life on the earth.

## 2. PRIMARY PRODUCTIVITY

The major source of organic compounds to the ocean comes from the primary production of marine plants. This primary production is the result of the growth of these organisms due to photosynthetic processes. It provides the basis of the marine food chain which culminates in fish and mammals. The plants remove $CO_2$ and micronutrients from the water and, using solar energy, convert them to complex organic compounds. The simple reaction is

$$CO_2 + H_2O \xrightarrow{\text{light}} CH_2O + O_2 \tag{56}$$

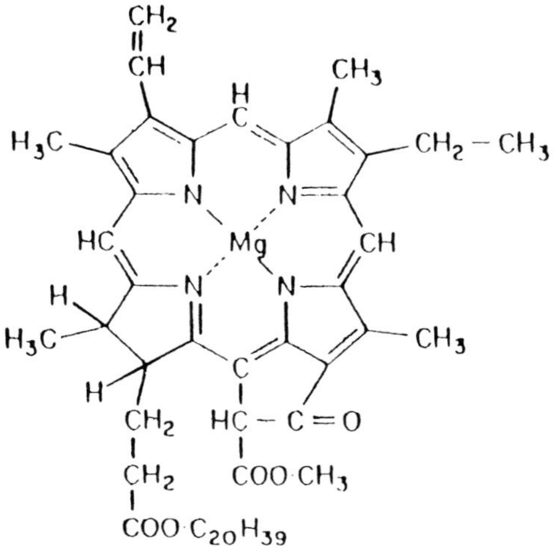

FIGURE 9.24. Sketch of chlorophyll-a.

Phytoplankton are the microorganisms responsible for the greatest primary production in the sea. Attached algae are important in shallow waters. Phytoplankton are free floating microscopic plants with only limited mobility that are distributed by ocean currents. They can be divided by size into

| | |
|---|---|
| Microplankton | 50 to 500 μm |
| Nanoplankton | 10 to 50 μm |
| Ultraplankton | 0.5 to 10 μm |

Since they live primarily by photosynthesis, they are *phototrophic organisms*.

Photosynthesis is a complicated process — solar energy is absorbed by phytoplankton cells and converted to biological energy stored in the form of organic compounds. The stages are:

1. Absorption of photons of light by photosynthetic pigments which are contained in the chromatophores. The principal pigments are chlorophylls. (The structure of chlorophyll-a is shown in Figure 9.24.) The resonating systems of double bonds stabilize the molecule and provide electrons that are easily excited to orbitals of higher energy when light is absorbed.
2. Part of the absorbed energy of these excited electrons is converted to chemical energy through a cyclic series of enzymatic reactions involving Cytochrome I, which leads to the production of high energy adenosine

triphosphate (ATP) from adenosine diphosphate (ADP) and orthophosphate (P)

$$ADP + P \rightarrow ATP \tag{57}$$

The remainder of the energy of the electrons is used in a series of enzymatic reactions of riboflavin phosphate and nicotinamide adenine dinucleotide phosphate (NADP)

$$4NADP + 2H_2O + 2ADP + 2P \rightarrow 4NADPH + O_2 + 2ATP \tag{58}$$

The protons are from water and the excited electrons are from the reduced form of NADP. The hydroxide from the water yields molecular oxygen and donates electrons to chlorophyll via the Cytochrome I chain.

3. The $CO_2$ is assimilated in a cyclic series of reactions utilizing the reducing action of NADPH and the phosphorylating power of ATP. These reactions, which can take place in the dark, lead to the production of carbohydrate ($CH_2O$)

$$CO_2 + 4NADPH + ATP \rightarrow CH_2O + H_2O + 4NADP + ADP + P \tag{59}$$

Combining the three equations gives

$$CO_2 + H_2O \rightarrow CH_2O + O_2 \tag{60}$$

Labeled $^{14}CO_2$ finds its way very quickly into compounds other than carbohydrates. Compounds such as fats and amino acids are synthesized from intermediates in the carbon cycle in addition to being produced from $CH_2O$.

The photosynthetic quotient (PQ = molecules of $O_2$ liberated/molecules of $CO_2$ assimilated) differs from 1.0 because of the production of other classes of compounds. PQ = 1.4 if only lipids are formed, PQ = 1.05 when amino acids are formed using $NH_3$, and 1.6 using $NO_3$. Values of 1.20 to 1.39 are common for natural populations provided there are adequate nutrients. At low light intensities, heterotrophic utilization of compounds such as acetate may take place. This may be advantageous because it requires less energy than to fix $CO_2$.

The energy required for metabolic processes is obtained by the oxidation of the photosynthesized organic compounds. This process is known as respiration. The simple overall reaction is

$$CH_2O + O_2 \rightarrow CO_2 + H_2O \qquad (61)$$

The respiration quotient (RQ = molecules of $CO_2$ liberated/molecules of $O_2$ assimilated) lies close to 1.0. Thus, fats and protein are not normally used. Respiration takes place in the light and dark at approximately the same rate. The rate is ~ 5 to 10% of the maximum rate of photosynthesis of cells grown under optimal conditions.

### 2.1. Phytoplankton Production

Both the amount and the rate at which phytoplankton are produced are important to consider with respect to availability to the food chain. The standing crop is the amount of phytoplankton at a given time (mg of C/$m^3$ of seawater). The *rate of primary production* (P) is defined as the weight of inorganic carbon fixed photosynthetically per unit time per unit volume (mg C/$m^3$ h) or under unit surface area (g C/$m^2$ d). The primary production is determined from measurements performed at 100, 10, and 1 m depth.

$$P = 1/5(2P_{100} + 2P_{10} + P_1)D/2\ NK \qquad (62)$$

where D = depth, light is 1% of surface, N = number of hours to sunset, and K = fudge factor (1.0 in the tropics).

### 2.2. Standing Crop or Biomass

The biomass is concentrated by centrifugation or filtration and the cells are counted by eye or coulter counter or by analyzing a chemical component of the cell. The concentration of chlorophyll-a, determined by fluorimetric methods, is often used as an indicator of biomass levels. The rate of consumption of micronutrient elements is also used to obtain integrated values of primary productivity over fairly lengthy periods. Since N and P can be regenerated and recycled several times a year, these values will tend to be minimal levels.

Primary Productivity is determined by measuring (1) the rate of $O_2$ liberated, and (2) the rate at which $CO_2$ is removed and converted into plant material. Methods using $O_2$ liberation and the uptake of $CO_2$ are discussed below.

## 2.3. $O_2$ Liberation

### Method of Measurement

A series of 300 ml glass stoppered bottles are filled with seawater and suspended at various depths throughout the photic zone. Dark or "light tight" bottles filled with the same seawater are suspended at the same depths. After 3 to 8 h, the water is analyzed for $O_2$. The increase of $O_2$ in the light bottles is a measure of the net photosynthesis. The gross productivity is obtained from the difference in the $O_2$ in the light and dark bottles (the loss of $O_2$ due to respiration). Some problems with the method are:

1. Insufficient sensitivity (~ 3 mg $C/m^2/h$) in low productivity waters (oligotrophic).
2. Cannot be used in polluted waters.
3. Results will be low when lipids or proteins are photosynthesized (e.g., diatoms).
4. Bacteria can cause bottle wall problems.
5. The growth may be affected by confinement.

## 2.4. Uptake of $CO_2$

### Method of Measurement

The best method of measuring primary productivity is using carbon 14. The 300 ml bottles are filled with seawater and 2 ml of bicarbonate solution with labeled $^{14}C$ (1 to 25 $\mu Ci$ depending upon productivity expected). The bottles are suspended at various depths or, less satisfactorily, placed in a thermostatic bath and illuminated with light from fluorescent lamps filtered to give the intensities and spectral distribution of a given depth. After 2 to 6 h, the seawater is gently filtered through a 0.45 $\mu m$ filter. The filters are washed with seawater or HCl vapor to remove calcareous inorganic carbon. The radiocarbon flux is counted with a scintillation counter to determine the activity. About 1 to 2% of the $CO_2$ may be exchanged in the dark (10% in tropical waters). The productivity is given by

$$\text{Productivity (mg C/m}^3\text{/h)} = 1.05\,(C_S - C_D)W/CN \qquad (63)$$

where W = total $CO_2$ in the sample (mg $C/m^3$), N = number of hours exposed, 1.05 = compensates for slow uptake of algal cells of $^{14}CO_2$ compared to $^{12}CO_2$, $C_S$ and $C_D$ are counting rates of light and dark bottles, and C = normalized counting rate of $^{14}C$ which has been added to the bottles. Strickland gives the full details of using this technique.

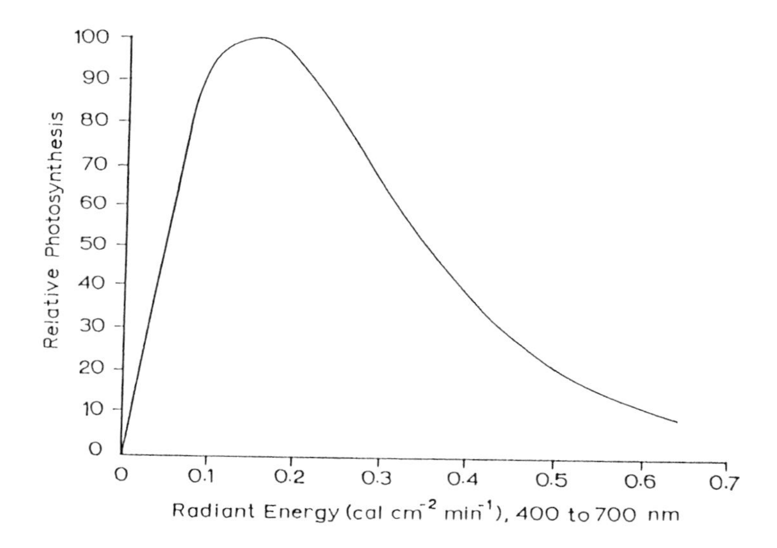


FIGURE 9.25. Relative photosynthesis as a function of the radiant energy.

Some problems with the method are:

1. Thought to measure net production (60 to 90% of gross production)
2. Extracellular metabolites may not be retained by the filter (30%)
3. Sample enclosure may cause changes
4. May miss diurnal effects
5. Measures only fixation of inorganic C

Laboratory measurements with cultures show reasonable agreement between estimates of primary productivity based on $O_2$ production, $CO_2$ assimilation, uptake of P, and the volume of packed cells obtained after centrifugation. The $O_2$ method gives more meaningful results for >24 h sampling and highly productive waters. $^{14}C$ is best for oligotrophic waters — after making corrections for extracellular products.

Factors affecting the growth of phytoplankton are given below.

**Light** — Two factors that influence light must be taken into consideration: (1) factors controlling the intensity and the spectral composition in the sea; and (2) species preference for a given intensity and wavelength. The amount of light reaching the surface oceans is controlled by altitude of the sun; cloud cover; wave length (370 to 720 nm); and reflection, absorption, and scattering. The rate of photosynthesis increases in proportion to the light intensity at low intensities (Figure 9.25).

Light saturation is reached at moderate intensities. At high intensities inhibition occurs perhaps as a result of the inhibition of chlorophyll production. Permanent inhibition occurs due to photo-oxidation at high light intensities.

**Temperature** — In the $-2$ to 30°C range, phytoplankton are rapidly killed at temperatures 10 to 15°C above the temperature at which they are adapted to live. Lowering the temperature slowly has less of an effect. Phytoplankton are best cultured at 5 to 10°C higher than natural conditions. The combined effects of temperature and light may determine the pattern of succession at a given latitude.

**Salinity** — Marine phytoplankton will grow at salinities as low as 15, some more successfully than at 35. *Stenohaline* organisms will only thrive in limited ranges of salinity, e.g., Peridinium balticum at S = 8 to 12 (Baltic). *Euryhaline* organisms can live over a wide range of salinities.

**Micronutrients and trace metals** — Organisms need N and P for healthy growth. A few organisms (e.g., cyanobacteria) can fix molecular N. Others need $NH_3$, $NO_2^-$, and $NO_3^-$; heterotrophs can use organic N. Some can use dissolved or particulate organic P. Most species can grow at N and P levels greater than 29 times the maximum values in the oceans. The minimum level of P is ~5 mg $PO_4^{3-}$/l, but may vary with different species. Heavy blooms in enclosed areas can be caused by excess N and P from sewage or fertilizer runoff. $SiO_2$ is needed for the growth of diatoms (minimum is ~50 mg Si/l in ocean waters except in subtropical waters). Trace metals (Fe, Mn, Mo, Zn, Cu, Co, V) are required for healthy growth. These metals are needed for proteins and enzymes, for example, ferrodoxin (Fe protein) and enzymes (Mn, Mo, Cu, Zn, Co). Phytoplankton can assimilate chelated forms of many elements. EDTA is frequently added to cultures to permit nontoxic forms of metals to be available without the risk of the precipitation of hydroxides. Recent work has indicated that shortages of Fe may limit the production of plankton in waters that have high levels of nutrients (N, P, and Si); for example, the North Pacific and southern ocean waters. The input of Fe from atmospheric sources could control the primary production in these regions.

**Organic factors** — Phytoplankton growth will not take place unless minute amounts of specific organics are available. Some organics, however, also inhibit the growth of algae. Vitamin $B_{12}$ and $B_1$ (thiamine) are growth promoters. Ascorbic acid and cystine may also be needed. Bacteria may produce some of these compounds.

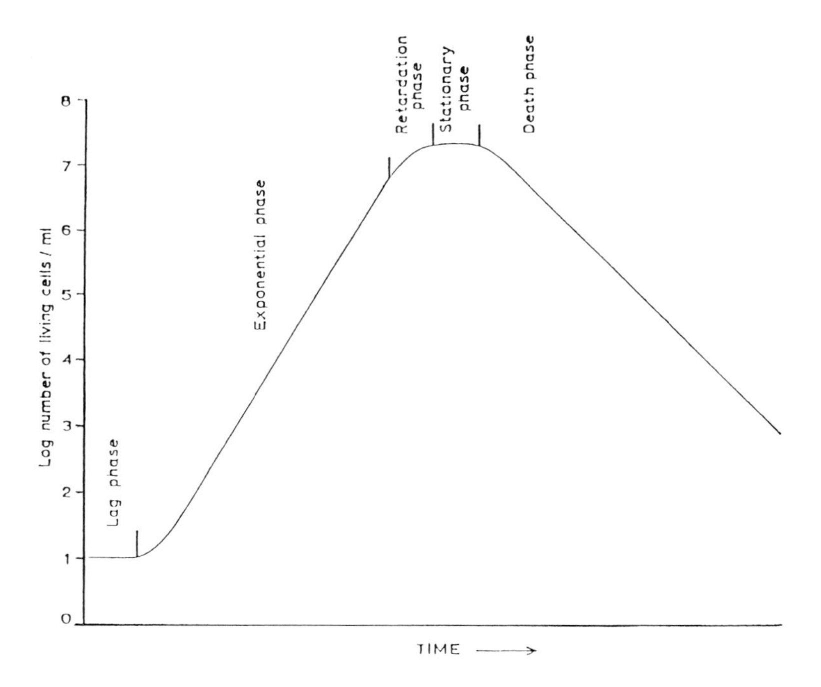


FIGURE 9.26. Idealized phytoplankton growth curve.

## 2.5. Growth and Distribution of Phytoplankton in the Sea

The growth kinetics of plankton is similar to that of bacteria which have been extensively studied. An idealized growth curve is given in Figure 9.26. This simple growth curve does not occur in the oceans due to physical and chemical factors as well as biological (grazing) and hydrographic (horizontal and vertical water movement) factors. Discussion of the various phases follows.

**Lag phase** — This delay of growth is thought to be due to a lack of balance in the enzymatic reactions in the cell due to growth promoters.

**Exponential phase** — This is the linear exponential growth phase where the number of cells (N) increase according to

$$\ln(N/N_0) = kt \qquad (64)$$

where $N_0$ is the initial number of cells, and k is related to the nature of the

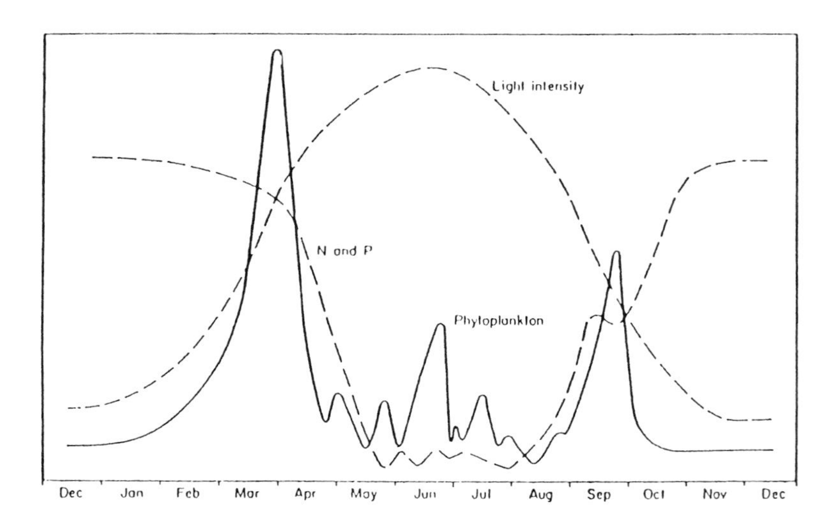


FIGURE 9.27. Seasonal variation of phytoplankton, nutrients and light in a typical northern temperate sea.

organism and conditions of growth. The half-time of growth, $t_{1/2}$, = ln2/k. In a culture the value of k = 0.1 to 1.0 h; in the sea, k = 0.09 to 0.015. This gives an average $t_{1/2}$ = 8 to 46 h.

**Retardation phase** — This is due to the reduction of the growth rate due to the depletion of nutrients, inhibition due to growth inhibitors, and reduction in photosynthesis due to shading.

**Stationary phase** — This is when there is no net increase in living cells, but metabolic activity still is not limited.

**Death phase** — This is when cells die at an exponential rate. Not all of the cells die, some go into a resting state and survive for long periods. In the oceans a bloom will dissipate due to grazing by zooplankton, sinking below the photic zone, horizontal movement, and toxins.

The seasonal variation of primary productivity is shown in Figure 9.27. The productivity peaks in the spring and fall and is related to the availability of N and P and appropriate light intensity.

The relative rates of photosynthesis as a function of depth in the summer and winter are shown in Figure 9.28. In the summer the maximum occurs at a depth where the transmission of light is between 25 to 50%. The low rates of photosynthesis that occur in the winter result in a compensation depth near the surface. In the open oceans the photosynthetic compensation depth is related to the availability of nitrogen from deeper waters. The geographical

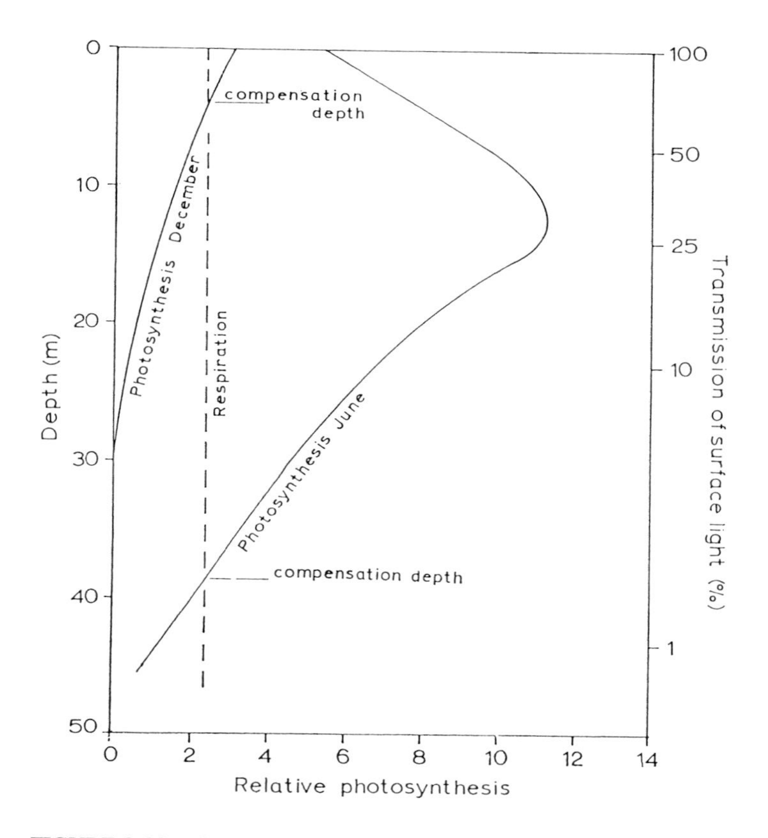


FIGURE 9.28. Integrated daily rate of primary productivity as a function of depth.

distribution of primary productivity is shown in Figure 9.29. It is highest in upwelling coastal regions and in the southern oceans. Some typical values for various regions are given in Table 9.5. The totals for the oceans range from mean values of 50 to 370 g C $m^{-2}$ $year^{-1}$, the higher values coming from $O_2$ utilization estimates. The annual values of 3 to 15 $\times$ $10^{10}$ metric tons of carbon can be compared to land estimates of 2 to 3 $\times$ $10^{10}$ metric tons of carbon.

In recent years, a number of workers have questioned the validity of the values of primary productivity determined by the carbon-14 technique. Part of the problem is related to the uncertainty of whether the technique is measuring net or gross primary production. Recent studies conclude that there is no evidence of errors unique in using the $^{14}C$ method to measure gross production. Thus, it means that the $^{14}C$ method provides an upper limit for net production.

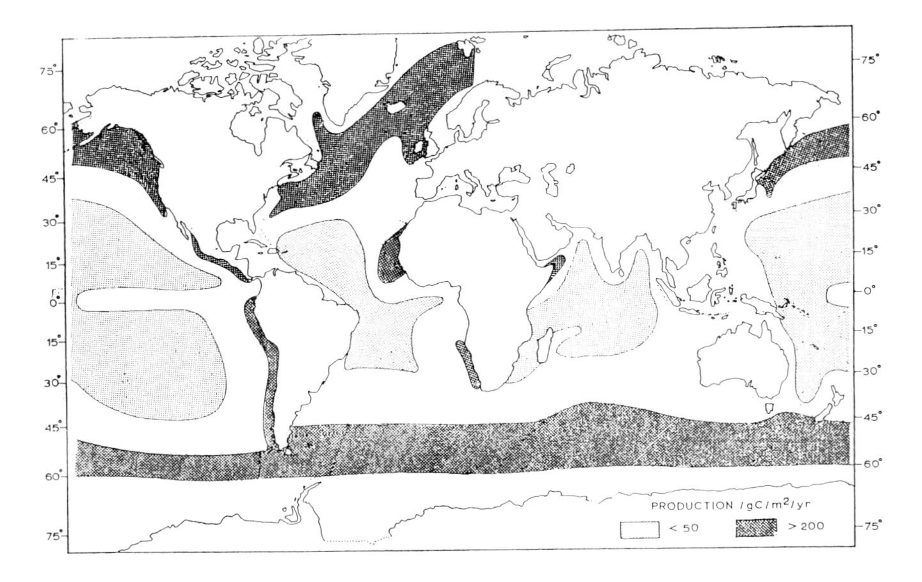


FIGURE 9.29. Geographical distribution of primary productivity in the oceans (g carbon $m^{-2}$ $year^{-1}$).

**TABLE 9.5**
**Primary Productivity in Ocean Waters**

| Area | $g\ C\ m^{-2}\ y^{-1}$ |
|---|---|
| Open ocean | 18—55 |
| Eq. Pacific Ocean | 180 |
| Eq. Indian Ocean | 73—90 |
| Upwelling areas | 180—3600 |
| Saragasso Sea | 72 (18—168) |
| Continental Shelf — NY | 120 |
| Continental Shelf — North Sea | 50—80 |
| Kuroshio | 18—36 |
| Arctic | 1 |
| Mean of all oceans | 50—370 |

Jenkins (WHOI) has measured oxygen utilization rates in the North Atlantic by combining tritium ($^3H$) and helium ($^3He$) measurements with apparent oxygen utilization (AOU). The $^3H/^3He$ ratio was used to date the waters

$$t = (1/18\ \text{year})\ \ln(1 + [^3He]/[^3H]) \qquad (65)$$

By plotting AOU vs. the age, he was able to determine the oxygen utilization rates. Below 100 m, he found a rate of 253 μmol $O_2$ consumed per $m^2$ per year. This means that 4.5 mol C $m^{-2}$ $year^{-1}$ or 55 g C $m^{-2}$ $year^{-1}$ is needed to reduce the oxygen levels. If this represents 80 to 90% of the amount being produced, the amount of primary productivity would be at least 60 g C $m^{-2}$ $year^{-1}$. This is much higher than the expected values for open oceans.

Since photosynthesis depends upon chlorophyll, the concentration of chlorophyll in seawater can be taken as an index of photosynthetic potential. This method is approximate since it is hard to distinguish between active and inactive chlorophyll in dead material. By using satellites, the chlorophyll in the waters can be used to get an idea of the primary productivity occurring over a wide area (Plate 24)*. If the satellite data is calibrated using direct measurements, one can improve the correlations between color and production. This method will continue to be used to study the primary production in ocean waters in the future.

The global primary production is shown in Figure 9.30. The highest levels occur in coastal upwelling areas. The oceanic values are compared to land production in Table 9.6.

## 3. HYDROTHERMAL VENT CHEMISTRY

In April, 1979 scientists aboard the submersible ALVIN discovered hot springs coming from the sea floor in areas of active volcanism. The area around the black smokers had deposits of metal-rich minerals and unique biological communities. A typical cluster of organisms around a hydrothermal vent is shown in Figure 9.31. Large clams, crabs, mussels, and worms are found fed by the hydrothermal fluids. These vent communities are thought to only persist for years to decades. Since the initial discovery in the Galapagos, more vent systems have been found along the East Pacific rise and more recently along the mid-Atlantic ridge. Two types of vents have been found: (1) warm vents with maximum exit temperatures of 5 to 23°C and flow rates of 0.5 to 2 cm $sec^{-1}$; (2) hot vents with maximum exit temperatures of 270 to 380°C and flow rates of 1 to 2 m $sec^{-1}$. Hot vents include white smokers ($\approx$ 300°C) and black smokers (350 $\pm$ 2°C).

* Plate 24 follows page 48.

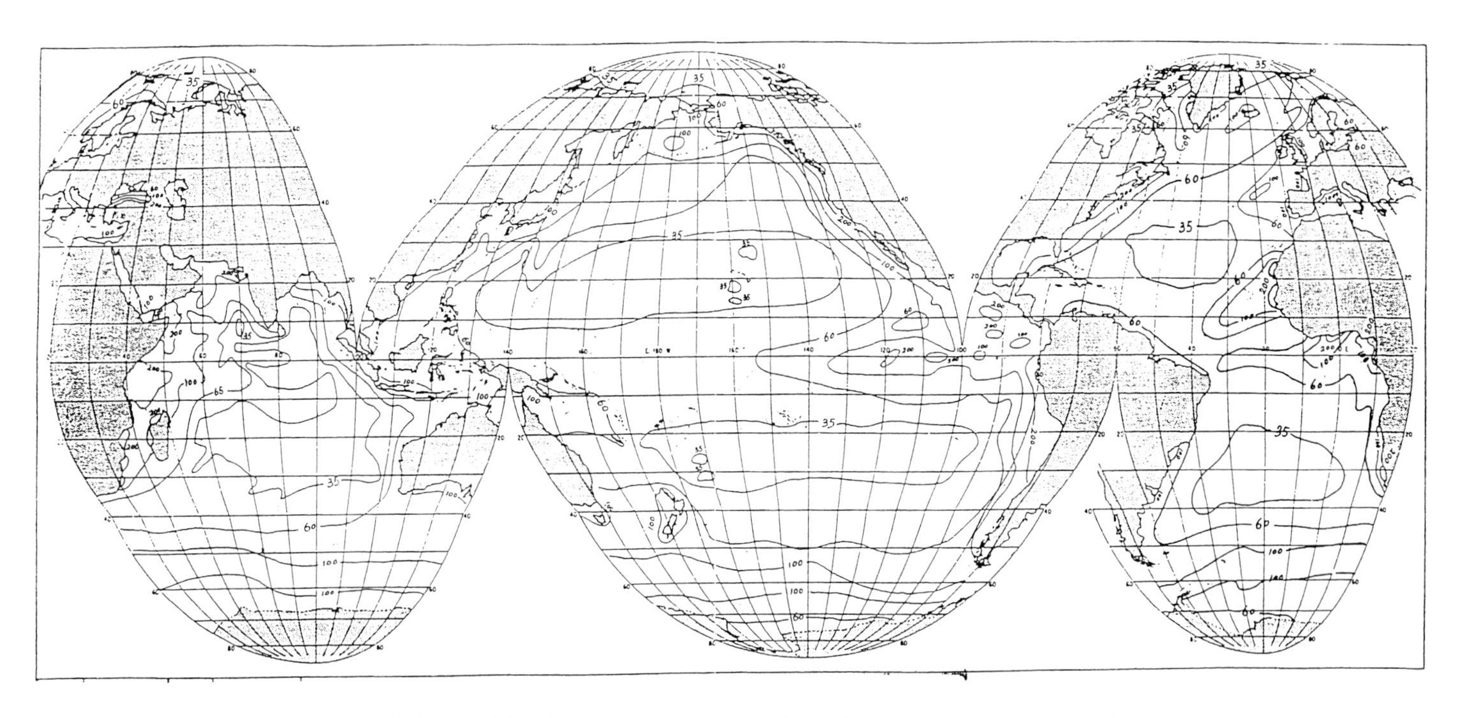


FIGURE 9.30. Global primary productivity in the oceans (mg carbon $m^{-2}$ $day^{-1}$).

## TABLE 9.6
## Comparative Global Primary Production

| | |
|---|---|
| Theoretical algal maximum | 27 g C · m$^{-2}$ · d$^{-1}$ |
| Rice field | 4 |
| Pine forest | 2 |
| Upwelling ocean | 2 |
| Antarctic Ocean | 1 |
| Neritic ocean | 0.2 (e.g., 20 mg C·m$^{-3}$ per 10 m photic zone) |
| Oceanic | 0.1 (e.g., 1.3 mg C·m$^{-3}$ per 77 m photic zone) |

FIGURE 9.31. A typical hydrothermal cluster of structures.

Bacteria are responsible for utilizing the geothermal energy from the vents. The process taking place is bacterial chemosynthesis. Like photosynthesis, the process involves the biosynthesis of organic carbon compounds from $CO_2$ with the source of energy being chemical oxidation (rather than light). Chemoautotrophy is the assimilation of $CO_2$ by bacteria with the ability to use reduced inorganic compounds as a source of energy (chemolithotrophy). The following equations illustrate the relationships between the two processes.

**TABLE 9.7**
**Electron Sources and Types of Chemolithotrophic Bacteria Potentially Occurring at Hydrothermal Vents**

| Electron donor | Acceptor | Organisms |
|---|---|---|
| $S^{2-}$, $S^0$, $S_2O_3^{2-}$ | $O_2$ | Sulfur-oxidizing bacteria |
| $S^{2-}$, $S^0$, $S_2O_3^{2-}$ | $NO_3^-$ | Denitrifying and sulfur-oxidizing bacteria |
| $H_2$ | $O_2$ | Hydrogen-oxidizing bacteria |
| $H_2$ | $NO_3^-$ | Denitrifying hydrogen bacteria |
| $H_2$ | $S^0$, $SO_4^{2-}$ | Sulfur- and sulfate-reducing bacteria |
| $H_2$ | $CO_2$ | Methanogeneic and acetogenic bacteria |
| $NH_4^+$, $NO_2^-$ | $O_2$ | Nitrifying bacteria |
| $Fe^{2+}$, ($Mn^{2+}$) | $O_2$ | Iron- and manganese-oxidizing bacteria |
| $CH_4$, CO oxidizing | $O_2$ | Methylotrophic and carbon monoxide-bacteria |

$$2CO_2 + H_2S + 2H_2O \xrightarrow{h\nu} 2[CH_2O] + H_2SO_4$$

(nonoxygenic photoautolithotrophy, purple and green bacteria) (66)

$$CO_2 + H_2O \xrightarrow{h\nu} [CH_2O] + O_2$$

(oxygenic photoautolithotrophy, green plants) (67)

$$CO_2 + H_2S + O_2 + H_2O \rightarrow [CH_2O] + H_2SO_4$$

(aerobic chemoautolithotrophy, bacteria) (68)

$$2CO_2 + 6H_2 \rightarrow [CH_2O] + CH_4 + 3H_2O$$

(anaerobic chemoautolithotrophy, bacteria) (69)

The possible electron donors used by bacteria are shown in Table 9.7. Some are the same as those used in anaerobic chemosynthesis ($NO_3^-$, $SO_4^{2-}$, or $CO_2$). The inorganic sources of energy are used to produce ATP.

$H_2S$ as the reduced compound coming from the vents is the source of the major electron donor. It is formed by the reduction of $SO_4^{2-}$ from seawater, coupled with the oxidation of $Fe^{2+}$ in basalt to $Fe^{3+}$. The $H_2S$ may also be formed from crystal basalts. Both mechanisms have been found in laboratory experiments at 300°C. The concentration of $HS^-$ (25 m*M*) is nearly the same

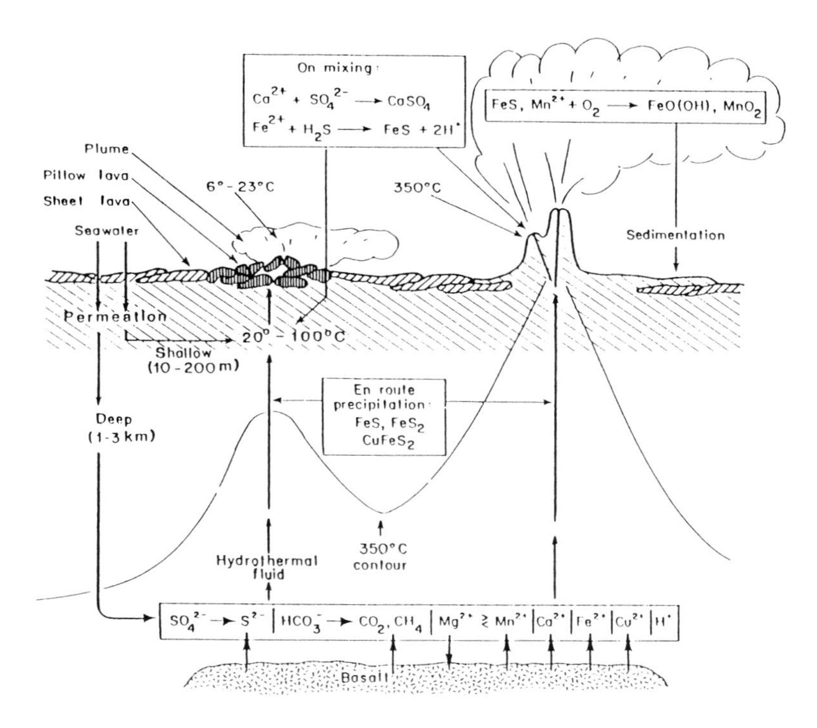


FIGURE 9.32. Schematic diagram showing the inorganic processes occurring at a vent site.

as $SO_4^{2-}$ in seawater (28 m*M*). This indicates that seawater circulating through the hydrothermal system reacts with an amount of rock nearly equal to its own mass. Since seawater can lose $SO_4^{2-}$ as $CaSO_4$ (anhydrite), little of the seawater $SO_4^{2-}$ may get to the hotter parts of the system where it can be reduced. Sulfur isotope work of the hydrovent waters indicates that $H_2S$ is mainly derived from the basalt. A diagram of the chemical processes occurring at a vent system is shown in Figure 9.32. The hydrothermal reactions associated with the fluid flow are shown in Figure 9.33.

Before we examine the chemistry of the hydrothermal fluids, we will examine the prefinding evidence for hydrothermal fluxes. The first evidence came from geophysical observations. Heat flow measurements, for example, were lower than the theoretically predicted values (Figure 9.34). The lost heat is attributed to the heat carried away by circulating seawater. Other geophysical evidence for hydrothermal fluxes comes from magnetic measurements and compression wave velocity measurements.

The second source of evidence for hydrothermal fluxes comes from experimental laboratory measurements. These laboratory measurements were

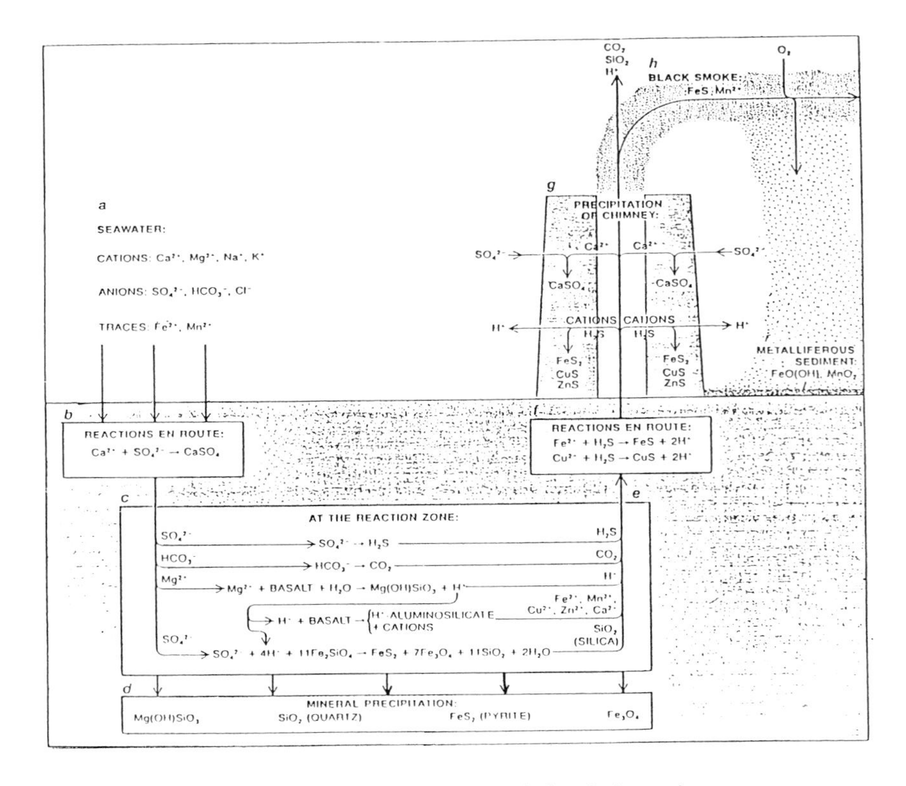


FIGURE 9.33. The hydrothermal chemical reactions.

made between basalt and seawater mixtures (at water/rock ratios of 10/1) in steel bombs. The measurements were made at low temperatures (70°C), high temperatures (150 to 300°C), and very high temperatures (300 to 350°C). The low temperature measurements showed losses of $Mg^{2+}$, $Na^+$, and $K^+$ from seawater and a release of $Cu^{2+}$ and $SiO_2$ by the basalt. The higher temperature measurements (water/rock ratio = 50/1 or 62/1) showed a loss of $Mg^{2+}$, $SO_4^{2-}$, and $Na^+$ from seawater and a release of $Ca^{2+}$, $H_2S$, $CO_2$, $SiO_2$, $K^+$, and $H^+$ from the basalt. The release of the $H^+$ causes the pH to decrease to values near 5.0. Slight increases of $Fe^{2+}$, $Mn^{2+}$, $Ba^{2+}$, $Al^{3+}$, and $Cu^{2+}$ were also found. In the very high temperature measurements (water/rock ratio = 50/1 or 62/1), higher concentrations of trace metals were released.

The third source of evidence for hydrothermal fluxes comes from geochemical field measurements. This evidence includes finding altered basalts, sulfide deposits, and hydrothermal precipitates. The iron and manganese rich hydrothermal precipitates found by Bostrum and Bonatti in 1966 on active ridge systems was a strong indication that high temperature reactions were

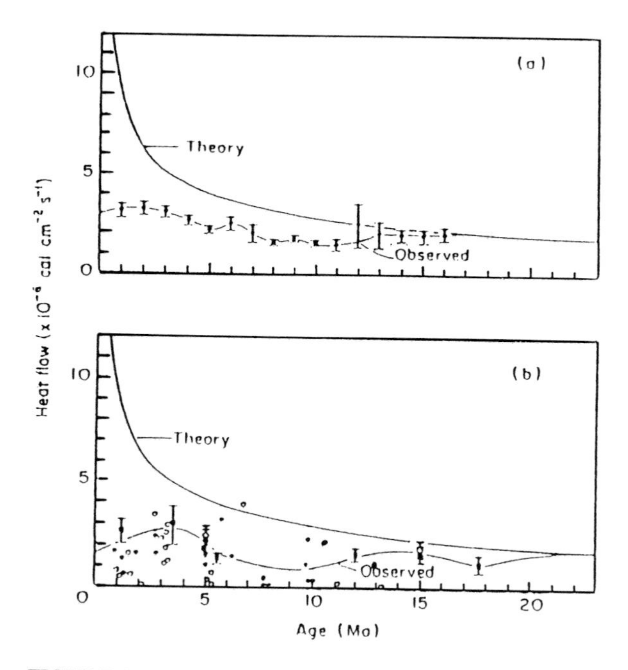


FIGURE 9.34. Comparison of the measured and theoretical heat flow as a function of the crustal age.

occurring. Earlier studies on the Red Sea hot brines also demonstrated that circulating seawater is heated by molten basalt.

The finding of hot springs in the Galapagos and on the East Pacific rise demonstrated that hydrothermal processes were occurring in the present oceans. More recent studies of hydrothermal vents have been made in Guaymas Basin (Figure 9.35). The Galapagos vents studies showed a linear increase of $SiO_2$ with temperature (Figure 9.36). As the $SiO_2$ increases, the $Mg^{2+}$ decreases (Figure 9.37). The $Mg^{2+}$ concentration is inversely proportional to the temperature (Figure 9.38). This has led to the use of $Mg^{2+}$ as a measure of the mixing of the hot hydrothermal fluids with seawater. The flux of $Ca^{2+}$, $Sr^{2+}$, $Li^+$, $K^+$, and $Rb^+$ coming from the Galapagos vents is shown in Figures 9.39 to 9.43.

The pH of the hydrothermal waters is close to 4.0 (Figure 9.44). The $SO_4^{2-}$ ion shows a loss in the hydrothermal waters as expected (Figure 9.45). The trace metals $Mn^{2+}$, $Fe^{2+}$, and $Zn^{2+}$ show increases in vent waters (Figures 9.46 to 9.48). The measurements of $Na^+$ and $Cl^-$ vs. $Mg^{2+}$ are not as conclusive (Figures 9.49 to 9.50). The comparison of the estimated composition of the 350°C end member and seawater are given in Table 9.8.

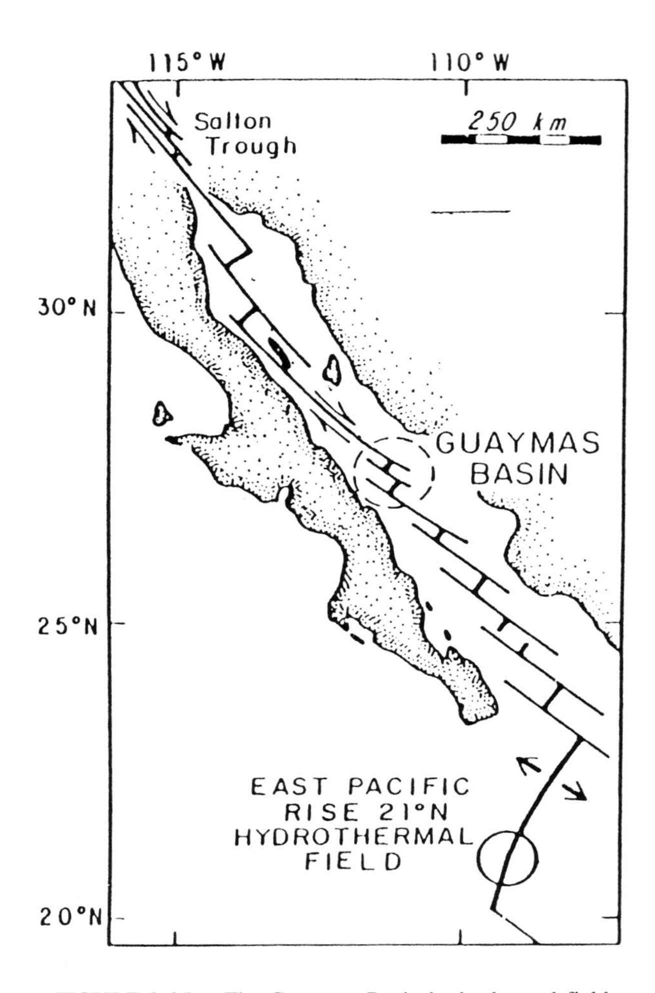


FIGURE 9.35. The Guaymas Basin hydrothermal field.

A comparison of the hydrothermal and river fluxes of elements is shown in Table 9.9. These comparisons indicate that for some elements the hydrothermal flux is the same order as the river flux.

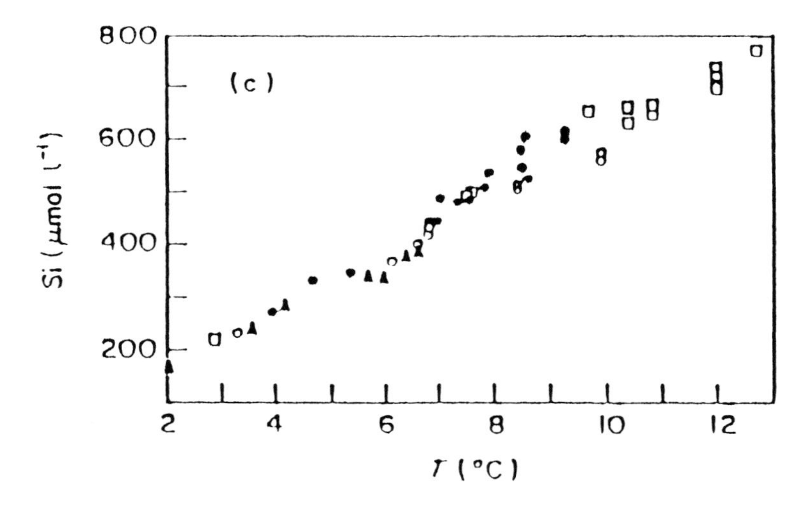


FIGURE 9.36. The concentration of $Si_2$ as a function of the temperature for the Galapagos Vent waters.

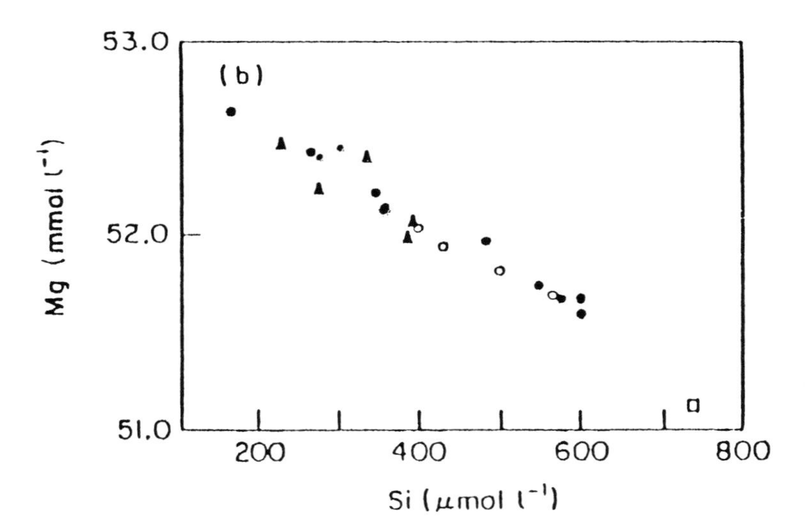


FIGURE 9.37. The concentration of magnesium as a function of the silica concentration.

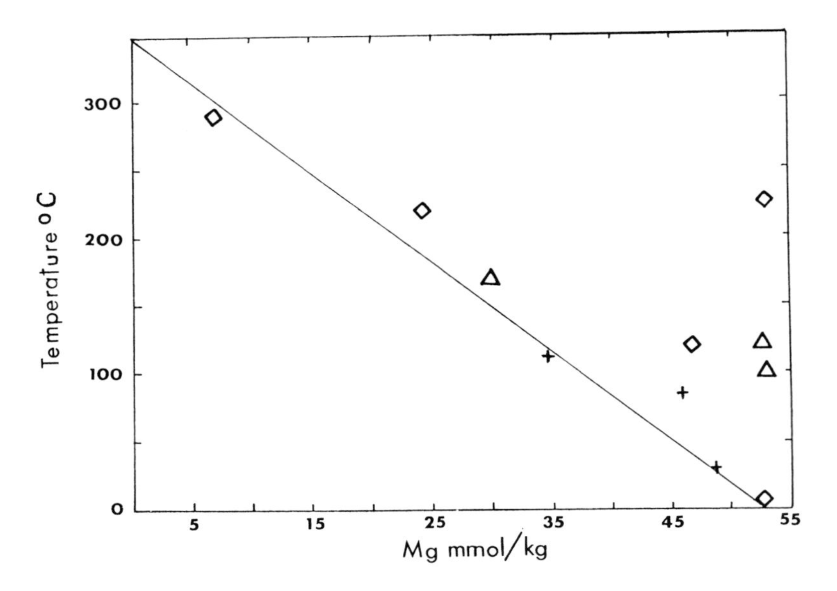


FIGURE 9.38. The magnesium concentration as a function of the temperature of vent waters.

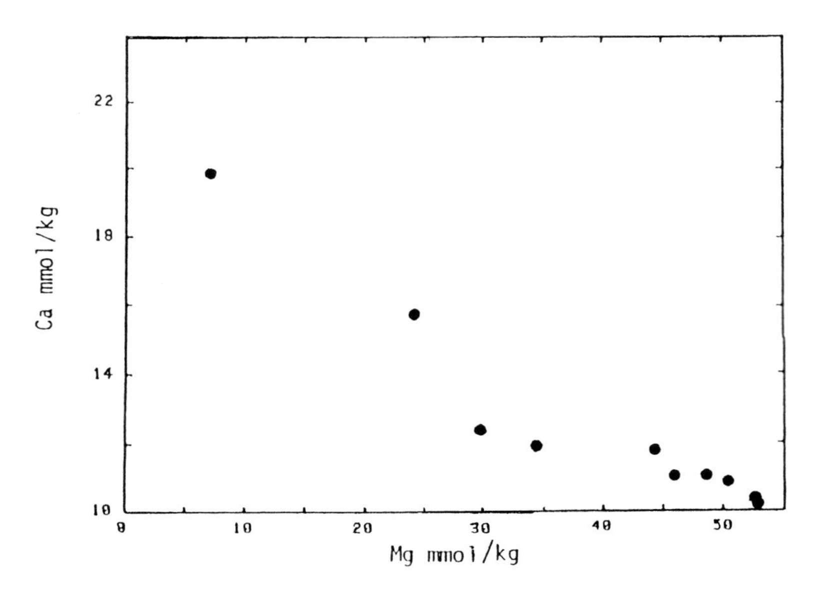


FIGURE 9.39. The calcium concentration as a function of magnesium in vent waters.

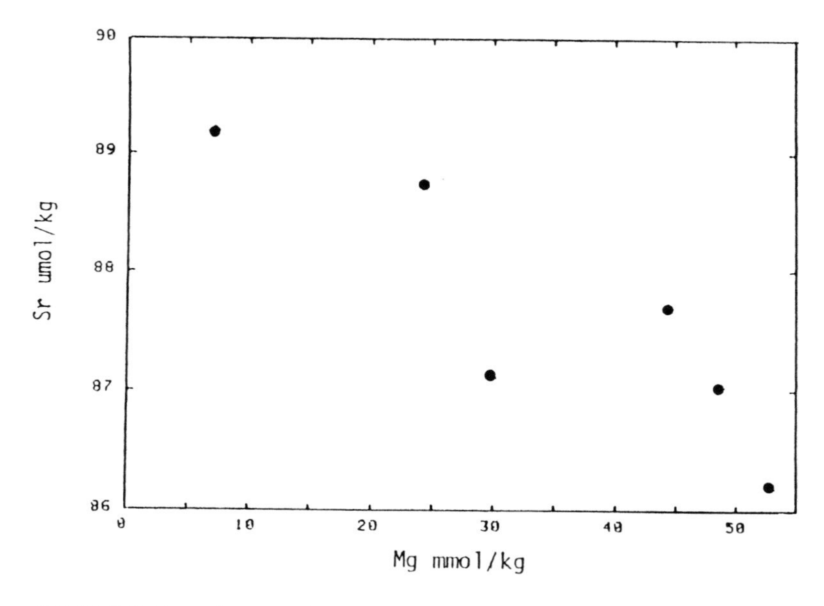


FIGURE 9.40. The strontium concentration as a function of magnesium in vent waters.

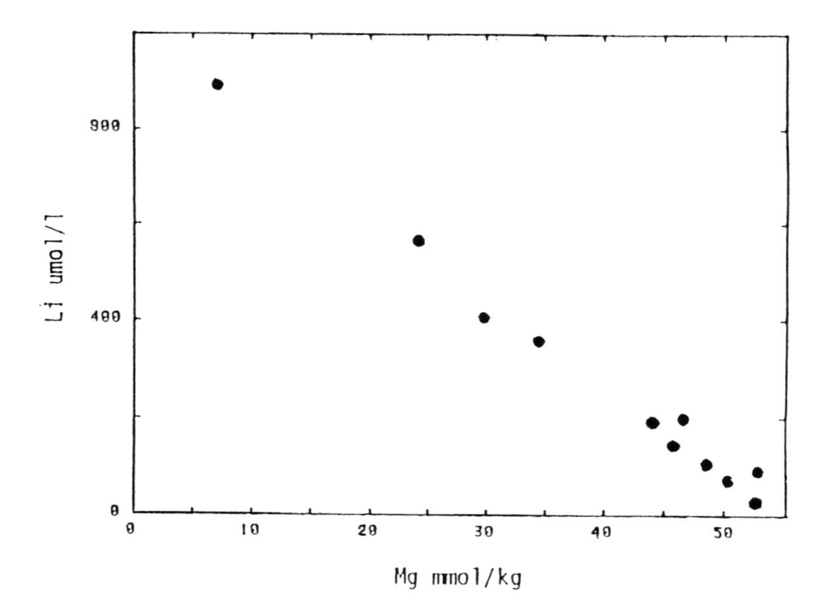


FIGURE 9.41. The lithium concentration as a function of magnesium in vent waters.

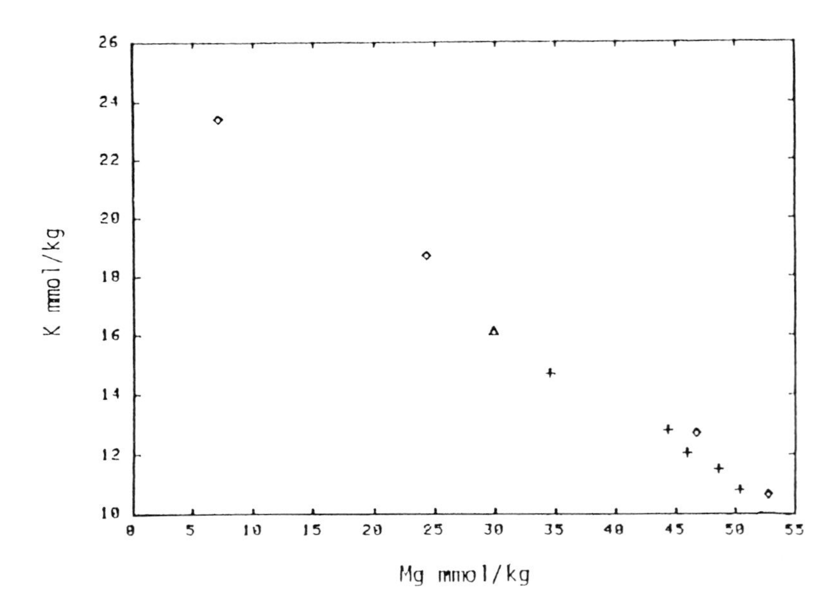


FIGURE 9.42. The potassium concentration as a function of magnesium in vent waters.

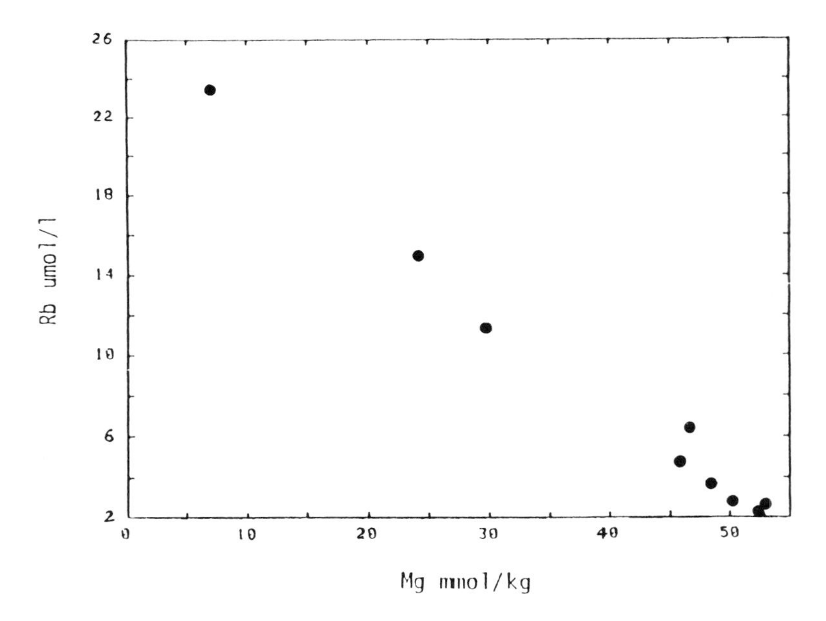


FIGURE 9.43. The rubidium concentration as a function of magnesium in vent waters.

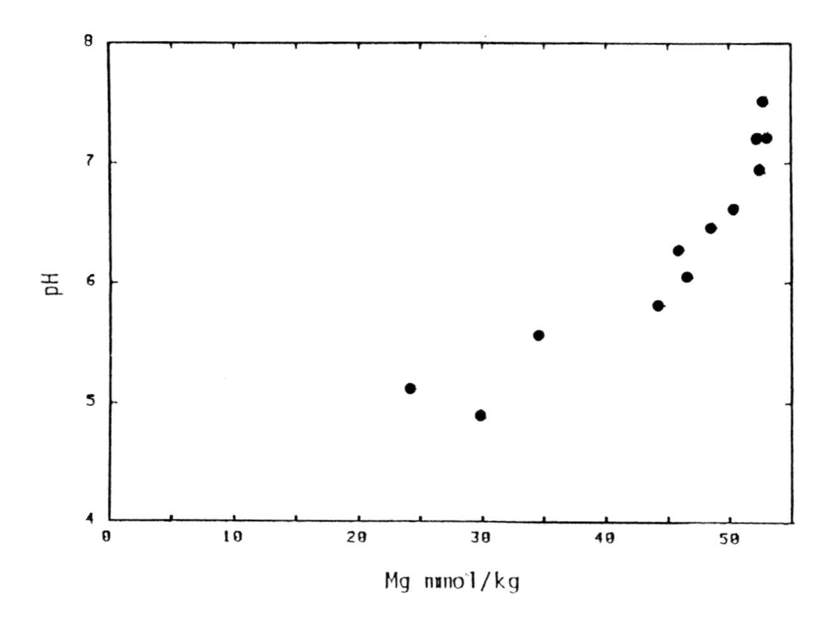


FIGURE 9.44. The pH as a function of magnesium in vent waters.

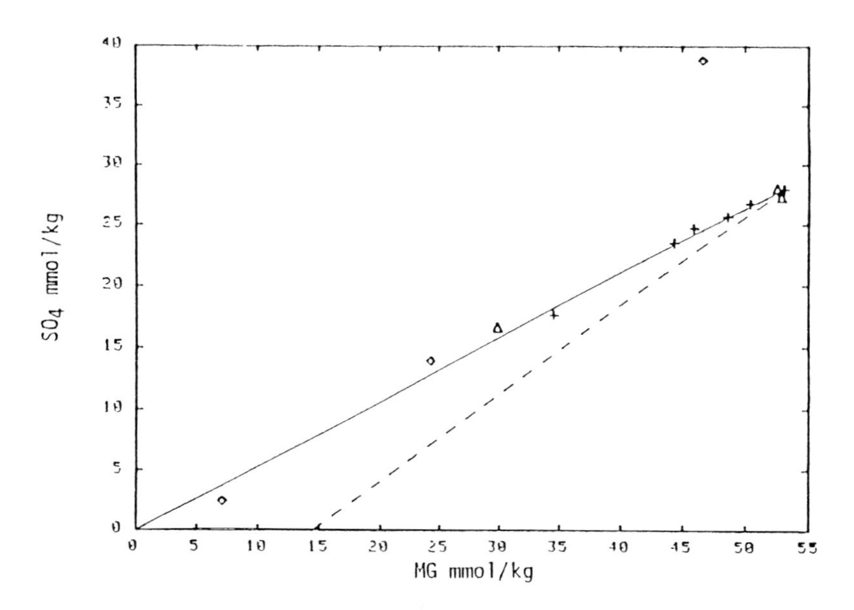


FIGURE 9.45. The sulfate concentration as a function of magnesium in vent waters.

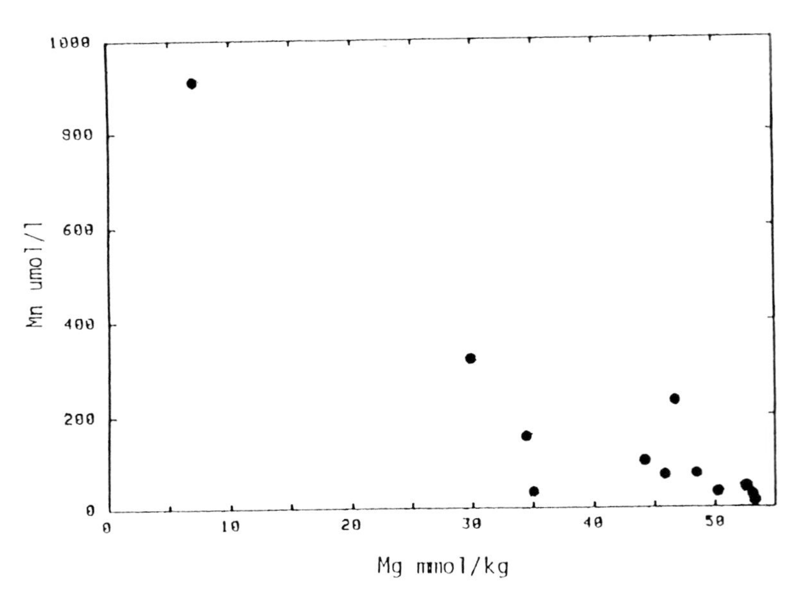


FIGURE 9.46. The manganese concentration as a function of magnesium in vent waters.

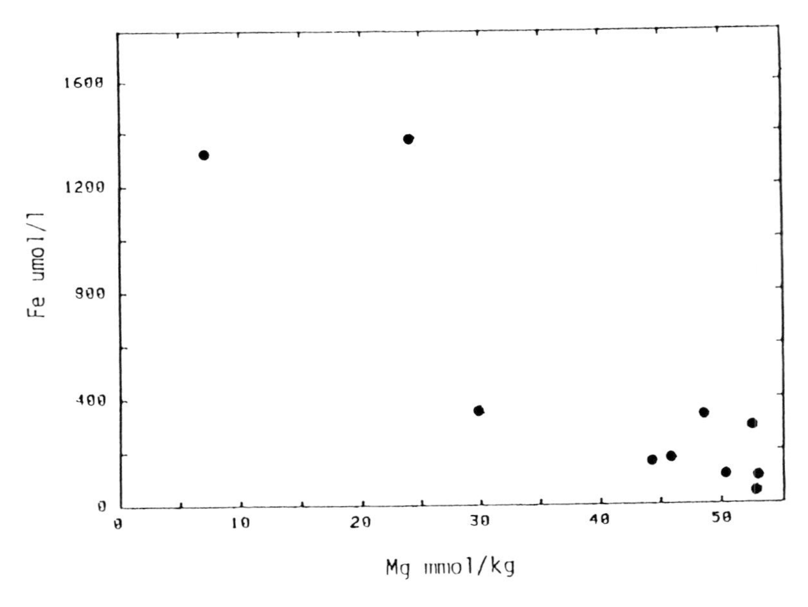


FIGURE 9.47. The iron concentration as a function of magnesium in vent waters.

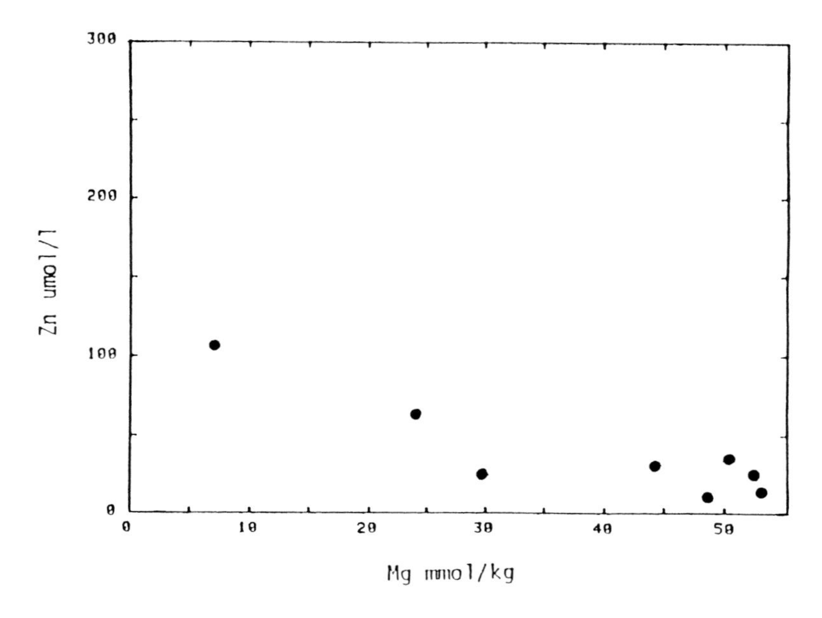


FIGURE 9.48. The zinc concentration as a function of magnesium in vent waters.

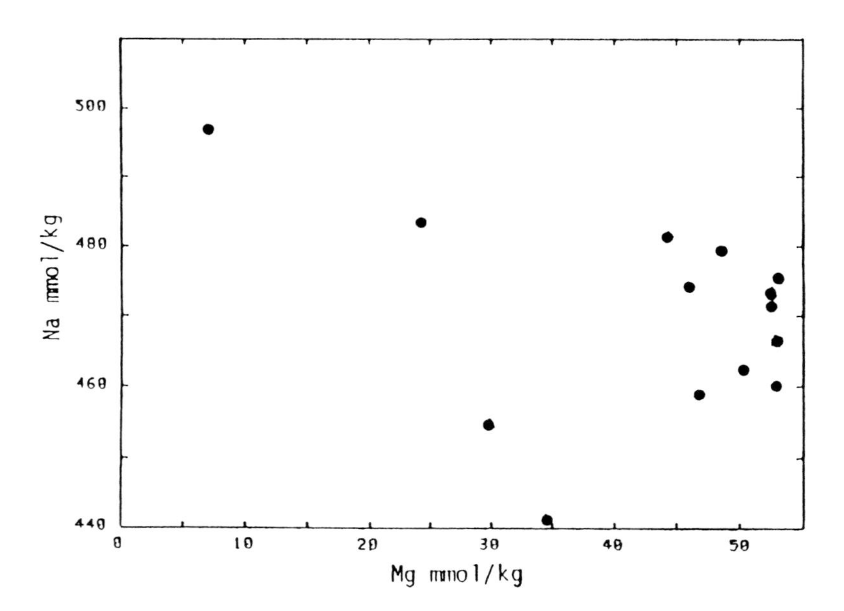


FIGURE 9.49. The sodium concentration as a function of magnesium in vent waters.

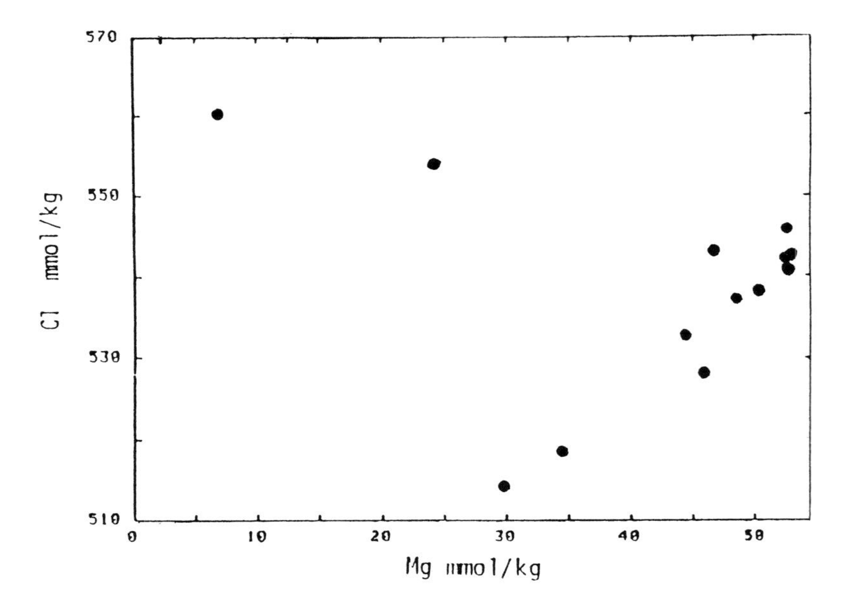


FIGURE 9.50. The chloride concentration as a function of magnesium in vent waters.

## 4. PHOTOCHEMICAL PROCESSES IN SEAWATER

In recent years there has been an increasing interest in the photochemical processes that occur in the surface waters of the oceans and other natural waters. Much of this work has been stimulated by the pioneering work of Zika (U. Miami) and Zafiriou (WHOI), who applied state of the art photochemistry to ocean systems. In this chapter we will briefly discuss the principles of photochemistry and the application of these principles to ocean systems.

**TABLE 9.8**
**Comparison of the Estimated Composition of a 350°C Hydrothermal End Member (based on extrapolation of the Galapagos data) With that Observed at 21N**

| | Galapagos | 21N | Seawater |
|---|---|---|---|
| Li ($\mu$mol kg$^{-1}$) | 1142–689 | 820 | 28 |
| K (mmol kg$^{-1}$) | 18.8 | 25.0 | 10.1 |
| Rb ($\mu$mol kg$^{-1}$) | 20.3–13.4 | 26.0 | 1.32 |
| Mg (mmol kg$^{-1}$) | 0 | 0 | 52.7 |
| Ca (mmol kg$^{-1}$) | 40.2–24.6 | 21.5 | 10.3 |
| Sr ($\mu$mol kg$^{-1}$) | 87 | 90 | 87 |
| Ba ($\mu$mol kg$^{-1}$) | 42.6–17.2 | 95–35 | 0.145 |
| Mn ($\mu$mol kg$^{-1}$) | 1140–360 | 610 | 0.002 |
| Fe ($\mu$mol kg$^{-1}$) | +[a] | 1800 | –[b] |
| Si (mmol kg$^{-1}$) | 21.9 | 21.5 | 0.160 |
| $SO_4^{2-}$ (mmol kg$^{-1}$) | 0 | 0 | 28.6 |
| $H_2S$ (mmol kg$^{-1}$) | +[a] | 6.5 | 0 |

*Note:* The ranges of the Galapagos results derive from the different trends for composition vs. heat observed between individual vent fields.

[a] Nonconservative to subsurface mixing.
[b] Seawater concentration not accurately known.

## 4.1. Principles

A photochemical process is a chemical reaction that takes place only under the influence of light. Photochemistry is concerned with the interaction of electromagnetic radiation and matter; it is at the crossroads of physics and chemistry. The light for photochemical processes includes the visible (700 to 400 nm), near ultraviolet (400 to 100 nm), and near infrared (700 to 1000 nm). This light is considered to be made up of individual photons of energy $E = h\nu$ (where h is Planck's constant, $6.63 \times 10^{-34}$ Js and $\nu$ is the frequency of light).

Only light which is absorbed can cause a photochemical effect (Grotthus-Draper law). Although this may seem to be obvious, it has been known as the first law of photochemistry for a long time. The primary photochemical reaction can be written in the form

$$M + h\nu \rightarrow M^* \qquad (70)$$

**TABLE 9.9**
**A Comparison of Hydrothermal and River Fluxes of Elements into the Oceans (mol/year)**

| | 21N | GSC[a] | River[b] |
|---|---|---|---|
| Li | $1.2 \rightarrow 1.9 \times 10^{11}$ | $9.5 \rightarrow 16 \times 10^{10}$ | $1.4 \times 10^{10}$ |
| Na | $-8.6 \rightarrow 1.9 \times 10^{12}$ | +, − | $6.9 \times 10^{12}$ |
| K | $1.9 \rightarrow 2.3 \times 10^{12}$ | $1.3 \times 10^{12}$ | $1.9 \times 10^{12}$ |
| Rb | $3.7 \rightarrow 4.6 \times 10^{9}$ | $1.7 \rightarrow 2.8 \times 10^{9}$ | $5.0 \times 10^{6}$ |
| Be | $1.4 \rightarrow 5.3 \times 10^{6}$ | $1.6 \rightarrow 5.3 \times 10^{6}$ | $3.3 \times 10^{7}$ |
| Mg | $-7.5 \times 10^{12}$ | $-7.7 \times 10^{12}$ | $5.3 \times 10^{12}$ |
| Ca | $2.4 \rightarrow 15 \times 10^{11}$ | $2.1 \rightarrow 4.3 \times 10^{12}$ | $1.2 \times 10^{13}$ |
| Sr | $-3.1 \rightarrow +1.4 \times 10^{9}$ | 0 | $2.2 \times 10^{10}$ |
| Ba | $1.1 \rightarrow 2.3 \times 10^{9}$ | $2.5 \rightarrow 6.1 \times 10^{9}$ | $1.0 \times 0^{10}$ |
| F | $-1.0 \rightarrow 2.3 \times 10^{9}$ | $2.5 \rightarrow 6.1 \times 10^{9}$ | $1.0 \times 10^{10}$ |
| Cl | $0 \rightarrow 1.2 \times 10^{13}$ | $-31 \rightarrow +7.8 \times 10^{12}$ | $6.9 \times 10^{12}$ |
| $SiO_2$ | $2.2 \rightarrow 2.8 \times 10^{12}$ | $3.1 \times 10^{12}$ | $6.4 \times 10^{12}$ |
| Al | $5.7 \rightarrow 7.4 \times 10^{8}$ | na | $6.0 \times 10^{10}$ |
| $SO_4$ | $-4.0 \times 10^{12}$ | $-3.8 \times 10^{12}$ | $3.7 \times 10^{12}$ |
| $H_2S$ | $9.4 \rightarrow 12 \times 10^{11}$ | + | |
| $\Sigma S$ | $-2.8 \rightarrow 3.1 \times 10^{11}$ | − | |
| Mn | $1.0 \rightarrow 1.4 \times 10^{11}$ | $5.1 \rightarrow 16 \times 10^{10}$ | $4.9 \times 10^{9}$ |
| Fe | $1.1 \rightarrow 3.5 \times 10^{11}$ | + | $2.3 \times 10^{10}$ |
| Co | $3.1 \rightarrow 32 \times 10^{6}$ | na | $1.1 \times 10^{8}$ |
| Cu | $0 \rightarrow 6.3 \times 10^{9}$ | − | $5.0 \times 10^{9}$ |
| Zn | $5.7 \rightarrow 15 \times 10^{9}$ | na | $1.4 \times 10^{10}$ |
| Ag | $0 \rightarrow 5.4 \times 10^{6}$ | na | $8.8 \times 10^{7}$ |
| Cd | $2.3 \rightarrow 26 \times 10^{6}$ | − | |
| Pb | $2.6 \rightarrow 5.1 \times 10^{7}$ | na | $1.5 \times 10^{8}$ |
| As | $0 \rightarrow 6.5 \times 10^{7}$ | na | $7.2 \times 10^{8}$ |
| Se | $0 \rightarrow 1.0 \times 10^{7}$ | na | $7.9 \times 10^{7}$ |

*Note:* +: gain; −: loss; na: not analyzed.

[a] GSC data is from Edmond et al. (MIT).
[b] River concentrations and fluxes are from either Edmond et al. (1979a,b) or Broecker and Peng (1982).

where M is a molecule that can absorb light (a chromophore), hν is the photon of light, and M* is the molecule in an "excited state". The extra energy (hν) and the properties of M* result in the photochemical process. The second law of photochemistry (Stark-Einstein law) states that an atom or molecule undergoing a photochemical process absorbs only a single photon.

$$\text{Reactants} + n\,h\nu \rightarrow \text{Products} \tag{71}$$

It is necessary to consider the yield in terms of the amount of light absorbed by the sample. For long irradiations it is often possible to obtain nearly quantitative yields like common chemical reactions (which can be characterized in terms of the yield of products formed or reactants used). The yield of a photochemical reaction can be limited by four factors:

1. The occurrence of competitive reactions

$$M^* \begin{array}{l} \nearrow A \\ \searrow B \end{array} \tag{72}$$

   each with its own yield.

2. The occurrence of further photochemical reaction products

$$M^* \rightarrow A \tag{73}$$

$$A + h\nu \rightarrow A^* \tag{74}$$

$$A^* \rightarrow B \tag{75}$$

   The maximum yield of A will thus depend upon the wavelength of irradiation, quantum yield, etc.

3. The existence of true photochemical equilibrium

$$M \underset{h\nu'}{\overset{h\nu}{\rightleftarrows}} A \tag{76}$$

   This equilibrium will depend on the irradiation wavelength.

4. The existence of a dark reverse reaction

$$M \underset{\text{dark}}{\overset{h\nu}{\rightleftarrows}} A \tag{77}$$

The quantum yield or efficiency of a photochemical reaction is defined as the number of molecules of product formed for each photon absorbed

$$\phi = \frac{\text{number of product molecules formed}}{\text{number of photons absorbed}} \tag{78}$$

This is the quantum yield of product formation. It is also possible to define the quantum yield of the disappearance of the reactant

$$\phi = \frac{\text{number of reactant molecules which disappear}}{\text{number of photons absorbed}} \tag{79}$$

The values of $\phi$ are generally less than 1.0, but can be greater than 1.0 for chain reactions.

The sequence of a photochemical reaction takes place in two steps. The first step starts with the absorption of light

$$M + h\nu \rightarrow M^* \text{ (absorption)} \tag{80}$$

The excited molecule can react by rearrangement

$$M^* \rightarrow P \tag{81}$$

or by reaction with another species N

$$M^* + N \rightarrow P \tag{82}$$

These steps involving M* are called the primary photochemical process. The products P may undergo further reactions which lead to final, stable products.

The excited molecule can lose its energy by the loss of heat

$$C^* \rightarrow C + \text{heat} \tag{83}$$

by the loss of light energy

$$C^* \rightarrow C + h\nu \tag{84}$$

or by an ionization

$$C^* \rightarrow C^+ + e^- \tag{85}$$

where $e^-$ is an electron. Secondary reactions of the primary photo products can frequently occur with the components of the solution. For example, when the solutions contain $O_2$, the free radical, $O_2^-$, can be formed. This radical can react as both an oxidizing and reducing agent.

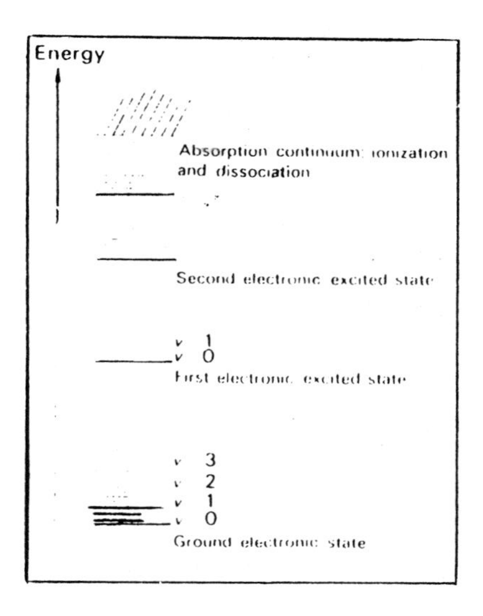


FIGURE 9.51. Repartition of electronic, vibrational, and rotational states in polyatomic molecules.

Before we discuss various photochemical reactions important in seawater, it is useful to briefly review the energy states of molecules. The types of excited molecules will depend upon the amount of energy absorbed. Rotational excitation of a molecule requires the smallest amount of energy. This results in a spinning of the molecule around a preferred axis; however, the molecule is not changed chemically. With higher energies, the molecule may be promoted to a vibrationally excited state. The energy is in the form of vibrations of various parts of the molecule; again, the molecule is not changed chemically. With higher energies, the molecule can be excited to an electronically excited state. One or more electrons can be promoted to higher energy orbitals.

Photochemical reactions occur from such electronically excited states of molecules. If a very high amount of energy is absorbed, a continuum is reached and the molecule is ionized or dissociated

$$M \xrightarrow{h\nu} M^+ + e^- \tag{86}$$

$$M \xrightarrow{h\nu} X + Y \tag{87}$$

The electronic states of a molecule are shown in Figure 9.51. In the electronic ground state as well as the various electronically excited states, there are

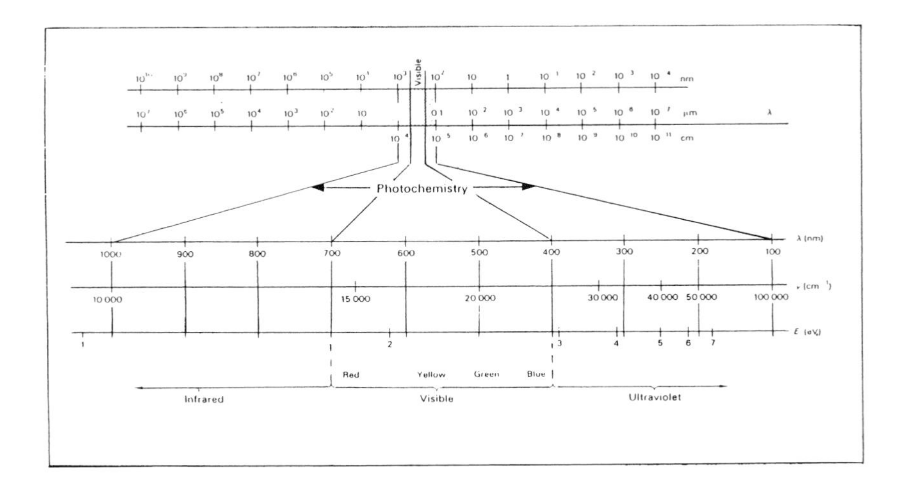


FIGURE 9.52. The electromagnetic radiations.

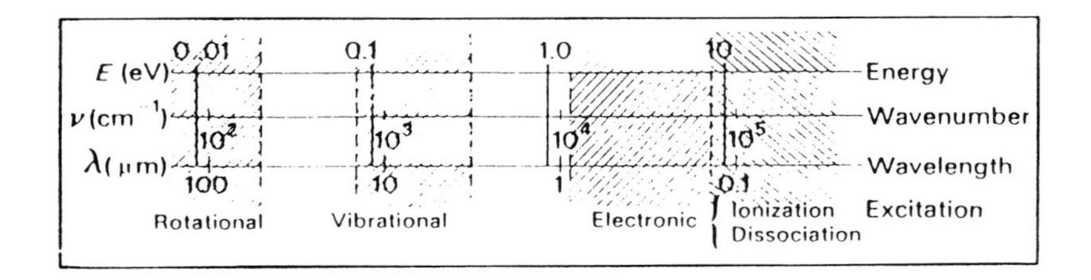


FIGURE 9.53. The energies of states and characteristics of radiation.

sublevels ($\nu$) containing various vibrational states and each vibrational state can have various rotational sublevels.

The energy of an atom or molecule is given in units of electron-volts, kilocalories, or kilojoules. One electron-volt is equal to $9.65 \times 10^4$ J. The frequency ($\nu$) of electromagnetic radiation is equivalent to a given energy ($E = h\nu$). Spectroscopists often use the wave-number rather than the frequency (wave number $= \nu/c$, where $c = 3 \times 10^8$ m s$^{-1}$, the speed of light). The wavelength is related to the frequency by

$$\lambda = c/\nu \tag{88}$$

The wave length is given in units of "Angstroms" ($1 \text{ Å} = 10^{-10}$ m) or the preferred unit of the nanometer ($1 \text{ nm} = 10^{-9} \text{ m} = 10 \text{ Å}$).

The relationship between energy and the various types of excitation are shown in Figure 9.52. These energies can be compared (Figure 9.53) to the energy of thermal agitation (RT = 144 J $mol^{-1}$) which is the same order of magnitude of translational and rotational energy. Although some of the low vibrational levels (v = 0) can be occupied, electronically excited states cannot be normally obtained from heat.

The absorption of a photon of electromagnetic radiation is possible if the energy corresponds to the energy separation between orbitals. Emission of a quantum of energy by an atom takes place when an electron jumps back from an outer orbital to an inner orbital. The frequencies of the light emitted are the same as the frequencies of the light absorbed. In the ground state, all the electrons fill the available orbitals in an order of increasing energy. An excited state has one or several electrons occupying higher energy orbitals, leaving one or several vacancies of lower orbitals. A ground state can only absorb light. An excited state can either emit light going to a lower or ground state. It can also absorb light and go upwards in energy to a higher excited state.

A schematic diagram of the photophysical process and its relationship to the various electronic states of an atom or molecule is shown in Figure 9.54. The absorption of light excites the molecule from the ground state ($S_0$) to an excited state ($S_n$). The excited molecule loses some of its energy and can be transferred to a lower singlet (spin paired) state. For an isolated atom, the emission wavelength is the same as the absorption wavelength. The emission from molecules is almost exclusively observed from the lowest excited state. The lifetime of this emission is usually very short ($10^{-8}$ s). The emission process of an excited molecule from a singlet state to the ground state is called fluorescence. The emission of many molecules is shifted to longer wavelengths (lower energy). This is shown for anthracene in Figure 9.55. Many molecules show a longer-lived emission at longer wavelengths. This emission can last for seconds. It comes from a special type of excited state known as a triplet (unpaired) and is called phosphorescence. Two electrons can share the same orbital only if they have different spin quantum numbers ($\uparrow\downarrow$). In an excited state one electron is in a singly occupied higher orbital, while it has left a vacancy in the lower orbital which is also singly occupied. These two electrons in different orbitals are quite independent and can have their spins antiparallel ($\uparrow\downarrow$) as in the ground state or parallel ($\uparrow\uparrow$ or $\downarrow\downarrow$). When the spins are parallel the excited state is called the triplet state. The emission from a molecule in the triplet state is called phosphorescence.

Delayed fluorescence arises from the production of an excited singlet state from the triplet or by chemical reactions. Singlet excited states can also be formed by triplet-triplet annihilation. This can result in delayed fluorescence if the diffusion of the excited molecules is rapid. An excited state that occurs

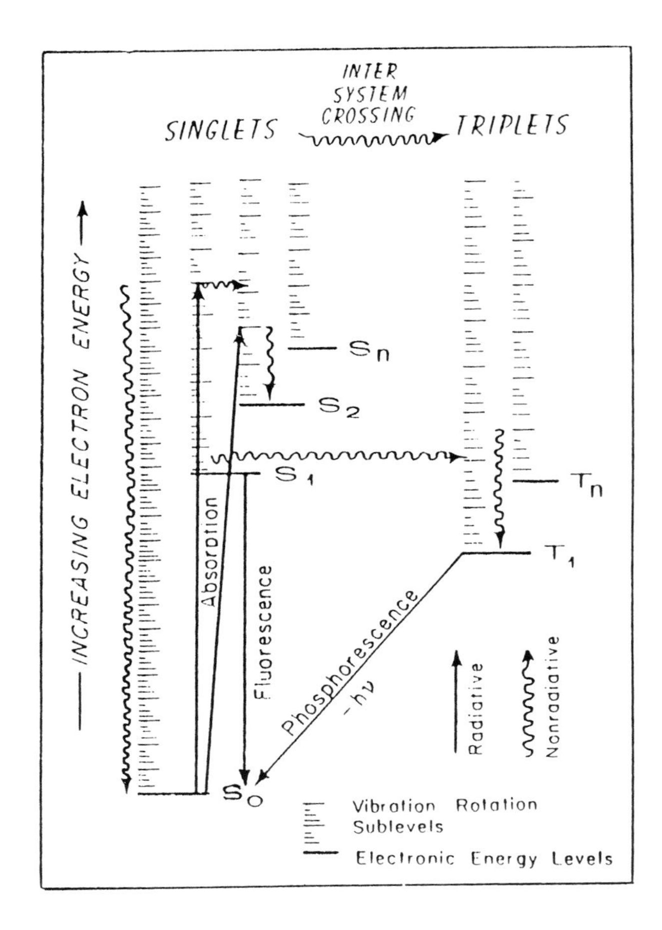


FIGURE 9.54. The photophysical processes and electronic states.

by chemical effects can also have a delayed fluorescence called chemiluminescence.

Photochemical processes in the surface waters of the oceans can be important because they can generate various redox transients. One, however, cannot rule out the formation of redox transients from precipitation (rainout) or gas phase deposition as well as biological processes (Figure 9.56). For photochemical processes to occur, it is necessary to have light of sufficient energy (usually UV). As shown in Figure 9.57, below a depth of 100 m little or no UV light is available. For photochemical processes to occur, chromophores must be available to absorb the light. Some of the photochemical processes that occur in natural waters are given in Table 9.10. The

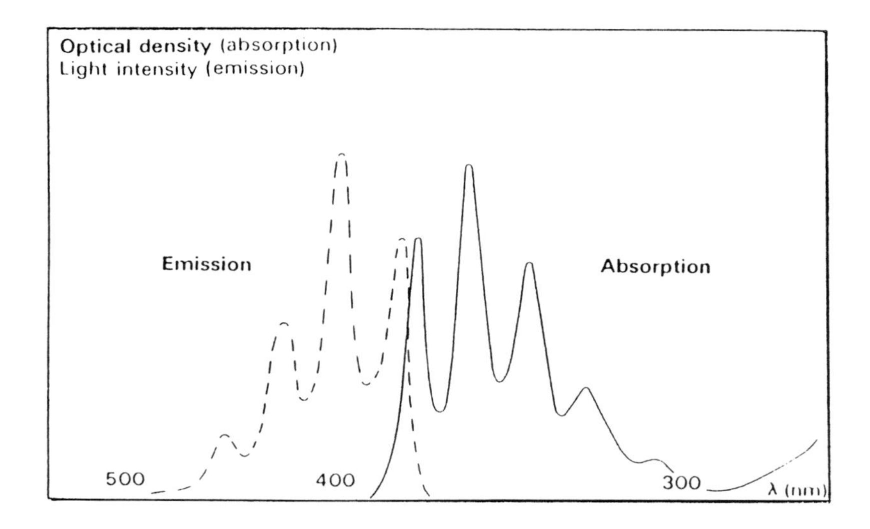


FIGURE 9.55. The fluorescence emission and absorption spectra of anthracene in solution.

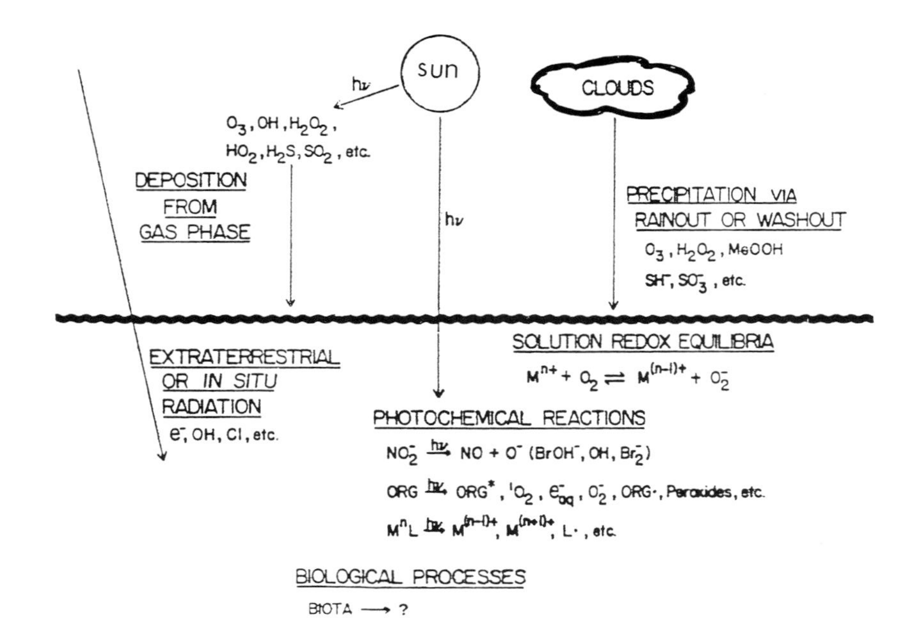


FIGURE 9.56. The primary sources of redox transients in the oceans.

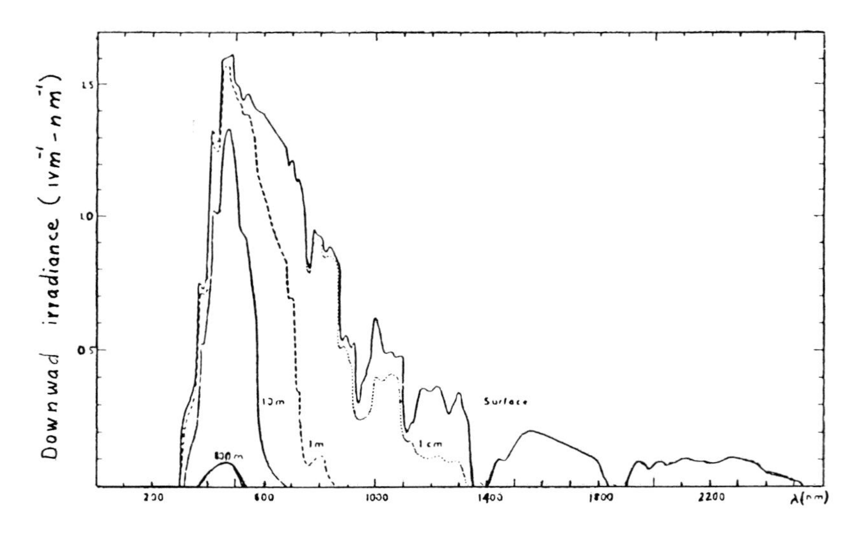


FIGURE 9.57. The spectra of light in the surface of the ocean.

**TABLE 9.10**
**The Diversity of Natural Water Photoprocesses**

| Environment | Substrates | Products | Probable mechanisms |
|---|---|---|---|
| Marine and fresh | Natural org. chromophores | C· + $HO_2$ | H atom transfer to $O_2$ or A |
| | | C· + AH· | |
| | | $C^+$ + $O_2^-$ | Energy transfer to $O_2$ |
| | | HOOH | Disproportionation of $HO_2$ |
| | $NO_2^-$ | NO + OH | Direct photolysis |
| | $Br^-$, $CO_3^{2-}$, RH | $Br_2^-$, $CO_3^-$, R· | OH Radical redox and abstraction |
| | R· | ROO· | $O_2$ Addition |
| Marine | $CH_3I$ | $CH_3^+$ | Direct photolysis |
| | $MnO_2$ (colloidal) | $Mn^{2+}$ | Unknown |
| | Cu(II) | Cu(I)Cl | $H_2O_2/O_2^-$ Reduction of Cu(II) |
| | Cu(II)L | Cu(I)Cl | Charge transfer to metal |
| | Fe(II) | Fe(III) + $H_2O_2$ | OH radicals ? |
| Freshwater | Fe(III)L | Fe(II) | $O_2^-$ ? |
| Surface films | $I^-$ | HOI, Oxidized organics | |
| Polluted waters | RH, ArH, $R_2$ | R=O, $RCO_2^-$, $R_2SO$ | Free radicals, direct photolysis, singlet $O_2$ |

## TABLE 9.11
## Photoreactions of Organic Compounds

| Chromophore | Products or effects |
|---|---|
| Humic, fulvic, etc. | Bleaching of absorption and fluorescence |
| | Production of singlet oxygen |
| | Fe(III) reduction |
| | Release of soluble P |
| | Oxidation of cumene via ROO and OH radicals |
| | Oxidation of phenolic groups to ArO and formation of $e^-$ and $O_2^-$ |
| | CO formation |
| | $H_2O_2$ formation (via $O_2^-$ ?) |
| Chlorophyll | Loss of chlorophyll |
| Vitamins | Loss of bioassay activity |
| Amino acids | ? |
| Glycine | COOH $C^{14}$ loss, HCHO 1 formation |
| $CH_3SSCH_3$ | $CH_3S$ |
| $CH_3I$ | $CH_3$ |
| Fatty acids | Particles, absorb., hydroperoxides |
| Aldehydes | RCO, R, CO |

## TABLE 9.12
## Photoreactions of Inorganic Compounds

| Compounds | Products or effects |
|---|---|
| $Fe^{2+}$ | $Fe^{3+} + H_2$ (Precambrian) |
| $Cu^{2+}$ complexes | $NH_3$ and HCHO |
| Fe-humates | $Fe^{2+}$, $CO_2$, $O_2$ |
| | $Fe^{2+}$, $PO_4^{3-}$ |
| $I^-$ | $I + e^-$ |
| $NO_2^-$ | $NO + OH$ |
| | Methionine destruction |
| $NO_3^-$ | $NO_3^-$ loss, $NO_2^-$ |

chromophores responsible for the initial absorption of light are thought to be humic and fulvic materials. A number of other organic and inorganic molecules may also be important and are given in Table 9.11 and 9.12.

The recent findings of hydrogen peroxide ($H_2O_2$) in surface seawaters (Figure 9.58) has strongly supported the notion that photochemical processes

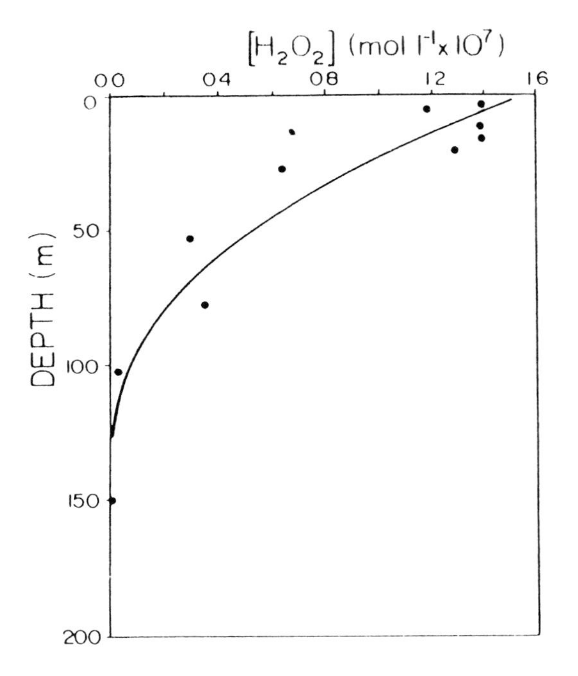


FIGURE 9.58. Profile of hydrogen peroxide in the Gulf of Mexico.

occur in the oceans. The sources of the $H_2O_2$ can result from three major processes.

$$\text{I. Org} + h\nu \rightarrow \text{Org}^* \tag{89}$$

$$\text{Org}^* \rightarrow \text{Org}^+ + e^- \tag{90}$$

$$O_2 + e^- \rightarrow O_2^- \tag{91}$$

$$O_2^- + H^+ \rightarrow HO_2 \tag{92}$$

$$HO_2 + HO_2 \rightarrow H_2O_2 + O_2 \tag{93}$$

$$\text{II. Org}^* + O_2 \rightarrow \text{Org}^+ + O_2^- \tag{94}$$

$$\text{Org}^* + \text{sub} \rightarrow \text{Org}^- + \text{sub}^+ \tag{95}$$

$$\text{Org}^* + \text{sub} \rightarrow \text{Org}^+ + \text{sub}^- \tag{96}$$

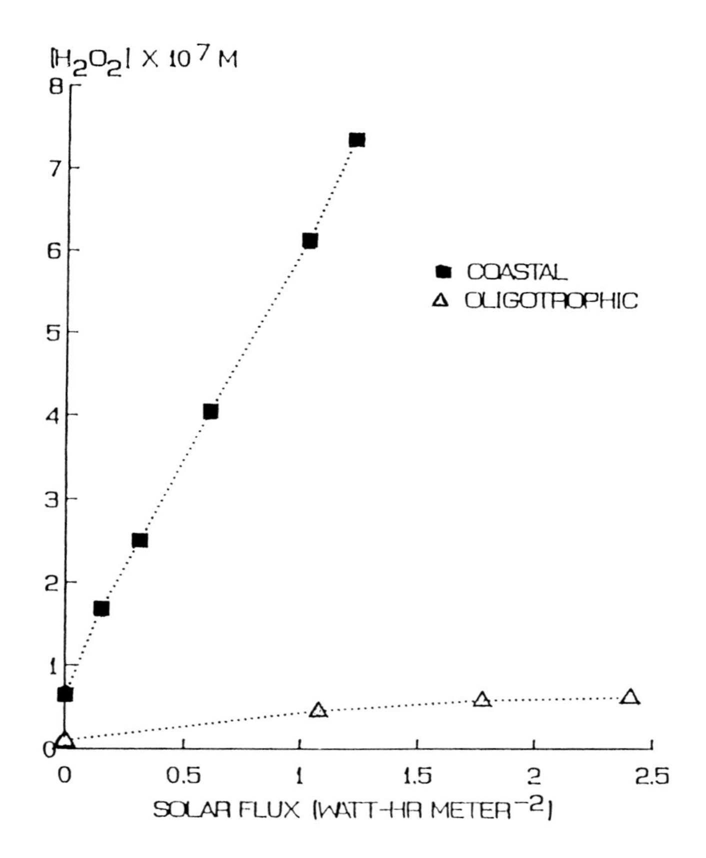


FIGURE 9.59. The photochemical accumulation of hydrogen peroxide in seawater.

$$Org^{+} + sub \rightarrow Org + sub^{+} \quad (97)$$

$$Org^{-} + O_2 \rightarrow Org + O_2^{-} \quad (98)$$

$$sub^{-} + O_2 \rightarrow sub + O_2^{-} \quad (99)$$

$$\text{III. } M^{n+} + h\nu \rightarrow M^{(n-1)} + L^{-} \quad (100)$$

$$M^{(n-1)} + O_2 \rightarrow M^{n+} + O_2^{-} \quad (101)$$

where sub is a substance able to give up or accept an electron. Recent studies by Zika and co-workers indicate that the formation of $H_2O_2$ is higher in coastal waters with higher humic concentration than in oligotrophic waters (Figure 9.59). The production rates also appear to be directly related to the concen-

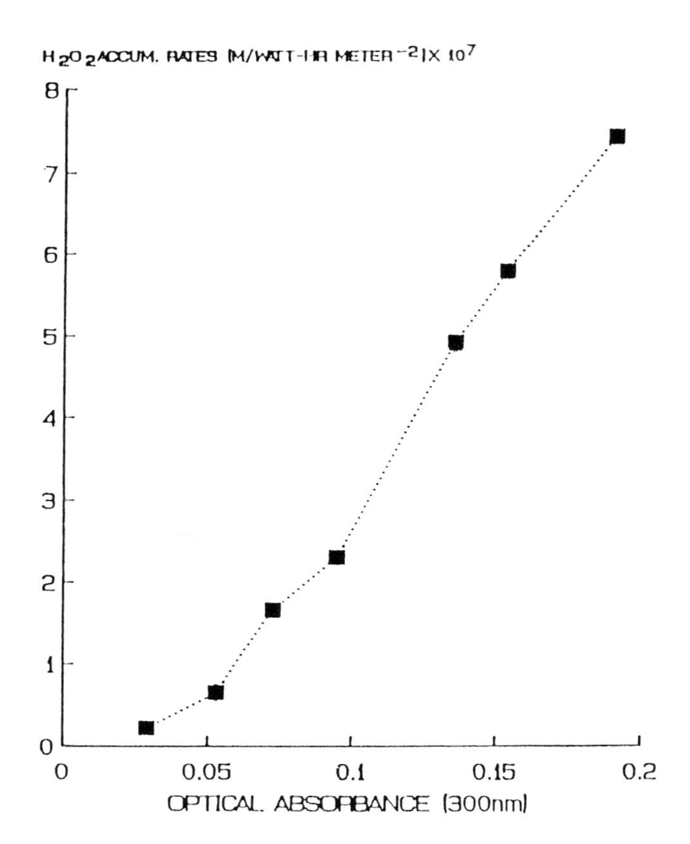


FIGURE 9.60. The accumulation rates for different seawaters as a function of the optical absorbance.

tration of the humic in the water (Figure 9.60) as measured by absorbance at 300 nm.

Since steady state levels of $H_2O_2$ are found at $2 \times 10^{-7}$ *M*, the decay of $H_2O_2$ must be slow. The diurnal variation shown in Figure 9.61 shows concentrations that vary from $6 \times 10^{-8}$ to $18 \times 10^{-8}$ *M* during the day. The half-time of $H_2O_2$ in Gulf Stream waters is about 120 h and about 12 h in coastal waters. Studies also indicate that the lifetimes are much longer in deep waters (1900 h at 250 m). These results indicate that the lifetime of the decay of $H_2O_2$ may be controlled by particles or biological processes. The effect of particles is shown in Figure 9.62. The filtration through a 0.2 μm filter slows down the decay.

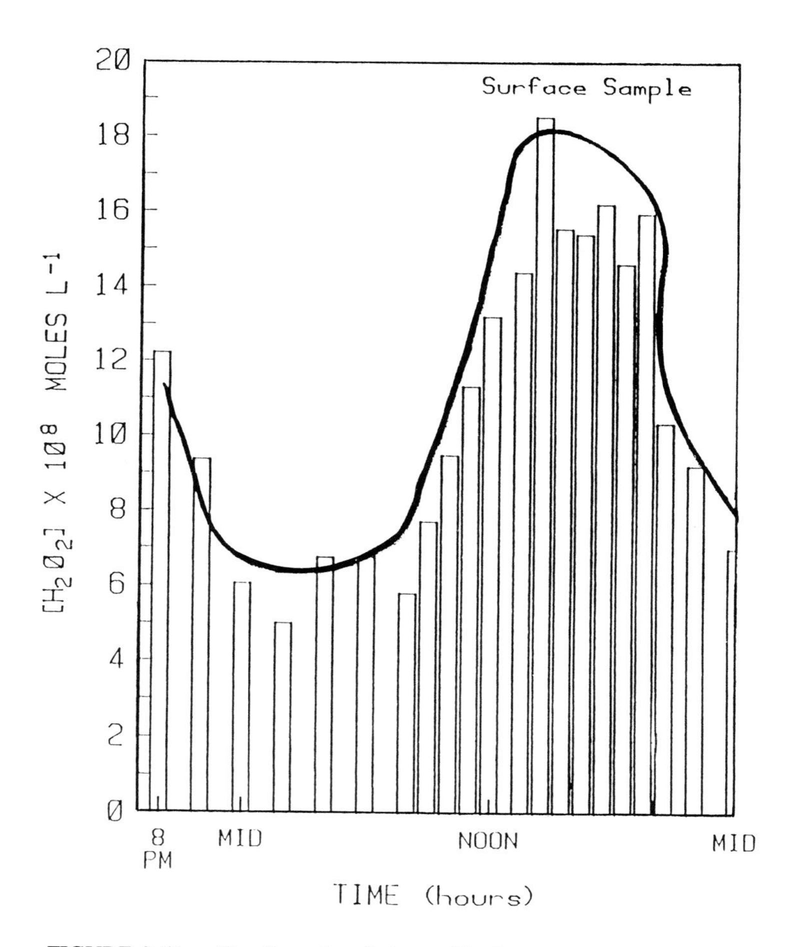


FIGURE 9.61. The diurnal variations of hydrogen peroxide in surface waters in the Bahamas.

Since autoclaving the seawater has a similar effect (Figure 9.63), the decay appears to be partly related to biological particles (living and dead cells). Further work is needed to elucidate the decay mechanism of $H_2O_2$ and the role that cells have on the formation and destruction of $H_2O_2$. At present, the formation of $H_2O_2$ appears to be abiotic. The decomposition is not affected by light and occurs in a matter of days. Enzymes as well as particles may be important. Although abiotic processes are small, they may be important in the open oceans.

The superoxide anion ($O_2^-$) is a key intermediate in the formation of $H_2O_2$

$$2O_2^- + 2H^+ \rightarrow H_2O_2 \qquad (102)$$

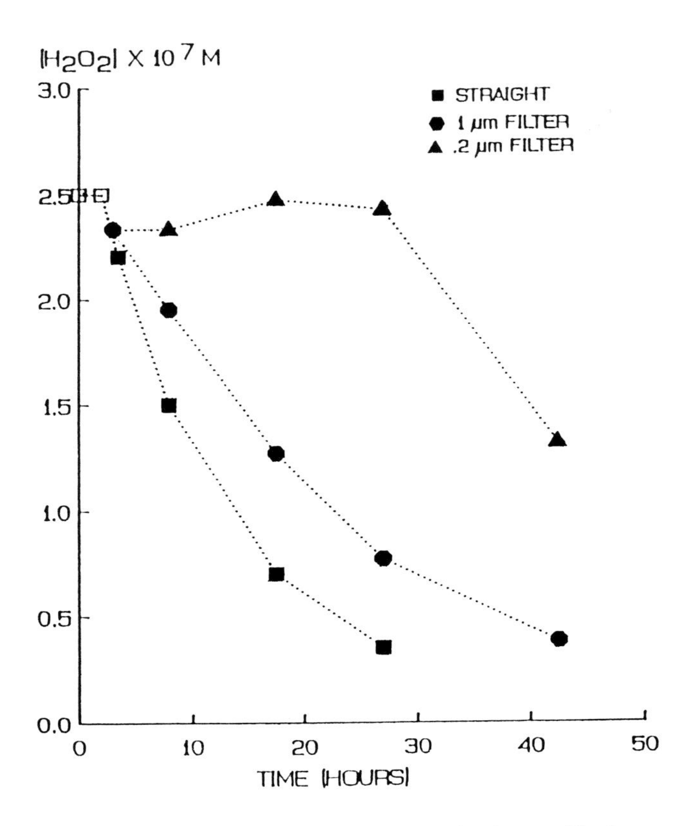


FIGURE 9.62. The effect of filtering on the decay of hydrogen peroxide in seawater.

This anion is a strong reductant and can result in the formation of reduced metals

$$Fe^{3+} + O_2^- \rightarrow Fe^{2+} + O_2 \tag{103}$$

$$Cu^{2+} + O_2^- \rightarrow Cu^+ + O_2 \tag{104}$$

Since $Cu^+$ has recently been measured in surface seawaters by Moffett (Figure 9.64), this reduction process may result in transient species. The oxidation of $Cu^+$ to $Cu^{2+}$ is prevented by the formation of chloro complexes that are resistant to oxidation

$$Cu^+ + Cl^- \rightarrow CuCl \tag{105}$$

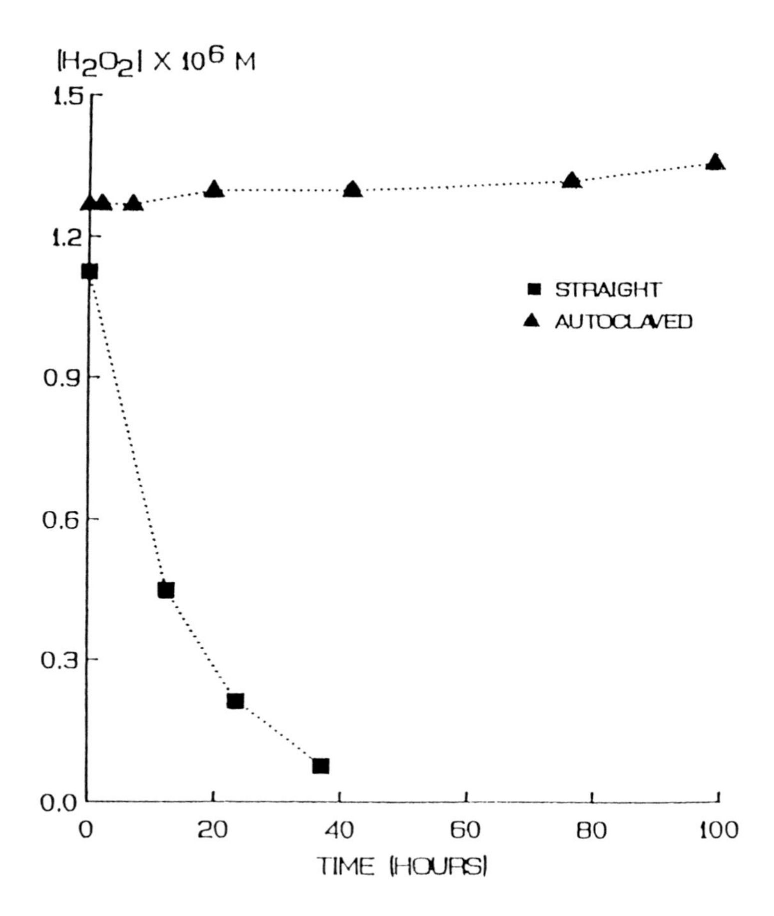


FIGURE 9.63. The effect of autoclaving on the decay of hydrogen peroxide in seawater.

$$CuCl + Cl^- \rightarrow CuCl_2^- \tag{106}$$

$$CuCl_2^- + Cl^- \rightarrow CuCl_3^{2-} \tag{107}$$

The lifetime of $O_2^-$ in seawater should be about 300 min if it decays only to form $H_2O_2$. Since the measured values are about 20 min, the reaction of $O_2^-$ with metals and perhaps nonmetals may be important in surface waters. Our entire view of redox processes in the surface waters of the oceans has changed as a result of these new findings of photochemically generated redox species.

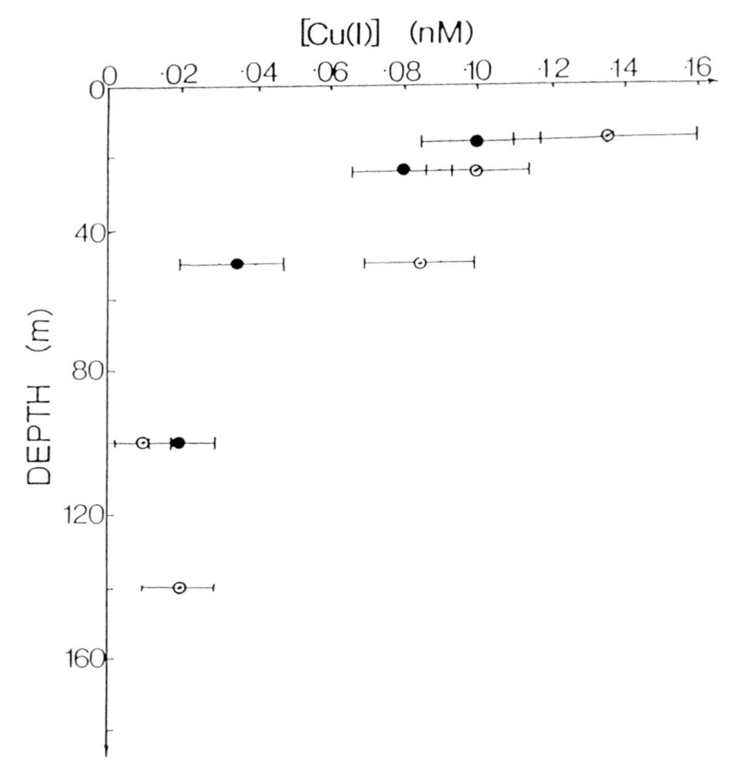


FIGURE 9.64. Profile of copper (I) in the surface waters of the Sargasso Sea.

## 5. ANOXIC WATERS

Anoxic basins were first described in detail by Richards in 1965. He defined anoxia as a condition where no dissolved oxygen is available. This condition can arise in natural waters when the consumption rate of oxygen exceeds the supply. The consumption is linked to the oxidation of organic matter by bacteria at a greater rate than the oxygen supply from the atmosphere. The supply of $O_2$ below the photic zone is dependent upon diffusion and advection. This normally occurs in enclosed basins where physical barriers (sills) and density stratification limits the advection of $O_2$ to the deep waters. There are two types of anoxic basins. The most common occurs because of a strong halocline (salinity gradient) which is the result of a net outflow of low salinity water from a positive estuary. The halocline prevents low salinity oxic waters from mixing with the high salinity deep waters. Examples of this type of basin are the Black Sea, the Baltic Sea, and many fjords such as the

**TABLE 9.13**
**Anoxic Basins**

| Name | Location | Bottom (m) | Sill (m) | Maximum $H_2$ ($\mu M$) |
|---|---|---|---|---|
| Goteland Deep | Baltic Sea | 249 | 60 | 20 |
| Cariaco Trench | Caribbean Sea | 1390 | 150 | 160 |
| Lake Nitinat | British Columbia | 250 | 4 | 250 |
| Saanich Inlet | British Columbia | 236 | 65 | 250 |
| Black Sea | Europe | 2243 | 40 | 350 |
| Framvaren | Norway | 350 | 2 | 8000 |

Framvaren in Norway. The second type of basin arises because of a strong thermocline preventing the mixing of surface and deep waters. The Cariaco Trench off the coast of Venezuela is an example of this type of basin. It is a deep trench with a maximum depth of 1400 m. The water is isohaline and isothermal from 600 m to the bottom. It is permanently anoxic below a depth of 350 m. The appearance of $H_2S$ above the thermocline is due to mixing. Both basins have a physical obstacle that prevents horizontal mixing of various water masses. In a fjord-type basin, a shallow sill prevents the salty seawater rich in $O_2$ from entering the basin and sinking to the bottom.

In recent years estuarine systems with deep basins like the Baltic and Chesapeake have experienced periodic anoxic behavior. This has been attributed to higher productivity in the surface waters due to increases of nutrients used as fertilizers and perhaps as acid rain. Some examples of anoxic basins are given in Table 9.13.

The loss of oxygen in a basin or sediment pore waters causes a series of reactions to occur in a given sequence that is shown in Figure 9.65. In an aerobic environment, $O_2$ is the electron acceptor and aerobic respiration occurs according to

$$(CH_2O)_{106}\,(NH_3)_{16}\,H_3PO_4 + 138O_2 \rightarrow 106CO_2 + 122H_2O + 16HNO_3 + H_3PO_4 \quad (108)$$

After the oxygen is depleted, denitrification occurs with $NO_3^-$ being the electron acceptor. The denitrification can be described by

$$(CH_2O)_{106}\,(NH_3)_{16}\,H_3PO_4 + 84.8HNO_3 \rightarrow 106CO_2 + 42.4N_2 + 148.4H_2O + 16NH_3 + H_3PO_4 \quad (109)$$

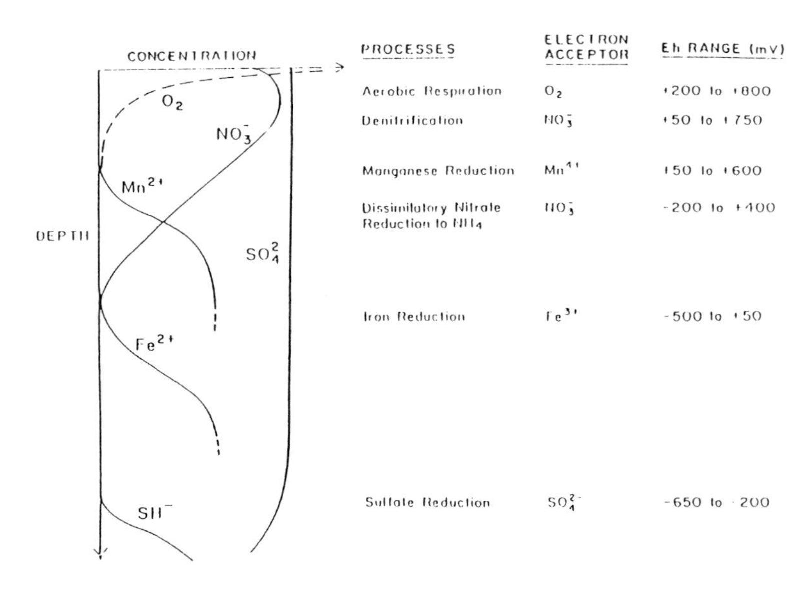


FIGURE 9.65. Schematic representation of chemical profiles in a stratified water column.

and

$$(CH_2O)_{106}\,(NH_3)_{16}\,H_3PO_4 + 94.4HNO_3 \rightarrow 106CO_2 + 55.2N_2 + 177.2H_2O + H_3PO_4 \quad (110)$$

$MnO_2$ reduction occurs next with $Mn^{4+}$ being the electron acceptor, followed by $NO_3^-$ reduction to $NH_4$ and reduction of Fe(III) to Fe(II). Since the concentration of these electron acceptors is not very high, these processes do not dominate the oxidation of organic material. The next electron acceptor is $SO_4^{2-}$ and sulfate reduction occurs according to

$$(CH_2O)_{106}\,(NH_3)_{16}\,H_3PO_4 + 53SO_4^{2-} \rightarrow 53CO_2 + 53HCO_3^- + 53HS^- + 16NH_3 + 53H_2O + H_3PO_4 \quad (111)$$

It should be noted that no oxidation occurs of $NH_3$ and the ratio of changes in $H_2S$ to $NH_3$ is 3.3.

The decomposition can also occur with $CO_2$ being the electron acceptor and the resultant formation of $CH_4$

$$(CH_2O)_{106}\ (NH_3)_{16}\ H_3PO_4 \rightarrow 53CO_2 + 53CH_4 +$$

$$16NH_3 + H_3PO_4 \qquad (112)$$

In the next sections, the distribution of various chemicals will be examined for some anoxic basins.

### 5.1. The Black Sea

The Black Sea is the largest and most studied anoxic basin. The present anoxic conditions began about 1500 to 2000 years after the influx of saline Mediterranean seawater into the basin. It is an estuarine basin of an area of 413,500 $km^2$, where precipitation and river runoff exceeds evaporation. There is a net outflow (12,600 $m^3/s$) of low salinity water compared to the inflow (6100 $m^3/s$) from the Mediterranean. This results in a permanent halocline as in the Baltic sea of thickness 100 to 240 m. The anoxic waters extend to a maximum depth of 2234 m. The oxygenated water from the Mediterranean Sea is unable to displace the anoxic bottom water. The low salinity surface waters leaving the basin are also unable to penetrate the more saline anoxic deep waters.

The depth of the transition zone between the oxic and anoxic waters varies from 100 to 250 m. Past studies showed an upper boundary of 150 m with a 50 m layer that contained both $O_2$ and $H_2S$. More recent studies in the summer of 1988 showed the interface has moved up to 100 m and showed a depth zone of 50 m without $O_2$ and $H_2S$ (Figure 9.66).

A number of workers have examined the balance of waters in the Black Sea. The annual estimates made by Fonselius are given in Figure 9.67, where R = river, p = precipitation, e = evaporation, O = outflow, and I is inflow. Complete renewal of the deep waters has been estimated to be 2500 years. Carbon measurements yield shorter lifetimes of about 800 years, presumably due to lateral mixing of salty waters from the Mediterranean through the Bosporus. A large portion of the exchange between the surface and deep waters occurs along the edge of the Black Sea.

The 1988 expedition to the Black Sea has resulted in new measurements of a number of chemical parameters. Some of the measurements were made to depths of 400 m with a pump system in real time. The resulting profiles are shown in Figures 9.68 and 9.69 for a station in the center of the Black Sea (Figure 9.70).

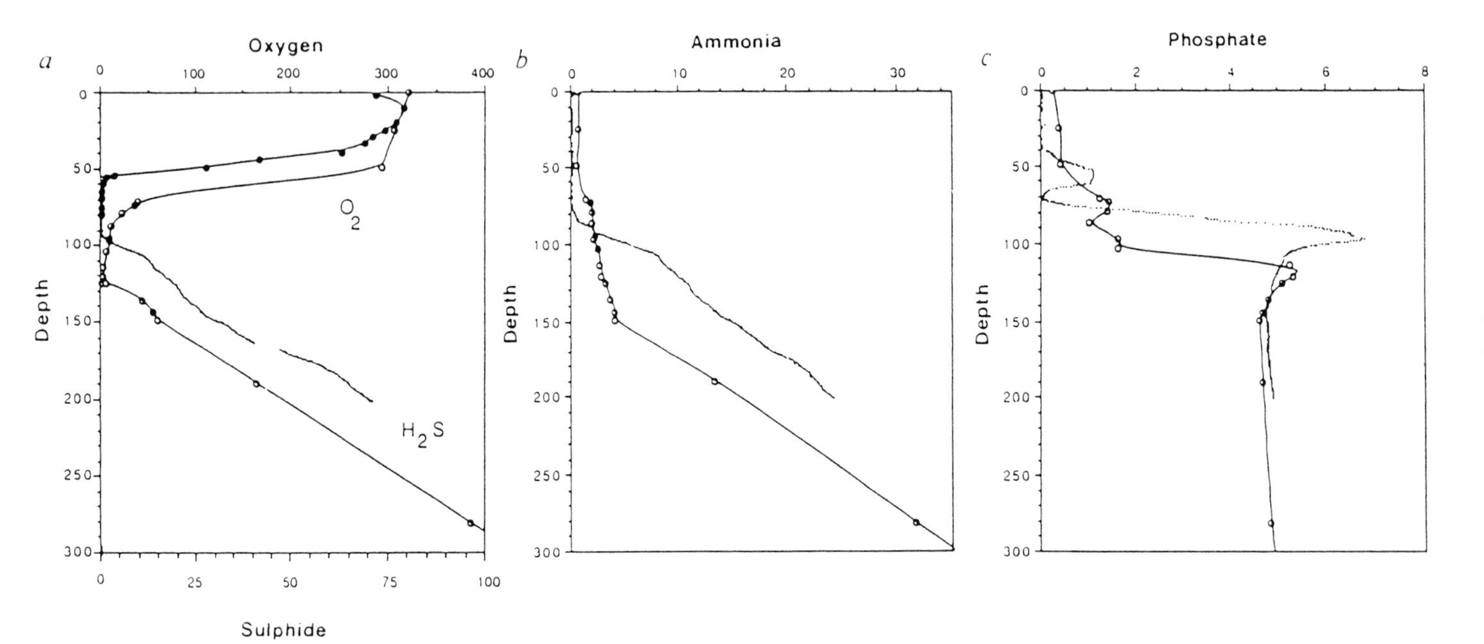


FIGURE 9.66. The changes in the oxic-anoxic interface in the Black Sea. Open circles are past studies; closed circles are recent studies.

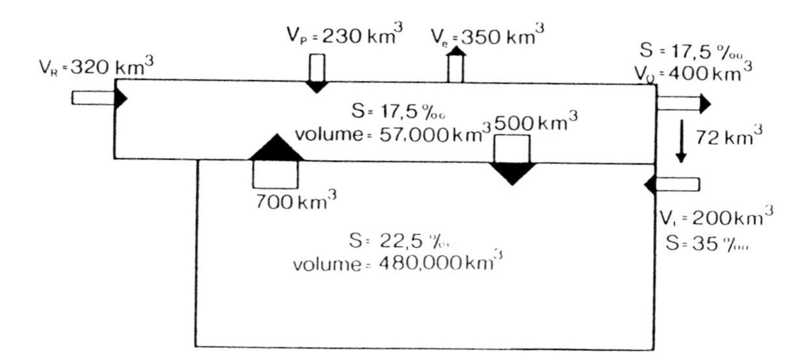


FIGURE 9.67. A box model of the Black Sea.

Recent measurements of trace metals at another station in the Black Sea have been made by Haraldsson and Westerlund (U. Gotenburg). Their results are shown in Figures 9.71 and 9.72. Mn, Co, and Fe go through a maximum near the interface due to the reduction of Mn(IV), Co(III), and Fe(III). Cu, Cd, and Pb show sharp decreases across the interface due to the low solubility of the sulfides of these metals. Ni shows little change across the interface because its concentrations are above the saturation levels in sulfidic waters.

## 5.2. Framvaren Fjord

The Framvaren is a fjord that is located in south Norway (Figure 9.73). The catchment area (31 km$^2$) is dominated by granitic rock. The rock consists of quartz (20%), microcline perthite (35%), oligoclas (36%), hornblende (5%), and accessory minerals. The fjord has a sill depth of about 2 m and a maximum depth of 180 m. It has extremely steep sides above and below the water (Figure 9.74). A major river entering the Lyngdalsfjord causes a flow of fresh water through the canal into the Framvaren. A vertical profile of salinity is shown in Figure 9.75. The surface waters have a salinity of 12, while the deep waters have salinities as high as 24. The salinity gradient prevents vertical mixing and the formation of $H_2S$ below a depth of about 18 m (Figure 9.76). The temperatures of the surface waters fluctuate from 0°C in the winter to 19°C in the summer. The deep waters have a uniform temperature of 7 to 8°C.

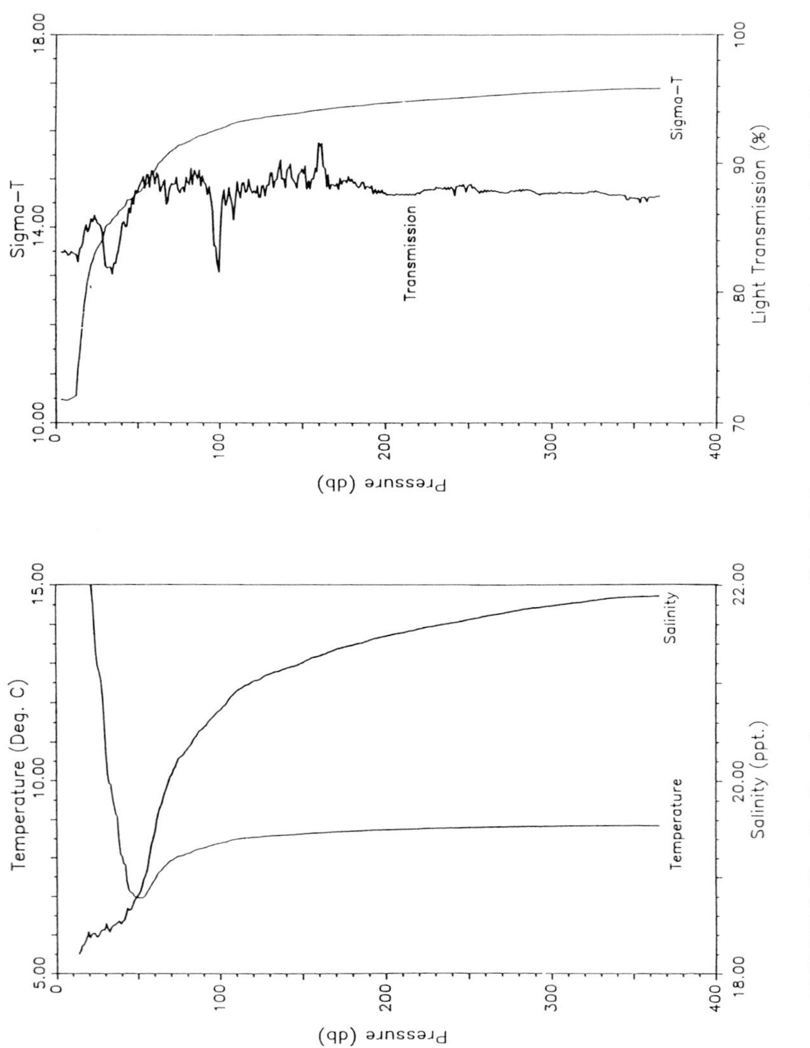


FIGURE 9.68. Profiles of temperature, salinity, light transmission, and sigma-t in the Black Sea.

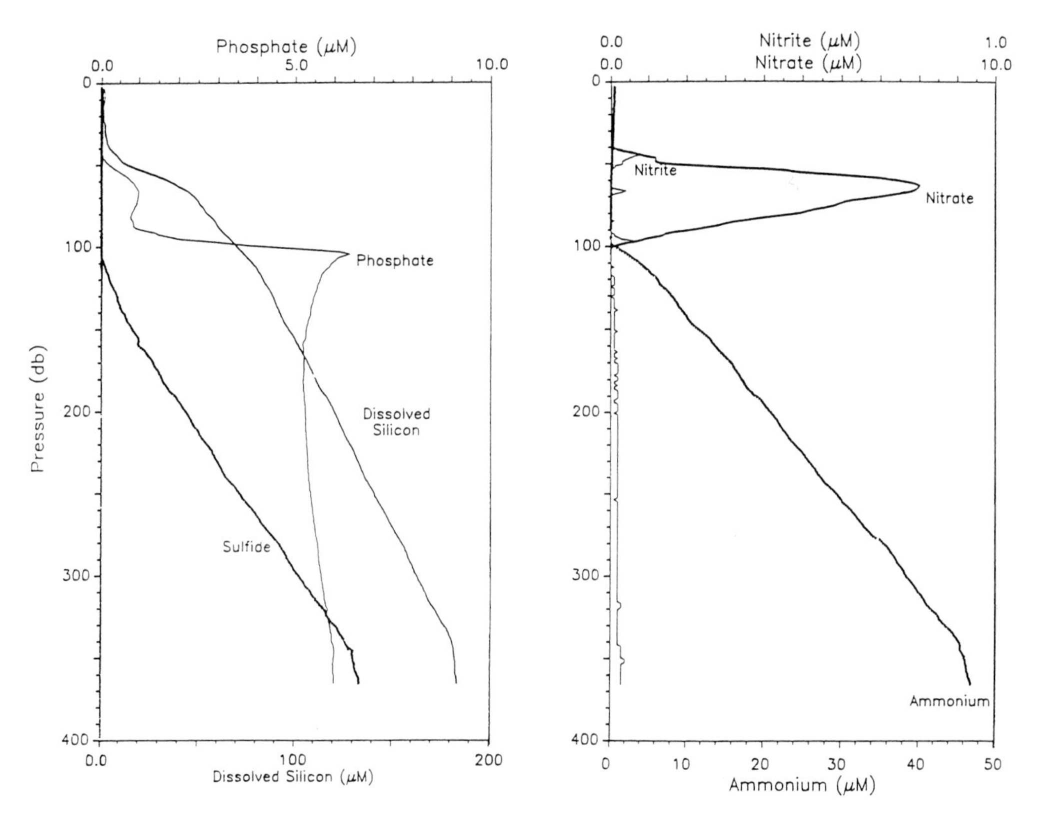


FIGURE 9.69. Profiles of silicate, phosphate, hydrogen sulfide, nitrate, nitrite, and ammonium in the Black Sea.

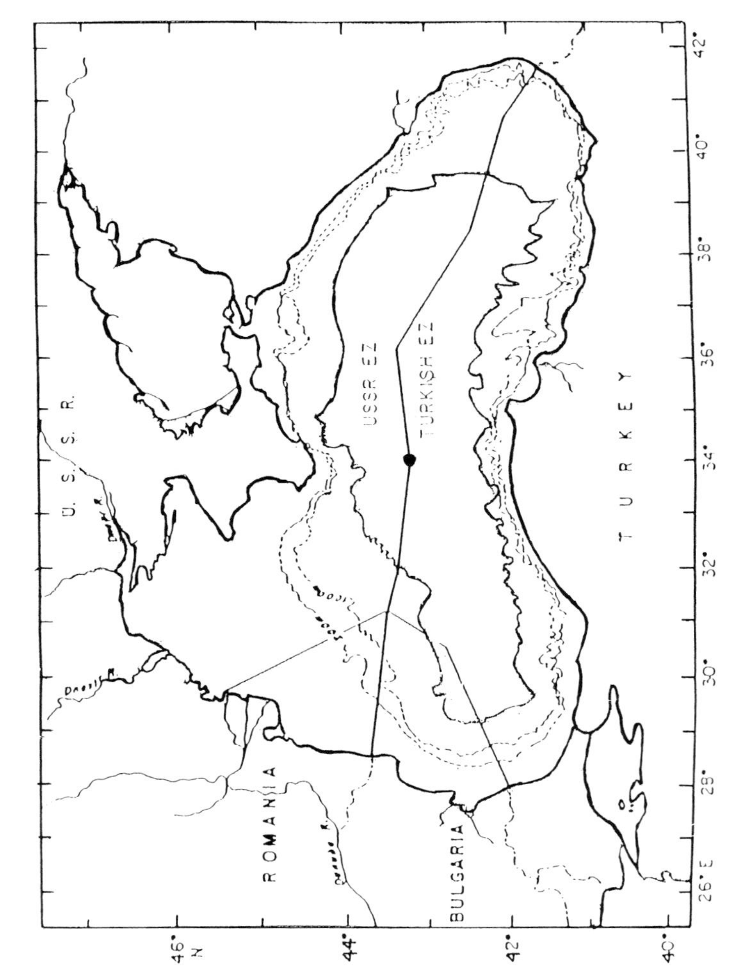


FIGURE 9.70. A sketch of Black Sea Station 5.

The deep basin has been anoxic for about 8000 years. The sill was dredged in 1850, resulting in the formation of a new layer of $H_2S$ (Figure 9.77). A vertical section of $O_2$ and $H_2S$ in the central basin is shown in Figure 9.78. A surface maximum is observed in $O_2$ due to the photosynthesis of phytoplankton. The $O_2$ goes to zero at about 18 m and the $H_2S$ increases to concentrations as high as 8 m*M* or 8000 $\mu M$. These values are the highest levels of $H_2S$ found in any anoxic basin.

The vertical distribution of ATP is shown in Figure 9.79 and indicates that bacteria are concentrated at the interface. There is also a concentration of particles at the interface as shown by the vertical distribution of transmission of light (Figure 9.80). The increase in the particles in the deep waters is related to the formation of metal sulfides and polysulfides. Framboidal pyrite ($FeS_2$) has been found in the deep waters (Table 9.14).

The concentration of metals in the Framvaren has been determined by Emerson and co-workers (U. Washington). Their results are shown in Figure 9.81. $Mn^{2+}$, $Fe^{2+}$, and $Co^{2+}$ show a maximum at the interface. $Ni^{2+}$ shows no changes across the interface. The metals $Cu^{2+}$, $Zn^{2+}$, and $Cd^{2+}$ show large decreases in the concentration at the interface due to the low solubility of their metal sulfides. More recent trace metal measurements have been made by Haraldsson and Westerlund.

### 5.3. Cariaco Trench

The Cariaco Trench (Figure 9.82) is an anoxic basin located in the Caribbean Sea north of Venezuela. It has a depth of 1400 m and a sill depth of 150 m. The distribution of $O_2$ and $H_2S$ across the interface is shown in Figure 9.83. Emerson and co-workers have measured trace metals in the Cariaco Trench. The results in Figure 9.84 are similar to the Black Sea and the Framvaren Fjord. Earlier measurements have shown the $CH_4$ in the anoxic waters (Figure 9.85). More recent measurements by Mopper and co-workers have been made on dissolved amino acids and amines. Measurements were also made on some low molecular weight carbonyl compounds and humic substances.

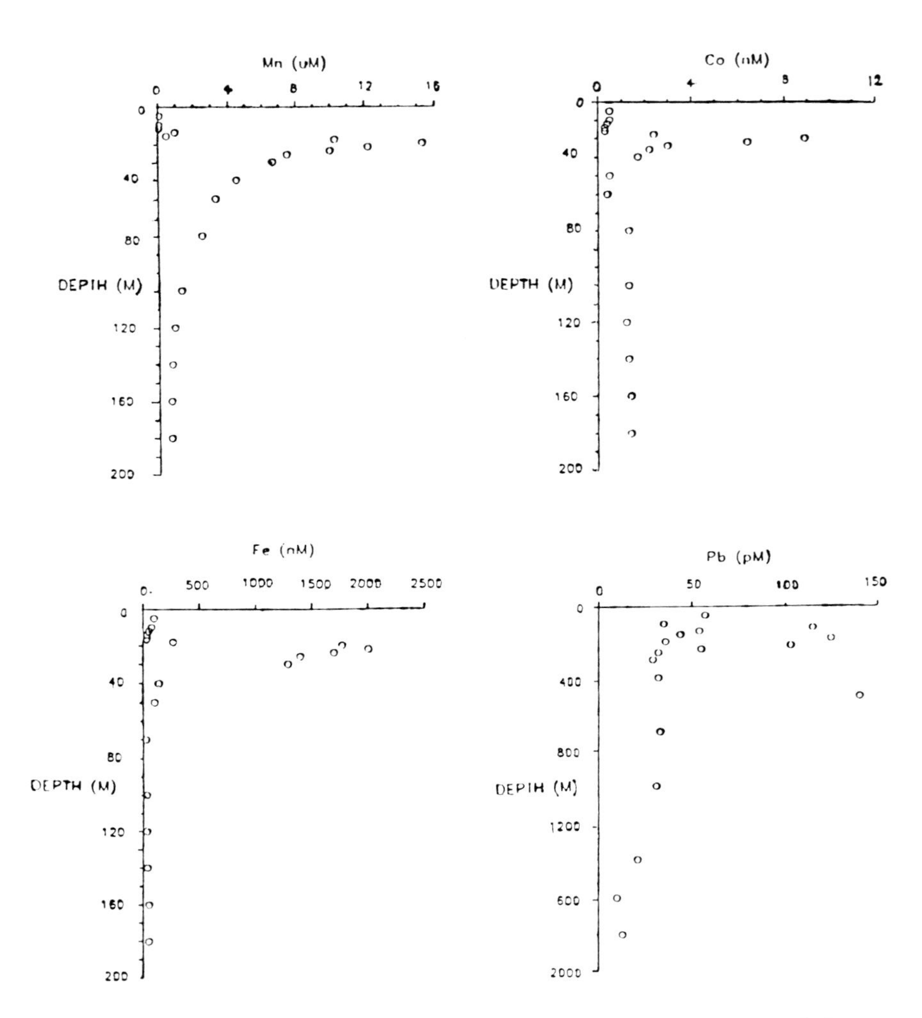


FIGURE 9.71. Profiles of manganese, cobalt, iron, and lead in the Black Sea.

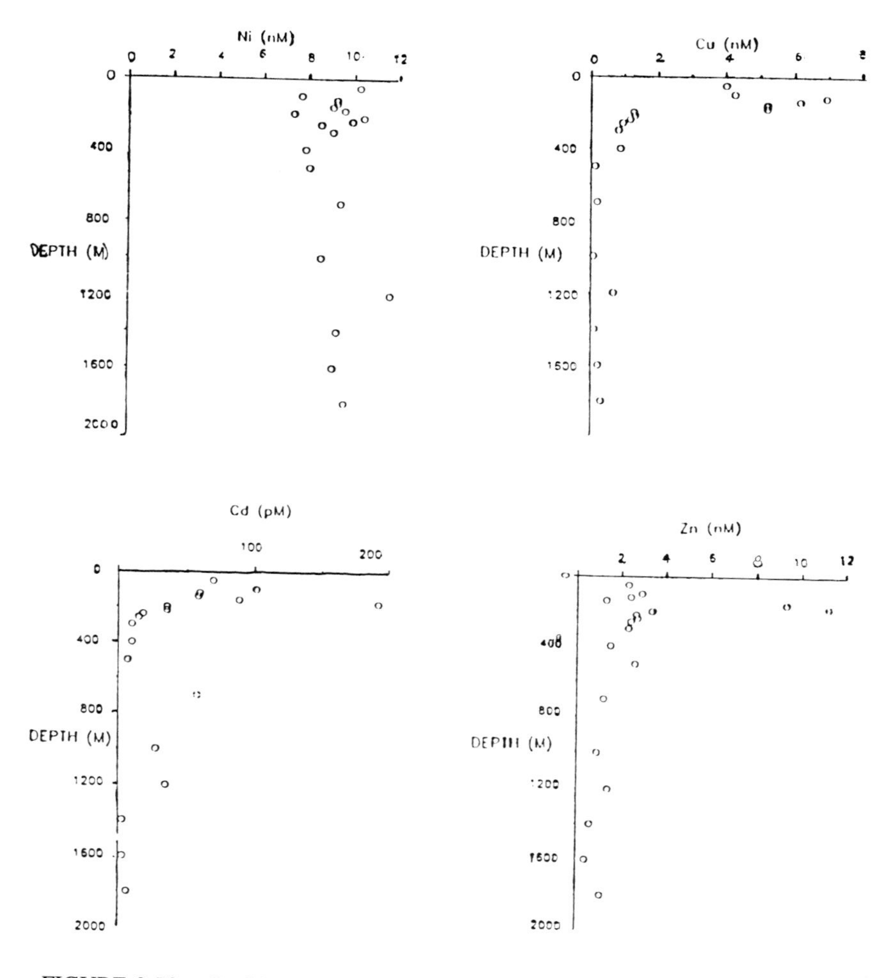


FIGURE 9.72. Profiles of nickel, copper, cadmium, and zinc in the Black Sea.

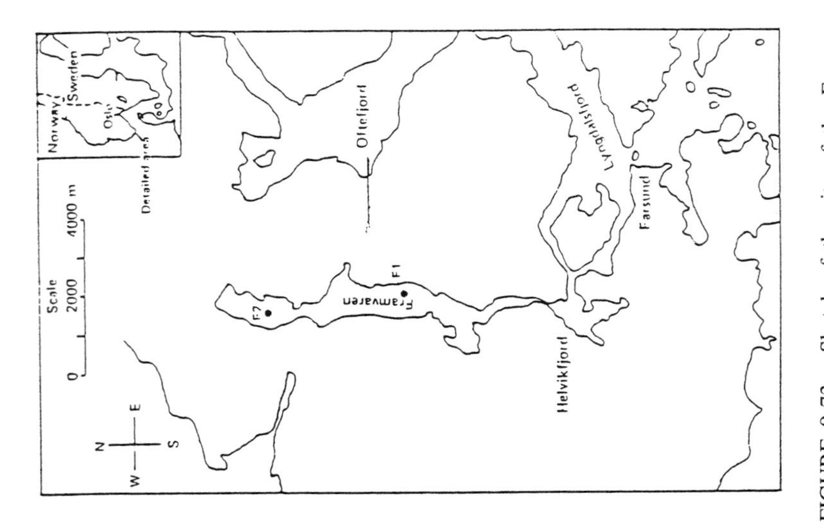


FIGURE 9.73. Sketch of the site of the Framvaren Fjord in Norway.

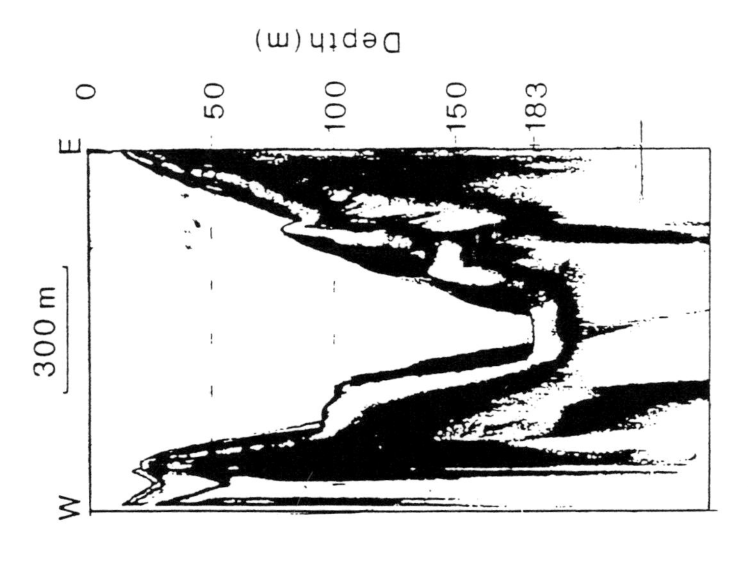


FIGURE 9.74. A cross-section of Framvaren Central Basin.

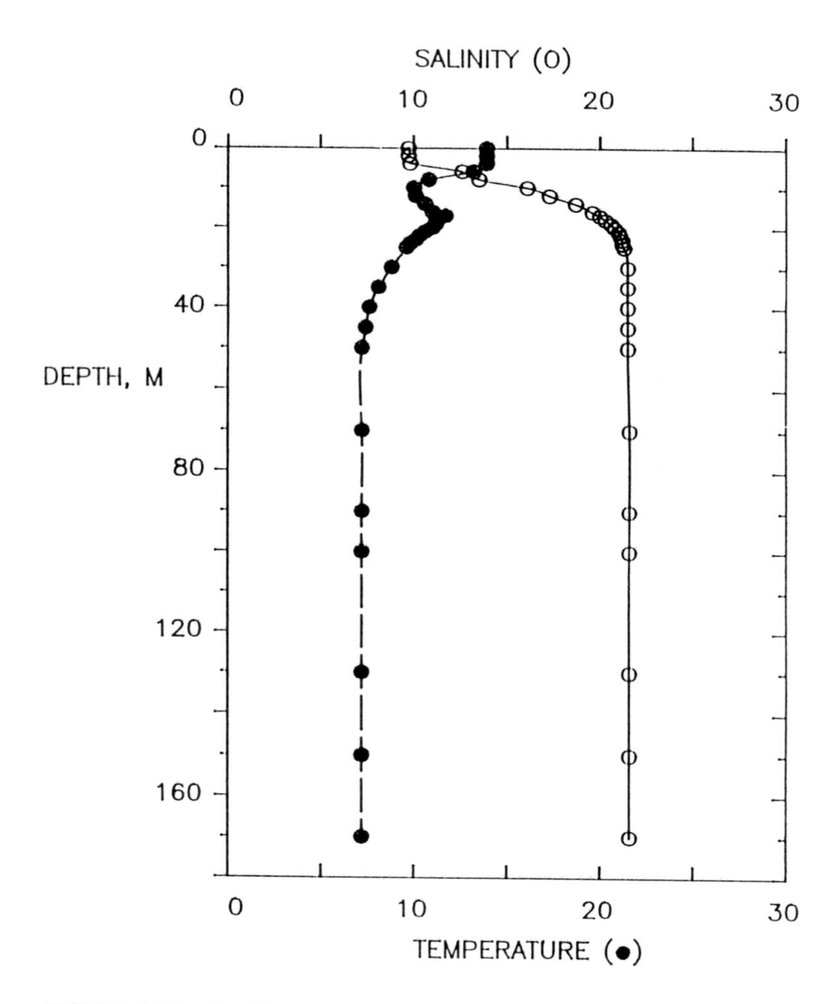


FIGURE 9.75. Profile of salinity and temperature in the Framvaren Fjord.

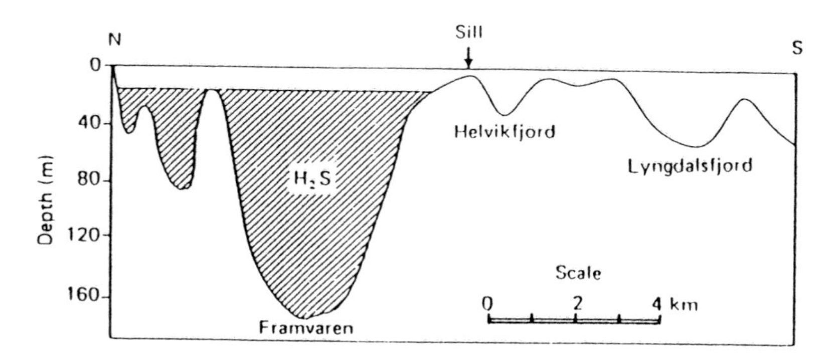


FIGURE 9.76. A longitudinal profile of the Framvaren Fjord.

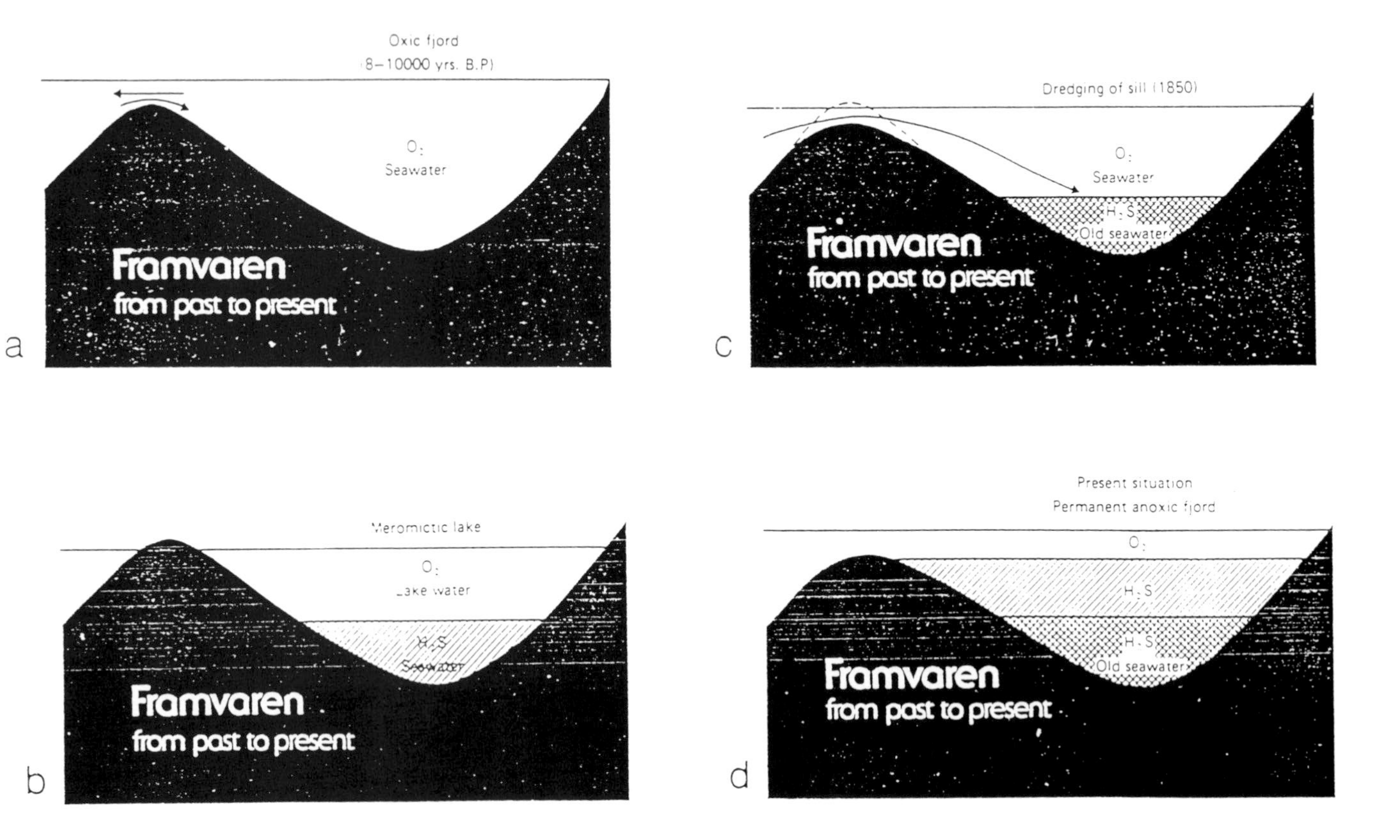


FIGURE 9.77. The historical development of the Framvaren Fjord.

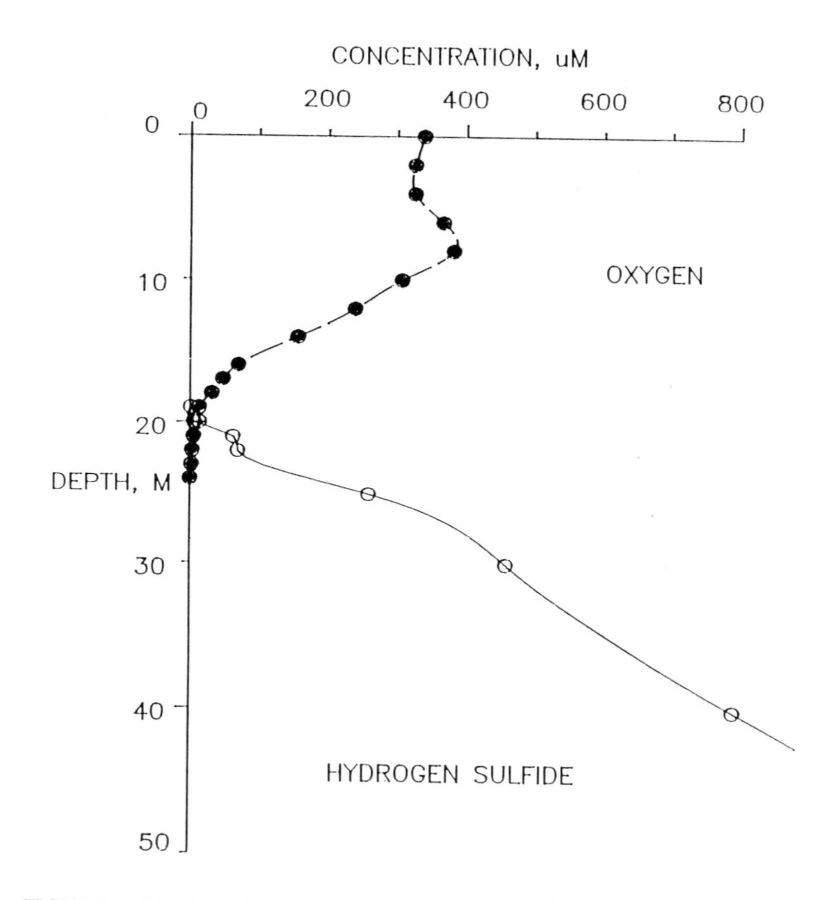


FIGURE 9.78. Profiles of oxygen and hydrogen sulfide in the Framvaren Fjord.

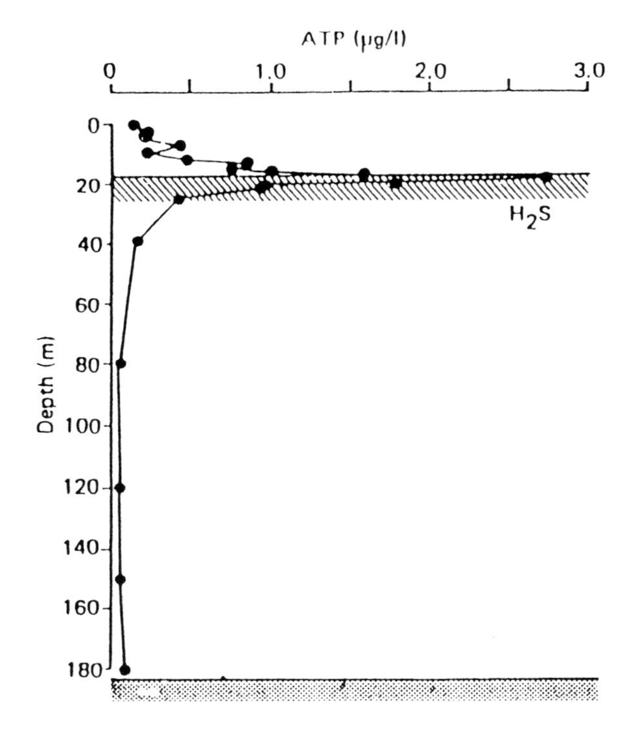


FIGURE 9.79. The vertical distribution of ATP in the Framvaren Fjord.

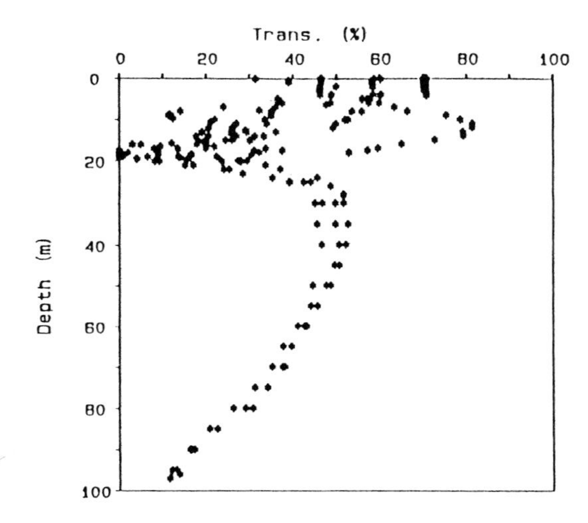


FIGURE 9.80. Transmission distribution in the Framvaren Fjord.

**TABLE 9.14**
**Framboidal Pyrite Composition (Average of 22 Framboids)**

| Element | wt% | SD |
|---|---|---|
| Cu | 0.5 | 0.08 |
| Zn | 0.3 | 0.11 |
| Mn | 1.0 | 0.16 |
| Ni | .03 | 0.03 |
| Co | .03 | 0.04 |
| Fe | 33.5—40.3 | |
| S | 36.2—45.5 | |

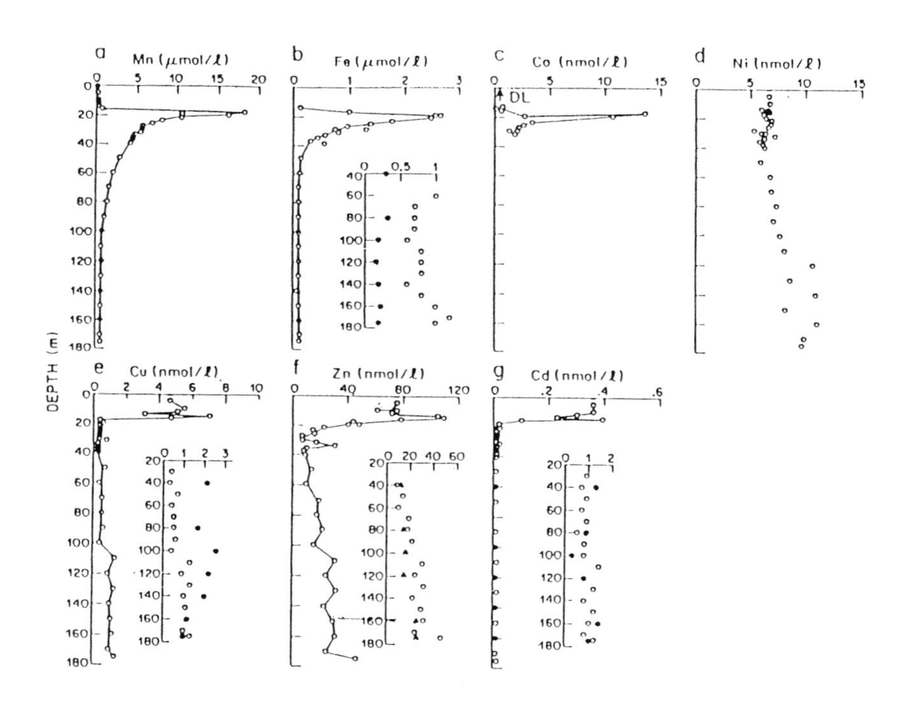


FIGURE 9.81. Profiles of metals in the Framvaren Fjord.

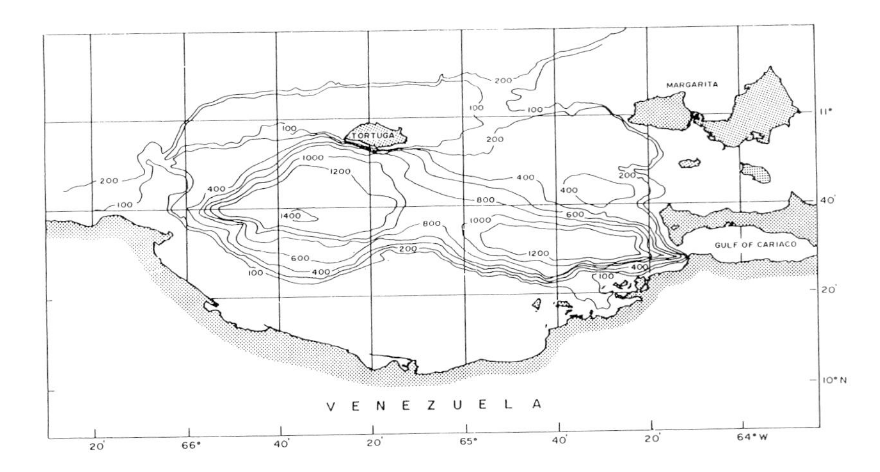


FIGURE 9.82. Sketch of the Cariaco Trench in the Caribbean Sea.

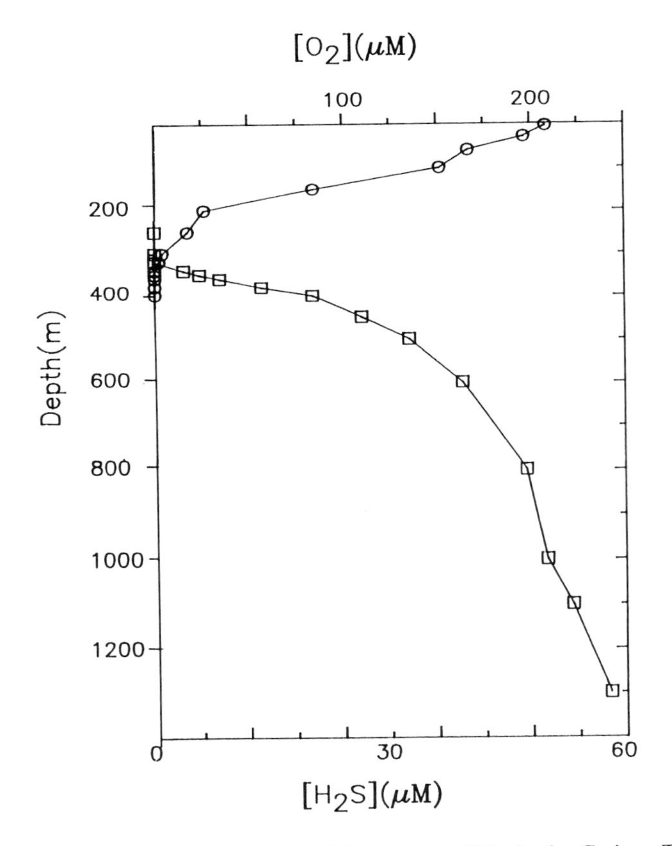


FIGURE 9.83. Profile of oxygen and hydrogen sulfide in the Cariaco Trench.

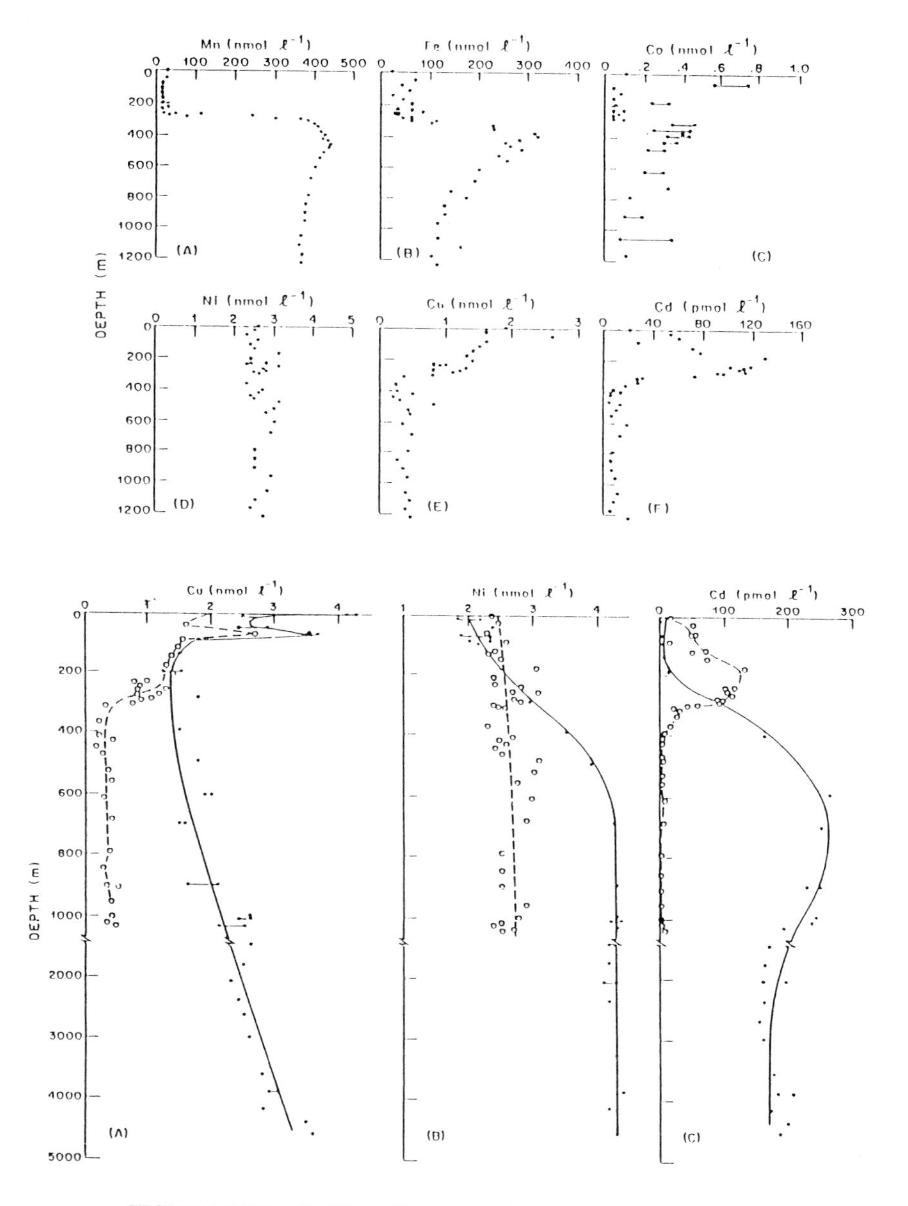


FIGURE 9.84. Profiles of trace metals in the Cariaco Trench.

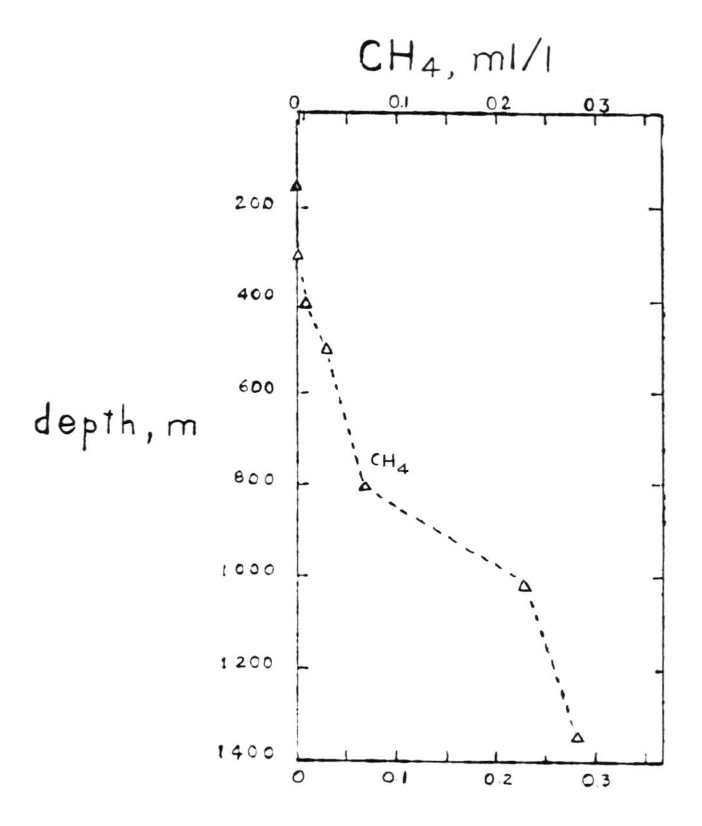


FIGURE 9.85. Vertical distribution of methane in the Cariaco Trench.

## 5.4. Saanich Inlet

The Saanich inlet is a fjord on the southeast side of Vancouver Island, British Columbia (Figure 9.86). It has a maximum depth of 200 m and a sill depth of 70 m. Water is trapped behind the sill in the late summer and winter. The basin stratifies and becomes anoxic below 130 m for about six months. In the fall, dense oxygenated water flows over the sill due to coastal upwelling. Emerson (University of Washington) and co-workers have recently studied the redox chemistry of this system. Nielson (U. Wisconsin) has studied the role of bacteria in the oxidation of $Mn^{2+}$. Profiles of $O_2$ and $H_2S$ are shown for stations A and E in the Figure 9.87. The cyclic behavior of $H_2S$ in Saanich inlet is shown in Figure 9.87. The profiles of metals across the interface are similar to what was found for other anoxic basins. Studies have been made on the nitrogen and iodine redox species across the oxic-anoxic interface in the Saanich Inlet.

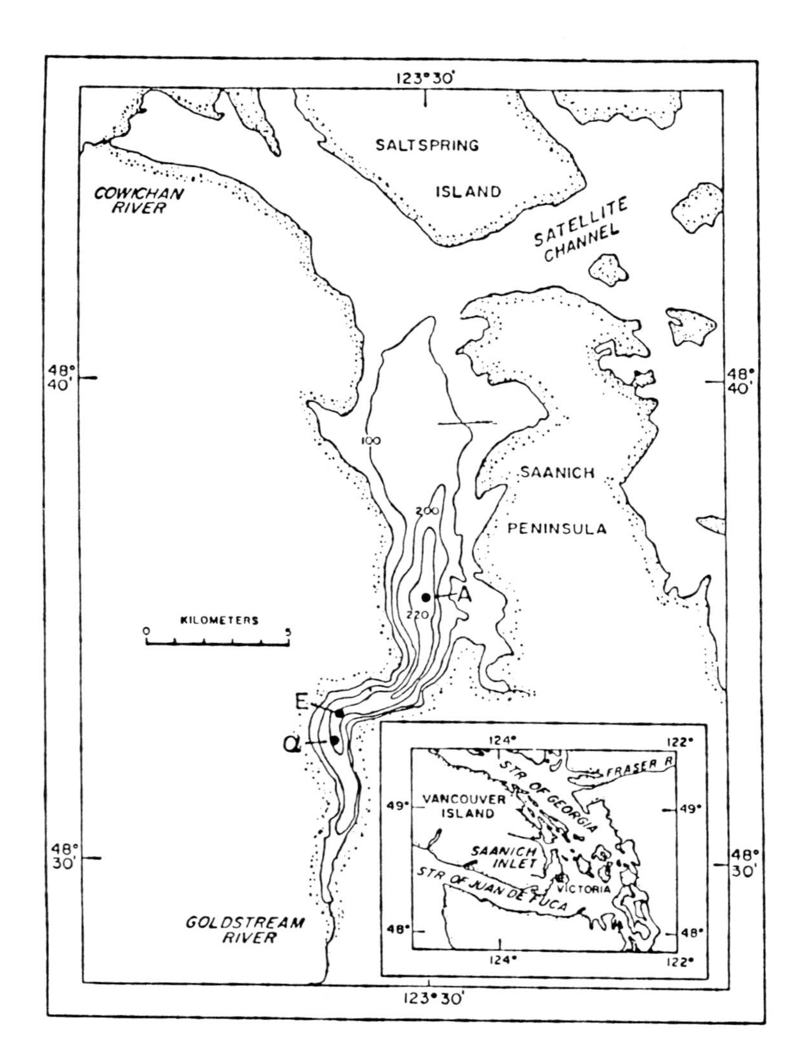


FIGURE 9.86. A sketch of the Saanich Inlet.

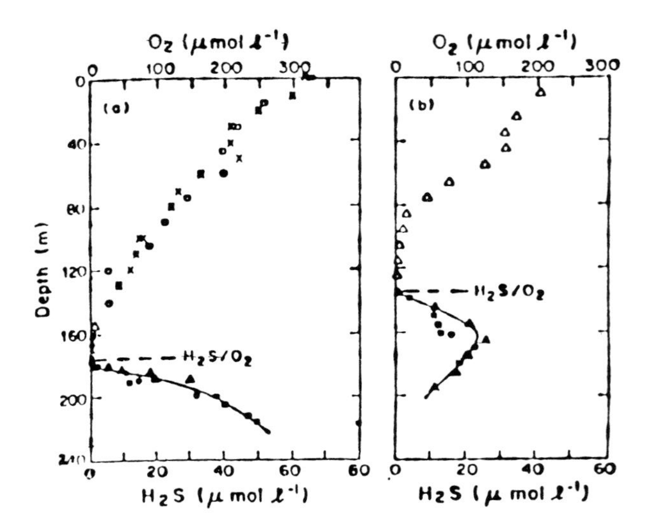


FIGURE 9.87. The $O_2$ and $H_2S$ interface at stations A and E in the Saanich Inlet.

## FURTHER READING

### Atmospheric Chemistry

Berg, W. W. and J. W. Winchester, Aerosol Chemistry of the Marine Atmosphere, chap. 38, *Chemical Oceanography,* Vol. 7, 2nd ed., J. P. Riley and G. Skirrow, Eds., Academic Press, New York, 173-231 (1978).

Chameides, W. L. and D. D. Davis, Chemistry in the Troposphere: Special Report, *Chem. Eng.,* Oct. 4, 1982.

Johnson, R. W. and G. E. Gordon, *The Chemistry of Acid Rain,* ACS Symp. 349, ACS Publishers, Washington, D.C., 337 pp. (1987).

Senfeld, J. H., *Atmospheric Chemistry and Physics of Air Pollution,* John Wiley & Sons, New York, 738 pp. (1986).

### Primary Productivity

Fogg, G. E., Primary Productivity, chap. 14, *Chemical Oceanography,* Vol. 2, 2nd ed., J. P. Riley and G. Skirrow, Eds., Academic Press, New York, 386-453 (1975).

Strickland, J. D. H., Production of Organic Matter in the Primary Stages of the Marine Food Chain, chap. 12, *Chemical Oceanography,* Vol. 1, 1st ed., J. P. Riley and G. Skirrow, Eds., Academic Press, New York, 478-610 (1965).

## Hydrothermal Vents

Jones, E. J. W., Sea-Follor Spreading and the Evolution of the Ocean Basins, *Chemical Oceanography,* Vol. 7, 2nd ed., J. P. Riley and G. Skirrow, Eds., Academic Press, New York, 1-74 (1978).

Rona, P. A., K. Bostrom, L. Lanbier, and K. L. Smith, Jr., *Hydrothermal Processes at Seafloor Spreading Centers,* NATO Conf. Ser. IV. Mar. Sci., 12, Plenum Press, New York, 796 pp. (1983).

Thompson, G., Hydothermal Fluxes in the Oceans, chap. 47, *Chemical Oceanography,* Vol. 8, 2nd ed., J. P. Riley and G. Skirrow, Eds., Academic Press, New York, 272-337 (1983).

## Photochemistry

Zafiriou, O. C., Natural Water Photochemistry, chap. 48, *Chemical Oceanography,* Vol. 8, 2nd ed., J. P. Riley and G. Skirrow, Eds., Academic Press, New York, 339-379 (1983).

Zika, R. G. and W. J. Cooper, *Photochemistry of Environmental Aquatic Systems,* ACS Symp. Ser. 327, ACS Publishers, Washington, D.C., 288 pp. 1879 (1989).

## Anoxic Basins

Deuser, W. G., Reducing Environments, chap. 16, *Chemical Oceanography,* Vol. 3, 2nd ed., J. P. Riley and G. Skirrow, Eds., Academic Press, New York, 1-37 (1975).

Grasshoff, K., The Hydrochemistry of Landlocked Basins and Fjords, *Chemical Oceanography,* Vol. 2, 2nd ed., J. P. Riley and G. Skirrow, Eds., Academic Press, New York, 456-597 (1975).

Richards, F. A., Anoxic Basins and Fjords, chap. 13, *Chemical Oceanography,* Vol. 2, lst ed., J. P. Riley and G. Skirrow, Eds., Academic Press, New York, 611-645 (1965).

# INDEX

## A


AAIW, see Antarctic Intermediate Water
Absorption, 143—146
Acetic acid, 276
Acid rain, 417, 427—430, 487
Acid-base equilibria reactions, 272—277
Acids, classification, 200, 276—278
Activity coefficient(s)
  acids, 204, 207
  calculations, 212—216
  electrolytes, 188, 190
  gas, 264, 265
  ions, 200—202, 212—213
  water dissociation, 273—275
Activity of water, see Water, activity
Additivity principle, 175
Adenosine diphosphate (ADP), 443—444
Adenosine triphosphate (ATP), 331, 337, 443—444, 493, 503
Adiabatic temperature, see Temperature, adiabatic
ADP, see Adenosine diphosphate
Adsorption, 143—146
Advection, 235, 250, 255—256, 259, 486, see also Plates 12—14
Aerobic respiration, see Respiration, aerobic
Aerosols, marine, 104, 107, 365
Age, Oceans, 49—51, see also Plates 9—11
Agulhas Current, 43
Air, composition, 228—231
Air injection, 244, 245—247
Air-sea interface
  gas exchange, 227, 271, 294, 305—307
  organic matter input, 364
  salt exchange, 104, 107—108
Alaska, 292
Alcohols, 408—409
Aldehydes, 406—407
Aldose, 367
Algae, 361, 382, 391, 407—408
Algal kerogens, see Kerogens
Alkali metals, 120, see also Metals
Alkaline metals, 120, see also Metals
Alkalinity, see also Calcium carbonate; Chlorinity; Conductivity; Density; Evaporation; pH; Salinity
  Baltic Sea concentrations, 86—87, 90, 91
  carbonate cycling, 283—284, 287—291
  specific, 298—299, 303
  surface water, 294, 298—300
  total, 298, 302, 304, 305, 307, 310, see also Plate 15
Alkanes, 378, 382—385
Alkenes, 378, 385—387
Allochthonous carbon, see Carbon, terrestrial-derived
Aluminum, ions, 125, 137, 138, 430
Amide bonds, see Peptides
Amino acids, 337, 372—375
β-Aminoglutaric acid, 376
Aminophosphonic acids, 324
Ammonia, ion
  analysis, 335
  anoxic water, reactions, 487—490, 494
  nitrogen fixation, 336—340
  oxidation states, 334
Ammonium, ions, 338, 340, 489, 494
Amylose, 368
Anions, see also Individual entries
  evaporation of, 101, 103
  ion pairing, 201—204
  organism affinity, 142
  rivers, 83—86
  speciation, 207
Anoxic basins, see Basins, anoxic
Anoxic sediment, see Sediment, anoxic
Anoxic water, see also Basins, anoxic; Deep water
  denitrification, 338, 339
  maxima/minima type profile, 138, 140—142
  nitrogen gas conversion, 246, see also Air injection
Antarctic
  bottom water, 33, 37, 45, 349
  circumpolar water, 41
Convergence
    Indian Ocean, 45
    Pacific Ocean, 41
    southern ocean circulation, 33, 36—38
  Diatom oozes, 346, 348
  Intermediate Water (AAIW)

Atlantic Central Waters, 30
Indian Ocean, 45
nitrogen oxide tracing, 350
oxygen concentration, 250—251, see also plates 12-14
Pacific Ocean, 40, 41
physical characteristics, 33, 36
sources, 38
ozone reduction, 438—442, see also Ozone
Anthracene, 475, 477
Antilles Current, 27
AOU, see Apparent oxygen utilization
Apparent molal property, see Molality, apparent
Apparent oxygen utilization (AOU), see also Oxygen
carbon dioxide, 289—290
determination, 253, 254, 255, 339, 344, 452—453
dissolved organic matter comparison, 358, 359
plant oxidation measure, 349, 350
Aragonite, see also Calcite; Calcium carbonate
Atlantic/Pacific Ocean profile, 311, 312
dissociation, 345
flux, 319
hydrogenous sediment, 10
solubility, 281—282
saturation levels, 309, 311, 317, 318
Archaebacteria, 389
Arctic
bottom water, 48
deep water, 48
deuterium concentration, 112
intermediate water, 31, 40, 250—251
zone, 33, 37
Arctic Sea, 46—49
Argon, gas, 233, 243, 246, 247, 252
Aromatic acids, 391
Aromatic hydrocarbons, see Hydrocarbons, aromatic
Arrhenius theory, 160
Arsenate, ions, 331
Artificial seawater, see Seawater, artificial
Astaxanthin, 402, 403
Atlantic Ocean, see also North Atlantic; South Atlantic
abyssal plain, 375
apparent oxygen utilization, 254, 256, 259
bottom water, 31, 32
calcium carbonate lysocline, 311, 313, 314, 316
carbon dioxide, atmospheric, 294—295
central waters, 29—30
central gyres, 43, 138, 142
coastal waters, 375—376
currents, 27—33, 43
density, 21, 23
depth profile
alkalinity, 298, 302—304
aluminum, 138
aragonite saturation, 311, 312
barium, 131, 132
cadmium/phosphate, 131
carbon dioxide, 294, 299, 307, 309
carbon monoxide, 260
deep water metals, 138, 141
lead, 133, 136
nickel/silicate/phosphate, 131, 135
nitrate, 338, 339, see also Plates 19, 20
nitrogen/phorphorus ratio, 342
nitrous oxide/oxygen, 339, 341
oxygen, 250, 251
pH, 294, 301
phosphate, 332—333, see also Plates 17, 18
saturation/compensation depth comparison, 313
silica, 132
silicate, 346—347
zinc/silicate, 131, 133
micronutrient ratios, 344
nitrogen oxide tracer, 350
nitrogen-phosphorus ratio, 341, 343
salinity profile, 17, 20, see also Plates 4—6
Atmosphere
carbon dioxide partial pressures, 291—294
layers of, 418
nitrogen input to oceans, 336
terrigenous organic matter, 364—366
trace metal input, 138, 142
Atoms, electronic configuration, 120, 122
ATP, see Adenosine triphosphate
Autoclaving, 483, 485

## B

Bacteria

chemosynthesis, 455—456
copper adsorption, 143—144
hydrocarbons in, 382, 383, 385, 386
lactic acid utilization, 390
Baffin Bay Current, 46
Baltic Sea
chlorinity, 87, 90, 91
composition, 91
depth profile
oxygen/pH, 86, 89
phosphate/nitrate, 86, 88
salinity temperature, 86, 87
estuary, 86
oxygen/hydrogen sulfide levels, 92, 95
river water comparison, 88—89, 92—94
Bar built estuary, 52, see also Estuaries
Barium, 131, 132
Basalt, 457, 458
Basins, 46—47, 50—54, 486—489
Benguela Current, 27, 43
Benthic algae, 361, 380, see also Algae
Bering Sea, 349
Bicarbonate, 63—64
Biodegradation, 378, 384, 398
Biogenic alkenes, 387, see also Alkenes
Biogenous sediment, 10, see also Sediment
Biomarkers
alicyclic hydrocarbons as, 384, 385
criteria, 381
cycle diterpenoids as, 401
fatty acids as, 388, 389
steroids as, 405
trace compounds as, 399
Biomass, indicators, 412, 445
Biosphere, affect on minor elements, 142—143
Biscayne Bay, 413
Bitumen, 392, 409
Bjerrum theory, 160, 189—191, 277, 278
Black Sea, see also Basins, anoxic
anoxic basin, 108, 138
box model, 489, 491
carotenoid reduction, 403
depth profiles
iron, 138, 140
manganese, 138, 141
oxic-anoxic interface, 489—491
physical properties, 489, 492
trace metals, 489, 491, 494—497
research station, 489, 495
stenol/stanol conversion, 405
Blooms, 160, 332, 346, see also Diatoms; Phytoplankton
Blue-green algae, 336, see also Algae
Boiling point, 162, 163, 177
Borate, 64, 67, 85
Boric acid, 64, 67, 91, 204, 280
Born model, 180—184, 187—188
Born-Haber cycle, 177
Bottom, oceans, see Seafloor
Bottom water, 31, 32, 378
Boundary layer model, 238
Box model, 489, 491
Branched fatty acids, see Fatty acids
Brazil Current, 27
Break in slope, 5
Brines, 52, 101—102, 106
Bromine, 63, 65, 91, 107
Bubbles, 104, 107, 233, 242, 243, see also Air injection
Buffers, 274—277
Bunsen coefficient, 232
*n*-Butane, 376

## C

Cadmium, depth profiles, 131, 491, 493, 497, 504, 506
*Calanus* spp., 380
Calcite, 10, 281—282, 309, 311, 345, see also Aragonite; Calcium carbonate
Calcium, ion
association constants, 326
Baltic Sea, 86, 88, 90
determination methods, 64
hydrothermal vent levels, 459, 462
silicon complexes and, 346
Calcium carbonate, see also Aragonite; Calcite
compensation depth, 313—318
dissolution, 290—291, 294, 298, 344—345
solubility, 308—310
Calcium oxalate, 64
Calcium sulfate, 101, 104
California Current, 40
Calomel electrode, 275
Canadian basin, 46—49
Canary Current, 27
Cape Lookout, 376
Capillary gas chromatography, 360
Capillary waves, gas exchange, 242
Carbohydrates
classification, 366—368
dissolved organic matter, 363

marine environment, 368—371
photosynthetic production, 444, see also Photosynthesis
Carbon, see also Dissolved organic carbon (DOC)
Black Sea, levels in, 489
dating, 50
dissolved organic matter assay, 364
inorganic/organic ratios, 291
organic, oxidation, 289
primary productivity measurement, 446—447, 451
terrestrial-derived, 363—365
Carbon dioxide, see also Carbonate system
Atlantic/Pacific oceans, profile, 309
cycling, 269—271, 292, 293
equilibria reactions, 277—282
fixation, 444—445
global climate changes, 430
partial pressure, 269—270, 287, 291—297, 305—306
phosphorus ratio, 344
primary productivity measurement, 271, 446—448
reservoirs, 271—272
total, 282, 284, 286, 307, 310, see also Plate 16
Carbon monoxide, 260, 261, 423
Carbon number, distribution, 381
Carbon tetrachloride, 436, 437
Carbonate, see also Calcium carbonate; Carbonate system
alkalinity, 283
compensation depth, see Carbonate compensation depth (CCD)
determination methods, 63—64
equilbria reactions, 277—282
surface-deep water transfer, 308
Carbonate compensation depth (CCD), 314, 316, 317
Carbonate system, see also Calcium carbonate; Carbonate; Carbon dioxide; Carbonic acid; pH
characterization, 282—283
direct measurement, 286—291
equilibria equations, 269
titration method, 284—286
Carbonic acid, 216, 278, 279, 283
Carbonyl compounds, see Aldehydes; Ketones
Carboxylic acids, 388—391
Cariaco Trench, see also Basins, anoxic
alkene isolation, 386
anoxic basin, 108, 138
carotenoid reduction, 403
depth profile
oxygen/hydrogen sulfide, 505
trace metals profile, 493, 506
dicarboxylic acids from, 391
location, 493, 505
nitrogen levels, 246
Carotane, 386
Carotene, 386, 402, 403
Carotenoids, see Carotane; Carotene
Cations
$d^0$ concentration, 120—121
$d^{10}$ concentration, 121, 123—124
evaporation of, 101, 102
ion-pairing, 201—204
organism affinity, 141
river components, 83, 84
speciation in seawater, 205—206, 208—211
Cavity formation, 266
CCD, see Carbonate compensation depth
Cellulose, 367, 368
Cement, 269—270
CFC, see Chlorofluorocarbons
Challenger expedition, 4, 39, 60
Chemical oceanography, first law, 60
Chemiluminescence, 476
Chemolithotrophy, 455—456, see also Hydrothermal vents
Chemosynthesis, 455—456, see also Hydrothermal vents
Chesapeake Bay, 427
Chloride, ion, 121, 123, 459, 468
Chlorin, 410
Chlorine, see also Chlorinity
ion exchange effects, 150
ozone loss, 433—435
salinity and, 73, 84
variation with depth, 65, 66
Chlorine monoxide, 439, 441
Chlorinity, see also Alkalinity, Chlorine; Conductivity; Density; Evaporation; pH; Salinity
alkalinity, 298—299, 300, 302, 303
Baltic Sea, 86—88, 90, 91
determination methods, 62—63, 65—70
salinity/conductivity, 73—76
*Chlorobium*, 385
Chlorofluorocarbons (CFC), 417, 430, 434—437

Chloroform, 392
Chlorophyll, 409, 410, 443, 445, 453
Chlorophyllide, 410, 411
Cholesterol, 404, 409
Chloroplatinate, 64
Chromophores, 476, 479
Circulation, 25—27, 33, 43
Circumpolar Current, 27, 36—37
Clathrate models, water, see Water, structure models
Clay minerals, 10, 150—151, 346, 348
Climate, 3, 23, 430
Clouds, 431
Cluster models, water, see Water, structure models
Cluster expansion theory, 189—192, see also Water
Coastal plain estuary, 52, see also Estuaries
Coastal sediments, 387, 407, see also Sediments
Coastal waters, 348, 481—482
Cobalt, depth profiles, 491, 493, 496, 504, 506
Colloids, 59—62
Colorimetric technique, 370
Compensation depth
Compression, temperature effects, 14, 167, 273
Conductance, electrical, 177
Conductance cell, 79—80
Conductivity, see also Alkalinity; Chlorinity; Evaporation; Density; pH; Salinity
  chlorinity determination, 63
  estuaries, 93, 94, 97, 99—100
  salinity, 74—81
Continental shelf, 5
Continental slope, 5
Continuum models, see Debye-Huckel theory; Bjerrum theory
Convergence zones, 15, 33, 37
Copecods, 402
Copenhagen Seawater, 74
Copper, ion
  bacterial adsorption, 143—144
  depth profile
    Black Sea, 491, 497
    Cariaco Trench, 493, 506
    Sargasso Sea, 484, 486
    Framvaren Fjord, 493, 504
  oxidation, 484—486
  speciation, 205—206, 212
  toxicity, 159—160
Core rings, 28, 29
Coriolis effect, 26, 36—37, 49
Cortisone, 404
Cosmogenous sediment, 10, see also Sediment
Costa Rican thermal dome, 40—42
Countercurrents, 33, 39, 40, 43, 44, see also Currents
Covalent bonding, 191, 195
CP, see Cross polarization
Cromwell Current, 39, 42—43
Cross polarization (CP), 395—397
Cross square rule, 220
Cross-square diagram, 192
Crude oil, 384, 386
Crystal molal volume, 188
Crystalline hydrates, see Hydrates, crystalline
Currents, 3—4, see also Individual entries
Cyanobacteria, 391, 448
Cyclic alkanes, see Alkanes
Cysteine, 413
Cystine, 413
Cytochromes, 409, 443, 444

## D

Dabob Bay, 371
Dalton's law of partial pressure, 228
DDT, 365—366
Dead Sea, 216, 217
Death phase, see Growth, phases
Debye-Huckel theory, 160, 189, 190
Decay, see Hydrogen peroxide, decay
Deep basin estuary, 52, see also Estuaries
Deep sea bottom, see Seafloor
Deep water, see also Anoxic water; Basins, anoxic
  air injection, 246—247
  Black Sea, 489
  calcium carbonate dissolution, 290—291, 312
  carbon dioxide, 294, 307, 309
  exchange, 49
  formation, 17, 18, 21, 33, 34, 38
  Framvaren Fjord, 491, 493, 504
  halocline effects, 486, 487
  hydrogen peroxide decay, 482, 484
  Indian Ocean, 45
  Mediterranean Sea, 51

metals, 138, 141
nitrogen levels, 338
nitrogen-phosphorus ratio, 341
oxygen concentration, 111—112, 250, 254
Pacific Ocean, 41
pH, 298, 301
phosphorus levels, 332
salinity, 17
silicon cycling, 346, 348
sources, 31
thermocline effects, 487
Dehydration, 278
Dehydroabietic acid, 401
Delayed fluorescence, see Fluorescence
Denitrification, 336—338, 487—488, see also Nitrification
Densimeter, 81—82
Density, see also Alkalinity; Chlorinity; Conductivity; Evaporation; pH; Salinity
chlorinity determination, 63
estuaries, 93, 99—100
measured vs. calculated, 98—100
Pacific Ocean, 41
Red Sea brine, 219, 222
salinity calculations, 75—76, 81
seawater, 80—83, 219
surface water, 21
vertical distribution, 11, 15, 16, 18
Deoxyribonucleic acid (DNA), 412
Deposits, maturity indicators, 382
Deuterium, 109—111
DFAA, see Dissolved free amino acids
D/H ratio, see Deuterium, hydrogen ratio
Diagenesis, 383, 398, 406, 410
Diatoms, 346—348, 385, 448
Dibasic acids, 276—278
Dicarboxylic acids, 391
Dicothermal layer, 14
Dielectric constant, 162, 184
Diffusion, 235—237
Dimethyl sulfide (DMS), 413
Diplotene, 401, 402
Dipole-dipole interactions, 164, 167, see also Water
Disaccharides, 366
Dissociation, photochemical reactions, 473
Dissolution, 102, 108, 263—266, 290, see also Individual entries
Dissolved free amino acids (DFAA), 374
Dissolved organic carbon (DOC)
amino acid levels in, 374
assays for, 360
concentration in seawater, 356—357, 362, 363
fatty acid levels, 389
Dissolved organic matter (DOM), 356—359, 377
Diterpenes, 386, 401
Diurnal variations, 11, 271, 294, 301, 482—483
DMS, see Dimethyl sulfide
DNA, see Deoxyribonucleic acid
DOC, see Dissolved organic carbon
DOM, see Dissolved organic matter
Downwelling, 50, 294, see also Upwelling; Volcanism
DPEP porphyrin series, 411

## E

Earth, surface topography, 5, 8
East Greenland Current, 46, 47, 49
East Pacific rise, 453, 459, 460, see also Hydrothermal vent
Echinenone, 402, 403
Eddy, 26
EDTA, see Ethylenediamine N,N,N′,N′-tetraacetic acid
EGTA, see Ethyleneglycol-bis(2-amino-ethyl)-N,N,N′,N′-tetraacetic acid
Eh, see Electrical potential
Ekman-induced upwelling, 17, see also Upwelling
El Chichon, 423, 426
El Niño, 40, see also Wind
Electric dipole, water, 162, 164, 166, see also Water
Electrical potential (Eh), 193—197
Electrolyte solutions, interactions, 160—161
Electromagnetic radiation, see Radiation, electromagnetic
Electron donors, 456
Electron states, 475—476
Electromotive force (emf), 274, 284, 285
Electrostatic bonding, 191, 195
Electrostatic partial molal entropy, 181, 184, 185, see also Entropy
Electrostriction, see Water, ion interactions
Elements, trace
anoxic waters, 138, 140—142
biological interactions, 139, 141—146

cations, 120—124
classification, 117—120
conservative, 130
depletion, 131, 133
electronegativity, 151—152
geochemical balance, 147—149
mid-depth maxima/minima, types, 137—140
nutrient type, 131, 132, 134—136
ocean formation model, 149—151
residence times, 124—129, 133, 151
seawater/phytoplankton concentration, 323—324
weathering, 152—154
emf, see Electromotive force
*Emiliana huxleyi,* 407
Energy
carbohydrates, 367
density, 18
geothermal, 455
photochemical reactions, 469—472
states of molecules, 473—475
English channel, 332, 338, 340
Enthalpy
gas dissolution, 263, 264
metal hydration, 180, 182, 184
sodium chloride, 178
Enthalpy of hydration, 174
Entropy
gas dissolution, 263, 264
metal hydration, 180, 183
metal ions, 174, 175
Environment, 375, 393
Equator
coriolis effect, 26
intertropical convergence zone, 423
salinity, 15, 17
surface temperature, 11
surface water, 375
total carbonate, 305, 307
Equatorial Atlantic, 33, 35
Equatorial current, 43
Equatorial undercurrent, see Cromwell Current
Equatorial waters, 111, 305
Ergosterol, 404
Erosion, terrestrial, 407
Essential oils, see Terpenes
Estuaries, see also Specific entries
alkene distribution, 386—387
anoxic, 487, see also Basins, anoxic
characteristics, 52—54
dissolved organic carbon levels, 363
humic substance analysis, 395, 396
Ethanol, 408
Ethyl ketones, 407
Ethylene, 385
Ethylenediamine N,N,N′,N′-tetraacetic acid (EDTA), 64, 124, 335, 348
Ethyleneglycol-bis(2-aminoethyl)-N,N,N′,N′-tetraacetic acid, 64
Etio porphyrin series, 411
Eubacteria, 389
Eucken polymer model, see Significant structure theory
Euphotic zones, 360—361, 374
Eurasian Basin, 46—47, 48, 49
Eutrophication, 427
Evaporation, see also Alkalinity; Chlorinity; Conductivity; Density; pH; Salinity
basins, 50—52, 101—105
Black Sea, 489, 491
Pacific Ocean, 40
salinity, 15, 17, 20, 72—73, 298
Evaporite, 52, 101
Exit coefficient, 238, 240, 241
Exponential phase, growth, see Growth, phases


## F


Falkland Current, 27
Fast ice, 49
Fatty acids, 364, 388—390
Fecal pellets, 146, 332
Ferric ion, see Iron
Fichtelite, 401
Fick's first law, 235—236, 238
FID, see Flame ionization detection
Filtration, 356, 360
Fish, 429—430
Fixation, nitrogen, see Nitrogen, fixation
Fjords, 54, 138, see also Framvaren Fjord
Flame ionization detection (FID), 360
Florida Current, 27
Flow injection, 331
Fluorescence, 475—476, 478
Fluoride, complexes, 120, 123
Fluorine, 63, 91
Food chain, marine, 442
Foraminifera, 299
Fossil fuel
biological origin, 410

burning and nitrogen oxide, 425
carbon dioxide levels, 269—270, 292
inanimate organic matter, 357—358
organic sources, 361
oxidation, 428
Framboidal pyrite, see Pyrite, framboidal
Framvaren Fjord, see also Basins, anoxic
central basin, 491, 499
depth profile
ATP distribution, 493, 503
light distribution, 493, 503
oxygen/hydrogen sulfide, 502
trace metals profiles, 493, 504
historical development, 493, 501
location, 491, 498
Frank and Wen flickering cluster model, see Cluster models
Free energy, 174, 175, 180, 181
Free radicals, 435—436, see also Ozone
Freezing, 15, 48, 108
Freezing point depression, 177
Freshwater, 111, 208—211, 491
Friction velocity, gas exchange, 242
Fructose, 370
Fulvic acid, 363, 392—393, 398, 479
Fungi, 384
Fuoss model, 191

## G

Galapagos, 459, 469, see also Hydrothermal vents
Gas, see also Specific entries
air-sea exchange, 235—242
atmospheric, 228—231
dissolution in seawater, 232—234
nonreactive, 240, 242—243
seawater sources, 234—235
solubility, structural aspects, 263—266
trace, 423, 424, 431—433
Gas chromatography-mass spectrometry (GC/MS), 360, 397
Gas film model, see Liss model
GC/MS, see Gas chromatography/mass spectrometry
Gel filtration, 393
Genes, 412
Geochemical markers, 405, 410
GEOSECS
Atlantic Ocean currents, 29, 34
nitrogen-phosphorus ratio measure, 343
ocean age, 49
oxygen concentration measure, 250, 252, 254
phosphorus measure, 332—333
Geothermal energy, 455
Germanium, 131, 134
Global cycle, 261
Global warming, 378, see also Methane
Glucan, 370
Glucose, 367, 370, 371
Glutamic acid, 337, 376
Glycerol, 388
Glycine, 372, 375
Glycosides, 368
Gran function, 285—286
Grass Lake, 393, 394
Greenhouse effect, 269, 417, 430—433
Greenland gyre, 46
Greenland Sea, 46
Grotthus-Draper law, 469—470
Ground state, electrons, 475
Growth, 323, 324, 449—450
Guaymas Basin, 459, 460
Gulf of Maine, 338, 340
Gulf of Mexico, 260, 262, 479—480
Gulf of Mexico Loop Current, 27
Gulf Stream, 26—29, 482—483
Gypsum, 101
Gyres
Atlantic Ocean, 27, 28, 33, 34
central, 138, 142
coriolis effect, 27
Indian Ocean, 43
Pacific Ocean, 39, 41
southern oceans, 33

## H

Halides, 62
Halocline, 17, 48, 54, 486, 489
Hawaii, 291—292
Heat
capacity, 167, 472
exchange, 14, 244
flow, 41, 457, 459
hydration, 177
Heavy water, 109—112
Helium, gas
exchange, 227
mid-depth profiles, 137, 140
saturation anomalies, 244, 247, 248
Hemicellulose, 368
Hemoglobin, 409

Heneicosahexane, 385
Henry's law, 232, 236, 239—240, 278, 307
*n*-Heptadecane, 382
High temperature combustion (HTC), 358
Hopanes, 384
HTC, see High temperature combustion
Humboldt Current, see Peru Current
Humic acid, 392—393, 398
Humic substances
  amino acids, 373
  analysis, 393—398
  characterization, 392—393
  hydrogen peroxide formation, 481—482
  light absorption, 479
  particulate organic matter, 361
  surface microlayer, 364
Humidity, 17, 229—230, 243
Humification, 398
Humin, 393, 398
Hummocks, 49
Hydrates, crystalline, 264
Hydration, see also Water
  Born model, 180—184
  carbon dioxide, 271
  structural model, 184—188
  thermodynamic functions, calculations, 174—175, see Enthalpy; Entropy; Free energy
Hydrocarbons
  aromatic, 387
  classification, 376—377
  high molecular weight, 379
  saturated, 379—385
  unsaturated, 385—387
  volatile, 377—378
Hydrochloric acid, 284
Hydrogen, 109—110, 260, 261
  carbon ratios, 397—398
  isotopes, 111
Hydrogen bonding, water, 162, 164—168, see also Water
Hydrogen peroxide
  decay, 482—485
  Framvaren Fjord profile, 493, 502
  ocean sources, 480—481
  rain, 423
  surface waters, 479—480
Hydrogen radicals, 420, see also Ozone
Hydrogen sulfide
  Baltic Sea concentrations, 89, 95
  chemolithotrophy, 456
  depth profile
    Black Sea, 489, 494
    Cariaco Trench, 493, 505
    Framvaren Fjord, 493, 502
    Saanich inlet, 507, 509
  vertical mixing effects, 487, 491, 500, 501
Hydrogenous sediment, 8, 10, see also Sediments
Hydrographic station, 11, 21, 291, 489, 495, 507
Hydrolysis, 144—146, 370, 372, 374
Hydroperoxyl radical, 422—425
Hydrophobic bonding, 264
Hydrostatic pressure, 243
Hydrothermal end member, 469
Hydrothermal fluid, see Hydrothermal vent, fluids
Hydrothermal vent, see also Downwelling; Upwelling; Volcanism
  chemical reactions, 457—459
  chemosynthesis at, 455—457
  community cluster, 453, 455
  fluids, 247
  geochemical cycle, 149
  mid-ocean ridge input, 137, 139
  silicon production, 346
  trace metal levels, 459—467, 468
  volcanism, 103
Hydroxides, 121, 123
Hydroxy acids, 390—391
Hydroxyl radicals, 421—423, 425, 435, see also Ozone

## I

Ice
  Arctic Sea, 46, 49
  deuterium concentration, 112
  inert compounds reactions, 441—442
  melting, 15, 33, 36, 48, 49
  phases, 169, 171
  structure, 165
Icebergs, Arctic Sea, 49
ICES, see International Council for the Exploration of the Sea
Ideal gas law, 228
IF, see Infrared energy
Igneous rocks, see Rocks, igneous
Indian Ocean, see also South Indian Ocean
  apparent oxygen utilization, 254, 258
  atmospheric carbon dioxide, 294, 297

circulation, 43—46
nitrate depth profile, 339, 341
micronutrient ratios, 344
Pacific Ocean comparison, 41
sub-oxic layer, 137
Inductance conductance cell, 81
Industrial effluents, 364, 431
Inert compounds, 441
Infrared analysis, 393
Infrared energy (IF), 417, 431
Inorganic compounds, photoreactions, 479
Institute of Ocean Sciences (IOC), 74
International Council for the Exploration of the Sea (ICES), 72
International Equation of State, 18, 22, 63, 101
International Geophysical Year, 29
Interstitial waters, 109, 375—376
Intertropical convergence zone (ITCZ), 423
Intrinsic partial molal volume, 182—183
IOC, see Institute of Ocean Sciences
Iodine, ion, 107
Ion-dipole interactions, 184, see also Water
Ion exchange, model, 150
Ion-ion interactions, see also Bjerrum Theory; Cluster expansion theory; Debye-Huckel Theory
ion-pairing, 201—212
models, 189—192
seawater, 161
specific interaction, 206, 212—216
state of ions, 193—198
types, 189, 192, 195
Ion pairing, see Ion-ion interactions, ion-pairing
Ion pairs, 201—203, 326, 327, 329
Ion-quadrupole interactions, 184, see also Water
Ion-water interactions, 161, 176, 184—188, see also Water
Ionic solution theory, see Ion-ion interactions, specific interaction
Ionization, 280—281, 346, 472
Iron, ions
depth profile
Black Sea, 140, 491, 496
Cariaco Trench, 493, 506
Framvaren Fjord, 493, 504
hydrothermal vent levels, 459, 466
oxidative state in seawater, 194—198
plankton growth, 448
Irradiation, see Ultraviolet (UV) irradiation
Irving-Williams order, 124, 141
Isobutane, 376, 377
Isohalines, 54
Isoprene, 379, 386
Isoprenoids, 380, 384, see also Alkanes; Alkenes
Isopycnals, 21, 254—256, 344
Isotherms, 11—13, 15
Isotopes, seawater content, 109—112, 382
ITCZ, see Intertropical convergence zone


## J


JGOFS, see Joint Global Ocean Flux Study
Joint Global Ocean Flux Study, 146
Joint Panel for Oceanographic Tables and Standards (JPOTS), 74, 76
JPOTS, see Joint Panel for Oceanographic Tables and Standards


## K


Kaolinite, 150—151
Kerogen, 398—399
Ketoglutaric acid, 337
Ketones, 406—408
Ketose, 367
Kurochio current, 26, 40
K-mica, 150


## L


Labrador Current, 27, 46
Lactic acid, 390
Lagoons, 101, 104
Lag phase, growth, see Growth, phases
Lake Tanganyika, 216, 217
Lakes, 429—430, see also Individual entries
Laminar film model, 235, 236
Land masses, 18, 43
Lanthanide metals, 120, see also Metals
Latitude
carbon dioxide input and, 271
ozone levels and, 440
salinity variation, 15, 17
temperature zones, 14
water density comparison, 18, 21, 22
Lattice heat, 177
Lavantine intermediate water, 51
Leaching, 363
Lead, ion, 130, 133, 136, 138, 491, 496
Liebig's law, 341—342

Ligands, 120, 121, 124, 160, 198
Light, see also Ultraviolet (UV) irradiation
  absorption and pH measure, 276
  Framvaren Fjord, 493, 503
  nitrous oxide reactions, 428
  photochemical reactions, 469—472, 474—476
  phytoplankton growth, 447
  polarized, plane rotation, 367
  spectra of ocean surface, 476, 478
  tropospheric chemistry, 418
  transmission profiles, 489, 492, 503
Lignin, 368, 395, 397
Limonene, 400
Lipids, 362, 365
Liquid junction, 274—276
Liss model, 238, 239
Lithium, ion, 459, 463
Lithogenous sediment, 10, see also Sediments
Lomonosov ridge, 46, 48
Lysocline, 313—314

## M

Magic angle spinning (MAS), 395—397
Magnesium, ion
  association constants, 326
  Baltic Sea concentrations, 90
  chlorophyll diagenesis, 410
  determination methods, 64
  hydrothermal vent levels, 459, 461, 462
  silicon complexes and, 346
Manganese, ion, see also Individual Manganese compounds
  depth profile
    Black Sea, 138, 141, 491, 496
    Cariaco Trench, 493, 506
    Framvaren Fjord, 493, 504
    Pacific, 133, 137, 139
  hydrothermal vent levels, 459, 466
  nodules, 10, 144
Manganese dioxide, 488, see also Manganese
Manganese hydroxide, 248—249, see also Manganese
Mariana Trench, 5
Marine snow, 360
MAS, see Magic angle spinning
Mass spectrophotometry, 109, 110, 330—331
Mass transfer coefficient, see Exit coefficient
MBTH, see 3-Methyl-2-benzothiazolinone hydrazone hydrochloride
MCHO, see Monosaccharides, total dissolved
Mediterranean Sea
  Atlantic Central Waters, 31
  Black Sea influx, 489
  closed basin, 50—51
  isotope content, 111
  salinity, 17, 18
Melting point, 162
Membrane, diffusion, 178
Mercury, 123
Mesopause, 418
Mesosphere, 418
Metals, see also Individual entries
  alkali, 127
  anoxic water concentrations, 138, 140—142
  biological interactions, 139, 141—146, 448
  cations, 120—121, 123—124
  conservative type profile, 130
  electronic configuration, 120, 122
  molal volume comparison, 187—188
  geochemical balance, 148—149
  hydrothermal flux, 457—459
  hydrothermal vent levels, 459—467, 468
  mid-depth, type profile, 137, 138, 139, 140
  nutrient type profile, 131—136
  porphyrins, 409, 410
  residence times, in seawater, 124—128, 129
  reduction, 484—485, 486
  surface enrichment, 131, 133, 136, 137
  thermodynamic activity, 160
  transition, 124
Meteor expedition, 4, 29
Meteorites, ferric, 10
Methane
  atmospheric, 423, 425, 431, 433
  depth profile
    Gulf of Mexico, 260, 262
    Cariaco Trench, 493, 507
  seawater concentrations, 377—378
Methanol, 408
Methionine, 413
Methogenesis, 378
3-Methyl-2-benzothiazolinone hydrazone hydrochloride (MBTH), 370

Methylchloroform, 436, 437
Methylheptadecane, 384
Methyl ketones, 407
Methyl mercaptan, 413
Methyperoxyl radicals, 422—423
Micronutrients, see Nutrients, phytoplankton growth
Mid-Atlantic Ridge, 103, 453, see also Ridge systems
Mid-Pacific Ridge, 137, 140, see also Ridge systems
Mindanao Trench, 15, 16
Minus-minus interactions, see Ion-ion interactions
Mixing, vertical
  Black Sea, 489
  estuaries, 52
  Framvaren Fjord, 491
  oceans, 11, 18
  thermocline, 338, 487
Mixture models, see Water, mixture models
Molal ionic strength, 190
Molality, see also Individual entries
  apparent, 216—218, 222—223
  ions, 190
  seawater, 68—69
  solution ions, 181—182
Mole fraction, 68, 228
Molecules, energy states, 473—476
Molybdate, 346—347
Molybdenum blue, 330—331, 347
Monosaccharides, 366—367, 370
Monoterpenes, 386, 400
Monsoon Current, 45
Monsoon winds, see Wind, monsoon
Moretane, 401, 402
Montmorillonite, 150—151

## N

NADP, see Nicotinamide adenine dinucleotide phosphate
NADW, see North Atlantic Deep Water
Nanoplankton, 371
Narragansett Bay estuary, 387
National Bureau of Standards (NBS), 110, 274—275
NBS, see National Bureau of Standards
Neon, gas, 247
Nernst equation, 274
Nickel, depth profile
  Atlantic/Pacific Ocean, 131, 135
  Black Sea, 491, 497
  Cariaco Trench, 493, 506
  Framvaren Fjord, 493, 504
Nicotinamide adenine dinucleotide phosphate (NADP), 444
Nitrate, see also Nitrogen; Individual Nitrogen compounds
  anoxic waters, 487—488
  apparent oxygen utilization, 349—350
  denitrification, 337—338
  depth profile
    Atlantic/Pacific Oceans, 338, 339, see also Plates 19, 20
    Baltic Sea, 86, 88
    Black Sea, 489, 494
  determination, 335
  low oxygen waters, 339, 341
  phytoplankton, 336—337
Nitric acid, 427, see also Acid rain
Nitrification, 336—338, see also Denitrification
Nitrilotriacetic acid, 124
Nitrite, see also Nitrogen; Individual Nitrogen compounds
  chlorine monoxide reactions, 441
  depth profile
    Black Sea, 489, 494
    Gulf of Maine, 338, 340
  determination, 335
Nitrogen, see also Nitrate; Nitrite; Nitrous oxide
  carbon ratios, 398
  content in proteins, 372—373
  depth profile, 357
  determination, 335
  distribution, 242, 336—340
  fixation, 336—338, 448
  oxidation states, 334
  phosphorus ratios, 340—345
  radicals, 420, see also Ozone
  reaction in troposphere, 424—425, 427
  saturation anomalies, 245—246
Nitrous oxide
  atmospheric increases, 436, 437
  depth profile
    Atlantic Ocean, 339, 341
    Pacific Ocean, 260, 263, 342
  hydroperoxyl radical reaction, 424—425
  mixing tracer, 350, 351
  ozone, 428, 435
*Nitzschia alba,* 385

NMR, see Nuclear magnetic resonance
Noble gases, see Individual entries
Nonideal gas behavior, 228—229
Normality, 69
North Atlantic, salinity, 17, see also Atlantic Ocean
North Atlantic Current, 27
North Atlantic Deep Water (NADW), see also Atlantic Ocean
  calcium carbonate saturation, 311, 312, 313
  Indian Ocean comparison, 45
  nitrogen oxide tracing, 350
  oxygen concentration, 251
  salinity, 18
  saturation anomalies, 247
  southern ocean deep water, 38
  water source, 31, 33, 49
North Equatorial Current, 27, 39, 44—45
North Pacific, see also Pacific Ocean
  Intermediate Water, 41, 251
  manganese depth profiles, 133, 137, 139
  salinity, 17, 18
  sub-Antarctic intermediate water, 18
  terrigenous carbon sources, 365
North Sea, 370
Northern Hemisphere, 423, 426
Norwegian Current, 27, 46
Norwegian Sea, 31, 46, 49, 386
Nuclear magnetic resonance (NMR) spectroscopy, 393, 395
Nucleic acids, 411—412
Nutrients
  phytoplankton growth, 323—324, 448
  trace distribution, 131—133, 135—136
  water mass tracer, 349—351

## O

$O_3$, see Ozone
Oceanography, 3
Ocean-rock partitioning, 151
Oceans, see also Individual entries
  bottom structure, 8
  carbon dioxide partial pressures, 291—294
  carbonate compensation depths, 315, 317, 318
  chlorinity values, 66
  currents, 27, see also plate 7
  deep water circulation, 41, 49—51
  elements in, 120
  geochemical cycle, 148—154
  nitrogen cycles, 335, 336
  photochemical processes in, 476—479
  physical characteristics, 5—10
  primary productivity, 451, 452
  salinity distribution, 11, 17—25
  surface, total carbon dioxide, 305
  temperature distribution, 11—16, see also Plates 1—3
  water mass tracers, 331
Odd over even predominance (OEP), 381—383
OEP, see Odd over even predominance
Olefins, 379, 385
Oligosaccharides, 366
Oozes, 346, 348
Opal, 346
Optical activity
  amino acids, 372
  carbohydrates, 367
  crude oil, 384, 386
  glycine, 375
  hydrogen peroxide, 482
  hydrocarbons, 377
Orca basin, 107
Organic carbon, see Carbon; Dissolved organic matter (DOM); Dissolved organic carbon (DOC); Organic matter; Particulate organic carbon (POC); Particulate organic matter (POM)
Organic matter
  deposition, 364—365
  dissolved, see Dissolved organic matter (DOM)
  inanimate as reaction modifiers, 357
  particulate, see Particulate organic matter (POM)
  photochemical reactions, 479
  seawater sources, 355, 362—366
  transport at air-sea interface, 364—365
Organosulfur compounds, 412—413, see also Sulfur
Orthophosphate, 443—444, see also Phosphate
Osmotic pressure, 177—178
Oxalic acid, 391
Oxic-anoxic interface
  Black Sea, 489—491
  Cariaco Trench, 493, 505
  Framvaren Fjord, 493, 503
  Saanich inlet, 507

Oxidation
estimation, 349—350
environment and metals, 144
nutrient levels and, 343—344
organic compounds, 257, 258
organic nitrogen, 337—338
phosphate determination, 330
states, 193—196, 334
sub-oxic layer, 137
surface enrichment, 133
Oxygen
deep water supply, 46, 486
depth profile
Atlantic/Pacific Oceans, 250, 251, 339, 341, 342, see also Plates 12—14
Baltic Sea, 86, 89, 92, 95
Black Sea, 489, 490
Cariaco Trench, 493, 505
Framvaren Fjord, 493, 502
Saanich inlet, 507, 509
determination, 249
dissolved, analysis, 248—249
distribution, 250—251
exchange and wind speed, 241
free radicals, 472
isotopes, 109, 111—112
minimum layer, see Oxygen minimum layer
ozone formation, 420—422, see also Ozone
phosphate/nitrate ratio, 344
primary productivity measurement, 446, see also Primary productivity
utilization, see Apparent oxygen utilization (AOU)
Oxygen mimimum layer
advection diffusion models, 255—256, 259
nitrate, 339
nitrogen-phosphorus ratio, 341, 343
nitrous oxide, 260, 263, 339, 341, 344
phosphate, 332
Ozone
chlorine, 433—439, 441
formation, 419—423, 425, 428
free radicals, 435—436
global trends, 438, 440
hole, 438, 441—442
UV irradiation, 438

## P

Pacific Central Waters, 40
Pacific Deep Water, 41
Pacific Equatorial water, 40, 41
Pacific Ocean, see also Apparent oxygen utilization (AOU); North Pacific
apparent oxygen utilization, 254, 257
carbohyrate composition, 371
carbon cycling, 291
circulation, 39—43
deep water, metals levels, 138, 141
depth profile
alkalinity, 298, 302—304
aluminum, 137, 138
cadmium, 131
calcium carbonate lysocline, 313, 314, 316
carbon dioxide, 294, 296, 299, 307, 309
dissolved organic matter vs. apparent oxygen utilization, 358, 359
germanium, 131, 134
helium, 137, 140
lead, 133, 136
molybdenum, 130
nickel/silicate/phosphate, 131, 135
nitrate, 338, 339, 341, see also Plates 19, 20
nitrogen-phosphorus ratio, 341, 342
nitrous oxide, 260, 263
oxygen, 250, 251
pH, 294, 301
phosphate, 131—333, see also Plates 17, 18
silica, 131, 134
silicate, 346, 347
tungsten, 130
nitrous oxide/oxygen relationship, 339, 342
salinity, profile, 17, 20, see also Plates 4—6
saturation anomalies, 247—250
sub-oxic layer, 137
surface waters, 319
vertical section, 349, see also Plate 22
upwelling, 40, 253, 254, 294, see also Hydrothermal vent; Upwelling; Volcanism
Pacific Sub-Arctic Water, 40
Pack ice, 49, see also Ice
Packing effects, 188, 263
Palmitic acid, 388, 389
Palmitoleic acid, 389
*Paracentrotus lividus,* 402

Paraffins, see Alkanes
Partial pressure
  air-sea interface model, 235, 236
  atmospheric gas, 228
  carbon dioxide, 269—270, 287, 291—297, 305—306
  carbonate characterization, 287—288
  dissolved gas, 232
  gas bubble effect, 242
  water vapor, 229—230
Particulate amino acids (PAA), 374—375
Particulate organic carbon (POC), 356—357
Particulate organic matter (POM), 356, 360, 361, 371
Pauling self-clathrate model, see Clathrate models
PCB, see Polychlorinated biphenyls
PCHO, see Polysaccharides, total dissolved
*n*-Pentadecane, distribution, 382
Pelgic sediment, 308, 311
Peptides, 372, 374
Permeability coefficient, see Exit coefficient
Periodic table, elements, 120
Peru, upwelling regions, 378, 408; see also Upwelling
Peru Current, 40
Petroleum, 377, 380, 409
Petroporphyrins, 409—411
pH, see also Alkalinity; Chlorinity; Conductivity; Density; Salinity
  ammonia dissociation, 334
  Baltic Sea profile, 89
  carbonate regulation of, 269, 282, 287—288
  clay exchange, 150
  hydrothermal vent levels, 459, 465
  ion state in seawater, 194, 196—198
  phosphoric acid, effect of, 327—329
  rain, 429
  seawater measurement, 272—277
  surface water, 294, 298, 300, 301
*Phaeodactylum* spp., 361
Phenol, 408
Pheophorbide, 410, 411
Pheophytin-a, 410
Phosphate, see also Phosphoric acid; Phosphorus
  carbonate ratios, 344—345
  conservative tracer, 350
  depth profile
    Atlantic/Pacific Ocean, 131, 135, 136, 333
    Baltic Sea, 86, 88
    Black Sea, 489—490, 494
    English Channel, 332
  determination/distribution in seawater, 324—327, 329, 330—333, see also Plates 17, 18
  esters, 324
Phosphomolybdic acid, 330
Phosphorescence, 475
Phosphoric acid, 327, 328, see also Individual Phosphorus compounds
Phosphorus, 323—329, 340, 357, see also Individual Phosphorus compounds
Photic zone, 486
Photoautolithotrophy, 456
Photochemistry
  copper oxidation, 484—485
  hydrogen peroxide accumulation, 480—485
  inorganic/organic compounds, 479
  laws, 469—471
  molecular excitation, 473—476
  oxygen lifetime, 485
  quantum yields, 471—472
  surface water, 476, 478
Photodissociation, 434—436
Photolysis, hydroxyl radical production, 421
Photons, 469—471, see also Light
Photosynthesis
  carbon dioxide cycling, 271—272, 292—293
  Framvaren Fjord, 493, 502
  organic matter source, 355
  oxygen concentration, 241, 250, 251, 253
  rate, factors affecting, 447—448
  stages of, 443—445
Photosynthetic compensation depth, 338—340, 450—451
Photosynthetic quotient, 444
Phototropic organisms, see Phytoplankton
Phytane, 381
Phytol, 380, 381, 386, 409, 410
Phytoplankton, see also Zooplankton
  alkanes from, 382
  amino acid levels, 374
  blooms, 160, 332
  carbohydrates, 368, 370, 371
  cholesterol production, 404

dimethyl sulfide release, 413
Framvaren Fjord, 493
growth phases, 449—450
metal adsorption, 144—146
nitrogen fixation, 336—338
nitrogen-phosphorus ratio, 340
organic matter constituent, 360—362
phosphate cycling, 332—333
photosynthetic compensation depth, 450—451
pristane in, 380
productivity, 443
stenol source, 406
Pinene, 400
Piston velocity, see Exit coefficient
Pitzer equation, 215—216, 223
Planck's constant, 469
Plankton, 370, 371, 390, 448
Plant wax, 385
Plants, 289, 292—293, 367—368, 371
Plus-minus interactions, see Ion-ion interactions, models
Plus-plus interactions, see Ion-ion interactions, models
POC, see Particulate organic carbon
Polar cap, 49
Polar Current, 43
Polar ice core, 378
Polar vertex, 441—442
Poles, surface temperature, 11
Pollution, 364, 377
Polychlorinated biphenyls (PCB), 365—366
Polynyas, 49
Polyphosphates, 330
Polysaccharides, 364, 366, 370
POM, see Particulate organic matter
Pore waters, 109, 327, see also Interstitial water
Porphyrins, 380, 381, 409—411
Potassium, ions, 64, 90, 459, 464
Potassium chloride, 76, 77
Potassium chromate, 62
Practical Salinity Scale, 76—78, 89, 97, 101, see also Salinity
Precipitation
Black Sea, effect of, 489, 491
Mediterranean Sea, 50—51
Red Sea, 51
redox transients, 476—477
salinity, 15, 20
seawater composition, 102
Pressure
acid-base equilibria, 272, 273, 280—282
ocean density, 18
phosphate ionization, 327, 329
water properties, effect of, 164, 167—169
Primary photochemical reaction, see Photochemistry, laws
Primary productivity, see also Phytoplankton
carbon dioxide availability, 269, 271
geographical distribution, 451, 452
global trends, 453—455, see also Plate 24
marine plants, 442
measurement, 446—448
Pacific Ocean, 41
seasonal variation, 356—357, 450
Pristane, 380, 381
Protein(s), 337, 361, 368, 372
Protons, 275, 418—421, 443
Purines, 411, 412
Pycnocline, 18, 26
Pyrimidines, 411, 412
Pyrite, 144, 493, 504


## Q


Quadrupole moment, 164, 166
Quantum yield, photochemical reactions, 471—472, see also Photochemistry
Quartz, 10


## R


Racemic mixtures, 372
Radiation, electromagnetic, 469, 474—475
Radiolarians, 346, 348
Radon, 133, 227
Rafting, 49
Rainfall, 17, 365, 423
Red Sea, 17, 18, 45—46, 51—52, 106, see also Brines
hot brine, 101, 219, 222, 459
Redfield model, 298, 343—344, 349
Redox transients, sources, 476—477, 507
Remineralization, 376
Repartition, polyatomic molecules, 473
Residence times, 49, 124—128, 150—151, 362
Resins, 401
Resistance, 239—240

Respiration, 445, 487
Retardation phase, growth, see Growth, phases
Retene, 401
Revelle factor (R), 305, 307
Reworking, 375
Ribonucleic acid (RNA), 412
Ribose, 371
Ridge systems, 5, 9, 49, 247, 248, see also Individual entries
Rivers
- Baltic Sea estuary, river comparison, 86—93
- Black Sea, effect of, 489, 491
- density calculations and seawater, 98—101
- element input to oceans, 125, 126, 336
- flow rate, 88—89, 94
- Framvaren Fjord, effect of, 491
- hydrothermal flux comparison, 468, 470
- major components, 83—86, 96, 216, 217
- organic matter input to oceans, 363—364
- runoff, 52, 88—89, 94
- salinity calculations, see Salinity, calculations
- silicon transport, 346, 348
- weathering products, 152, 154

RNA, see Ribonucleic acid
Rocks, 125, 147—149, 152—154, 491
Ross Sea, 37
Rubidium, ion, 459, 464

## S

Saanich inlet, 507, 508
St. Lawrence River, 98—101
Saline, contents, 60
Salinity, see also Alkalinity; Chlorinity; Conductivity; Evaporation; Density; Practical Salinity Scale; Salts
- anoxic waters, 486—487
- Atlantic Ocean, 27, 30—31
- calculations, 72, 73, 97
  - estuaries, 89, 92—93, 97—98
  - evaporation technique, 72, 73
- chlorinity relationship, 73—74, 76, 93
- conductivity relationship, 74—81, 93—94, 97—101
- density relationship, 75—76, 81, 93, 100
  - conductivity and, 94, 96, 98, 99
- depth profile
  - Baltic Sea, 86, 87
  - Black Sea, 489, 492
  - Framvaren Fjord, 491, 500
- determination methods, 78—83
- dissolved organic matter and, 363
- estuaries, definition, 52—54, 363
- Mediterranean Sea, 50—51
- nitrogen oxide, 350
- ocean distribution, 5, 11, 17—25
- oxygen concentration, 254
- Pacific Ocean, 40
- phytoplankton growth, 448
- Red Sea, 45—46, 51
- southern ocean water, 33, 36, 37
- surface water, 15, 19
- water mass tracer, 349

Salinometers, 75, 80, 81, see also Salinity, determination methods
Salt marsh, 367, 371
Salt wedge estuary, 54
Salting coefficient, 234, 264—266
Salting out, 233, 364
Salts, see also Individual entries
- additivity effects to water, 216, 218, 219
- evaporation effects, 101, 105
- river sources, 152, 154

Samoa, 292
Santa Barbara Basin, 391
Sargasso Sea, 28, 370, 371, 484, 486
Satellites, ring formation, 29, see also Plate 8
Saturated hydrocarbons, see Hydrocarbons, saturated
Saturation, 101, 104, 244, 245, 246, 313
Scale particle theory, 266
Scavenging, 143—144
Sea floor, 5, 8, 148, 247
Sea salt, 67, 219—223, see also Individual entries
Seamounts, 5
Seas, 6, 7, see also Individual entries
Seasons, variations in
- carbonate cycling, 271, 292, 293
- nitrogen, 338, 340
- organic matter, 356
- phosphorus, 332
- silicon, 348
- temperature, 14

Seawater, see also Basins; Estuaries; Oceans; Rivers

acid-base equilibria, 272—277
air interface, see Air-sea interface
amino acid levels, 374
analysis, historical background, 60—62
artificial, 70, 71, 176—178
brines, 101—102, 106
calcium carbonate, see Calcium carbonate
carbohydrate levels, 370
carbon dioxide cycling, 305, see also Carbon dioxide; Carbonate system
chlorinity, see Chlorinity
composition and stoichiometry, 65—70
conductivity, see Conductivity
density, depth correlation, 18, see also Salinity
determination methods, components, 62—65
dissolution in, 102, 108, 232—235
elements in, see Cations; Elements; Metals
evaporation of, 101—105
freezing, 108, see also Ice
geochemical balance, 147—154
hydrogen peroxide, decay, 481—485
interstitial waters, 109
ion pairing, see Ion-ion interactions, ion-pairing
ions, state of, 193—198
isotopes, 109—112, 382
mixing, 487, see also Mixing, vertical
natural, 476, 478
organic matter concentration, 355, 362—366
physical properties, 216—223
photochemical processes, 476, 478
quality, 364
residence times, elements in, 124—128
salinity in, see Salinity
speciation, see Speciation
stratified column, 487—488
type recognition, 21, 23
volcanism, see Volcanism
Sediment, see also Individual entries
alkane distribution, 383—384
components, 8, 10
cycling in, 125—128, 147—148, 375, 387
pore waters, 487
shelf, 133, 137, 138, 142, 406
slope, 406
trap assay, 361
water interface, 227, 378
Sedimentary rock, 398—399, see also Rocks
Sedimentation rates, 126
Selenium, 131, 136
Sensing, remote, 29, see also Plate 8
Serine, 375
Sesquiterpenes, 400
Setchenow salting out, 233
Sewage, 364
Shallow water, 131, 356
Shelf sediment, see Sediment, shelf
Shore, 5
Sigma-t, 18, 22, 23, 489, 492
Significant structure theory, 172—173
Silica, 131—132, 134, 136, 348; see also Individual Silicon compounds
Silicate, see also Individual Silicon compounds
depth profile
Atlantic/Pacific Oceans, 131, 133, 135
Black Sea, 489, 494
theory, 149—151
Silicon, 345—349, see also Plates, 21, 22; Individual Silicon compounds
Silicon dioxide, 448, 459, 461, see also Individual Silicon compounds
Sills, 487, 491, 493, 507
Silver chromate, 62
Slope sediment, see Sediment, slope
Smog, 417
SMOW, see Standard mean ocean water
Sodium, ion
Baltic Sea concentrations, 90
chlorinity, 65—68
determination methods, 64—65
hydrothermal vent levels, 459, 467
Sodium acetate, 276
Sodium chloride
ion-ion interactions, 192
phosphoric acid forms in, 327, 328
salting coefficients, 265, 266
synthetic seawater, 176—178
Sodium thiosulfate, 249
Soil, 395, 396
Solar radiation, 419, 421, 430, 431, see also Sun
Solubility, 312, 345
Somali Current, 45
Sound velocity, 82, 167
South Atlantic, see also Atlantic Ocean; North Atlantic

barium profile, 132
basins, 31, 32
central water, 45
circulation, 27, 33, 35
silica profile, 132
South Equatorial Current, 27, 39, 43—45
South Indian Ocean, 43, see also Indian Ocean
South Pacific bottom water, 41, see also Pacific Ocean
South Pole, 292, 438
Southern Hemisphere, 423
Southern oceans, 33, 36—38
*Spartina alterniflora,* 370
Speciation
elements, 118—119
metals, 198, 199, 202, 412, 413
phosphoric acid, 327—330
seawater
free anions, 207
free cations, 205—206, 208—212
minor components, 204—205
Spectrum, visual, humic substances, 393, 394
Sponges, 348, 402
Squalene, 386, 401—402
Stability constants, 123, 124
Stagnant film model, 235, 236
Standard mean ocean water (SMOW), 110—111
Standard temperature and pressure (STP), 229, 232
Standing crop, 445, see also Phytoplankton
Stanols, 405, 406
Stanones, 405
Starch, 368
Stark-Einstein law, 470—471
Starvation, 380
Static stability, 18
Stationary phase, growth, see Growth, phases
Steady state model, element circulation, 148
Stearic acid, 388
Stenols, 405, 406
Steranes, 384, 405, see also Steroids
Steroids, 379—380, 404
Sterols, 405
Stone Harbor, 393, 394
Storms, 242, see also Air-sea interface
STP, see Standard temperature and pressure
Stratified estuaries, 53—54, see also Estuaries
Stratopause, 418
Stratosphere, 418—420, 428
Strontium, ion, 459, 463
Structural hydration model, 184—188
Sub-Arctic convergence, 40
Sub-Antarctic zone, 36
Sub-Antarctic Intermediate Water, 17, 18
Subtropical convergence, 29—31, 36, 37, 43
Sub-oxic layer, 137
Succinic acid, 391
Sugars, see Individual entries
Sulfate, ion, see also Individual Sulfur compounds
Baltic Sea concentrations, 86, 91
determination methods, 63
hydrothermal vent levels, 459, 465
reduction, 488
Sulfide, 108, 458, see also Individual Sulfur compounds
Sulfur, 112, 413, 428—430, see also Acid rain; Indivudal Sulfur compounds
Sulfur dioxide, 428, see also Fossil fuels; Individual Sulfur compounds
Sulfuric acid, 429, see also Individual Sulfur compounds
Summer, 241, 450, see also Seasons
Sun, 25, 88, 89, see also Solar radiation
Superoxide anion, 483—484, see also Hydrogen peroxide; Hydroxyl radicals; Ozone
Supersaturation, gases, 244—246, 250, 252, 253
Surface enrichment, 131, 133
Surface microlayer, 364, 375, 390
Surface mixed layer, 14, 251
Surface renewal model, 238
Surface sea slicks, 408
Surface water, see also Air-sea interface
adenosine triphosphate (ATP), 331
alkalinity, 298—300, 302—304, see also Plate 15
carbon dioxide exchange, 305, 306
circulation and convergence, 33, 36, 37
copper ions, 484—485
Framvaren Fjord, 491, 500
gas solubility, 233
hydrogen peroxide, 479—481, see also Hydrogen peroxide
isotopes, 112

nitrate levels, 338
nitrogen-phosphorus ratio, 341
oxygen concentration, 250, 252—254
pH, 294, 298, 300, 301
phosphate levels, 327, 331, 332
productivity levels, 487
photochemical reactions, 476—478
salinity/temperature profile, 18, 19
silicon levels, 346—348

## T

TCHO, see Total dissolved carbohydrates
TDS, see Total dissolved sugars
Temperature
adiabatic and water stability, 14, 21
acid-base equilibria, 273
atmospheric layers, 418
Baltic Sea, 86, 87
Black Sea, 492
carbonate cycle, 279, 282
distribution in oceans, 11—16
Framvaren Fjord, 491, 500
gas solubility, 244—247
high, reactions, see Hydrothermal vent
interstitial water composition, 109
Pacific Ocean profile, 250, 253
phytoplankton growth, 448
potential, 15, 16, 350, 351
salinity profiles, see also Salinity
Arctic Sea, 48
Indian Ocean, 45
Pacific Ocean, 40—42
use of, 21, 23, 25, 79
South Atlantic bottom water, 31, 32
southern ocean surface water, 33, 36, 37
subsurface maximum, 48
water properties, effect of, 164, 167—169
Terpenes, 379, 386, 400
Terpenoids, 399—400
Terrestrial-derived carbon, see Carbon, terrestrial-derived
THAA, see Total hydrolyzable amino acids
Thermal agitation, 474—475
Thermal radiation, trapping, 430—434, see also Plate 23
Thermoclines, 11, 14, 15, 40, 338, 487
Thermohaline, 26
Thermosphere, 418
Thin film model, see Laminar film model
Thin layer chromatography (TLC), 360
Thiols, 413
Tidal currents, 52
Titration calorimeter, 62—63
TLC, see Thin layer chromatography
TOC, see Total organic carbon
Total carbon dioxide, see Carbon dioxide, total
Total dissolved carbohydrates (TCHO), 368—371
Total dissolved polysaccharides (PCHO), 371
Total dissolved sugars (TDS), 370
Total hydrolyzable amino acids (THAA), 374
Total organic carbon (TOC), 360, 362, 363
Trace elements, see Elements, trace
Tradewinds, see Wind, trade
Transamination, 337
Transfer velocity, 306—308, see also Exit coefficient
Transient Tracers in the Ocean (TTO), 29, 33, 34
Transition metals, see Metals, transition
Trench, ocean, 5, 14, see also Individual entries
Triglycerides, 388—389
TRIS, see Tris (hydroxymethyl) aminomethane
Tris (hydroxymethyl) aminomethane, 276, see also Buffers
Triterpanes, 386, 401, see also Squalene
Tritium, 33, 34, 109, 349, 452
Tropics, 17, 20, see also Plates 4—6
Tropopause, 418
Troposphere, 418, 424—425, 427, 430
TTO, see Transient Tracers in the Ocean
Turpentine, 400

## U

UCM, see Unresolved complex mixture
Ultraviolet (UV) irradiation, 330, 358, 419, 438, 476—470
Undersaturation, 128—129
Uniformist models, 166, 168
Unresolved complex mixture (UCM), 360
Unsaturated fatty acids, see Fatty acids
Unsaturated hydrocarbons, see Hydrocarbons, unsaturated
Upwelling, see also Downwelling; Hydrothermal vent; Mixing, vertical; Volcanism

oceans, 40, 50
oxygen concentration, 253, 254
phosphate levels, 332
primary productivity, 294, 338, 451, 452
Saanich inlet, 507
silicon concentrations, 348
surface temperature, 11, 15
vertical circulation, 33
UV irradiation, see Ultraviolet (UV) Irradiation

V

van der Waals equation, 228, 229
van't Hoff equation, 233
Vapor pressure, 177, 232
VCF, see Volume concentration factor
Velocity section, 43
Vents, types, 453
Vertical advection velocity, 255
Vertical circulation, see Circulation, vertical
Vertical eddy diffusion coefficient, 255
Vertical mixed estuary, 53
Vertical mixing, see Mixing, vertical
Vibrating flow densimeter, see Densimeter
*Vibrio anginolyticus,* 159—160
Viscosity, water, 164, 168
Vitamins, 403, 448
Volatilization, 377, 378, 389
Volcanism, 103, 336, 423, 425, see also Downwelling; Hydrothermal vent; Mixing, vertical; Upwelling
Volume concentration factor (VCF), 144—145

W

Walfish Ridge, 31
Walvis Bay, 405, 406
Water
activity, 178—179
bonding, 162, 166
crystal molal volume, 188
ion interactions, 174—178, 180—184
hydration, 184—188
solvent properties, 161—164, 167
structure models, 165—166, 168—173, 176
Water vapor, 229—230, 431
Wax esters, 408
Weathering, 152, 154, 348
Weddell Sea, 37
West Greenland Current, 46
Westerlies, see Wind
Western Pacific, 294, 298
Wheatstone Bridge, 80
Wind
-driven ocean circulation, 25, 26, 33, 43
field and salinity, 17
monsoon, 44
speed, 236—238, 240, 241
trade, 26—27, 39, 44
transfer velocity effects, 306—307
Winkler technique, 248—249
Winter, 241, 450, see also Seasons
WOCE, see World Ocean Climate Experiment
World Ocean Climate Experiment (WOCE), 4
World river, see Rivers

X

Xanthophylls, 402, 403
Xenon, gas, 244

Y

Yeast, 384
Young's rule, 217—218, 220—223

Z

Zinc, ion
depth profiles
Atlantic/Pacific Oceans, 131, 133
Black Sea, 491, 497
Framvaren Fjord, 493, 504
hydrothermal vent levels, 459, 467
Zinc uranyl acetate, 64
Zooplankton, see also Phytoplankton; Primary productivity
alkanes from, 382
alkene distribution, 385
cholesterol production, 404
particulate organic matter, 361
phosphate levels, 332
pristane in, 380
stenol source, 406
total dissolved carbohydrate levels, 371
wax esters from, 408